THE GEOGRAPHY OF THE WORLD ECONOMY

The Geography of the World Economy provides an in-depth and stimulating introduction to the globalization of the world economy. The book offers a consideration of local, regional, national and global economic development over the long historical term. The theory and practice of economic and political geography provide a basis for understanding the interactions within and among the developed and developing countries of the world. Illustrated in color throughout, this new edition has been completely reworked and updated to take account of recent significant changes in the world economy.

A new companion website also accompanies the book, with additional resources for each chapter including multiple choice and short essay questions and links to relevant websites. Figures and tables are also available for download located at www.routledge.com/cw/knox

The text is signposted throughout with an glossary of key terms, and is richly illustrated with full-color maps, diagrams and illustrations. It is ideal for upper level university undergraduates and for post-graduates in a variety of specializations including geography, economics, political science, international relations and global studies.

Paul Knox, University Distinguished Professor and Senior Fellow for International Advancement, Virginia Polytechnic Institute and State University, USA.

John Agnew, Distinguished Professor, University of California, Los Angeles, USA.

Linda McCarthy, Associate Professor, University of Wisconsin-Milwaukee, USA.

THE GEOGRAPHY OF THE WORLD ECONOMY

Sixth Edition

Paul Knox, John Agnew and Linda McCarthy

Routledge
Taylor & Francis Group

LONDON AND NEW YORK

First edition published in 1989 by Edward Arnold
Fifth edition published in 2008 by Hodder Education

Sixth edition published in 2014
by Routledge
2 Park Square, Milton Park, Abingdon, Oxon OX14 4RN

and by Routledge
711 Third Avenue, New York, NY 10017

Routledge is an imprint of the Taylor & Francis Group, an informa business

British Library Cataloguing in Publication Data
A catalogue record for this book is available from the British Library

Library of Congress Cataloging in Publication Data
Knox, Paul L.
The geography of the world economy/Paul Knox, John Agnew and
Linda McCarthy.
   pages cm
   Includes bibliographical references and index.
   1. Economic geography. 2. Economic history. I. Agnew, John A.
   II. McCarthy, Linda (Linda Mary) III. Title.
   HF1025.K573 2014
   330.9–dc23   2013027243

ISBN: 978-0-415-83128-4 (hbk)
ISBN: 978-1-44-418470-9 (pbk)
ISBN: 978-0-203-77518-9 (ebk)

Typeset in Gill Sans and Sabon
by Florence Production Ltd, Stoodleigh, Devon, UK

Printed and bound in India by Replika Press Pvt. Ltd.

Contents

List of figures   viii

List of tables   xi

List of boxes   xiii

Acknowledgements   xiv

PART 1
ECONOMIC PATTERNS AND THE SEARCH FOR EXPLANATION   1

1 The changing world economy   3
    1.1 Studying the world economy   5
    1.2 Economic organization and spatial change   6
    1.3 Spatial divisions of labor   12
         Key sources and suggested reading   19

2 Global patterns and trends   20
    2.1 What "economic development" means   23
    2.2 International patterns of resources and population   29
    2.3 International patterns of industry and finance   45
         Summary   59
         Key sources and suggested reading   60

3 Geographical dynamics of the world economy   61
    3.1 History of the world economy   62
    3.2 States and the world economy   66
    3.3 "Market access" and the regional motors of the world
         economy   79
         Summary   91
         Key sources and suggested reading   91

PART 2
RISE OF THE CORE ECONOMIES   93

4 Preindustrial foundations   95
    4.1 Beginnings   95
    4.2 Emerging imperatives of economic organization   100
    4.3 Emergence of the European world-system   100
         Summary   113
         Key sources and suggested reading   115

5 Evolution of the core regions   116

    5.1   The Industrial Revolution and spatial change   116
    5.2   Machinofacture and the spread of industrialization in Europe   117
    5.3   Fordism and North American industrialization   125
    5.4   Japanese industrialization: Two economic miracles   130
    5.5   Emergence of "organized" capitalism   137
    5.6   Principles of economic geography: Summarizing lessons from the industrial era   142
        Summary   144
        Key sources and suggested reading   144

6 Globalization of economic activities   145

    6.1   Transition to advanced capitalism   145
    6.2   Patterns and processes of globalization   159
        Summary   174
        Key sources and suggested reading   174

PART 3
SPATIAL TRANSFORMATION OF CORE AND PERIPHERY   177

7 Spatial reorganization of the core economies   179

    7.1   The context for urban and regional change   180
    7.2   Spatial reorganization of the core economies   182
    7.3   Old industrial spaces   194
    7.4   New industrial spaces   196
    7.5   Regional inequality in core economies   205
        Summary   211
        Key sources and suggested reading   212

8 Dynamics of interdependence: Transformation of the periphery   213

    8.1   Colonial economies and the transformation of global space   213
    8.2   Economic mechanisms of enmeshment and maintenance in the colonial world economy   218
    8.3   Influence of colonial administration on interdependence   226
    8.4   Mechanisms of cultural integration   228
    8.5   Changing global context of interdependence   230
    8.6   Alternative models of development?   242
        Summary   243
        Key sources and suggested reading   244

9 Agriculture: The primary concern?   245

    9.1   Agriculture in the periphery   246
    9.2   Land, labor, and capital   251
    9.3   Rural land reform   260
    9.4   Capitalization of agriculture   262
    9.5   Science and technology in agriculture   270
        Summary   272
        Key sources and suggested reading   273

10 Industrialization: The path to progress? 274

    10.1 National and global stimuli to industrialization 275
    10.2 Limits to industrialization in the periphery 280
    10.3 Geography of industrialization in the periphery 285
    10.4 Rise and fall of the Soviet model of industrialization 299
    10.5 China's rise in the world economy 304
        Summary 310
        Key sources and suggested reading 312

11 Services: Going global? 313

    11.1 Defining and theorizing services 315
    11.2 National and global stimuli to the growth of services 318
    11.3 Services outsourcing: Benefits and drawbacks for all? 319
    11.4 Limits to service export growth in the semi-periphery and periphery? 322
    11.5 Geography of services 325
    11.6 Variety in the internationalization of services 332
        Summary 347
        Key sources and suggested reading 349

PART 4
ADJUSTING TO THE WORLD ECONOMY 351

12 International and supranational institutionalized integration 353

    12.1 Economic change and geopolitics 353
    12.2 International and supranational institutionalized integration 356
    12.3 Spatial outcomes of economic integration 362
        Summary 377
        Key sources and suggested reading 378

13 Reassertion of the local in the age of the global: Regions and localities within the world economy 379

    13.1 Regionalism and regional policy 380
    13.2 Nationalist separatism 386
    13.3 Grassroots reactions 391
        Summary 396
        Key sources and suggested reading 397

14 Conclusion 398

        Key sources and suggested reading 401

Glossary 403
Bibliography 423
Index 453

Figures

1.1	The inter-relationships surrounding economic organization and spatial change	6
1.2	Major features of economic change in the world's developed economies	8
1.3	Employment outsourcing and insourcing, United States	14
1.4	Basic elements of commodity chains	15
2.1	The world-system: core, semi-periphery and periphery	22
2.2	GDP per capita (PPP, constant 2005 international dollars)	22
2.3	Gap in growth of GDP per capita (PPP, constant 2005 international dollars)	23
2.4	Happy planet index, 2012	27
2.5	Change in human development index (HDI) in regions as a percentage of potential	27
2.6	Human development index (HDI)	28
2.7	Human development index (HDI), 2011	28
2.8	North American shale plays	31
2.9	Estimated shale gas resources in 14 regions	31
2.10	Marine "dead zones"	31
2.11	World energy consumption by fuel, historical and projected	35
2.12	The world's cultivable land	37
2.13	Stress on freshwater supplies, 1995 and 2025	38
2.14	Ecological footprint	39
2.15	Share of agricultural products in world merchandise exports, 1950–2009	40
2.16	The demographic transition	41
2.17	Demographic transition map of the world	42
2.18	Remittance flows	43
2.19	Remittance flows, top countries, 2010	44
2.20	Share of a country's nationals with a university degree living in an(other) OECD country	45
2.21	Growth in volume of world merchandise trade and GDP, 2005–13 (annual % change)	50
2.22	Trade, imports and exports	51
2.23	Intra- vs. inter-regional connectedness of major exporters	52
2.24	Index of commodity concentration of exports, 2011	53
2.25	Government debt of euro-zone countries, 2011 (Maastricht limit: 60%)	55
2.26	Iceland and the global financial crisis	57
3.1	Shifting fortunes in the world economy (after Maddison)	62
3.2	Corruption perceptions index, 2011	63
3.3	The increasing pace of the world economy	69
3.4	F.D.I. inflows, developed, developing, and transitioning* countries (*includes South-East Europe and Commonwealth of Independent States)	69
3.5	United States military expenditure	76

3.6	The product life cycle model and possible effects on U.S. production and trade (after Vernon, 1966)	83
3.7	Cheaper transport and communications costs on the global highway	88
4.1	The urbanization of the classical world	99
4.2	Plan of a medieval manor	103
4.3	The rise of merchant capitalism and the changing space-economy	105
4.4	Towns and cities of the Hanseatic League	105
4.5	The emergence of a European-based world-system	108
4.6	Transoceanic rim settlements of the mercantile era	110
4.7	Colonialism and urban settlement patterns	114
5.1	Output growth in Western Europe, 500–1990	118
5.2	Europe in 1875	121
5.3	Core and periphery in Europe	124
5.4	The American Manufacturing Belt in 1919 (after Conzen, 1981: 340, Figure 9.13)	128
5.5	Index of manufacturing production for selected countries	134
6.1	Forces in the deindustrialization of the United Kingdom: dramatic loss of competitiveness (1978–83) and consequent import penetration	148
6.2	Internet users	149
6.3	Boeing's 787 Dreamliner global production system	165
6.4	The changing global geography of clothing manufacturing	168
7.1	The foreign automobile industry in the United States	184
7.2	Research and development (R&D) in the USA by U.S. affiliates of foreign companies and R&D performed abroad by foreign affiliates of U.S. TNCs	185
7.3	Geography of an Indian offshore services provider	187
7.4	The system of world cities	191
7.5	Employment by sector in the United Kingdom	195
7.6	Employment shares, by economic sector, USA	196
7.7	French competitiveness clusters	200
7.8	Regional inequality across the European Union	206
7.9	An index of deprivation in England in 2010, by district level (average score)	207
7.10	The relationship between interregional inequality and levels of economic development, as posited by Williamson (1965)	208
7.11	Regional trends in per capita incomes in the United States	209
7.12	Myrdal's model of regional cumulative causation	210
8.1	Geographical extent of European political control, 1500–1950	214
8.2	Long waves and colonization: the two long waves of colonial expansion and contraction	216
8.3	The Atlantic system, 1650–1850	219
8.4	Geographical distribution of British foreign investment, 1860–1959	220
8.5	Colonization of Africa: (A) 1850; (B) 1914	220
8.6	World steamship routes, by volume of trade, 1913	221
8.7	Telegraph system in Asia and Africa, 1897	222
8.8	Development of roads and railways in the River Plate region of South America, 1885–1978	223
8.9	New states of the world since the Second World War	231

8.10 Exports between and from developed and less developed countries, 2001–2010 — 236
8.11 Soviet and U.S. spheres of influence, 1982 — 238
9.1 External migration flows in Sub-Saharan Africa — 257
10.1 Manufacturing as a percentage of GDP, 2010 — 286
10.2 Three "tests" of industrialization, 1975 and 2004 — 287
10.3 The geography of growth in manufacturing output, 1995–2005 — 289
10.4 Share of certain NIEs in total LDC exports, 2005–06 by value — 289
10.5 Labour processes in three manufacturing industries — 290
10.6 Locally owned electronics industry plants in Malaysia, 1999 — 291
10.7 Share of U.S. and European companies' outsourcing contracts with an offshore element, 2003–12 — 293
10.8 Manufacturing establishments of the Japanese electronics industry in Asia, 1995 — 297
10.9 Economic and demographic growth in China — 305
10.10 Percentage of China's GDP from industry and services — 307
11.1 Changing share of employment in primary, secondary and tertiary sectors of the economy — 314
11.2 Service employment and GDP per capita, selected countries — 314
11.3 The world's largest service providers — 315
11.4 Service encapsulation — 318
11.5 Submarine fiber-optic cable network — 323
11.6 Percentage of workers in services — 327
11.7 Informal economy and level of development — 327
11.8 Average annual percentage growth in services — 328
11.9 Service production — 328
11.10 Exports and imports of services, 2010 — 330
11.11 International tourist arrivals and receipts, 2006 — 337
11.12 The world's major stock markets — 343
11.13 Major world areas of offshore banking — 343
11.14 Global R&D service providers, 2012 — 347
12.1 Selected supranational integration agreements — 358
12.2 Enlargement of the European Union — 368
12.3 Regional policy in the European Union, 2007–2013 — 374
13.1 The Thatcher government's rolling back of regional aid, 1979–84 — 385
13.2 Governmental decentralization of Russia, 1993 — 389
13.3 Geographical clustering of export shares in Italy, by province, 1985–1999 — 392
13.4 How plant size in the United States has shrunk — 393

Tables

1.1	Technology systems and the evolution of the world economy	11
1.2	Global foreign exchange market turnover (daily averages, US$ billions)	17
2.1	World manufacturing	46
2.2	World's top non-financial TNCs ranked by foreign assets, 2010	47
2.3	Types of special economic zone	48
3.1	The continuum of geographical incorporation into the world economy	65
3.2	The geographical development of the world economy in the nineteenth century	66
3.3	The institutions and ideological basis of the world's dominant capitalisms	72
3.4	The old and the new pillars of world trade	81
3.5	"Hymer's stereotype", in which the space–process relationship takes the form A→B→C	87
3.6	Spatial transaction costs versus externalities: six scenarios	88
5.1	Growth rates in Europe	123
6.1	Percentage of value of shipments accounted for by the four largest companies in selected manufacturing industries in the USA	156
6.2	Inter- and intra-regional merchandise trade, 2011 (US$ billions)	160
6.3	United States trade in goods and services with the European Union (US$ billions)	161
6.4	Contrasts in production and labor: Fordism versus flexible production	163
7.1	Contrasts in governance: Fordism versus flexible production	182
7.2	World Internet users	199
7.3	Propulsive industries and new industrial spaces	203
8.1	Stock of foreign capital investment held by Europe, 1825–1915 (US$ billion)	220
8.2	World exports by region, 1876–1937, percent of total	224
8.3	World imports by region, 1876–1937, percent of total	224
8.4	Barter terms of trade and world prices for primary product commodity groups	232
8.5	Adaptations to global capitalism	243
9.1	Food production per capita for selected countries (2004–2006 = 100)	249
9.2	Land, labor, capital, and markets: haciendas and plantations	252
9.3	Index of total (and per capita) gross agricultural production: selected African countries (2004–2006 = 100)	255
9.4	Average farm size, 2010	258
9.5	A typology of land reforms	261
9.6	Selected export crops in some Latin American countries (metric tons)	266
10.1	Labor conditions in global manufacturing	278
10.2	Changing geography of manufacturing employment: paid employment in manufacturing (millions of people)	279

10.3   International variations in the concentration of incomes (higher figures indicate greater inequality)   284

10.4   Manufacturing plants outside the USA owned or leased by top US electronics corporations, 2013   292

10.5   Stages in South Korea's export-oriented industrial development   296

10.6   China Mobile Limited's cellphone global production system   298

10.7   The BRIC road to economic growth   309

11.1   Deciphering the outsourcing terminology   321

11.2   Changing employment in services as a percentage of total employment   326

11.3   EPZs targeting services   331

11.4   Wal-Mart retail stores   334

11.5   World's largest financial TNCs, 2011   341

11.6   E-finance, 2009   342

11.7   Major international business process outsourcing (BPO) corporations   345

12.1   Average tariff levels (percent) for selected countries, 1988 and 2009–2011   359

Boxes

1.1	Outsourcing and global commodity chains	14
1.2	Barbie: American icon and global product	16
2.1	HIV/AIDS and the impact on development in Sub-Saharan Africa	24
2.2	Indicators of well-being	26
2.3	E-waste and the digital divide	33
2.4	Migrant workers' remittances	43
2.5	From darling to disaster: Iceland and the global financial crisis	56
3.1	Neoliberalism	77
3.2	The semiconductor industry and the workings of the market-access regime	84
5.1	Core and periphery in Europe	124
5.2	The growth of the U.S. manufacturing belt	127
5.3	Regional dimensions of Japanese industrialization	135
6.1	Technological breakthroughs and the information economy	150
6.2	Fast fashion and IT: Zara responds rapidly to changing customer demand	154
6.3	The changing geography of the clothing industry	167
6.4	The myth of the new industrial districts of the Third Italy?	170
6.5	Coming to America?	173
7.1	The Sunbelt	188
7.2	Agglomeration and the "relational turn" in economic geography: Motorsport Valley	189
7.3	The digital divide	198
7.4	The demise of the Celtic Tiger	202
7.5	Hollywood and the cultural economy of cities	204
7.6	National economic development and regional inequality	208
8.1	The Asian financial crisis	240
9.1	The coffee commodity chain	248
9.2	From free trade to fair trade?	250
9.3	Agribusiness and the developed countries	263
9.4	The great land grab?	265
9.5	Science and rice	271
10.1	Dreaming the BRIC future	308
11.1	Bucking the Fisher-Clark thesis in China? Concurrent growth of manufacturing and services	320
11.2	A day in the life of a call center worker in India	325
11.3	Wal-Mart®	334
11.4	Abu Dhabi, a tourist Mecca?	338
11.5	India's competitive advantage in BPO	346
12.1	Genetically modified foods and U.S.–EU trade	375
13.1	International terrorism	389

Acknowledgements

The authors would like to thank the following organizations for permission to use the material listed:

Figure 1.4 from "Shifting governance structures in global commodity chains, with special reference to the Internet," by Gereffi, published in *American Behavioral Scientist* 44, 2001, with permission of Sage Publications.

Figure 2.8 from *Energy in Brief: What is shale gas and why is it important?* by E.I.A. (U.S. Energy Information Administration, U.S. Department of Energy) 2012.

Figure 2.9 from *Analysis and Projections: World shale gas resources: An initial assessment of 14 regions outside the United States* by E.I.A. (U.S. Energy Information Administration, U.S. Department of Energy) 2011.

Figure 2.10 from *Aquatic Dead Zones* by NASA, Earth Observatory, 2008.

Figure 2.13 from *Vital Water Graphics: An overview of the state of the world's fresh and marine waters* by UNEP (United Nations Environment Program) 2002.

Figure 2.14 from *Living Planet Report 2010: Biodiversity, biocapacity and development* by WWF (World Wide Fund for Nature) 2010.

Figure 2.18 from *Human Development Report 2009: Overcoming barriers: Human mobility and development* by UNDP (United Nations Development Program) 2009.

Figure 2.20 from *Poverty Reduction and Social Development: Migration and the brain drain phenomenon* by OECD (Organization for Economic Cooperation and Development) nd.

Figure 2.21 from *World Trade 2011, Prospects for 2012: Trade growth to slow in 2012 after strong deceleration in 2011* by WTO (World Trade Organization) 2012.

Figure 3.2 from *Corruption Perceptions Index 2011* by Transparency International, 2012.

Figure 4.1 from *An Introduction to Urban Historical Geography*, 1983, by Carter, with permission from Routledge.

Figure 4.2 from the *Historical Atlas*, 1923, by Shepherd, with permission from the Perry-Castañeda Library Map Collection, University of Texas at Austin.

Figure 5.2 from *Peaceful Conquest: The industrialization of Europe*, 1760–1970, 1981, by Pollard, with permission of Oxford University Press.

Figure 6.1 from "Industrial restructuring: an international problem," by Hamilton, published in *Geoforum* 15, 1984, with permission of Elsevier.

Figure 6.3 from "Boeing's outsourcing for the 787 Dreamliner," published in *Seattle Times*, 2006, 29 September, with permission of Seattle Times.

Figure 7.2 from *Science and Engineering Indicators 2012*, 2012, by National Science Board, with permission of National Science Foundation.

Figure 7.3 from *The Offshore Services Value Chain*, 2010, by G. Gereffi and K. Fernandez-Stark, with permission of G. Gereffi, Duke University, North Carolina, Center on Globalization, Governance, and Competitiveness (CGGC) http://www.cggc.duke.edu/pdfs/CGGC CORFO_The_Offshore_Services_Global_Value_Chain_March_1_2010.pdf

Figure 8.8 from "The River Plate countries," by Crossley, published in Blakemoor and Smith (eds) *Latin America: Geographical perspectives*, 2nd edition, 1983, with permission of Routledge.

Figure 12.3 from *Regional Policy—Inforegio: Cohesion policy 2007–2013* by the European Commission.

Every effort has been made to trace copyright holders of material reproduced in this book. Any rights not acknowledged here will be acknowledged in subsequent printings if notice is given to the publisher.

We are indebted to many of our colleagues for their advice at various stages in the conception and preparation of this book, and in its current revision. We would particularly like to recognize Stuart Corbridge, Raymundo Cota, Bob Dyck, Richard Grant, Larry Grossman, Naeem Inayatullah, D. Michael Kirchoff, Soo-Seong Lee, Andrew Leyshon, Ragnhild Lund, Sallie Marston, Ezzeddine Moudoud, Pritti Ramamurthy, Bon Richardson, Susan Roberts, David J. Robinson, Freddy Robles, Mark Rupert, David Short, Barney Warf, and Colin Warren for their contributions.

Part 1

Economic patterns and the search for explanation

In the first part of this book, we introduce the scope and complexity of our subject, establish the salient patterns in the world's economic landscapes, and review alternative theoretical approaches to understanding the development of these patterns. Chapter 1 provides the orientation for the book by outlining the relationships between the organization of the world economy and spatial change. In Chapter 2, the major dimensions of the world's contemporary landscapes are described. We identify dominant and recurring patterns and note the major exceptions to these patterns. Both the patterns and the exceptions raise a number of critical questions about process and theory in economic geography. For example: "How should the development process be conceptualized?," "What are the processes that initiate and sustain spatial inequalities?," and "Why are economic activity and prosperity spread so unevenly?" These questions are pursued in Chapter 3, where we outline a broad theoretical framework for understanding the interdependence of the world economy and its spatial components.

Picture credit:
Linda McCarthy

Chapter 1

The changing world economy

Picture credit: Linda McCarthy

The perspective of this book is global. Although local, regional, and national circumstances remain important, what happens in any given locale is increasingly influenced by its role in and relationship to systems of production, trade, and consumption that are global in scope.

In the 1970s, only a few **less developed countries** (LDCs) had opened their borders to trade and investment. About one-third of the world's labor force lived in countries with centrally planned economies. Another third lived in countries insulated from international markets by protective trade barriers and currency controls.

Today, more than 7 billion people populate our planet, and most live in countries that have been integrated into global markets. Three population blocs—China, India, and the republics of the former Soviet Union—account for more than 40 percent of the world's labor force and are important participants in the global market. Many other countries such as the **newly industrializing economies** (NIEs) of Brazil, Hong Kong, Singapore, South Korea, and Taiwan have also become vital contributors to the world economy.

More than two decades ago, Robert Reich, former U.S. Secretary of Labor, underscored the significance of the rapid pace of globalization:

> We are living through a transformation that will rearrange the politics and economics of the coming century. There will be no national products or technologies, no national corporations, no national industries. There will no longer be national economies, at least as we have come to understand that concept. . . . As almost every factor of production—money, technology, factories, and equipment—moves effortlessly across borders, the very idea of a U.S. economy is becoming meaningless, as are the notions of a U.S. corporation, U.S. capital, U.S. products, and U.S. technology.
>
> (Reich, 1991: 3, 8)

People are not only increasingly interconnected; they are interdependent as the following narratives generated by the **World Bank** illustrate:

> Joe lives in a small town in southern Texas. His old job as an accounts clerk in a textile firm, where he had worked for many years, was not very secure. He earned $50 a day, but promises of promotion never came through, and the firm eventually went out of business as cheap imports from Mexico forced textile prices down. Joe went back to college to study business administration and was recently hired by one of the new banks in the area. He enjoys a comfortable living even after making the monthly payments on his government-subsidized student loan.

> Maria recently moved from her central Mexican village and now works in a U.S.-owned factory in Mexico's *maquiladora* sector. Her husband, Juan, runs a small car upholstery business and sometimes crosses the border during the harvest season to work illegally on farms in California. Maria, Juan, and their son have improved their standard of living since moving out of subsistence agriculture, but Maria's wage has not increased in years; she still earns about $10 a day but does not complain because she has heard rumors that the company is considering moving the factory to China.

> Xiao Zhi is an industrial worker in Shenzhen, a Special Economic Zone in China. After three difficult years on the road as part of China's floating population, fleeing the poverty of nearby Sichuan province, he has finally settled with a new firm from Hong Kong that produces garments for the U.S. market. He can now afford more than a bowl of rice for his daily meal. He makes $2 a day and is hopeful for the future.

The complex relationships revealed in this anecdote would have been unthinkable 30 years ago.

Although the outcomes in this tale are positive—Joe secured a position at a bank and earns a comfortable living; Maria and her family improved their standards of living; and Xiao Zhi escaped poverty—not everyone benefits from globalization. Although the world is increasingly *flat* with capital crossing borders in nanoseconds in the pursuit of the highest rate of return and corporations locating operations where labor markets, tax codes, and regulatory regimes are most favorable, it is not necessarily increasingly *fair*.

In a hypothetical sequel to the World Bank narratives, the bank where Joe worked leveraged its portfolio with uncollateralized debt and had to close its doors at the height of the global financial crisis. With no income and a mortgage that exceeded the value of his house, Joe was forced to file for bankruptcy, but he still owes $830 per month on his student loans. He was fortunate enough to find a part-time job at a Wal-Mart warehouse, but with a one-way commute of 56 miles and gas prices nearing $4 per gallon, he cannot afford health insurance.

Juan was detained by border patrol and has joined approximately 390,000 other illegal immigrants incarcerated indefinitely in U.S. detention centers. The recession bloated inventories and decreased the demand for textiles produced in the factory where Maria worked. When she was laid off, she was forced to withdraw her son from school, move in with relatives, and return to subsistence farming.

And poor Xiao Zhi contracted a skin infection when he was forced to handle chemicals in the factory without protective gloves. The floor manager fired him when he could no longer maintain the required pace of production. After many failed attempts to secure another job in Shenzhen, he returned to Sichuan province but remains hopeful that the herbal remedies prescribed by the village doctor will heal him sufficiently so, one day, he will be able to earn $2 per day again.

1.1 STUDYING THE WORLD ECONOMY

How can one make sense of these stories? On the surface, cause-and-effect seem straightforward—the demand for textiles declined; Maria lost her job—but in the undercurrent one discovers a complex array of forces that have broad and dramatic effects that can often produce surprising and unexpected results. Deciphering the impact of those forces, interpreting their local, regional, and national implications and how they alter the contours of the economic landscape is the job of the economic geographer.

What are the implications of the Arab Spring; persistently high unemployment rates in EU countries such as Italy and Spain; the AIDS pandemic in Africa; continued environmental degradation in China; increased immigration to the EU from countries in North Africa and the Middle East; growing income inequality and increasingly polarized political landscape in the United States; technological advances such as **fracking** that enable the extraction of previously unprofitable carbon fuels; and the greater magnitude and frequency of natural disasters?

In addition to these headline-grabbing phenomena, what are the local, regional, and national implications of less newsworthy but equally profound changes in the world economy such as **resource grabbing** in Africa by **developed countries**, the rapid spread of genetically modified organisms (GMOs) in agricultural production, and technological advances that have enabled cost-effective 3-D printing?

How can we interpret the significance of specific changes that have been occurring in the world's economic landscapes: The **deindustrialization** of traditional manufacturing regions (for example, the Rustbelt around the Great Lakes in the United States, northern England, the Ruhr region in Germany), the economic revival of formerly "lagging" regions (for example, New England, Bavaria), the spread of **branch plants** in the towns and cities of some NIEs (for example, Taipei, Seoul), the emergence of high-technology complexes (for example, Silicon Valley in California, the Research Triangle in North Carolina), the consolidation of global financial and corporate control functions in a few cities (London, New York, Tokyo), and the unprecedented rates of urbanization in China's coastal regions?

Our task is to develop an understanding of the general economic forces and socioeconomic relationships within the world economy and of the unique features that represent local and historical variability.

But first we need to clarify the use of the terms "general" and "unique" as well as a third term, "singular":

- *General*: Widespread phenomena, such as migration or colonialism.
- *Unique*: Distinctive phenomena—where there are no other instances of it—but its distinctiveness can be explained by a particular combination of general processes and individual responses. An example of this would be the migration streams prompted by the famine in Ireland in the mid-1800s. The general processes that precipitated the famine were environmental (potato blight) and governmental (laissez-faire policy); in response, many people emigrated, including to the United States.
- *Singular*: Distinctive phenomena that cannot be accounted for by combinations of general processes and individual responses. An example would be the growth of the automobile industry in Detroit. With no established pattern of manufacturing, automobile manufacture could have developed in any number of cities; but Detroit was Henry Ford's home town (his father had emigrated to Michigan from Ireland to flee the famine), and he put his ideas into practice there.

With these concepts in hand, we can begin to map some of the interrelationships between economic organization and spatial change.

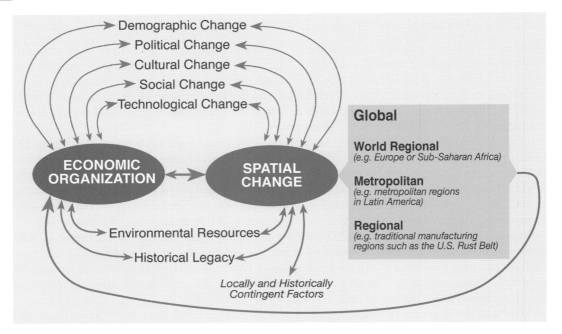

Figure 1.1 The inter-relationships surrounding economic organization and spatial change

Figure 1.1 shows that economic organization, while critical to spatial change, is implicated with demographic, political, cultural, social, and technological change. Many important interactions also emerge, for example, between political and cultural change and between locally contingent factors and spatial change.

All these direct, indirect, and interaction effects are important to developing a holistic understanding of spatial change: *They are implicated in and account for the general and the unique*. The task of the economic geographer is to unravel these relationships to develop a coherent and comprehensive explanatory framework.

To accomplish this goal, first we must gain a clear perspective on the central relationship between economic organization and spatial change. In the next section, we outline the most important aspects of this relationship and introduce several concepts that we will refer to throughout the book.

1.2 ECONOMIC ORGANIZATION AND SPATIAL CHANGE

At the most basic level, the idea of economic organization approximates to the concept of **mode of production:** The way in which societies organize productive activities to advance and reproduce their socioeconomic life. The theoretical and historical identification of modes of production is a difficult and controversial matter, but five major forms of economic organization are commonly recognized:

1. subsistence
2. slavery
3. feudalism
4. capitalism
5. socialism.

These broad categories also can be broken down into more specific forms of economic organization. For example, scholars have often found it useful to differentiate between **merchant capitalism** (or **mercantilism**), industrial (or **competitive**) capitalism, **organized capitalism,** and **advanced** (or globalized) **capitalism.**

What distinguishes these forms of economic organization are differences in the relationships between the **factors of production** such as land and other natural resources, labor, and physical and human capital. Under slavery, for example, the laborer is private property and may be bought and sold similar to any other instrument of production. Under **feudalism** (or rank redistribution), peasant laborers are legally tied to specific tracts of land. They may own some of the instruments of production, but the land and a percentage of the product of their labor is the property of the feudal lord. Under capitalism, the laborers own no instruments of production, but they are free to sell their labor power.

Different forms of economic organization are also characterized by different *forces of production* (for example, technology, machinery, means of transportation) and *social formations* (with specific proportions of participants from various social classes).

The economic "logic" of these different forms of economic organization results in substantially different forms of spatial organization. Where feudalism translates into a patchwork of self-sufficient domains with little trade and, therefore, few market centers, merchant capitalism requires a highly developed system of market towns and an inherent tendency to colonize new territories to amass the wealth and resources necessary to sustain ever expanding markets.

In contrast to feudalism and mercantilism, industrial capitalism requires spatial restructuring that enables the exploitation of new energy sources, development of increasingly efficient production techniques, and the adoption of new forms of corporate organization. Mining and manufacturing towns appear, and whole regions, such as the manufacturing cities around the Great Lakes or the Ruhr region of Germany, become specialized in certain kinds of industrial production.

The "classic" sequence of transformation from one form of economic organization to another runs from **subsistence economies** through slavery, feudalism, mercantilism, **industrial capitalism,** and **advanced capitalism.**

This sequence is also distinctly European. The rise of capitalism in much of Europe and the subsequent drive to acquire resources to propel economic growth led to different sequences of development in other regions. In North America, capitalism was imposed directly on the subsistence economies of Native American communities. In Japan, feudalism was uprooted suddenly by state-sponsored industrial capitalism. In Russia, an embryonic industrial capitalism was displaced by a **socialism** that soon gave way to state capitalism. Today, in many lesser developed countries, aspects of multiple forms of economic organization may also coexist. As a result of these variations, important regional differences have emerged in the world economy.

Spatial change and further regional differentiation also occurs with the evolution of forms of economic organization. So a regional agricultural landscape must be seen as one of a number of possible realizations rather than a straightforward reflection of a particular form of economic organization. Each economic landscape should be interpreted, therefore, as the product of broad economic forces interacting with local social, cultural, political, and environmental factors: A product of the *general* and the *unique*.

THE EVOLUTION OF CAPITALISM

The evolution of the capitalist system of economic organization has been perhaps the most important influence on the development of the world's economic landscape. Its evolution can be traced through three broad phases: Competitive, organized, and globalized capitalism.

Competitive capitalism

The earliest phase began in the late 1700s in the United Kingdom. It spread through much of northwestern Europe and North America, and continued until the end of the nineteenth century. This phase of competitive capitalism was the heyday of free enterprise and laissez-faire economic development. Competition between small family businesses characterized the markets, and there were few constraints or controls imposed by governments or public authorities (see Figure 1.2).

In the earlier years of this phase, the dynamism of the system rested on the profitability of agriculture and, increasingly, manufacturing and machinofacture, which involved industrial production based less on handicraft and direct labor power than on mechanization, automation, and intensively used skilled labor.

Manufacturing boosted the wealth of NIEs, and their collective prosperity was further consolidated through imperialism, which ensured supplies of raw materials and markets for manufactured goods. Gradually, competition led to consolidation. Some businesses prospered and expanded their operations while less nimble, well-capitalized, or adept entrepreneurs saw their businesses contract and eventually become absorbed by their more successful counterparts.

As companies expanded operations to serve regional and national markets rather than exclusively local ones, business owners also experimented with new organizational structures. Labor markets became more organized as wage norms spread. And as private interests acquired ever greater wealth, the need for governments to regulate public affairs and mediate between increasingly powerful interests become apparent.

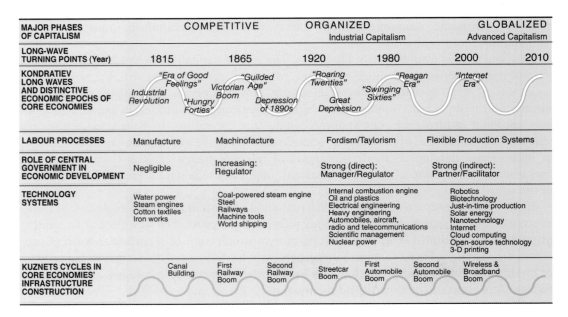

Figure 1.2 Major features of economic change in the world's developed economies

Near the end of this phase of capitalism, the United States also surpassed the United Kingdom as the leading industrial economy.

Organized capitalism

By the early 1900s these trends had altered the nature of the capitalist enterprise so significantly that a new phase—organized capitalism—was demarcated.

In the early decades of the twentieth century, profitability became increasingly dependent on new labor processes. Fordism, named after the automobile manufacturer Henry Ford, ushered in the era of mass production using assembly-line techniques. Frederick Winslow Taylor, an engineer and early critic of Fordism, also outlined the principles of scientific management (often known as Taylorism) which became central to the efficiency movement in manufacturing.

During this period, mass production lowered the costs of many goods, and higher wages and sophisticated advertising techniques fuelled mass consumption. In turn, mass consumption and production initiated the race to find ever more efficient production processes and untapped markets.

A hallmark of this period was the emergence of a workable relationship between business interests and labor unions. Unions had grown in size and strength in the progressive era, and constituted another increasingly important element of "organization." Government also expanded the scope of its activities in part to mediate the relationship between organized business and labor.

The market failures that triggered the Great Depression of 1929–1934 undermined the legitimacy of classical economic liberalism and led to its eclipse in the New Deal era by an egalitarian liberalism that relied on the state to manage economic development and soften the unwanted side-effects of free market capitalism.

In this expanded role, government assumed responsibility for the management of the national economy and the organization of various dimensions of social well-being. This type of economic policy, which seeks to mitigate the deleterious effects of private market activity in aggregate through active fiscal and monetary policy, is known as Keynesianism after the eminent economist John Maynard Keynes.

Globalized capitalism

After the Second World War, another important transformation in the nature of capitalism began to take place and led to a third major phase: Advanced or globalized capitalism.

This period is characterized by a shift away from industrial production and toward services, particularly sophisticated financial and business services, as the basis for profitability within the more developed economies. Labor-intensive manufacturing declined although manufacturing *production* continued to expand in these countries as sophisticated, technology-intensive manufacturing processes gained prominence.

The globalizing of the economy also meant that large transnational corporations (TNCs) were able to outmaneuver the national scope of governments and labor unions and contributed to a destabilization of the "organized" relationship between business, labor, and government. By the mid-1990s, the world's largest TNCs accounted for two-thirds of international trade, and the largest ten reported total income that exceeded that of the world's 100 poorest countries.

Meanwhile, Fordism became a victim of its own success as many markets were saturated with low-cost goods. As profit margins in conventional markets narrowed, many enterprises chased revenues by catering to specialized market niches. Such specialization required flexible production systems. The overall result has sometimes been labeled disorganized capitalism for its distinct contrast to the orderly interdependence of business, labor, and government in the system of organized capitalism.

One of the driving forces behind growth during this phase of capitalism has been the global **information** (or knowledge) **economy**, a form of production and management where productivity and competitiveness rely heavily on knowledge generation and on gaining access to and the rapid assimilation of new information.

A second critical driver of this era has been the ubiquity of high technology. In particular, the Internet has affected nearly every facet of the economy and sparked **disruptive innovation** that has radically altered the dynamics of marketplaces and industries such as news and media, music, publishing, and advertising. Although the first phase of euphoric investment in Internet technologies culminated in the bursting of the dot.com bubble in 2000, from its ashes emerged a number of robust and dynamic businesses such as Google, eBay, and Amazon.com that continue to transform the economic landscape. With the expansion of broadband and wireless technologies, cloud-computing, open source development tools, and the advent of Web 2.0, numerous platform technologies and social media enterprises such as YouTube, Twitter, Facebook, Flickr, and WordPress continue to change how people, companies, and institutions collaborate, communicate, and compete with each other.

There is some question, however, about the sustainability of technological change as a driver of economic growth. We may have reached a technological plateau in which we continue to exploit yesterday's ideas rather than develop new ones. For example, 80 percent of total growth in U.S. GDP between 1950 and 1993 came from the application of previously discovered ideas plus huge investments in education and scientific research that cannot be easily repeated in the future. The overall rate of innovation from medieval times to the present peaked in the nineteenth century and has gone downhill since. So, we should be careful not to presume that somehow technological change will necessarily continue to endlessly create more growth.

TECHNOLOGY AND ECONOMIC DEVELOPMENT

> The opening up of new markets, foreign or domestic, and the organizational development from the craft shop and factory to such concerns as U.S. Steel illustrate the same process of industrial mutation ... that incessantly revolutionizes the economic structure *from within*, incessantly destroying the old one, incessantly creating a new one. This process of Creative Destruction is the essential fact about capitalism.
>
> (Schumpeter, 2010/1943: 73)

Coined by the Austrian economist Joseph A. Schumpeter, the term **creative destruction** captures the essence of the capitalism system—the relentless drive to innovate in the competition for markets—perhaps better than any other concept.

Creative destruction is at the heart of the broad structural shifts that occur in the development of **technology systems** (see Table 1.1). The transportation system provides perhaps the starkest example of creative destruction at work. With each advancement new markets opened, industries developed, and cities were built, but not without costs. The railroads that opened the West in the United States also effectively closed the canal systems that had been built in the previous decades.

Technologies also destroy old jobs even as they create new ones. Typically, more primitive technologies are associated with higher levels of labor inputs than are more sophisticated ones. So as technology has become more central to economic development, employment, particularly in limited or semi-skilled jobs, has suffered. The newest technologies today require "leaner" and more skilled labor than did the older ones. By way of example, in 1900 nearly 350,000

Table 1.1 Technology systems and the evolution of the world economy

I: Water (around 1785 in England)

- Water power and steam engines
- Cotton textiles, pottery, and iron working
- Transportation systems (e.g. river systems, canals, turnpike roads)

II: Steam transport (late 1820s)

- Coal-powered steam engines
- Steel
- Railroads
- Global shipping
- Machine tools

III: Steel and electricity (late 1870s)

- Internal combustion engine and automobiles
- Oil and plastics
- Electrical and heavy engineering
- Radio and telecommunications
- Airplanes

IV: Fordist (around 1915 in the United States)

- Nuclear power
- Durable goods and consumer industries
- Aerospace industries
- Electronics
- Petrochemicals

V: High technology (late 1970s–present)

- Microprocessors
- Biotechnology
- Robotics
- Broadband and wireless systems
- Genetic engineering
- Nanotechnology
- Internet technology (e.g. cloud-computing, open-source applications)

people were employed as blacksmiths or carriage and harness manufacturers. Today, thanks to the internal combustion engine, few earn a living in these professions, but millions of people work as auto mechanics, long-haul truckers, and taxi drivers, and millions of others are employed in production, sales, and manufacturing jobs related to the auto industry.

The evolution of systems does not only impact the markets for goods, services, and labor. As one system is eclipsed by another, so different regions are favored or disadvantaged. A city such as Chicago, which aggressively invested in railroads, saw its population increase fourfold in the span of a decade in the mid-1800s. From this small **initial advantage**, Chicago became the dominant city in the Midwest, the hub of transcontinental trade. By the time it supplanted Philadelphia as the second largest city in the United States, it had developed a diverse economy with leading corporations in industries as diverse as iron and steel, garment production, publishing, banking, insurance, and mail-order retail. During this rapid period of growth, it also surpassed older and initially larger cities such as St. Louis, which depended on and were invested in steamboat commerce.

As this example illustrates, changes in technology are crucial to understanding the geographical path dependence of economic activities, that is, the historical relationship between present economic activities associated with a place and its past experience. As new technologies eclipse old ones, industries—and sometimes entire industrial regions—are "dismantled" (or, at least, neglected) as investors shift capital to fund the creation of new centers of profitability and employment.

1.3 SPATIAL DIVISIONS OF LABOR

The evolution of capitalism has also brought about changes in the spatial division of labor. The division of labor within and between firms and over space is not fixed; rather, it responds to changes in the historical-structural context in which firms operate.

For example, during the Fordist period in countries such as Britain and the United States, the basic division of labor was organized primarily within regional parts of the national economy. Plant, firm, and industry were national phenomena. They were organized around national markets and industries, and they created national social (class) divisions. Although capital, labor, and technology were often imported and exported, these factors of production were subject to intensive regulation by national governments.

The internal geography of a national economy such as Britain's reflected its position in the international division of labor. In the 1930s Britain specialized in certain key manufacturing industries such as coalmining, iron and steel manufacturing, and shipbuilding. Previous investments in these industries and the increasing returns to scale and external economies of scale they generated defined Britain's trading patterns. Elsewhere, different industries, often newer, mass-production ones based on larger firms, also took root. As a result, trade reflected cumulative competitive advantages in sectors where each had a "head start."

The locational consequences in the British case are laid out by Massey (1984: 28–29) as follows:

> It was the United Kingdom's position as an imperial power, its early lead in the growth of modern industry, and its consequent commitment to free trade and its own specialization in manufacturing *within* this international division of labor, which enabled the rapid growth, up to the First World War, of these major exporting industries. The spatial structures that were established by those industries were those where all the stages of production of the commodity are concentrated within single geographical areas. The comparatively low level of separation of functions within the process of production, and the relatively small variation in locational requirements between such potentially separable functions, were not sufficient to make geographical differentiation a major attraction.

In other words, the spatial division of labor of key industries within national economies was based largely on different regional industrial specializations. And agglomeration was a major feature of economic organization across a number of manufacturing industries.

Similarly, in the United States during this period, the northeast contained a vast array of specialized manufacturing clusters—steel in Pittsburgh, automobiles in Detroit, chemicals in Wilmington, and photographic equipment in Rochester—as well as regional areas of specialization in agricultural (for example, tomatoes and sweet corn in the Garden State of New Jersey, low-bush blueberries in Maine, maple syrup in Vermont), and raw materials (for example, stone quarries in the "granite state" of New Hampshire). In this respect, places and regions could readily be associated with specific products.

GLOBALIZATION AND CHANGING SPATIAL DIVISIONS OF LABOR

Under the conditions of flexible production, such regional specialization has been challenged and undermined to a considerable degree. Spatial divisions of labor are now structured in a variety of ways depending on the needs and characteristics of particular industries. In addition to (1) *regional specialization* and (2) *regional dispersal* (which has characterized consumer services such as stores, restaurants, and hospitals, and some manufacturing industries such as shoe production and food processing), four additional spatial divisions of labor can be identified:

1. *Three-tier regional functional separation*
   a. management and research activities in major metropolitan regions
   b. skilled labor in "old" manufacturing areas
   c. unskilled labor in regional peripheries (to exploit lower wages and non-unionized labor forces).
2. *Two-tier regional functional separation*
   a. management and research activities in major metropolitan regions
   b. semi-skilled and unskilled labor in regional peripheries.
3. *Regional and global functional separation*
   a. management, research, and skilled labor in advanced industrial regions
   b. unskilled labor in the global periphery.
4. *Divisions by areas of growth and decline*
   a. some areas characterized by investment, technical change, and job expansion
   b. other areas characterized by stagnant and progressively less competitive production and job loss.

These new spatial divisions of labor have been possible because transportation and communications technologies have created an environment in which firms can decentralize activities associated with primary production yet maintain central control. A firm can remain headquartered in New York, Zürich, or Hamburg, but locate manufacturing facilities in a location such as Chennai, India, or the Monterrey-Nuevo Laredo corridor in Mexico and reap the benefits of non-union labor forces, easier access to concentrated regional markets, and favorable regulatory environments.

Under this new international division of labor (NIDL), investment and production are no longer organized primarily around national economies. The process of production—most obviously in the examples of the automobile, electronics, and software industries—is now global. Components or specific services are sourced from multiple suppliers in different countries and assembled in several locations (see Box 1.1).

In fact, many products no longer have any obvious nationality. It is difficult to distinguish some "U.S." from some "Japanese" cars, for example, now that U.S. car companies import vehicles under "their" names from Japan, and Japanese companies now manufacture cars in the United States (for example, Honda in Marysville, Ohio, and Toyota in Blue Springs, Mississippi). Even manufacturing the Barbie doll, an all-American icon, includes operations in a number of countries (see Box 1.2).

Box 1.1 Outsourcing and global commodity chains

International outsourcing affects millions of people (see Figure 1.3) and is a controversial topic, particularly during periods of high or rising unemployment such as the Great Recession that followed the global financial crisis of 2008. But neoclassical economists see international outsourcing as a beneficial and natural consequence of free trade:

> In February 2004, when N. Gregory Mankiw, a Harvard professor then serving as chairman of the White House Council of Economic Advisers, caused a national uproar with a "textbook" statement about trade, economists rushed to his defense. Mankiw was commenting on the phenomenon that has been clumsily dubbed offshoring (or offshore outsourcing)—the migration of jobs, but not the people who perform them, from rich countries to poor ones. Offshoring, Mankiw said, is only "the latest manifestation of the gains from trade that economists have talked about at least since Adam Smith . . . More things are tradable than were tradable in the past, and that's a good thing." Although Democratic and Republican politicians alike excoriated Mankiw for his callous attitude toward U.S. jobs, economists lined up to support his claim that offshoring is simply international business as usual.
>
> (Blinder, 2006:1)

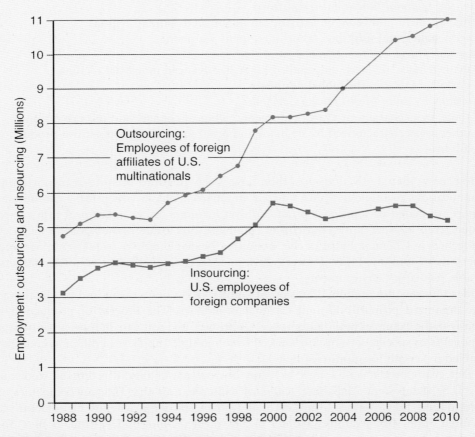

Figure 1.3 Employment outsourcing and insourcing, United States

Source: Adapted from Mankiw and Swagel (2006: 27, Figure 2)

Producer-driven commodity chains

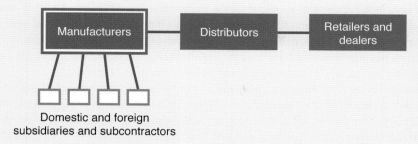

Buyer-driven commodity chains

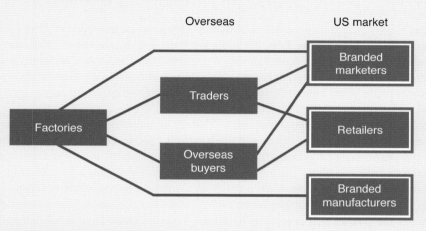

Figure 1.4 Basic elements of commodity chains
Source: Based on Gereffi (2001: 1619, Figure 1)

The world economy is constituted through numerous **commodity chains** (see Figure 1.4) that crisscross the globe. Commodity chains link the supply and processing of raw materials, the production of components, and the assembly and distribution of finished products in vast global systems. As we will see in subsequent chapters, these global assembly systems are increasingly important in shaping economic landscapes.

Global assembly systems also provide manufacturers several advantages.

First, a global assembly system for standardized products can maximize economies of scale. Second, it enables corporations to take greater advantage of the full range of geographical variations in costs for production and assembly. Basic wages in manufacturing industries, for example, are between 25 and 75 times higher in advanced industrial countries than in some LDCs. With a global assembly system, labor-intensive work can be done where labor is cheap, raw materials can be processed near their source of supply, and assembly can be done close to major markets. Finally, a global assembly system means that companies are less dependent on a single source of supply for specific resources, thereby reducing its vulnerability to industrial troubles and other disturbances.

Box 1.2 Barbie: American icon and global product

The famous (and impossible) physique of the Barbie doll says "Made in America" but the box it comes in says "Made in China." Tracing Barbie's production path raises interesting questions about how its place of origin can be identified and how the globalization of production ties together disparate locations in the world economic core and the periphery (a topic that will also be covered in greater detail in the next chapter).

Although Barbie is an American icon and a team of over 100 designers, beauticians, tailors, and sculptors at the headquarters of Mattel Corporation in El Segundo, California collaborate on her spring, fall, and holiday collections every year, Barbie has *never* been made in the United States. The first doll was produced in Japan in 1959. As costs rose in Japan, production was moved to other sites in Asia including Taiwan, Hong Kong, and the Philippines. Following a strike in 1988, Mattel closed its two Philippine factories resulting in the loss of 4,000 jobs.

Mattel closely guards its proprietary manufacturing process; however, in 1996 *Los Angeles Times* staff reporter, Rone Tempest, did some sleuthing and discovered the following about Barbie: She is made from ethylene, refined oil imported from Saudi Arabia, which is turned into pellets by a firm in Taiwan. Barbie's nylon hair comes from Japan. Her cardboard packaging is made in the United States. The manufacturing and packaging is managed from Hong Kong.

The production story begins, however, in Mattel's commodity management center where information about commodity prices and wage rates is used to decide on the best locations to buy the plastic resins, the cloth, the paper and other materials, and bring them together at a final point of assembly.

At one time, Japan and Taiwan were the main toymakers to the world economy. As their economies diversified into more capital-intensive production, they became the suppliers of the plastics that previously had come from the United States and Europe. At that time, production shifted to lower wage sites such as China, Thailand, and Indonesia (Foek, 1997).

Making Barbie is extremely labor intensive. Workers must operate plastic molds, sew clothing, and paint the details on the dolls. A typical Barbie requires 15 separate paint stations. Machines cannot perform these tasks. So the two Barbie plants in China employ about 11,000 workers, mainly unmarried women between 18 and 23 from poor regions of interior China brought to work at the factories for two to five years (Tempest, 1996).

So, Barbie is made in China. In the trade ledgers—where country trade deficits and surpluses are defined—Barbie is one of its exports. But a number of firms in different countries contributed to its production and reaped profits from the final product. Tempest estimated that Chinese firms and workers obtained only about 35 cents out of the $2 export value placed on each doll. In contrast, Barbie retails in the U.S. for $12–18.

In recent years, global sales for Barbie have fluctuated somewhat. Nevertheless, Barbie continues to account for $3 billion annually in retail sales for Mattel, and the secondary market for all things Barbie remains hotter than her "Barbie pink" toenail polish.

For many transnational corporations, national markets for capital, labor, and plant and office location exist only as parts of global **commodity chains**. Even small firms now have the opportunity to operate globally, outsourcing web development and design, production and packaging, customer service, and nearly any other facet of its business while competing for customers in local, regional, and global markets. So these "new" conditions cannot be solely identified with multinational corporations.

The pace of economic globalization has accelerated since the late 1960s. Between 1961 and 1976, for example, the number of employees of German firms outside Germany increased tenfold. The number of firms with foreign operations doubled during the same period of time. Today, the 30 largest corporations headquartered in Finland employ more than 50 percent of their employees outside Finland. German and Finnish firms have generally been less willing to expand foreign operations compared to U.S. and British firms, so these figures indicate something of a lower bound among countries with long histories of industrialization.

Paralleling and stimulating this trend has been the emergence of international devices for steering capital beyond national control (for example, the **eurodollars** in circulation outside the United States, see p. 54) and the **offshore financial centers** (see p. 340) that, rather like some city-states in the past, now service the new international division of labor. Table 1.2 highlights the rapid growth in the size and depth of the global foreign exchange market, where daily volumes now exceed to 25 percent of U.S. GDP. Some small countries such as Luxembourg and Switzerland have successfully cashed in on the world economy to the extent that they now have median household income levels higher than those of the "old" national manufacturing economies such as the U.K. and Germany.

National economies, therefore, are no longer the sole building blocks of the world economy. For an increasing proportion of agricultural and manufactured commodities and for some services, production and markets have become worldwide. This shift has had important

Table 1.2 Global foreign exchange market turnover (daily averages, US$ billions)

	1989	1992	1995	1998	2001	2004	2007	2010
Spot transactions[1]	317	394	494	568	386	631	1,005	1,490
Outright forwards[2]	27	58	97	128	130	209	362	475
Foreign exchange swaps[3]	190	324	546	734	656	954	1,714	1,765
Estimated gaps in reporting	56	43	53	61	26	107	128	144
Total 'traditional' turnover	590	820	1,190	1,491	1,198	1,901	3,209	3,874

1 Single outright transactions involving the exchange of two currencies at a rate agreed on the date of the contract for value or delivery (cash settlement) within two business days.
2 Transactions involving the exchange of two currencies at a rate agreed on the date of the contract for value or delivery at some time (more than two business days) in the future.
3 Transactions involving the actual exchange of two currencies on a specific date at a rate agreed at the time of conclusion of the contract (the short leg), and a reverse exchange of the same two currencies at a date further in the future and at a rate agreed at the time of the contract (the long leg).

Source: Based on Bank for International Settlements (2010: 7, Table B1)

consequences for the spatial distribution of economic activities both globally and within countries.

Globally, it has given rise to the growth of newly industrializing economies such as South Korea and Brazil. It has also contributed to a significant polarization of income and wealth. According to the World Bank (2010), the average per capita income in the richest 20 countries is 47 times that of the poorest 20. Within the "core" of advanced industrial national economies, the new international division of labor has led to a reorientation in employment away from manufacturing to services and a massive restructuring of regional economies. In Britain, for example, three sorts of local area have fallen victim to the loss of traditional manufacturing industries and the failure of new ones to replace them:

1. the centers of nineteenth-century industrialization in the north of England, south Wales, and central Scotland
2. the inner cities of London and other large metropolitan areas with concentrations of poor people and few of the unskilled jobs that they used to fill
3. the centers of the growth industries (specifically vehicles and engineering) of the 1950s and 1960s in the West Midlands and northwest of England.

We will draw on this framework throughout the remainder of the book as we analyze and describe the geography of the world economy. In the next chapter, we establish the major dimensions of the contemporary economic landscapes within the world economy. In Chapter 3, we outline a comprehensive global historical framework that serves as the context for the rest of the book.

In Part 2, we trace the emergence of three of the world's core economies—Europe, North America and Japan—and follow their paths towards increasing scale and complexity. Part 3 focuses on the rest of the world and pays special attention to the spatial transformations that have occurred as a consequence of colonialism and global capitalism that emanated from the core economies as well as the role of agriculture and manufacturing industries in economic development and spatial change.

Finally, in Part 4, we examine some of the reactions to the emergence of ever larger and more powerful economic forces that have come to characterize the world economy. In particular, we describe the spatial consequences of transnational political and economic integration and decentralist reactions: Nationalism, regionalism, and grassroots movements towards economic democracy.

SUMMARY

In this introductory chapter, we provided the orientation for the book by outlining the relationships between the organization of the world economy and spatial change. We stressed how studying the world economy involves developing an understanding of the general economic forces and socioeconomic relationships within the world economy and of the unique features that represent local and historical variability. We introduced the interrelated concepts of economic organization and spatial change and discussed the five major forms of economic organization commonly recognized: Subsistence, slavery, feudalism, capitalism, and socialism. We established some of the basic ideas and outcomes associated with globalization including technology, economic development, and changing spatial divisions of labor.

KEY SOURCES AND SUGGESTED READING

Blinder, A.S., 2006. Offshoring: The next Industrial Revolution? *Foreign Affairs*, March/April.

Brynjolfsson, E. and McAfee, A., 2011. *Race against the Machine: How the digital revolution is accelerating innovation, driving productivity, and irreversibly transforming employment and the economy*. Lexington, MA: Digital Frontier Press.

Cowen, T., 2011. *The Great Stagnation: How America ate all the low-hanging fruit of modern history, got sick, and will (eventually) feel better*. New York: Dutton.

Hughes, A. and Reimer, S. (eds.), 2004. *Geographies of Commodity Chains*. New York: Routledge.

Johnston, R.J., Taylor, P.J., and Watts, M. (eds.), 2002. *Geographies of Global Change: Remapping the world*. Cambridge, MA: Blackwell.

Mankiw, N.G. and Swagel, P., 2006. *The Politics and Economics of Outsourcing*, Working Paper 12398, National Bureau of Economic Research.

Massey, D., 1984. *Spatial Divisions of Labor*. London: Methuen.

O'Loughlin, J., Staeheli, L., and Greenburg, E. (eds.), 2004. *Globalization and its Outcomes*. New York: Guilford Press.

Schumpeter, J.A., 2010. *Capitalism, Socialism and Democracy*. New York: Routledge.

Storper, M. and Scott, A.J. (eds.), 1992. *Pathways to Industrialization and Regional Development*. London: Routledge.

Venables, A.J., 2006. Shifts in Economic Geography and Their Causes, *Federal Reserve Bank of Kansas City Economic Review* 31, 61–85.

Wallerstein, I., 1991. *Geopolitics and Geoculture. Essays on the changing world-system*. Cambridge: Cambridge University Press.

Chapter 2

Global patterns and trends

Picture credit: Linda McCarthy

Geography is about local variability within a general context.

(R.J. Johnston, 1984: 444)

In this chapter, we describe the major dimensions of the contemporary economic landscape. We identify dominant and recurring patterns and note the major exceptions to these patterns. In other words, we are concerned primarily with characterizing the *general context* referred to by Johnston in our opening quote. To the extent that we identify exceptions and contradictions, we are also concerned with *local variability*.

In subsequent chapters, our objective will be to uncover the processes that have contributed to these patterns—both the general and the locally distinctive or unique. As we will demonstrate, from the interaction of the unique with the general, distinctive economic regions emerge.

The dominant components of economic geography at the global scale are most often cast in terms of *core–periphery* differences. Meier and Baldwin (1957) were perhaps the earliest writers to attempt a conceptual description of this core–periphery structure on a global scale. They noted that a country is at the center of the world economy:

> [I]f it plays a dominant, active role in world trade. Usually such a country is a rich, market-type economy of the primarily industrial or agricultural-industrial variety. Foreign trade revolves around it: It is a large exporter and importer, and the international movement of capital normally occurs from it to other countries.

In contrast, they posited, a country could be considered peripheral:

> [I]f it plays a secondary or passive role in world trade. In terms of their domestic characteristics, peripheral countries may be market-type economies or subsistence-type economies. The common feature of a peripheral economy is its external dependence on the center as the source of a large

proportion of imports, as the destination for a large proportion of exports, and as a lender of capital.

(Meier and Baldwin, 1957: 147)

By the 1960s international socioeconomic inequalities had become more rather than less pronounced. A virtual avalanche of critical writings appeared claiming that the prosperity of the developed countries in the world economy (the United States, Europe, and Japan, in particular) was based on the *under*development of LDCs. The latter could not "follow" the previous historical experience of developed countries, it was argued, because their underdevelopment was a structural requirement for development elsewhere. By means of unequal trade, the exploitation of labor, and profit extraction, the less developed countries were becoming *increasingly* rather than decreasingly impoverished.

The **world-system** theory, developed by Immanuel Wallerstein (1984), took this disequilibrium into account. According to this perspective, the entire world economy can be seen as an evolving market system in which an economic hierarchy of countries—a *core, periphery* and *semi-periphery*—is the product of the long-wave economic rhythms that dominate the dynamics of the system. The forces of these economic long waves meant that the composition of each category is variable, meaning that countries can move from one hierarchical level to another (for example, core to periphery or periphery to semi-periphery).

The labels "core" and "periphery" are used by Wallerstein to refer to the dominant *processes* operating at particular levels in the hierarchy. Some of the key characteristics include the following:

- *Core*: Relatively high wages, advanced technology, and a diversified production mix
- *Periphery*: Low wages, more rudimentary technology, and a simple production mix
- *Semi-periphery*: A mix of core and periphery processes in place; exploitation of peripheral countries and by core countries.

Figure 2.1 represents an attempt to capture the current composition of the three categories based on countries' total **gross domestic product** (GDP) and GDP per capita. GDP is an estimate of the total value of all materials, foodstuffs, goods, and services that are produced by a country in a particular year. To standardize for countries' varying sizes, the statistic is normally divided by total population, which gives an indicator, per capita GDP, which provides a reasonable yardstick of relative levels of economic development. **Gross national income** (GNI), a similar measure, includes the value of income from abroad—flows of profits or losses from overseas investments, for example.

Countries with high scores on total GDP and GDP per capita are likely to be politically strong states and have large internal markets and predominantly high-wage, capital-intensive production—all, theoretically, defining characteristics of core status. Conversely, countries with low national economic output are likely to be weak states and have predominantly low-wage, labor-intensive production.

These assumptions are rather sweeping, and the allocation of individual countries to particular categories is inevitably somewhat arbitrary. Additionally, countries with intermediate scores for total GDP and GDP per capita comprise quite a diverse group. Semi-peripheral countries include resource-exporting countries such as Saudi Arabia and South Africa and NIEs such as Mexico, Brazil, Hong Kong, and Singapore; as well as more recent NIEs such as China, India, Malaysia and Thailand. The semi-periphery also includes European countries with less developed countries such as Greece and Portugal; the formerly socialist countries of eastern Europe; and Russia.

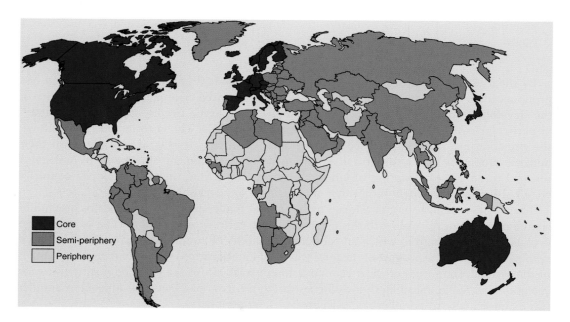

Figure 2.1 The world-system: core, semi-periphery and periphery

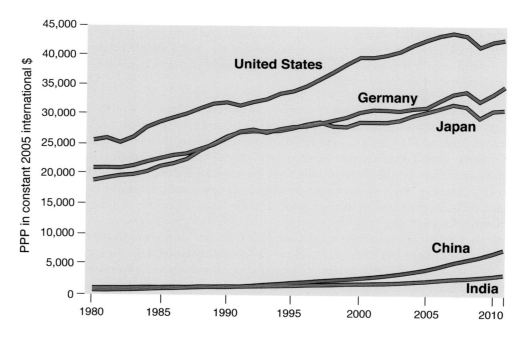

Figure 2.2 GDP per capita (PPP, constant 2005 international dollars)

Source: Based on online data from World Development Indicators Database (WDI) 2012. Washington, DC: World Bank Group

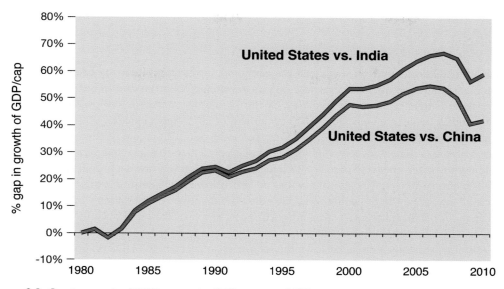

Figure 2.3 Gap in growth of GDP per capita (PPP, constant 2005 international dollars)

Source: Based on online data from World Development Indicators Database (WDI) 2012. Washington, DC: World Bank Group

Categorizing countries in this manner is not simply an exercise in alternative classification; it reflects a particular conception of the dynamics of the world economy. Economic prosperity is unevenly distributed across countries. Indeed, the absolute gap in income between the richest and poorest countries continues to widen. Consider the differences between China and India, the largest LDCs, and Germany, Japan and the United States, the economically largest DCs (see Figure 2.2). Between 1980 and 2010, the absolute difference in per capita GDP between China and the United States (measured in constant 2005 dollars) grew by more than 40 percent, while the gap between India and the United States increased by nearly 60 percent (see Figure 2.3).

2.1 WHAT "ECONOMIC DEVELOPMENT" MEANS

Major international economic cleavages not only reflect differences in prosperity but also different forms of economic organization, kinds of resource base, demographic characteristics, political system, and roles in the system of international specialization and trade.

Defining and measuring "economic development" is, therefore, problematic. As we have seen, there are strong grounds for thinking in terms of *under*development rather than development as far as LDCs are concerned, since the term "development" implies a trajectory of improvement in relative and absolute terms.

Increasingly, world leaders, nongovernmental organizations (NGOs), and scholars frame development in broader terms that include consideration for social well-being. Narrow economic definitions, while admirably precise, illuminate only one dimension of the picture. They encompass changes in the amount, composition, rate of growth, distribution, and consumption of resources, but they do not extend to the effects these changes have on people's lives. These broader considerations such as access to food and clean drinking water, healthcare, primary and secondary education, housing, an adequate level of security, and protection of

Box 2.1 HIV/AIDS and the impact on development in Sub-Saharan Africa

The HIV/AIDS epidemic has weakened the economic performance of many countries, but the impact has been the most dramatic on less developed countries and, in particular, the poorest members of those communities. Of the 31–35 million people worldwide who were estimated to be infected with HIV/AIDS at the end of 2010, approximately 23 million lived in Sub-Saharan Africa (UNAIDS, 2011). Despite having only 12 percent of the world's population, this region accounts for more than 68 percent of HIV-infected people.

In recent years, the pace of new HIV infections has fallen worldwide, including in 22 Sub-Saharan countries. The annual number of deaths from AIDS-related causes has also declined due largely to the widespread introduction of antiretroviral therapy. It is estimated that over 700,000 deaths were averted in 2010 as a result of the rapid scale-up of treatment efforts (UNAIDS, 2010).

Although these trends are positive, the long-term impact of HIV/AIDS on Sub-Saharan countries will be significant. Nearly 90 percent of the more than 16 million children orphaned as a result of HIV/AIDS live in the region. The impact on basic education has been dramatic with school enrolments declining even as the number of school-aged children increases. In South Africa, one in five teachers is infected with HIV. In Tanzania, an estimated 45,000 additional teachers are needed to meet the basic educational needs of school-aged children.

In countries such as Mozambique, Botswana, Namibia, and Zimbabwe, the agricultural workforce is anticipated to be 20 percent smaller in 2020 than it would have been in the absence of the disease (UNAIDS, 2006). It has also been estimated that the combined impact of AIDS-related healthcare expenditures, absenteeism, and lowered productivity reduces company profits in the region by at least 6 to 8 percent (UNAIDS, 2003).

Estimating the economic impact of HIV/AIDS is not easy; however, conservative calculations suggest the effect will likely reduce the GDP growth rate of many Sub-Saharan countries by 1.5 percent per year. Although this decline may seem small, over 25 years, the cumulative impact will be economies that are 31 percent smaller than they would have been otherwise in the absence of the disease (Greener, 2004).

civil rights are more clearly reflected in the United Nations Millennium Development Goals and "2015 End Poverty" campaign:

1. Reduce the number of people suffering from hunger and poverty by half.
2. Achieve universal primary education.
3. Eliminate gender disparity in primary and secondary education.
4. Reduce by two-thirds the mortality rate of children under five.
5. Reduce by three-quarters the maternal mortality rate.
6. Halt and begin to reverse the spread of HIV/AIDS and other diseases.
7. Integrate the principles of sustainable development into country policies and programs.
8. Develop open trading and financial systems that are nondiscriminatory.

Seen in this light, development is a normative concept; it involves values and standards that enable the comparison of a particular situation against a preferred one. In this respect, development can properly be evaluated only in terms of how it meets the human needs and values of those impacted *from their perspective*. It also follows that although "development" implies economic, social, political, and cultural transformations, these elements should be seen not as ends in themselves but as means for enhancing the overall quality of human life.

Recognizing the limitations of various economic measures as indicators of well-being, the United Nations Development Program (UNDP) established the Human Development Index (HDI). This index includes three dimensions—income, education, and health—and four indicators—life expectancy at birth, mean years of schooling, expected years of schooling, and GNI per capita—to estimate the relative welfare of countries.

Clearly, this indicator fails to provide a complete picture of human development, neglecting as it does political aspects like government corruption, human or worker rights, and political freedoms; but, like other indices of well-being (see Box 2.2) it reflects several quality-of-life dimensions beyond income and production, and as such offers a useful measure for comparing the relative progress of countries and regions.

Over the last decade the HDI has risen across all regions—developed and less developed—though at variable rates. Figure 2.5 illustrates these differences by providing a comparison of the rate of change in the HDI for OECD countries and selected less developed regions—East Asia and the Pacific, Latin America and the Caribbean, and Sub-Saharan Africa—as a percentage of the potential for improvement.

Framing improvements in terms of potential provides an important perspective for assessing the relative success (and failure) of efforts to enhance human development. Consider a relatively high human development country with an index value around 0.8 (for example, Bahrain or Portugal) and a relatively low development country with an index value of 0.4 (for example, Malawi or Côte d'Ivoire). While a 0.04 increase is only a 5 percent improvement for the former (0.8 to 0.84), it is a 10 percent improvement for the latter (0.4 to 0.44). However, given that the HDI is a normalized index with a range from 0 to 1, the potential for improvement for the high development country is significantly less than for the low development country (0.2 vs. 0.6). In other words, Malawi and Côte d'Ivoire have far greater room for improvement. A 0.04 increase in the HDI for Portugal and Bahrain reflects a 20 percent "capture" of the total improvement possible (0.04 of a possible 0.2), a relatively significant increase; but the same improvement reflects merely a 6.7 percent "capture" of the total improvement possible for Malawi and Côte d'Ivoire.

As Figure 2.6 illustrates, Sub-Saharan countries had the greatest potential for improvement; however, as shown in Figure 2.5, their rate of improvement lagged OECD countries in other less developed regions. The gap in the HDI between Sub-Saharan Africa and OECD countries narrowed slightly—from approximately 49 percent (0.37 to 0.75, respectively) in 1980 to 52 percent (0.46 to 0.87, respectively) in 2010—but, clearly, the disparity remains significant.

Figure 2.7 depicts the global pattern of human development in 2011. Many countries in South and Central America and East Asia have moved beyond the basic threshold of development. In contrast, a number of countries in South Asia and sub-Saharan Africa continue to have very low levels of development. Of the countries with index values for 1990, only four countries—Democratic Republic of the Congo, Lesotho, Swaziland, and Zimbabwe—had lower HDI scores in 2011.

Although statistics on development are often illuminating, we are concerned not only with the "ends" but also the "means." As a result, our perspective is necessarily broad and situates economic geography as the essential and dynamic core of human geography. In the next section,

Box 2.2 Indicators of well-being

By Bart Yavorosky

Despite their frequent use, economic indicators are not without limitations. For example, output that increases GDP today often also generates externalities, or costs, such as pollution, soil erosion, and depleted natural resources that are not fully reflected in current prices but may impose serious constraints on future growth. Similarly, although the production aspects of services, such as education and healthcare, are captured in economic statistics, the quality-of-life dimensions are not.

In an effort to overcome these shortcomings, a number of indicators have been developed over the last few decades. The Gross National Happiness Index (www.grossnationalhappiness.com) includes assessments of psychological well-being, health, education, cultural and ecological diversity and resilience, good governance, community vitality, and living standards. Similarly, the OECD's Better Life Index (www.oecdbetterlifeindex.org) incorporates 11 dimensions of well-being ranging from education and civic engagement to work–life balance and personal safety. The Measure of Economic Welfare broadens GNI by including the cost of economic "bads" such as pollution. This emphasis on balancing the costs and benefits of economic activity is also reflected in the Genuine Progress Indicator, which incorporates factors such as the loss of leisure time and the psychological cost of unemployment.

The Happy Planet Index (www.happyplanetindex.org) draws on only three variables—life satisfaction, life expectancy, and ecological footprint—and measures the relative efficiency of countries in fostering long, happy lives for inhabitants without robbing future generations of similar opportunities. The 2012 report found that no country achieved high and sustainable well-being (see Figure 2.4); of the nine countries that came closest, eight were in Latin America and the Caribbean. The United States, with its large ecological footprint, ranked 105th, and the highest ranked country in Europe—Norway—was 29th.

Although more robust than many economic indicators, these quality-of-life measures fail to reflect the diversity of experiences of people within countries. Wealthy people—whether in Ghana or Germany, Canada or Colombia—experience life quite differently compared to their counterparts at the other end of the economic spectrum who often lack access to basic necessities, live in high-crime communities, receive inadequate schooling and health services, and likely have few employment opportunities outside the $10 trillion black market that is the fastest growing segment of the world economy. This divergence is most evident in cities such as Chicago where the upscale shops of the Magnificent Mile stand in stark contrast to the gangland violence of the South Side. In this local snapshot of the "core–periphery" pattern, the impediments to upward mobility for certain segments of the population stand in stark contrast to the wealth of others.

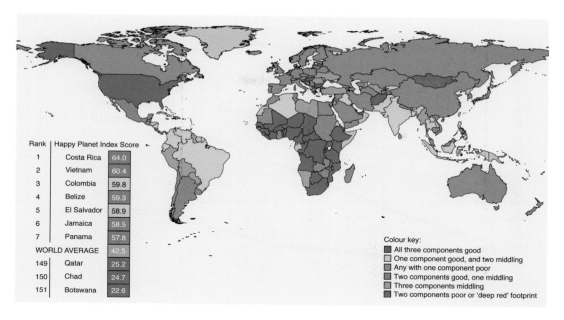

Figure 2.4 Happy planet index, 2012

Source: Based on New Economics Foundation (2012: 12, Figure 5)

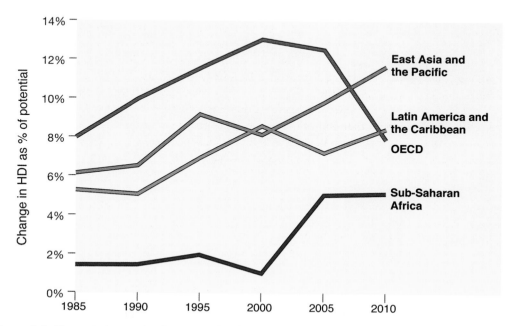

Figure 2.5 Change in human development index (HDI) in regions as a percentage of potential

Source: Based on online UNDP International Human Development Indicators at http://hdrstats.undp.org/en/tables/

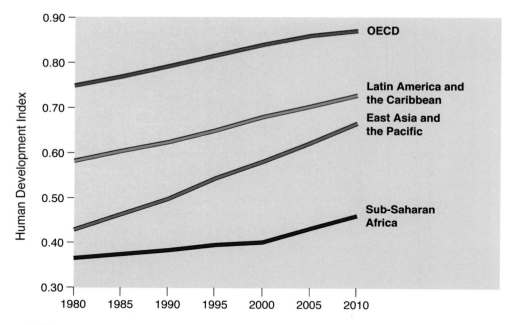

Figure 2.6 Human development index (HDI)

Source: Based on online UNDP International Human Development Indicators at http://hdrstats.undp.org/en/tables/

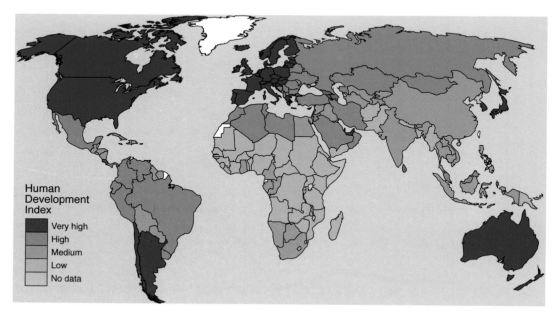

Figure 2.7 Human development index (HDI), 2011

Source: Based on online UNDP data at http://hdr.undp.org/en/data/map/

we begin our examination of some of the global patterns that reflect the "means" of transformation: Patterns of resources, population, manufacturing, trade, investment, aid, and debt. We then summarize the "ends," or net outcomes, in terms of an overall typology of socio-economic development.

2.2 INTERNATIONAL PATTERNS OF RESOURCES AND POPULATION

The distribution of natural resources has played an important role in the patterns of international economic activity and development. Not only are key resources such as energy, minerals, fresh water, and arable land unevenly distributed, but the *combination* of specific resources in particular countries and regions makes for a complex mosaic of opportunities and constraints. A lack of resources may be remedied through international trade (Japan provides the prime example, see p. 130), but for most countries the resource base remains an important determinant of development.

An exceptionally high proportion of the key nonrenewable natural resources are concentrated in the United States and Canada, Russia, China, South Africa, and Australia. Geology and physical geography play an important role in this concentration; but political instability in much of ex-colonial Africa, Asia and Latin America—instability that has hindered resource exploration and exploitation—and technological innovation also contribute to the disparity. In other words, the natural resource base of a country, while technically finite, also depends on a range of economic, political, and technological factors.

For example, U.S. Environmental Information Agency (EIA) data indicate that in 2009, Russia (27 percent), Iran (16 percent), and Qatar (14 percent) accounted for 57 percent of proven reserves of natural gas. The United States accounted for a paltry 4 percent of proven natural gas reserves. However, proven reserves merely reflect the estimated quantities of energy sources that analyses of engineering and geologic data indicate with reasonable certainty are recoverable *under existing operating and economic conditions.* Proven reserves provides one measure of a country's natural resource base, one that may differ significantly from other broader measures such as **undeveloped technically recoverable resource (UTRR) base, unproven reserves,** and **undiscovered resources** that can have a significant impact on the future development of a country.

As Figure 2.8 illustrates, the estimated shale reserves in North America alone are immense. Initial estimates also suggest that shale reserves are not solely a North American phenomenon (see Figure 2.9). By 2035, nearly 50 percent of natural gas production in the United States will likely be from the unconventional production of shale gas, up from 16 percent in 2009, as a result of technological advances in horizontal drilling and hydraulic fracturing—**fracking**—that enabled the development of previously unprofitable shale formations such as the Marcellus.

While technical innovation can enrich a country by expanding its economically recoverable natural resource base, it also means that resource-dependent countries such as Bolivia, Chile, Guinea, Guyana, Liberia, Mauritania, Sierra Leone, Surinam, and Zambia may face significant risks associated with technological change.

ECONOMIC DEVELOPMENT AND THE ENVIRONMENT

The rate of exploitation of some natural resources may also be a cause for concern. The sheer scale and capacity of the world economy means that humans are now capable of altering the environment at the global scale. The "footprint" of humankind extends to more than four-

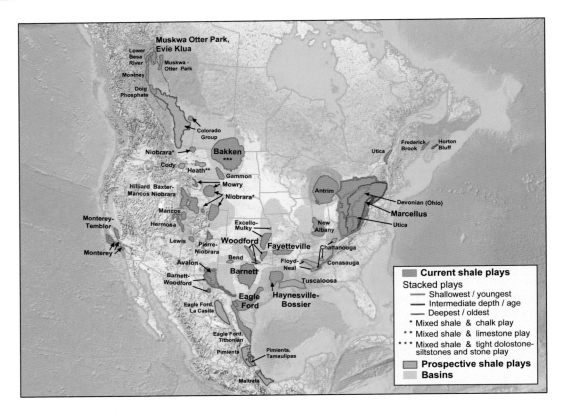

Figure 2.8 North American shale plays

Source: Based on EIA (U.S. Energy Information Administration) (2012, Figure 2) http://www.eia.gov/energy_in_brief/article/about_shale_gas.cfm

fifths of the Earth's surface, and many of the important issues facing modern society are the consequence—intended and unintended—of human modifications of the physical environment.

For example, clearing land for settlement, mining, and agriculture provides livelihoods and homes for some but alters physical systems and transforms human populations, wildlife, and vegetation. The inevitable by-products of economic development—garbage, air and water pollution, hazardous wastes, and so forth—place enormous demands on the capacity of physical systems to absorb and accommodate them.

Although names such as Fukushima, Deepwater Horizon, Chernobyl, Bhopal, Exxon *Valdez*, and Three Mile Island evoke the most vivid and notorious images of our ability to damage the planet, perhaps more devastating has been the relatively slow but persistent alteration of the planet's ecosystems as a result of human activity:

- The growing ubiquity of marine "dead zones," areas that cannot sustain life because they have been depleted of oxygen by algae blooms caused by the run-off of fertilizers and animal manure (see Figure 2.10).
- The worldwide depletion of topsoil that has resulted primarily from overgrazing, unsustainable agricultural activities, and deforestation.
- An accelerated loss of biodiversity due to habitat destruction, pollution, industrialization and agricultural production, and the fragmentation of forests.

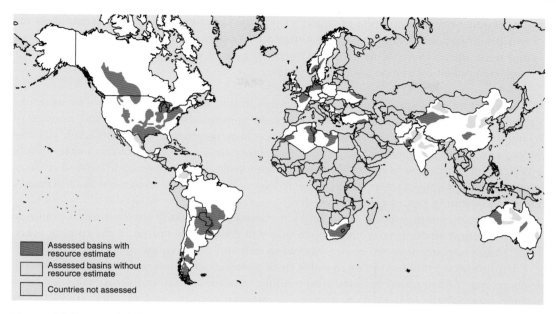

Figure 2.9 Estimated shale gas resources in 14 regions

Source: Based on EIA (U.S. Energy Information Administration) (2011a, Figure 1) http://www.eia.gov/analysis/studies/worldshalegas/

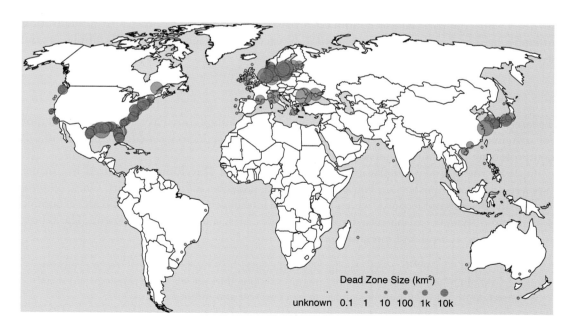

Figure 2.10 Marine "dead zones"

Source: Based on NASA, Earth Observatory (2008) (website) http://earthobservatory.nasa.gov/IOTD/view.php?id=44677

- The exponential accumulation of waste, such as the Great Pacific Garbage Patch, likely the world's largest dump, containing an estimated 100 million tons of slowly degrading plastic and toxic debris that circulates in the northern gyre of the Pacific Ocean and enters the ecosystem when consumed by marine wildlife.

The effect of these problems is neither insignificant nor inconsequential. For example, although the rate of deforestation has decreased somewhat—from an annual average of 16 million hectares from 1990–1999 to 13 million hectares from 2000–2009 (FAO, 2010)—the economic and ecological impact of commercial foresting, particularly in sub-tropical South America and Africa has been considerable. It has resulted in the loss of livelihood of local inhabitants, the silting of reservoirs, damage to hydroelectric plants, an increase in flash floods, and the devastating loss of genetic diversity.

These threats are greatest in the world's periphery, where daily environmental pollution and degradation amounts to a catastrophe that will continue to unfold in the coming years. These trends also intensify the contrast between rich and poor regions (see Box 2.3).

Environmental problems are inseparable from processes of demographic change, economic development, and human welfare. Additionally, environmental problems will inevitably become increasingly enmeshed in matters of national security and regional conflict. The spatial interdependence of economic, environmental, and social problems means that some parts of the world are effectively ecological time bombs.

The prospect of civil unrest and mass migrations resulting from the pressures of rapidly growing populations, deforestation, soil erosion, water depletion, air pollution, disease epidemics, and intractable poverty is real. These issues should also sound an alarm for citizens in developed countries whose continued prosperity depends on processes of globalization uninterrupted by large-scale environmental disasters, unmanageable mass migrations, or the breakdown of stability in the world-system.

SUSTAINABLE DEVELOPMENT

The implications of global warming and environmental despoliation have increased the clamor for **sustainable development**—economic development that seeks to meet current needs without compromising the ability to meet future needs.

In 1987 the World Commission on Environment and Development, chaired by former Norwegian Prime Minister Gro Harlem Brundtland, issued an influential report, *Our Common Future* (the Brundtland Report), which stressed the interdependence of ecological and economic systems, and made a strong plea for the principles of sustainable development. The report focused on two central and integrated concepts:

> The concept of "needs," in particular the essential needs of the world's poor, to which overriding priority should be given; and the idea of limitations imposed by the state of technology and social organization on the environment's ability to meet present and future needs.
>
> (World Commission on Environment and Development, 1987)

These ideas, although eminently common sense at first blush, are hardly uncontested. Confrontation between the DCs and the LDCs over the implications of sustainable development largely derailed the Earth Summit, the United Nations Conference on Environment and Development held in Rio de Janeiro in 1992.

Box 2.3 E-waste and the digital divide

By Bart Yavorosky

Although skeptics questioned the potential demand for Apple products in the LDCs, sales in China account for about 15 percent of the company's revenue. The rapid growth of China's middle class has fueled the demand for information technology and illustrates the increasingly unequal distribution of benefits and burdens in the world economy.

Like Barbie (see Box 1.2), the iPhone may be an American icon, but at least 90 percent of its more than 1,000 parts are manufactured outside the United States. Apple uses more than 150 suppliers, including many in China. An Apple internal audit revealed that more than one-fifth of these facilities failed to comply with requirements to prevent **involuntary labor**; two facilities experienced explosions that resulted in four dead and 77 injured workers; and more than 100 of these suppliers failed to properly store, move, or handle hazardous chemicals.

The negative impacts of the digital revolution are not confined to the factory floor; they also extend down the supply chain. Materials used in electronic equipment, such as tantalum, tin, and tungsten, often comes from conflict regions such as the Democratic Republic of the Congo. And 97 percent of the rare earth elements in components such as LED backlights are mined in China where refinement costs are much lower than in the DCs due to lax environmental regulations.

Although problematic, these issues pale in comparison to the human and environmental costs associated with the disposal of digital technology. Electronics is the fastest growing waste stream in the DCs. In the United States, hundreds of thousands of cell phones and computers are discarded every day, and some NGOs have calculated that between 50 and 80 percent of e-waste generated in the United States is exported overseas. This material ends up in places such as Guiyu in China where shipping containers arrive daily despite Chinese ratification of the Basel Ban Amendment to the Basel Convention that bans the import of e-waste.

The laborers in Guiyu are a stark contrast to the upwardly mobile workers in Shanghai and Shenzhen. Most toil in tiny shops and open-air fire pits, salvaging copper, lead, and gold, and earning perhaps $1.50 a day. The air in parts of Guiyu has the highest concentrations of dioxins in the world. Soil samples have revealed unacceptably high levels of cadmium, chromium, copper, iron, lead, and tin.

Despite the attention that some NGOs have brought to places such as Guiyu, the number of these electronic graveyards is expected to multiply. Between 2010 and 2025 the volume of e-waste is projected to triple, and LDCs, particularly those in Africa and Asia with the world's fastest growing economies and some of its largest populations, will be responsible for generating two-thirds of it. E-waste illustrates how one person's waste can become another person's wealth. It also reflects global, regional, and intra-national flows and divisions of labor that define the **digital divide**.

Economic development in industrialized countries has been powered to a significant extent by fossil fuels. Although renewables such as wind, solar, geothermal, biomass, and hydro-electricity have been the fastest growing source of energy, they account for only about 10 percent of global energy consumption (EIA, 2011b).

More importantly, as long as the market price of fossil fuels does not reflect the full cost of production or consumption, so that negative externalities continue to exist, conventional carbon-based fuels such as coal and oil will likely remain significantly cheaper than renewable alternatives, and so provide the most direct path to economic development.

For example, China led the world in investment in clean energy in 2010 at $54.4 billion (compared to $34 billion for the United States and $3.3 billion for the United Kingdom) (Pew Charitable Trusts, 2013). However, U.S. Environmental Information Agency (EIA) data indicate that coal continued to provide nearly 70 percent of the energy consumed in China. In 2000, China consumed 1,239 short tons of coal. Nine years later, its consumption had increased 180 percent to 3,474 short tons.

As this brief description demonstrates, sustainable development means different things to different people. To some, it means regulating economic systems so the benefits of development are distributed more equitably (if only to prevent poverty from causing environmental degradation). To others, it means reorganizing societies to improve education, healthcare, and social welfare and, as a result, raising environmental awareness and sensitivity to improve the quality of life of all. To the approximately 315,000 citizens of the Maldives, an island country with a median altitude of 1.5 meters (4.9 feet) above sea level, it may mean the difference between having a homeland and becoming a permanent diaspora.

Clearly, perspective plays an important role. Citizens of DCs may want lesser developed countries to be more responsible stewards of the environment, to curb deforestation, top soil erosion, and the profligate extraction of natural resources, but protecting the environment may also seem like a luxury to those who are among the estimated 925 million people who are undernourished (FAO, 2011). It may also seem unreasonable and inequitable to the 1.3 billion citizens of China who aspire to own 2.28 cars per household (Noor, 2008), 2,505-square feet houses (U.S. Census Bureau, 2013), and consume 185 pounds of meat per year as do their U.S. counterparts (USDA, 2010).

Perhaps the most significant obstacle to sustainable development is simply the inadequacy of institutional frameworks. Sustainable development requires economic, financial, and fiscal decisions to be fully integrated with environmental and ecological decisions. Such decisions require a weighing and reconciliation of interests that are often contradictory and policies with outcomes that can be estimated tenuously at best.

The implications of inaction are almost certainly significant—rising sea levels that swamp coastal cities, increased desertification and incidences of wildfires and severe weather phenomena such as tornadoes, hurricanes, drought, and monsoons—but coordinating a response, one that effectively balances the competing interests of differently situated peoples and countries, remains elusive.

As the United Nations Conference on Sustainable Development (Rio +20) demonstrated, in the 20 years since the first Earth Summit, perhaps the most poignant statement that can be made about sustainable development is that it remains an embarrassing contradiction in terms. And what is often referred to as Plan B, geoengineering the atmosphere to reflect sunlight away from the planet and reverse global warming, although perilous, increasingly assumes the appearance of Plan A by default, particularly for countries most at risk, such as the Maldives.

THREE ESSENTIAL RESOURCES: ENERGY, ARABLE LAND, AND WATER

Energy and arable land play an important role in shaping the economic geography of the world. As Figure 2.11 illustrates, natural gas, oil, and coal, barring the rapid development of low-cost alternatives, will be the major energy sources to power the world in coming years. Most DCs are reasonably well off in terms of energy *production*, the major exceptions being Japan and parts of Europe. Most LDCs, by way of contrast, are energy poor. The major exceptions are the Persian Gulf states, Nigeria, Angola, Venezuela, Algeria, Libya, and Kazakhstan—all major oil producers—and Indonesia, which has significant natural gas and coal production. This uneven distribution of nonrenewable energy resources plays an important component in world trade. Mineral fuels, including oil, coal, gas, and refined products, account for more than 14 percent of world trade (CIA, 2012).

For many LDCs, the cost of energy imports represents a significant burden. Consider, for example, the predicament of the small island LDCs, a collection of 39 island countries from Africa, the Caribbean, and the Pacific and Indian Oceans, united by their small size, isolation, lack of natural resources, and vulnerability to the effects of climate change and natural disasters. These countries are often considered the most petroleum-dependent ones on the planet. For example, countries such as the Cook Islands, named after Captain Cook who sighted them in the 1770s, Tonga, and Samoa import up to 95 percent of the oil required to meet energy needs at a cost that far exceeds the total value of their exports. Running persistent current account deficits, these countries are dependent on foreign aid and, increasingly, remittances from family members who have immigrated to other countries.

Needless to say, few LDCs can afford to consume energy on the scale of the DCs, so the pattern of commercial energy *consumption* tends to mirror the fundamental core–periphery cleavage of the world economy. For example, despite having a population one-fourth as large, the United States consumed 440 percent more energy than India in 2009. Compared to the countries of Africa combined, U.S. Energy Information Administration data indicates that energy consumption was 590 percent greater despite having a population less than one-third in size. OECD countries accounted for less than 18 percent of the world's population in 2009 (United Nations, 2010); however, U.S. Environmental Information Agency (EIA) data indicate that those countries consumed more than 47 percent of the world's primary energy.

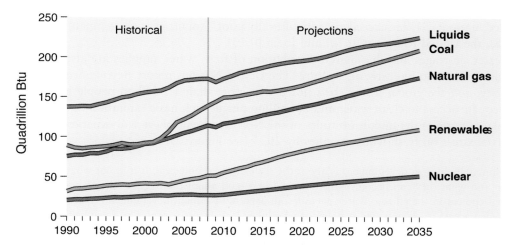

Figure 2.11 World energy consumption by fuel, historical and projected

Source: Adapted from EIA (U.S. Energy Information Administration) (2011b: 2, Figure 2)

It should be noted that these figures do not reflect the use of firewood and other traditional fuels for cooking, lighting, heating, and, sometimes, industrial needs. The Food and Agriculture Organization (FAO) of the United Nations estimates that wood accounts for approximately 9 percent of global energy consumption; however, more than 2 billion people, a disproportionate number residing in less developed countries, are dependent on firewood for cooking and heating. In 2011 the *Economist* reported that six of the ten fastest growing economies in the world were African countries, but wealth generated through resource extraction in countries such as Nigeria, Angola, and Botswana have done little to alleviate extreme poverty or the necessity for the poorest people to gather firewood for basic needs and suffer the health consequences of cooking over open fires.

Although it has been assumed that the collection of firewood causes considerable deforestation, recent studies have shown that non-forest sources such as dispersed woodlands and roadsides provide up to two-thirds of firewood. While there are insufficient data to assess the sustainability of firewood use, barring significant technological advances in fields such as artificial photosynthesis that offers the potential of providing low-cost, carbon-neutral fuels on a small scale to the world, dependence on wood energy will likely not abate and may increase substantially with population growth.

The distribution of arable land represents another important environmental influence on international economic differentiation. More than half of the earth's land surface cannot sustain traditional forms of cultivation. Figure 2.12 provides an approximation of the world's cultivable land. It excludes mountainous regions as well as those areas with poor soil, insufficient growing seasons, too little precipitation, or that cannot be converted easily to farmland because it is used for grazing or forestry or has been conserved. Agriculture is not completely absent from the unshaded areas of the map; rather, farming in these regions is marginal.

Clearly, the distribution of the world's arable land is highly uneven, concentrated primarily in Europe, west-central Russia, eastern North America, the Australian littoral, Latin America, parts of India, eastern China and parts of Sub-Saharan Africa (although it should be noted that some of these regions may be marginal for farming as a result of marshy soils or other adverse conditions; while irrigation, for example, can extend the local frontier of productive agriculture in other areas). World Bank data indicate that arable hectares per person in 2011 in countries such as Australia (2.14), Canada (1.25), the Russian Federation (0.85), and the United States (0.51) far exceeded "agriculturally poor" countries such as Japan (0.03), China (0.08), the United Kingdom (0.10) and India (0.13).

Water shortages and the uneven distribution of fresh water supplies are also increasingly imposing limits to intensive agriculture. The United Nations report *Beyond Scarcity: Power, poverty, and the global water crisis* noted that more than 1.1 billion people in LDCs have inadequate access to clean water and 2.6 billion people lack basic sanitation. It further warns that large parts of the middle latitudes of the Northern hemisphere will experience significant levels of stress on fresh water supplies by 2025 (see Figure 2.13):

> Ultimately, human development is about the realization of potential. It is about what people can do and what they can become—their capabilities—and about the freedom they have to exercise real choices in their lives. Water pervades all aspects of human development. When people are denied access to clean water at home or when they lack access to water as a productive resource their choices and freedoms are constrained by ill health, poverty and vulnerability. Water gives life to everything, including human development and human freedom.
>
> (UNDP, 2006: 2)

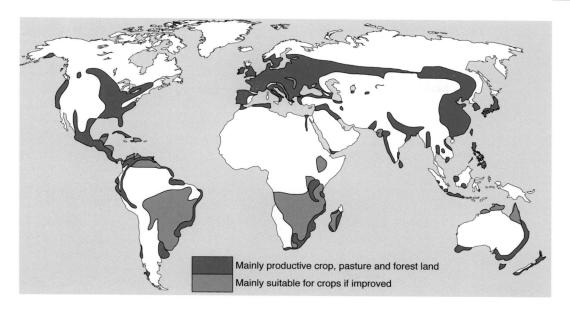

Mainly productive crop, pasture and forest land

Mainly suitable for crops if improved

Figure 2.12 The world's cultivable land

Source: Updated from FAO (1995) http://www.fao.org/docrep/v4200e/v4200e00.htm

We also have to bear in mind that not all arable land is equal. The carrying capacity of land, which is the maximum population that can be maintained in a place with rates of resource use and waste production that sustain the long-term productivity, vary considerably.

A companion concept is the **ecological footprint** of a population, which measures human pressures on the natural environment from the consumption of resources and assimilation of waste. It changes in proportion to population size, average consumption per person, and the resource intensity of the technology used in the production of goods consumed. The ecological footprint is measured in "area units," in which one area unit is equivalent to one hectare of biologically productive land with world average productivity. As land varies in productivity, a hectare of highly productive cropland would represent more "area units" than the same amount of less productive grazing land. Figure 2.14 shows the intensity of the ecological footprint across the world in 2007. Intensity increases with greater population densities, higher per capita resource consumption, and lower resource efficiencies. The World Wide Fund for Nature (WWF) has found that the ecological footprint of the world's population has been increasing steadily since the 1970s. In *Living Planet Report 2012*, the WWF estimated that we use the equivalent of 1.5 planets to support our current activities. This overshoot depletes the Earth's natural capital and renewable capacity and, therefore, cannot continue indefinitely.

AGRICULTURAL PATTERNS AND THE FOOD QUESTION

These issues shift our attention from the abstract to reality, and to a consideration of the world agricultural map and the global food situation. Both of these are rather different in configuration from the patterns of arable land and carrying capacity described earlier.

We must also recognize that the actual pattern of world agriculture is merely one of a vast number of possible realizations of the global agricultural resources. It reflects a variety

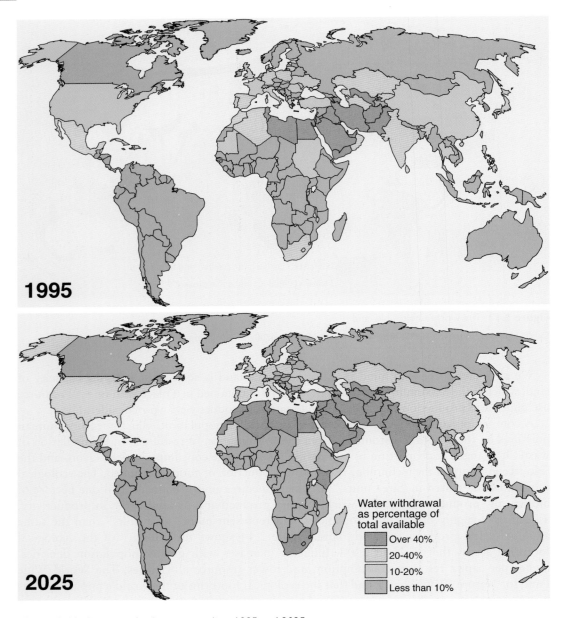

Figure 2.13 Stress on freshwater supplies, 1995 and 2025

Source: Based on United Nations Environment Programme (UNEP) 2002.

of interpretations, at different times, of environmental possibilities, desirable products, and marketable opportunities—all influenced, in turn, by prevailing land-tenure systems, wealth, levels of and access to technology, and global power politics. Most succinctly, it reflects the economic history of the world.

The mosaic of world agricultural regions shows a high degree of specialization. At the same time, the broader international division of labor means that some countries depend much more on agriculture for employment and income than others. In countries such as Liberia, Comoros,

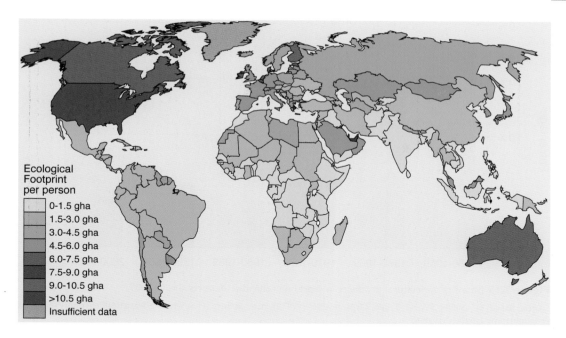

Figure 2.14 Ecological footprint

Source: Based on WWF (World Wide Fund for Nature) (2010: 36, Map 3)

Cambodia, Ghana, and Nepal, agriculture employs roughly 60–80 percent of the labor force and accounts for 30–75 percent of GDP. In DCs such as Canada, France, and the United States, agriculture typically employs less than 3 percent of the labor force and accounts for only about 2 percent of GDP (CIA, 2012).

This specialization results in a large volume of trade in agricultural produce. The biggest exporters of food, however, are not LDCs. A handful of DCs—Australia, Canada, France and the USA—with technologically advanced and highly productive agricultural sectors dominate global cereal production.

In recent years world trade in food has grown rapidly and there have been some significant changes in the pattern of trade. Until the 1950s the primary flow was food grains into Europe. These flows have now declined significantly. The major source of grains remains Canada and the United States, but the destinations are Japan and, increasingly, middle-income countries as categorized by the World Bank, such as Mexico. Over the last 60 years, however, the share of agricultural products in global exports has shrunk dramatically, from almost 40 percent to under 10 percent (see Figure 2.15).

Gross inequalities in the consumption of food, one of the most basic of all human needs, are an important corollary of these patterns and flows. Of the estimated 925 million people around the world who are undernourished, 98 percent live in LDCs (FAO, 2011). Malnutrition was the underlying contributing factor in over one-third of the approximately 6.9 million children under the age of five who died in 2011 (WHO, 2012). The fact that many of these children lived in countries that are net *exporters* (by value) of foods is a telling indictment of the world economic system.

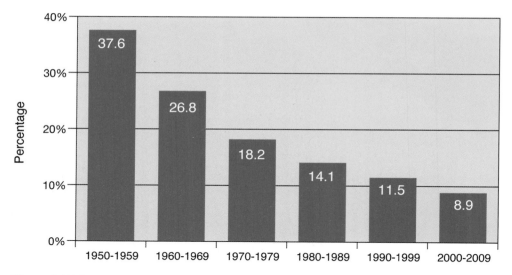

Figure 2.15 Share of agricultural products in world merchandise exports, 1950–2009

Source: Based on online World Trade Organization (WTO) data at http://stat.wto.org/StatisticalProgram/WSDBStat ProgramHome.aspx?Language=E

INTERNATIONAL DEMOGRAPHIC PATTERNS

The geography of population and the dynamics of population change are closely interrelated with patterns of economic development. Population density, fertility, mortality, and migration often directly reflect economic, social, and political conditions. They also can be important determinants of economic change and social well-being. Human resources are vital to economic development, but in the wrong set of circumstances they can be a liability rather than an asset. Although one cannot easily unravel cause and effect, understanding the broad context is important.

In global terms, population growth dominates the broad context. The World Bank estimates that the world population is increasing by approximately 1 million people every five days. According to the United Nations estimates, the current population of just over 7 billion is likely to grow to 9.2 billion by the year 2050. The population of the 49 least developed countries is expected to nearly double to 1.7 billion. The population in the rest of the less developed world is also expected to increase but at a slower rate, from 4.8 billion in 2009 to 6.2 billion in 2050. In contrast, the population of the DCs is expected to increase only slightly from 1.23 billion to 1.28 billion.

The demographic transition

This core–periphery contrast reflects differences in fertility and mortality rates. In turn, these rates relate to differentials in the **demographic transition** associated with the broad sweep of economic development and social change. Scholars conventionally portray this transition in three stages, to which we added a tentative fourth stage (see Figure 2.16).

In the first stage, populations exhibit high birth rates and high and fluctuating death rates with net growth rates around 1 percent. In the second stage, death rates fall sharply as people have access to more nutritious food, medicine, and medical care. Birth rates also fall, but the decrease in fertility lags largely because it takes time for social and cultural practices to respond

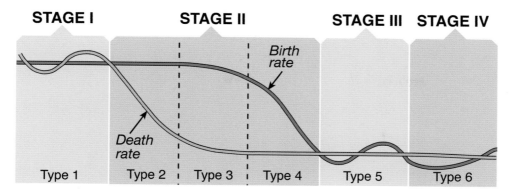

Figure 2.16 The demographic transition

to changing circumstances. The combination of these two factors results in an explosive increase in population. Most DCs experienced this stage during the nineteenth century. In the third stage, death rates even off at a low level; birth rates remain low but fluctuate and net growth rates return to around 1 percent. Quite a few LDCs, including many of the NIEs (Latin American countries such as Chile, as well as East Asian countries such as China, Singapore, South Korea, and Thailand) have already transitioned into this low growth stage (Stage III Type 5). The fourth tentative stage that we added captures those DCs, such as Germany and Japan, that are experiencing low birth rates but rising death rates due to an aging population (Stage IV Type 6?).

In contrast, quite a few African countries, including Angola, Niger, and Rwanda, are at the beginning of the demographic transition with relatively high death rates suppressing the rate of natural increase (Stage I Type 1). Of enormous concern are the southern Sub-Saharan countries, such as Botswana and Zimbabwe, which have slipped back from the most explosive phase of the growth stage to the beginning of the demographic transition as deaths rates skyrocketed due to having more than 15 percent of their population infected with HIV/AIDS.

While not all countries should be expected to follow the demographic transition path, it is useful to identify whether a country may be entering the critical second stage of rapid population expansion and so has the major part of its population growth ahead of it; whether it is in the middle of the population "explosion;" or whether it is on the verge of completing the growth stage.

Accordingly, the United Nations has suggested a threefold division of the second stage (see Figure 2.16). Figure 2.17 shows how the countries of the world fit into this classification system. Some countries in Africa, such as Gabon, Namibia, and Senegal, are experiencing the most explosive phase of the growth stage (Stage II Type 2). Quite a number of other Latin American and Asian countries—Colombia, Mexico, Peru, Venezuela, the Philippines, and Malaysia—seem to be in the final phase of the growth stage (Stage II Type 3). A number of countries have transitioned into the final slow growth stage, including Brazil, Costa Rica, the Cayman Islands, Jamaica, and Vietnam (Stage II Type 4).

Migration

Migration plays another important role in population change. International labor migration has been a vital part of the world economic system ever since the Industrial Revolution in the nineteenth century. The International Labor Office (ILO) estimated the total number of

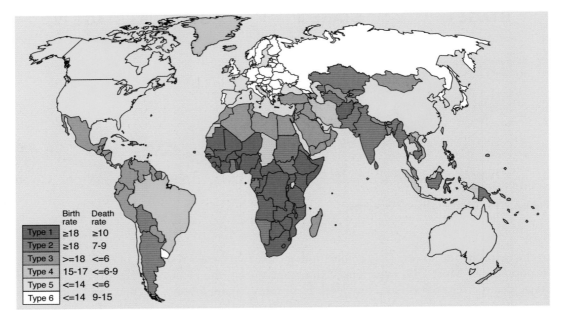

	Birth rate	Death rate
Type 1	≥18	≥10
Type 2	≥18	7-9
Type 3	>=18	<=6
Type 4	15-17	<=6-9
Type 5	<=14	<=6
Type 6	<=14	9-15

Figure 2.17 Demographic transition map of the world

Source: Based on online World Bank 2010 World Development Indicators data at http://data.worldbank.org/data-catalog/world-development-indicators

economically active international migrant workers at 105.4 million in 2010 with a comparable number of dependants accompanying them. The top immigration destinations are the United States, the Russian Federation, Germany, Saudi Arabia, Canada, and the United Kingdom; while the top emigrant countries are Mexico, India, the Russian Federation, China, and Ukraine (World Bank, 2011).

The United Nations estimated that there were 42 million immigrants living in the United States in 2010. The Pew Research Center estimated that as many as 11 million of those immigrants were in the country illegally from Mexico. Many of the members of the European Union provide an interesting contrast to the American situation. Assuming a constant participation rate, the size of the European labor market, absent migration, is anticipated to shrink from 227 million in 2007 to 201 million in 2025 and 160 million in 2050. This shift in demographics reflects a combination of reduced birth rates and an aging population (Munz *et al.*, 2007). As a result, maintaining the size of the workforce requires a net inflow of 1.5 million immigrants per year.

Two of the most significant concerns about migration are its impact on wages and salaries and its drain on public resources. While studies have shown that the impact of migration on local workers' earnings may be positive or negative, the impact is small in the short and long run (UNDP, 2009). In contrast, estimating the impact of immigrants on public services and resources is difficult at best with results highly dependent on expenditures and revenues included in the calculation, the composition of the immigrant population (for example, the age, level of experience, education, and fertility rates of immigrants). However, a synthesis of available research strongly suggests the hyperbole about immigrants—legal and otherwise—draining public budgets is unfounded (ILO, 2010) (see Box 2.4).

Box 2.4 Migrant workers' remittances

An increasingly important aspect of globalization is the flow of funds from international migrant workers to their home countries. These remittances are usually to family members in LDCs and, as unilateral transfers, they do not create any future liabilities such as debt servicing or profit transfers. Workers' remittances tend to move counter-cyclically with the economy in recipient countries in that migrant workers increase their support to family members during down cycles of economic activity back home offsetting some of the family income lost due to unemployment or other crisis-induced reasons. This process enables remittances to serve as a stabilizer by smoothing out large fluctuations in the national income over phases of the business cycle.

LDCs received recorded remittances in excess of $372 billion in 2011, a total that exceeded the combined sum of aid provided by DCs (World Bank, 2012). Figure 2.18 illustrates the flow of remittances from DCs to LDCs. The United States is by far the largest single source of remittances (followed by Saudi Arabia and Switzerland). South and East Asia receive the bulk of inflows with India and China accounting for more than one-fourth of all recorded remittances (see Figure 2.19). Unrecorded remittances increase the net inflow of funds, but by how much remains an open question. Estimates range from less than 5 percent for countries such as the Philippines, Dominican Republic, and Guatemala to more than 50 percent for Sub-Saharan and South Asian countries such as Bangladesh (Freund and Spatafora, 2005).

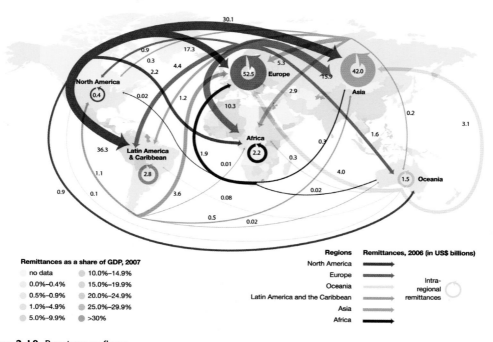

Figure 2.18 Remittance flows

Source: Based on UNDP (2009: 73) http://hdr.undp.org/en/media/ HDR_2009_EN_Chapter4.pdf

Another important facet of immigration is the flow of highly skilled laborers— physicians, engineers, natural and computer scientists, and mathematicians—the so-called *brain drain*. The principal recipients of these streams have been the United States, Canada, Britain, and Australia. As depicted in Figure 2.20, the brain drain disproportionately impacts Sub-Saharan and Central American countries. Typically, the brain drain results when students and professionals opt not to return home after the completion of educational courses or training programs in developed countries. The absolute number of people involved in these streams is

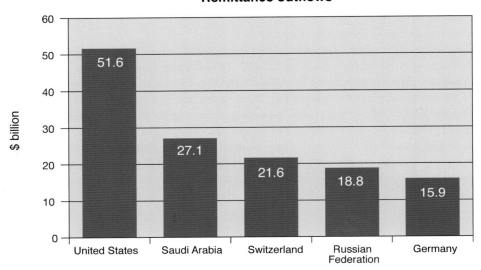

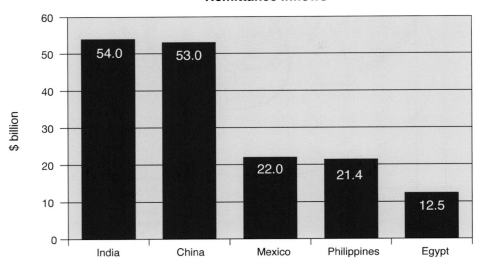

Figure 2.19 Remittance flows, top countries, 2010

Source: Based on online remittance data from the World Bank http://econ.worldbank.org/WBSITE/EXTERNAL/EXTDEC/ EXTDECPROSPECTS/0,,contentMDK:22759429~pagePK:64165401~piPK:64165026~theSitePK:476883,00.html

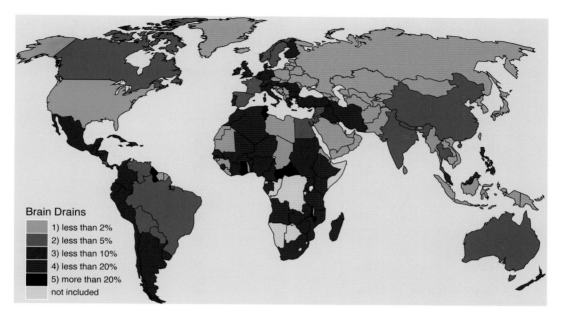

Figure 2.20 Share of a country's nationals with a university degree living in an(other) OECD country

Source: Based on OECD (n.d.) http://www.oecd.org/dev/poverty/ migrationandthebraindrainphenomenon.htm

insignificant, but the economic implications of the relative gains and losses of highly skilled personnel can be substantial.

Meanwhile, the demographic transition continues to flood the labor markets of most LDCs. Without the option to move into unsettled territories—an option that represents a crucial difference between the current experiences of LDCs compared to the historical experiences of DCs—the demographic transition in less developed countries has been a channeling of people along *rural-to-urban migration* streams. Shortage of land and a combination of real and perceived advantages of urban life push young, reproductively active rural populations into cities. As a result, the rate and scale of urbanization in LDCs represents yet another important contrast between core and periphery countries. The United Nations estimated that approximately 827 million people lived in slums in 2010. In Sub-Saharan Africa over 60 percent of urban dwellers, more than 190 million people, live in slums (UN Centre for Human Settlements, 2012).

2.3 INTERNATIONAL PATTERNS OF INDUSTRY AND FINANCE

Similar to the agricultural map of the world, the international mosaic of industrial production and employment is highly complex and includes a great deal of specialization in particular activities. Once again, the important aspect to stress is the overall framework.

First, the core economies account for almost 55 percent of world manufacturing value added (MVA) while Brazil, Russia, India, and China (the BRIC countries) account for nearly another fourth of manufacturing output. Three countries—the United States, China, and Japan—produced 47 percent of global MVA in 2009 (Table 2.1).

One important trend these numbers fail to illustrate is shifts in the types of manufacturing different countries increasingly engage in. As LDCs with large and relatively inexpensive labor

Table 2.1 World manufacturing

Rank	Country	MVA, 2009	MVA, 2009 as share of world (%)	2000–2009, annual growth rate (%)	Manufacturing as share of GDP, 1990 (%)	Manufacturing as share of GDP, 2009 (%)
1	United States	1,674.2	18.8	0.8	18	13
2	China	1,612.3	18.1	11.5	33	32
3	Japan	905.5	10.1	–0.7	27	18
4	Germany	567.9	6.4	–0.8	28	19
5	Italy	311.3	3.5	–2.3	23	16
6	France	253.6	2.8	–0.7	21	11
7	United Kingdom	217.6	2.4	—	23	11
8	Brazil	216.9	2.4	1.7	25	16
9	South Korea	208.8	2.3	5.4	27	28
10	India	190.3	2.1	8.0	17	15
11	Spain	172.4	1.9	–1.0	—	13
12	Russian Federation	154.8	1.7	—	—	16
13	Mexico	147.1	1.6	–0.2	21	18
14	Indonesia	142.2	1.6	4.4	21	25
15	Turkey	93.5	1.0	3.2	23	18

Source: Based on World Bank online World Development Indicators data http://data.worldbank.org/

forces capture the market for labor-intensive manufacturing, DCs have shifted production to more sophisticated, automated processes.

For example, U.S. Bureau of Labor Statistics data indicate that while manufacturing output increased in the United States from 2000 to 2009, the number of people employed in manufacturing *decreased* by 30 percent. China and the United States had nearly equal shares of global MVA in 2009; however, Chinese factories employed nearly seven times as many workers. In the early 1900s the typical factory worker in the United States was a first- or second-generation immigrant who may or may not have earned a high school degree. Today, an increasing number of workers on the "factory floor" are engineers who manage operations through highly complex, computerized systems. In contrast, a far greater number of Chinese factory workers look much more similar to U.S. manufacturing employees a century ago. Rather than immigrating from eastern and southern Europe, they have migrated from rural China to cities and special economic zones in cities such as Shantou and Shenzhen in the Guangdong province.

Although advanced industrial economies continue to account for the majority of global manufacturing, several important trends should be noted. When China nosed ahead of the United States in manufacturing production in 2011, it ended an approximately 110-year run for the United States as the global leader in the production of goods. In France, Germany, Italy, Japan, and the United Kingdom, manufacturing as a percentage of GDP declined significantly as a percentage of GDP from 1990 to 2009. Annual growth rates in manufacturing for each of those countries were negative in the first decade of the twenty-first century.

In contrast, growth rates in LDCs remained robust during that period with China and Vietnam topping 10 percent per year and 17 other countries (including India, Pakistan, Peru, Poland, Singapore, South Korea, Thailand, Turkey, and Ukraine) tallying increases in excess of 3 percent per year. We will examine the reasons for these shifts in Chapters 5 to 7, where we discuss the evolution and transition of the world's developed countries. As we will see, this shift has involved an *increasing degree of international interdependence* throughout the world economy.

These shifts are also part of a globalization of economic activity that has emerged as the overarching component of the world economy. As we will see in Chapters 6 and 10, corporate strategies, particularly the strategies of large transnational corporations (TNCs) have created this globalization of economic activity. For the moment, however, it is sufficient to take note of the magnitude of the phenomenon.

Table 2.2 World's top non-financial TNCs ranked by foreign assets, 2010

Rank	Corporation	Country	Industry	Foreign assets ($ million)	Total assets ($ million)
1	General Electric	USA	Electrical and electronic equipment	551,585	751,216
2	Royal Dutch Shell	Netherlands/ UK	Petroleum	271,672	322,560
3	BP	UK	Petroleum	243,950	272,262
4	Vodafone Group	UK	Telecommunications	224,449	242,417
5	Toyota Motor Corporation	Japan	Motor vehicles	211,153	359,862
6	Exxon Mobil Corporation	USA	Petroleum	193,743	302,510
7	Total SA	France	Petroleum	175,001	192,034
8	Volkswagen Group	Germany	Motor vehicles	167,773	266,426
9	EDF SA	France	Utilities	165,413	321,431
10	GDF Suez	France	Utilities	151,984	246,736
	Average				*327,745*

Source: Based on UNCTAD (2011: annex Table 29 additionally available on UNCTAD website: http://unctad.org/Sections/dite_dir/docs/WIR11_web%20tab%2029.pdf

Table 2.3 Types of special economic zone

Type of zone	Development objective	Physical configuration	Typical location	Eligible activities	Markets	Examples
Free trade zone	Support trade	Size < 50 hectares	Ports of entry	Entrepôt and trade-related activities	Domestic, re-export	Colon Free Zone, Panama
Traditional EPZ	Export manu-facturing	Size < 100 hectares; only part of the area is designated as an EPZ	None	Manufacturing, other processing	Mostly export	Karachi EPZ, Pakistan
Hybrid EPZ	Export manu-facturing	Size < 100 hectares; only part of the area is designated as an EPZ	None	Manufacturing, other processing	Export and domestic market	Lat Krabang Industrial Estate, Thailand
Freeport	Integrated development	Size > 100 km^2	None	Multi-use	Domestic, internal and export markets	Aqaba Special Economic Zone, Jordan
Enterprise zone (urban free zones)	Urban revitalization	Size < 50 hectares	Distressed urban or rural areas	Multi-use	Domestic	Empowerment Zone, Chicago
Single factory EPZ	Export manufacturing	Designation for individual enterprises	Countrywide	Manufacturing, other processing	Export market	Mauritius, Mexico, Madagascar

Source: Based on World Bank (2008: 10, Table 2)

One striking measure of the importance of transnational corporations in the world economy is the size of their annual turnover in comparison with the GNI of entire countries. By this yardstick, all of the top 50 global TNCs—including ExxonMobil, General Motors, Ford Motor Company, General Electric, Mitsubishi, IBM, Nestlé, Unilever, BP, Royal Dutch/Shell, Siemens, and Toyota Motor Corporation—carry more economic clout than many smaller LDCs. The median total assets of the ten largest TNCs are larger than the economies of countries such as South Africa, Portugal, Singapore, and Ireland (see Table 2.2). In fact, only 27 countries have higher GNIs than the median total assets of the ten largest TNCs.

Collectively, the 500 largest U.S. corporations now employ an overseas labor force as big as their domestic labor force. Similar statistics apply to the largest Japanese corporations. These overseas labor forces are spread among different parts of the world, but it is in LDCs and, in particular, the NIEs where the most rapid growth has been taking place. More than 30 percent of manufacturing by Japanese firms happens overseas. The foreign-made share of output for Toshiba exceeds 50 percent. For companies such as Fuji Xerox and Yamaha Motor, that share accounts for more than four-fifths of production with much of it relocated to countries in the region such as South Korea, Thailand and Taiwan (*Economist*, 2010b).

The governments of LDCs have sought to take advantage of transnational corporations' need for cheap labor by setting up **export-processing zones** (EPZs)—see also Chapter 10— adaptations of free trade zones in which favorable investment and trade conditions are created by waiving excise duties on components, providing factory space and warehousing at subsidized rates, allowing tax "holidays" of up to five years, and suspending foreign exchange controls. EPZs are merely one type of special economic zones that range in size, objective, and markets (see Table 2.3). One of the primary objectives of these zones is increasing employment opportunities. Currently, special economic zones are estimated to provide direct employment for over 68 million people, nearly 90 percent of those jobs are in Asia and the Pacific (World Bank, 2008, Table 15).

At a more general level, governments everywhere have responded to the "global reach" of transnational corporations by intensifying their involvement with supranational economic and political organizations such as the European Union (EU), the Association of South East Asian Nations (ASEAN) and the North American Free Trade Agreement (NAFTA). We will be reviewing the changing role of the state in the context of the internationalization of the world economy in Parts 2 and 3. In Part 4, we will explore supranational reactions to the internationalization of the world economy.

PATTERNS OF INTERNATIONAL TRADE

The fundamental structure of international trade has been based on a few **trading blocs** with most trade taking place *within* these blocs. Membership in these trading blocs is principally the result of the effects of (1) distance, (2) the legacy of colonial relationships, and (3) geopolitical alliances. For most of the period from 1950 to 1990, for example, four trading blocs dominated international trade:

1. western Europe including some former European colonies in Africa, South Asia, the Caribbean and Australasia
2. North America and some Latin American countries
3. the countries of the former Soviet Union
4. Japan along with other East Asian countries and the oil-exporting countries of Saudi Arabia and Bahrain.

In contrast, other countries have always exhibited a high degree of **autarky** from the world economy in that they do not contribute significantly to the flows of world trade. Typically, these countries are smaller LDCs such as Armenia, Fiji, Nepal, Nicaragua, and Swaziland. All LDCs combined account for about 1 percent of all world trade flows (exports plus imports) in industrial goods (UNCTAD, 2011).

But the geography of trade has been changing rapidly in response to the several factors:

1. *Innovation*: Improvements in transport, communications, and manufacturing technology have diminished the importance of the "classical" distance-based factors that have underpinned traditional trading blocs.
2. *Politics*: The trend toward political as well as economic integration in Europe, the increasingly important role played by China in the world economy, and the continuing trend away from isolationism on the part of the United States.
3. *Internationalization*: The globalization of economic activity has created new flows of materials, components, information, and finished products. TNCs account for about 70 percent of world trade.

Since 1990, the dissolution of the trading bloc of the former Soviet Union has left a tri-polar framework for international trade: North America; the European Union; and Japan, China, and the Asian NIEs. It is no coincidence, of course, that this tri-polar framework centers on the countries that constitute the core regions of the world economy. Since the 1950s, world trade has tripled in volume. As Figure 2.21 illustrates, world trade has grown significantly faster than the growth in world real GDP over the last 20 years, deepening economic integration and global interdependency.

The three most important developments in global trade over this period have been the increased role played by LDCs, particularly China; the growing importance of regional trade;

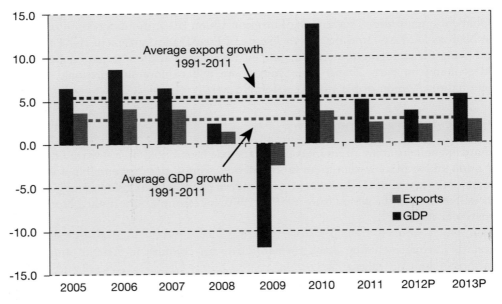

Figure 2.21 Growth in volume of world merchandise trade and GDP, 2005–13 (annual % change)

Source: Based on WTO (World Trade Organization) (2012a) http://www.wto.org/english/news_e/pres12_e/pr658_e.htm

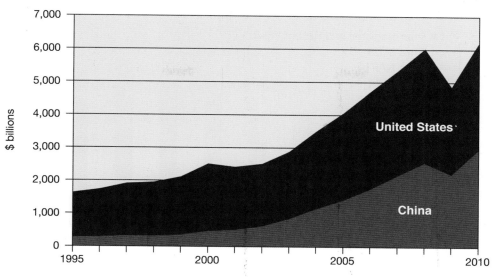

Figure 2.22 Trade, imports and exports

Source: Based on online data from UNCTAD, UNCTADstat online data: International Trade in Goods and Services http://unctadstat.unctad.org/ReportFolders/ reportFolders.aspx

and the downward shift in technology that has enabled emerging economies to develop robust export markets.

In the early 1970s LDCs accounted for approximately one-fourth of world trade. In 2010, their share exceeded 40 percent. Much of this rapid growth can be attributed to China (and to a lesser extent other East and South Asian countries) which has industrialized at a breakneck pace and has aggressively pursued foreign trade. In the late-1990s, China had one-fifth of the trade volume (imports and exports) of the United States. By 2010, its trade volume was only 8.3 percent less than the United States (see Figure 2.22), and it had surpassed all other developed countries in global trade importance.

Intra-regional trade has also become an increasingly important facet of world trade. While the level of *inter*-regional trade has not increased significantly as a percentage of global GDP since 1980, *intra*-regional trade has doubled during that period with particularly strong growth in Asia (see Figure 2.23).

A striking, albeit perhaps not surprising, aspect of the regionalization of trade is the continued dependence of LDCs for trade with developed countries. UNCTAD data indicate that, for example, the United States is the central focus for the exports of most Central American countries. Similarly, France, Italy, and Spain accounted for 32 percent of Algeria's exports in 2010 compared to only 3 percent for all other North African countries combined. The unsurprising aspect of this dynamic is that the combined economies of Tunisia, Egypt, Libya, and Morocco are approximately 6 percent the size of the combined economies of France, Italy, and Spain. Normalizing for this disparity—that is, holding the size of the combined economies constant—the share of Algerian exports destined for North African countries is nearly 50 percent larger than the share shipped to France, Italy, and Spain.

An important aspect of the "flattening" of the world economy has been the decreasing dependence of smaller, peripheral countries in their trading relationships with developed countries. For example, while France, Italy, and Spain accounted for one-third of Algeria's exports

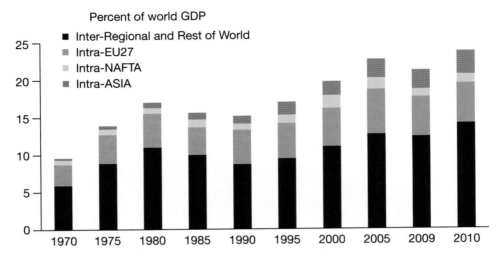

Figure 2.23 Intra- vs. inter-regional connectedness of major exporters

Source: Adapted from IMF (2011: 8, Figure 3) http://www.imf.org/external/np/pp/eng/ 2011/061511.pdf

in 2010, that total was down from over 50 percent a decade earlier. During that period, the share of Algerian exports destined for the United States increased from 16 to 24 percent. Similarly, China went from less than one-tenth of 1 percent of Algerian exports to over 2 percent. Although this level of exports is relatively insignificant, shifting to the import side of the trade equation, the story becomes more interesting. UNCTAD data indicate that the share of total imports held by France, Italy, and Spain dropped from 46 percent in 1996 to 31 percent in 2010. Similarly, the share of total imports from the United States was halved during that period. In contrast, the share of imports from China rose from barely more than 2 percent to over 11 percent. In other words, intra-region trade has increased, particularly among DCs, but internationalization—the combination of innovation in communication, the sharp reduction in the cost of information, the rise of transnational corporations, and the political embrace of free trade—appears to have reduced the dependence of LDCs on specific DCs that had previously benefited from geographical or colonial ties.

Another aspect of **dependency** is the degree to which a country's export base is diversified. Figure 2.24 shows one measure of dependency: The index of commodity concentration of exports. Countries with low values on this index have diversified export bases. They include Turkey, Thailand, Poland, and China, as well as most of the developed countries. At the other extreme are LDCs where the manufacturing sector is poorly developed and the balancing of national accounts and the generation of foreign exchange depends on the export of one or two agricultural or mineral resources: Angola, Azerbaijan, Chad, Congo, Iran, Iraq, Libya, and Nigeria, for example.

While some LDCs have a high concentration in exports, others have increasingly diversified since the mid-1990s. For example, Uganda's concentration index has dropped from 0.74 in 1995 to 0.19 in 2010. While that index level still reflects a high degree of concentration (Uganda is the leading exporter of coffee in Africa and coffee accounts for approximately one-fourth of its exports)—and far less than the diversification achieved by developed countries such as the United States (0.08), France (0.09), Italy (0.05), and Japan (0.13)—it demonstrates the effect of government efforts (as well as the help provided by foreign aid) to invest in and consistently broaden the economy.

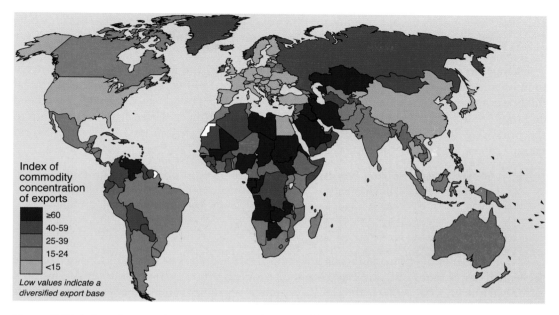

Figure 2.24 Index of commodity concentration of exports, 2011

Source: Based on data from UNCTAD, UNCTADstat online data: International Trade in Goods and Services: Concentration & diversification indices of merchandise exports and imports by country: exports http://unctadstat.unctad.org/TableViewer/tableView.aspx?ReportId=120

PATTERNS OF INTERNATIONAL FINANCE AND BUSINESS SERVICES

The spatial organization of world production and trade closely mirrors patterns of international finance and business services. By the beginning of the twenty-first century the United States and European Union accounted for close to three-quarters of global **foreign direct investment** (FDI) inflows and nearly 80 percent of outflows. However, the globalization of economic activity has modified the longstanding dominance of flows of foreign direct investment between these developed countries. In 2010 the U.S. and EU accounted for only 43 percent of inflows and 56 percent of outflows. The BRIC countries along with Hong Kong and Singapore increased their share of global FDI significantly, doubling their share of inflows to 26 percent and tripling their share of outflows to 18 percent during that period.

Until the early 1970s U.S.-based transnational corporations accounted for about two-thirds of the total outflows of foreign direct investment. About four-fifths of these investments were directed towards Canada and the most advanced industrialized economies of Europe. By the 1990s U.S. transnational corporations' share of the total had dropped to less than half, while foreign direct investment by Japanese, Canadian, and German corporations had increased significantly. Meanwhile, another source of foreign direct investment had begun to show up: Transnational corporations based in the NIEs. *Forbes* Global 2000 Leading Companies list in 2012 indicated that approximately 20 of the top 100 corporations in the world are now based in NIEs. Samsung Electronics (South Korea) and China Mobile are larger than AT&T and IBM; and Banco do Brasilia and Banco Bradesco of Brazil are larger than Goldman Sachs and Bank of America based on sales, profits, assets, and market value.

Along with these changes in the *sources* of investment have been changes in *destination*. The advanced industrial countries still absorb most of the inflows, but the globalization of economic activity has brought significant flows of capital into the NIEs; seven—Brazil, Chile, China (including Hong Kong), India, Mexico, Saudi Arabia, and Singapore—accounted for 60 percent of inflows among LDCs.

Changes in international investment have been contingent on other changes in the pattern of international finance. The **Bretton Woods Agreement** of 1944 set the pattern of international finance in the first part of the post-war period. During this period exchange rates were fixed and the U.S. dollar served as the convertible medium of currency with a fixed relationship to the price of gold. But, as the position of the United States deteriorated in terms of world manufacturing and trade, the system came under pressure.

> The result was that the Bretton Woods system crumbled. In particular, by the late 1960s fixed exchange rates effectively disappeared and every domestic currency became convertible into every other. Exchange rates "floated," and, as a result, all domestic currencies became a medium that could be bought and sold and out of which a profit could be made
>
> (Thrift, 1989: 34).

Meanwhile, a pool of **eurodollars** had developed—U.S. dollars held in banks located outside the United States (not to be confused with the European Union currency, the euro). This pool expanded quickly after 1971 as the U.S. government began financing its budget deficit with its own currency, effectively flooding the world with dollars and fuelling worldwide **inflation**.

Two years later, the eurodollar market swelled further as oil-producing countries rapidly acquired reserves of dollars as a result of the quadrupling of petroleum prices in the wake of the embargo by the **Organization of Petroleum Exporting Countries (OPEC)**. From the combination of floating currencies and the creation of a large market in eurodollars, a new, more sophisticated system of international finance emerged. Consequently, an expansion and internationalization of key business services such as stock exchanges, futures markets, and banks accompanied these new patterns of investment.

As developed countries became less competitive in labor-intensive industrial production, they increasingly relied on these services to earn foreign currency and balance national accounts. This system of international finance has produced giant financial conglomerates including Citigroup, UBS, HSBC, and Deutsche Bank that dominate world financial affairs. As Barnet and Cavanagh (1994: 17) noted:

> Twenty-four hours a day, trillions of dollars flow through the world's foreign-exchange markets as bits of data traveling at split-second speed. No more than 10 per cent of this staggering sum has anything to do with trade in goods and services. International traffic in money has become an end in itself, a highly profitable game.

But as the global debt crisis demonstrated, this interconnection also amplifies the potential domino effect of crises and can imperil not only these institutions but entire industries, countries, and the world economy. The subprime mortgage "meltdown" in the United States not only affected U.S. financial institutions. Its repercussions were felt across Europe and Asia. An estimated $2.2 trillion was written off on toxic assets and bad loans. The crisis triggered a sell-off in world financial markets and a global recession. In the initial years after the onset of the crisis, the reverberations continued to be felt as the debt burden of Greece, Ireland, Spain, Portugal, and Italy (Figure 2.25) increased the possibility of government default and the implosion of the euro.

Additionally, not all of the needs of this "casino economy" can be met by conventional financial and business services. Offshore financial centers have emerged to meet the need for secrecy and the desire for shelter from taxation and regulation; islands and micro-states such

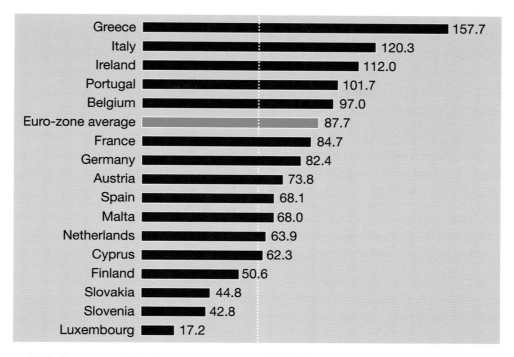

Figure 2.25 Government debt of euro-zone countries, 2011 (Maastricht limit: 60%)

Source: Adapted from Spiegel Online (website) http://www.spiegel.de/fotostrecke/graphics-gallery-the-most-important-facts-about-the-global-debt-crisis-fotostrecke-71636–5.html

as the Bahamas, Bahrain, the Cayman Islands, the Cook Islands, Luxembourg, Liechtenstein, and Vanuatu have become specialized nodes in the geography of worldwide financial flows. The chief attraction of these offshore financial centers is simply that they are less regulated than financial centers elsewhere. They provide low-tax or no-tax settings for savings and havens for undeclared income and "hot" money. They also provide discreet markets for transactions of currencies, bonds, loans, and other financial instruments without drawing the attention of regulating authorities or competitors. In an effort to curb tax evasion, the U.S. Internal Revenue Service initiated three voluntary disclosure programs (in 2009, 2010, and 2011) that netted $4.4 billion in previously unpaid taxes, an amount that almost certainly represents a small fraction of the income sheltered in offshore accounts.

INTERNATIONAL TRADE AND THE DEBT TRAP

The structured inequality of the world economy has led to a chronic problem of international debt. The role inherited by most LDCs within the international division of labor has been one of producing **primary commodities** for which both the **income** and **price elasticity of demand** are low. In contrast, the income and price elasticity of manufactured goods and high-order services (the specialties of developed countries) are both high. As a result, the **terms of trade** are stacked against the producers of primary commodities. No matter how efficient primary producers may become or how affluent their customers are, the balance of trade will be tilted against them. Quite simply, they must run in order to stand still.

Box 2.5 From darling to disaster: Iceland and the global financial crisis

By Bart Yavorosky

The global financial crisis produced a number of high-profile failures such as Lehman Brothers, Merrill Lynch, and Bear Sterns, but none illustrates the interconnectedness of the world economy quite as strikingly as Iceland.

For much of its history, Iceland had a "cod" economy. The ubiquitous fish of the Atlantic was its primary source of wealth, its main export, and its only impetus for "wars" (conflicts with the United Kingdom over the right to fish the waters off its mainland). In the span of a few years leading up to the financial crisis, however, Iceland became the envy of many other DCs.

Mimicking the profitable short-term lending practices of financial institutions in other countries, Icelandic commercial banks began offering low-interest, long-maturity loans at the start of the 2000s. Easy credit fueled a construction boom and a spike in real estate prices. By refinancing their mortgages, Icelanders converted their paper wealth into cash and increased personal consumption (see Figure 2.26 A). In an effort to curb inflation (see Figure 2.26 B), the central bank raised short-term interest rates to 15 percent. This rate was attractive to foreign investors who borrowed money abroad (for example, from Japan) at low rates and invested in high-returning Icelandic assets. This "carry trade" increased net inflows of FDI nearly eightfold (see Figure 2.26 C) and enabled Icelanders to embark on a foreign-denominated debt spending spree.

Deregulation of Iceland's financial sector also enabled its three main banks—Glitnir, Kaupthing, and Nyi Landsbanki—to expand operations abroad by acquiring commercial banks and securities brokerage houses in Europe and the United States. As bank revenues grew, Iceland's stock market soared (see Figure 2.26 D). By 2007 these banks had accumulated assets totaling more than eight times Iceland's GDP and were funding as much as 75 percent of their operations through short-term debt.

A country that traditionally had prided itself on prudence and thrift became one of the fastest growing economies in the world and an emblem of conspicuous consumption. When the global financial crisis hit and credit markets seized up, Iceland's banks could no longer acquire the capital to operate; given the size of their debt, Iceland's central bank could not act as a lender-of-last-resort. With no other options, the government seized the banks, sending its stock market tumbling, decimating the krona, and devastating its real estate market. The fishing families who had eked out modest livings for generations by trolling the North Atlantic for cod had become investors who, for a few brief years, earned but subsequently lost millions riding the waves of the global financial markets.

A prudent counterstrategy for these countries is to change their role in the international division of labor, moving away from the specialization in primary commodities and towards a diversified manufacturing base; that is, pursuing **import substitution**. Unfortunately, this strategy is easier to execute on paper than in practice. Establishing a diversified manufacturing base requires vast amounts of start-up capital. With the terms of trade running against them, these countries have a difficult time acquiring the requisite capital. The only alternative, short of striving for self-sufficiency—for example, Tanzania and Myanmar (Burma)—or opting out

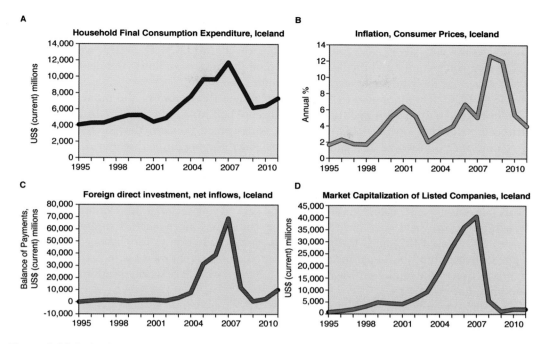

Figure 2.26 Iceland and the global financial crisis

Source: Based on online data from World Bank, World Development Indicators at http://data.worldbank.org/data-catalog/world-development-indicators

of the capitalist world economy altogether—for example, Cuba and North Korea—is raising the capital through loans. If the capital is invested in economic development projects that do not yield sufficient returns, further loans must be undertaken in order to service the original debt and to finance new development projects. This syndrome of constantly borrowing to fund "development" has come to be known as the **debt trap**.

Servicing long-term debts (associated with meeting costs of both interest charges and repayments) has become a significant burden for some countries. Total debt service on **International Monetary Fund (IMF)** loans exceeded 35 percent of exports of goods, services, and income for Latvia, Jamaica, Kazakhstan, and Tajikistan for example, and in excess of 30 percent for Turkey and Ukraine in 2011 according to World Bank data.

The debt problem has been seriously aggravated by patterns of international capital flows that have been stimulated by changing money market conditions rather than by the geography of trade. In order to avoid the punitive "taxation" of profits and savings by the high inflation rates caused by the combination of trade imbalances and debt, the domestic creditors of LDC governments (anyone holding its currency or its interest-bearing debt) tend to move their money abroad. The result is **capital flight**.

But capital flight should not be considered strictly a problem for LDCs. For example, in the first quarter of 2012, the central bank of Spain, Banco de España, reported that €97 billion, or roughly 10 percent of Spain's GDP, exited the country over concern of the government's ability to respond to the economic and financial crises.

A second significant problem for LDCs occurred in the 1970s when world monetary reserves increased twelvefold with the availability of OPEC petrodollars and as a by-product of the inflation that accompanied the break-up of the Bretton Woods system. The international banking system, awash with funds, began to "recycle" the surpluses that oil companies had deposited with them. Bankers from developed countries suddenly found a willingness to lend to LDC governments. They also found eager borrowers in LDCs. The result was that many LDC governments committed themselves to capital projects, wage bills, and debt repayments that far exceeded their financing capacity through taxes.

By 1981 LDC debt amounted to $739 billion. When **Reaganomics** and **monetarism** drove up interest rates in the early 1980s the debt "burden" became a debt "crisis." In 1982 Mexico threatened to default on its debts, triggering wide-spread concern that such action could snowball to the point where international financial stability might be threatened. Recognizing this possibility (and realizing that their long-term interests depended not only on international financial stability but also on reasonably healthy markets in LDCs), creditors in the developed countries allowed Mexico to reschedule its debts.

This pattern of events has since been repeated several times. Brazil, for example, borrowed so much money in the 1970s and 1980s that it could no longer meet its interest payments. Between 1983 and 1989 the IMF bailed it out on the condition it adopt austerity measures designed to curb imports. These measures included a 60 percent increase in petroleum prices and a reduction of the minimum wage to $50 a month, which gave workers half the purchasing power they had in 1940.

The Philippine government, meanwhile, has had to reschedule its debt five times with the Paris Club, a group of 19 creditor countries. However, by 2010 the Philippines carried foreign debt of approximately $60 billion, but it had reduced its debt–service ratio to less than 20 percent of its annual export earnings.

By 2005 the accumulated debts of LDC countries had risen to over $2.5 trillion, leading to calls for lending countries to provide debt relief. In 2000 the United States and Britain began to cancel LDC debts to the tune of $435 million and $1.43 billion respectively. Then, at the meeting of G8 countries (USA, Japan, Germany, UK, Canada, France, Italy, and Russia) in 2005, agreement was reached to write off the entire $40 billion debt of 18 highly indebted LDCs to the World Bank, the International Monetary Fund, and the African Development Fund.

Despite these efforts, by the end of 2010, in the wake of the global financial crisis, the World Bank reported that the cumulative external debt of LDCs had grown to $4 trillion. Addressing this magnitude of LDC debt will require continued substantial and sustained efforts on the part of developed and less developed countries.

PATTERNS OF INTERNATIONAL AID

The debt issue leads us logically to the question of aid. Large-scale movements of aid began shortly after the Second World War with the **Marshall Plan**, financed by the United States to bolster war-torn European allies whose economic weakness, it was believed, made them susceptible to communism. During the 1950s and 1960s, as more LDCs gained independence, aid became a useful tool in western and Sino-Soviet Cold War offensives to establish and preserve political influence throughout the world.

By the late 1960s the list of donor countries had expanded beyond the superpowers to include smaller countries such as Austria, Denmark, and Sweden, whose motivation in aid giving can be seen as more philanthropic than strategic. In addition, there was a greater geographic dispersal of aid, thanks largely to the activities of multilateral financial agencies such as the IMF and World Bank.

Nevertheless, the geography of aid still has a strong political flavor. Bilateral aid from some countries reflects localized political aspirations and colonial ties. So much British and French aid is directed towards former African colonies; Japanese aid is disbursed largely within Asia; and aid from the OPEC countries has been directed mainly towards the "frontline" Arab countries.

More significantly for the LDCs, the end of the Cold War, together with the balance of payments difficulties of several developed countries, has ensured that levels of aid have diminished. Whereas official development assistance from OECD countries amounted to nearly 0.5 percent of their total GNI in 1965, it had fallen to 0.3 percent by 2011 with only Sweden (1.02 percent), Norway (1.00 percent), Luxembourg (0.99 percent), Denmark (0.86 percent), and Netherlands (0.75 percent) meeting the United Nations official development assistance (ODA) target of 0.7 of GNI.

OECD data indicate that the most striking decrease in ODA as a percentage of GNI has come from the United States. It had maintained a ratio in excess of 0.5 percent for much of the 1960s but it fell to 0.19 percent at the end of the energy crisis in the 1970s and to around 0.1 percent when Republicans controlled Congress during the Clinton administration in the 1990s. It then rose steadily during the Bush administration and has been approximately 0.2 percent during the Obama administration. It should also be noted that the dollar value of aid from the United States continues to exceed the aid provided by the two next highest donor countries combined—Germany and the United Kingdom. In contrast, Japan has consistently maintained official development assistance at a level between 0.2 percent and 0.3 percent since 1960.

At these levels, aid cannot seriously be regarded as redressing core–periphery inequalities. Contrariwise, because most aid is "tied" in some way to donor countries' exports or to specific military, educational, or cultural projects, it has been argued by many that, insignificant as it is in relative terms, it is sufficient to reinforce the initial advantage of the donors.

SUMMARY

In this chapter, we described the major dimensions of the contemporary economic landscape. We identified dominant and recurring patterns associated with resources, population, industry and finance and noted the major exceptions to these patterns. Other important points include the following:

- The world economy can be seen as an evolving market system comprising an ever changing economic hierarchy of countries—*core*, *periphery*, and *semi-periphery*.
- Defining and measuring "economic development" is problematic. Increasingly, development is being framed in broader terms that include consideration for social well-being.
- Sustainable development requires economic, financial, and fiscal decisions to be fully integrated with environmental and ecological decisions. Such decisions require a weighing and reconciliation of interests that are often contradictory and policies with outcomes that can be estimated tenuously at best.
- The intensity of changes in patterns of economic activity and development are associated with the globalization of economic activity under the influence of the strategies of transnational corporations (TNCs).
- A striking, although perhaps not surprising, aspect of the regionalization of trade is the continued dependence of less developed countries for trade with developed countries.

KEY SOURCES AND SUGGESTED READING

Barnet, R.J. and Cavanagh, J., 1994. *Global Dreams. Imperial corporations and the new world order.* New York: Simon & Schuster.

Beyon, J. and Dunkerley, D. (eds.), 2000. *Globalization: The reader.* New York: Routledge.

FAO, 2011. *The State of Food Insecurity in the World.* Rome: FAO.

Held, D. and McGrew, A. (eds.), 2000. *The Global Transformations Reader.* Malden, MA: Blackwell.

Hutton, W. and Giddens, A. (eds.), 2000. *Global Capitalism.* New York: New Press.

Lechner, F.J. and Boli, J. (eds.), 2000. *The Globalization Reader.* Malden, MA: Blackwell.

Meier, G.M. and Baldwin, R.E., 1957. *Economic Development.* New York: John Wiley & Sons.

Mittelman, J.H., 2000. *The Globalization Syndrome: Transformation and resistance.* Princeton, NJ: Princeton University Press.

O'Meara, P., Mehlinger, H.D., and Krain, M. (eds.), 2000. *Globalization and the Challenges of a New Century.* Bloomington: Indiana University Press.

Thrift, N., 1989. The geography of international economic disorder, in R.J. Johnston and P.J. Taylor (eds.), *A World in Crisis?*, 2nd edn. Oxford: Blackwell, 16–78.

UNCTAD, 2011. *World Investment Report: Non-equity modes of international production and development.* New York and Geneva: United Nations.

Wallerstein, I., 1984. *The Politics of the World-Economy.* Cambridge: Cambridge University Press.

Chapter 3

Geographical dynamics of the world economy

Picture credit: Linda McCarthy

As we saw in the last chapter, the 1960s marked the beginning of a more interdependent world economy. Many firms operating in western Europe and the United States reorganized their operations and improved their profitability by subcontracting some of their activities and internationalizing their production facilities.

But, from another point of view, a world economy was already in existence and simply undergoing a process of "globalization." Beginning in the sixteenth century, but undergoing its greatest expansion and intensification in the nineteenth and twentieth centuries, a world economy had evolved from localized economic systems. As it became progressively more integrated, covering ever wider geographical areas and a greater number of economic activities (for example, resource extraction, capital investment, trade in manufactures, services, etc.), it underwent shifts in its mode of operation as well as shifts in the relative importance of different world regions (see Figure 3.1).

Viewed from this larger historical context, the changes that began in the late 1960s are merely the most recent manifestation of this evolutionary process.

We have a threefold objective for this chapter:

1. to sketch the historical development of the world economy
2. to pinpoint the geographical effects of state regulatory and macroeconomic actions
3. to identify the main causes and consequences of the current geographical reorganization of the world economy.

We provide a framework for understanding economic landscapes, outline the historical context for the emerging trends in world economic geography examined in Part 4, and give an overview of the important theoretical trends in the field of economic geography: *Geopolitical economy*, that is, the impact of states on the working of the world economy and the geography of economic activities.

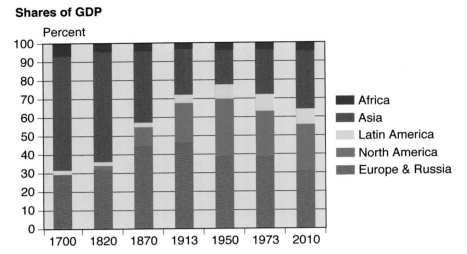

Figure 3.1 Shifting fortunes in the world economy (after Maddison)

Source: Updated from Venables (2006: 63, Chart 1)

3.1 HISTORY OF THE WORLD ECONOMY

Six basic features of the world economy can serve to organize our account of its history.

1. THE SINGLE WORLD MARKET

The world economy consists of a single global market. Within this market, the price of products is not fixed but set as a result of competition among producers. Price-setting markets, in this modern sense, took many years to become established, particularly since labor did not become a commodity (**wage labor**) until the late eighteenth century. Indeed, that point in time marks an important qualitative change in the nature of the world economy. European industrialization intensified world trade and produced increasingly complex markets for raw materials and manufactured goods. Extra-European investment and trade also became much more important for the growth of Europe after the late 1700s. Major waves of international migration in the nineteenth and late twentieth centuries testify to the periodic increase in geographical scope of the pools of labor on which firms and regions can draw to fuel their growth.

2. THE STATE SYSTEM

The world economy has always had a territorial division between political states. This division both pre-dates and grew along with the geographical expansion of the contemporary world economy. States can protect and stimulate **infant industries** and encourage the development of domestic production through **tariffs**, trade quotas (by restricting access to local markets of foreign-made goods), and financial incentives. The result is a competitive system in which each state attempts to simultaneously protect itself from and leverage to its advantage the world market.

Obtaining competitive advantage does not always require conventional protectionist measures such as tariffs on competing products. Numerous historical examples attest to the importance of the mediating role of the state and other institutions: From the government

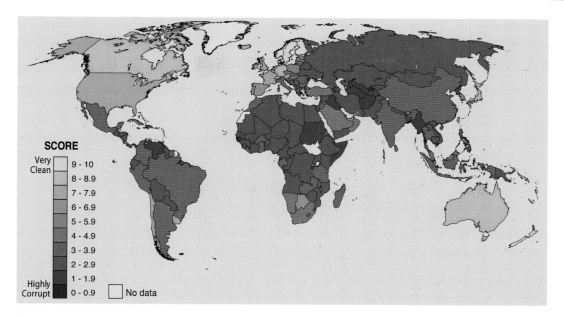

Figure 3.2 Corruption perceptions index, 2011

Source: Based on Transparency International, 2012 (website) http://www.transparency.org/cpi2011/results

stimulus that resulted in canal building in the United States in the early nineteenth century and massive spending on military goods since the Second World War; through Japan's disciplined conquest of market share in a wide range of industries after 1945; to China's recent investment in solar technology, which marginally reduces its dependence on nonrenewable energy sources but, more importantly, has enabled its large-scale, highly integrated solar panel manufacturers to operate profitably and capture market share from previously dominant Western suppliers as the price of photovoltaic cells continues to drop.

In some cases, for example in the Nordic model adopted by Norway, Sweden, Finland, and Denmark, social policy is intertwined with economic and state policy. These economies are characterized by pro-growth features such as minimal market regulation and a high degree of openness, that is, few barriers to free trade, but they also include an elaborate social safety net, universal healthcare, free education, and the strong protection of property and individual freedoms. These policies foster social cohesion and transparency, minimize corruption (Denmark, Finland, Norway, and Sweden rank in the top six of Transparency International's corruption perception index; see Figure 3.2), and enable the type of collective risk taking and openness to change that has become increasingly important to adapting to evolving global economic and market conditions.

3. THE THREE GEOGRAPHICAL TIERS

The modern world economy has developed a basic three-tiered geography as it has expanded to cover the globe (see Figure 2.1). The early world economy consisted of Europe and those parts of South and Central America under Spanish and Portuguese rule. The rest of the world was an external arena, essentially outside the workings of the world economy. Eventually, the rest of the world became incorporated by way of trade, colonization, or imperialism. So the world economy came to consist of a core (western Europe at first, joined later by the United

States and Japan), periphery, and semi-periphery as described at the beginning of Chapter 2. **Uneven development** at a global scale, therefore, is not a recent phenomenon or a mere by-product of the world economy; it is one of the basic features of the world economy.

However, even as it expanded to incorporate ever greater parts of the world, the world economy was not without change. Countries have moved between the tiers; for example, the rise of the United States and Japan, and decline of Spain and Portugal. Similarly, the semi-periphery, which is composed of countries such as Singapore, Malaysia, South Korea, and the BRIC countries and where a mix of core and periphery processes are at work, have also become increasingly important with globalization.

4. TEMPORAL PATTERNS AND HEGEMONY

The contemporary world economy has followed a number of cyclical patterns of growth and stagnation. In the first chapter, we identified and reviewed some of the main cycles of the world economy (see Figure 1.2). States can exploit or suffer from cyclical effects depending on their productivity, commercial supremacy, ability to restrict competition from rivals, and capacity to respond to crises.

Britain dominated the world economy through a mixed strategy of formal and informal imperialism (empire building and extensive investment outside its empire). The United States then dominated by sponsoring a set of international economic institutions—the IMF, International Bank for Reconstruction and Development (the World Bank), and the General Agreement on Tariffs and Trade (GATT), and its successor, the World Trade Organization (WTO)—and international and regional security institutions—the United Nations and the North Atlantic Treaty Organization (NATO)—in order to foster free market capitalism after the Second World War.

The term **hegemony** is often applied to instances of dominance such as those of Britain and the United States. This concept is contentious, but most scholars would accept the criterion that the leading state cannot be simply the strongest militarily; it must also have the economic and cultural power to set and enforce the rules of international conduct that it prefers. U.S. hegemony, therefore, not only signifies a shift in the identity of the hegemonic power from a previous one, it also implies a shift in the institutions and practices that the United States has brought to the world by virtue of its dominant position. These included mass production/consumption (Fordism), limited state welfare policies, electoral democracy based on weak mass political parties, and government economic policies directed towards stimulating private economic activities.

5. INCORPORATION, SUBORDINATION, AND RESISTANCE

One danger in focusing on the concept of world economy is that local histories can be deprived of their integrity and specificity. In fact, people resist or adapt to incorporation into the world economy rather than simply accept or succumb to it, and different parts of the world have reacted differently to the expansion of the world economy.

Hall (1986) provides a very useful typology of world-economy impacts. Along a continuum of patterns of incorporation he distinguishes a "weak" pole of areas external to the world economy (external arenas) and a strong pole of fully fledged dependent peripheries. Between these extremes are areas where contact has been slight (contact peripheries) and an intermediate category of marginal peripheries (see Table 3.1). The processes involved as an area shifts from the status of an external arena to a dependent periphery are also indicated in Table 3.1. Market

Table 3.1 The continuum of geographical incorporation into the world economy

Type of periphery	The continuum of incorporation			
	None	Weak	Moderate	Strong
	External arena	Contact periphery	Marginal periphery	Dependent periphery
Market articulation	None	Weak	Moderate	Strong
Impact of core on periphery	None	Strong	Stronger	Strongest
Impact of periphery on core	None	Low	Moderate	Significant

Source: Based on Hall (1986: 392, Diagram 1)

articulation refers to the nature of the capital and product flows between the expanding world economy and an area undergoing absorption. At the weak pole of the continuum, the areas are only slightly connected to the world economy, with the primary flow of influence from the core to the periphery. At the strong pole of the continuum, the exchange involved is important to core development. Although influence flows in both directions, net product and capital flows generally favor the core.

Movement along the continuum is contingent rather than inevitable, both in terms of the pace and the eventual degree of dependence. The pace of transition towards strong incorporation depends on the strength of the state engaged in expansion and the nature of the world economy at the time. For example, in the sixteenth century Spanish expansion led less immediately to effective incorporation and the spread of market exchange than did British expansion in the nineteenth century. Plunder and religious zeal were more important to sixteenth-century Spain than "bringing to market." In the nineteenth century, market exchange was effectively internationalized under British hegemony as production for the market everywhere replaced the mere exchange of commodities (see Table 3.2). The British national economy had become the "locomotive" of the world economy. But as its markets in Europe became more competitive, it was pushed to widen its markets elsewhere. The British Empire played an important role in this expansion. The internationalization of the British economy in the nineteenth century was a crucial element in the quickening pace and increased strength of incorporation worldwide (see Chapter 8).

6. ALTERNATIVE ADAPTATIONS

Finally, every part of the world has had its own particular relationship to the evolution of the world economy. In the case of the United States, for example, the existence from an early period of two contrasting and incompatible modes of socioeconomic organization within one territorial state—a plantation agriculture based on African slave labor in the South and a "classic" capitalist or free enterprise economy in the north—was uniquely American. Its heritage, in terms of the relative regional underdevelopment of the south and racially polarized politics wherever there are concentrations of African-Americans, continues to this day. Similarly, the industrialization and subsequent "unionization" of the Northeast has made the South, with its "business-friendly" state regulations and legislation, such as right-to-work laws (which secures

Table 3.2 The geographical development of the world economy in the nineteenth century

Stage:	Developed	Developing		Underdeveloped	
Factor intensity:	Capital	Labour	Land	Land	Labour
1800	Britain	Europe		USA	India
1840	Britain	Europe	USA	Latin America Australia Canada	India China
1870	Britain Europe		USA	Australia Latin America Canada Africa	China
1900	Britain Europe USA		Australia Canada Argentina Mexico South Africa	Latin America Africa	India China

Source: Based on Hansson (1952: 49–82)

the right of the worker to decide whether to join or provide financial support to a union), an increasingly attractive destination for corporations seeking to locate (or relocate) within the United States.

Likewise, the apartheid system of racial categorization and control in South Africa was a peculiarly South African response to the history of European settlement and economic exploitation in southern Africa. The nature of relations between the state and the economy also differs significantly among such nominally capitalist states as Britain, Italy, France, and Germany. These differences reflect the historical development of connections with the world economy and alternative approaches to maintaining international competitive advantage. The whole process of development and underdevelopment is frequently mediated geographically through the actions of state-level regulation. Japan (see Chapter 5) and China (see Chapter 10) offer fascinating examples of this process. Also, the most important experiment in providing a consciously designed *departure* from the guiding principles of the world economy was the model of economic development established in the Soviet Union in the aftermath of the Russian Revolution of 1917 (see Chapter 5).

3.2 STATES AND THE WORLD ECONOMY

The emergence of the world economy in Europe coincided with, and was dependent on, the consolidation of territorial states within Europe and the emergence of the Westphalian System. The Peace of Westphalia (1648), ending the 30-year war in Europe, established several key principles that have had a lasting impact on the world:

- the sovereignty of states and the fundamental right of political self-determination
- legal equality between states
- nonintervention of one state in the internal affairs of another state.

These principles are still relevant today, which explains why the system of states is often referred to as the Westphalian System. From approximately 500 political entities in Europe in 1500, no more than 25 territorial states existed in 1900. Modern territorial states serve as the political framework of the contemporary world economy. So the expansion of the world economy has been accompanied by a parallel expansion of the interstate system as the sole legitimate form of political, military, and administrative organization.

But the state is not a standardized entity. It includes different institutions (varying representative assemblies, bureaucracies, police forces, militaries, etc.), each a product of particular histories that adapted to shifts in national development and position within the world economy. The relative power of each state can be thought of in terms of three dimensions:

1. relative to one another
2. relative to its inhabitants
3. relative to the evolving world economy (Harris, 1986: 145–169).

The first—the hierarchy of states—has shown considerable variation historically, for example, in the rise and decline of Britain and the Soviet Union; the rise of the United States; and the growing power of China, India, and the **Asian Tigers**—Hong Kong, Singapore, South Korea, and Taiwan. The second—the power of states relative to their inhabitants—seems to grow incessantly, particularly in the more developed countries, with respect to regulation, taxation, and surveillance. However, the third—the power of states relative to others in the world economy, such as TNCs—appears to be declining in general, and to decline more as countries develop economically.

STATES AND THE EVOLVING WORLD ECONOMY

The latest phase in the evolution of the world economy, labeled earlier (see Figure 1.2) by the term globalized capitalism, has involved the erosion of national economies as the basic building blocks of the world economy. A transnational element has been ascendant in the form of the growth of a global market that is supplied by firms that organize their production and distribution without much reference to national boundaries. This "global shift" in production has given rise to an explosion of foreign direct investment (FDI) and to the emergence of trade *within* large firms as the most rapidly expanding component of total world trade. Perhaps as much as 40 percent of U.S. imports, for example, are intra-firm transactions involving parts and goods purchased by U.S. subsidiaries of multinational firms or from foreign subsidiaries of U.S. corporations (Hellerstein *et al.*, 2006).

Transnational corporations, therefore, are major engines in the growth of world trade. Chapters 7, 8, 10, and 11 document the impact of this global shift on the geography of the world economy. However, an important point that must be underscored is that, rather than marking the eclipse of national boundaries, the globalization of production has occurred in large part *because* of them. *The strategies of transnational corporations are designed to exploit national differences in labor forces, market conditions, regulatory environments, and macroeconomic (fiscal and monetary) conditions.* Without this variation, the attraction of shifting investments and relocating production would be reduced. The challenge to states is to compete in this more integrated and interdependent economic environment as effectively as possible.

One must also be cautious not to exaggerate the extent that the globalization of production has increased the power of global firms. Many "global" companies are still strongly attached to their home countries. For example, non-national board membership on transnational corporations, although steadily increasing, remains low. European and North American companies lead the trend towards global boards. In Japanese companies, in contrast, "foreign directors are as rare as British sumo wrestlers" (*Economist*, 1993a: 69). This bias manifests itself in the activities of even the most "global" firms. When total sales decline, home markets tend to be protected at the expense of foreign ones. When firms expand abroad they continue to rely heavily on suppliers from their home countries. Antitrust laws and nationalism still make it difficult for foreign firms to expand through takeovers. Many states and trading blocs (such as the European Union) have industrial and trading policies that serve to protect and enhance their national economies by encouraging domestic and discouraging foreign investment (for example, Airbus Industrie, the producer of the European Airbus is a consortium of aerospace firms from different European countries that collaborates to compete with U.S. producers such as Boeing).

Britain provides an interesting contrast to these protectionist policies. Between 2000 and 2010 foreigners spent nearly $1 trillion to acquire 5,400 British companies including Jaguar and Land Rover by Tata (India) and the purchase of the venerable chocolatier, Cadbury, by the U.S. company, Kraft Foods. But it should be noted that the British also spent U.S. $750 billion to acquire 6,000 foreign firms during this same period.

The sale of Cadbury, in particular, is also instructive from another perspective: 80 percent of its business and 8 percent of its employees resided outside the United Kingdom when it was purchased (*Economist*, 2010a), raising questions about whether the emphasis is more appropriately placed on the "multi" or "national" part of its operations and the extent that British government policies can and, perhaps more importantly, should seek to align and simultaneously advance the interests of UK-based multinational corporations and the broader interests of its citizenry.

INTERNATIONAL FINANCIAL SYSTEM

Perhaps the most significant factor contributing to the reduction of state power relative to the world economy has been the ever increasing importance of the international financial system. All forms of capital have become more mobile. Corporate and financial market decisions are made more frequently and within much tighter time schedules than was only recently the case. Decisions about employing and shedding labor are made weekly and monthly compared to quarterly (see Figure 3.3). Commodity, currency, and equity markets in world cities operate around the clock. Banking has become a global industry with the reduction of national and local barriers to operations and capital movements.

Flows of capital have become increasingly important drivers of the world economy. The volume of flows has increased exponentially since the 1970s. Although developed countries remain important recipients of FDI inflows, developing and transitioning countries accounted for 52 percent of inflows in 2010 (Figure 3.4). When FDI is broken down by economic activity, services had been the most important sector leading up to the financial crisis. But in the post-crisis period, manufacturing accounted for 48 percent of FDI, while services, led by the financial industry, experienced a sharp decline and only accounted for 30 percent of FDI. During the crisis, FDI in the primary sector also declined sharply, but it had recovered to pre-crisis levels by 2009–2010 (UNCTAD, 2011).

Government policies towards FDI have been liberalized worldwide. The privatization of state-owned industries has driven much of this change; however, pressure from multilateral institutions such as the IMF and the World Bank, which insist on increased openness to FDI

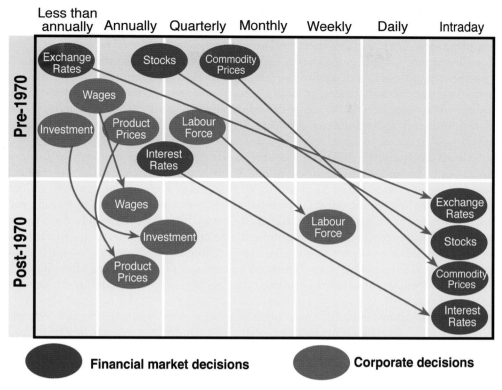

Figure 3.3 The increasing pace of the world economy

Source: Updated from *Economist* (1983: 11, Figure 2)

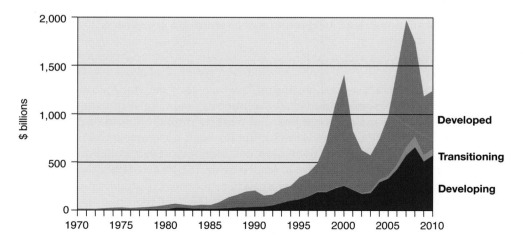

Figure 3.4 F.D.I. inflows, developed, developing, and transitioning* countries (*includes South-East Europe and Commonwealth of Independent States)

Source: Based on data from UNCTAD, UNCTADstat online data: Inward foreign direct investment flows http://unctadstat. unctad.org/ReportFolders/reportFolders.aspx

in return for favorable loan conditions, has also been instrumental. In some LDCs, the IMF and the World Bank have cooperated to intervene in domestic politics through **structural adjustment programs**. These policies are stipulations that must be met to receive new loans or to obtain lower interest rates on existing loans. Their intent is to ensure that the money lent will be spent in accordance with the overall goals of the loan, reducing the borrowing country's macroeconomic imbalances, promoting economic growth, and generating income in order to pay off accumulated debt. Structural adjustment programs are based on fiscal and monetary restraint and a combination of deregulation and the liberalization of national markets.

Even the most economically powerful countries are now less autonomous and more vulnerable to both foreign ownership of economic assets and shocks emanating from world financial markets. One commentator has proclaimed that global financial integration signifies the "end of geography" (O'Brien, 1992). The relative impotence of states to deftly "right the ship" following the financial crisis suggests that O'Brien may be more correct than not at least in certain respect. However, important institutional and cultural barriers remain to the free movement of capital. For example, knowledge about financial opportunities is still tied to knowledge of the local investment "scene" to some extent, and this knowledge often depends on having access to local social networks and relationships developed through interaction and common social bonds.

Financial markets are undoubtedly more open than they were when the Bretton Woods system of pegged exchange rates prevailed (see Chapter 2). Under that system, free trade and regulated finance ultimately proved incompatible; once free trade became more significant in the 1960s, the latter proved impossible to sustain. Confidence in a government's ability to maintain a fixed exchange rate (such as that of Britain in 1967) began to wane as capital controls proved incapable of neutralizing international capital movements. Flexible exchange rates after 1971 have proved relatively advantageous for large, relatively closed economies such as those of the United States and Japan. Some economists argue, however, that the costs of floating currencies have become increasingly high for smaller, open economies. According to this view, excessively volatile exchange rates disrupt domestic policymaking and reduce the ability of firms to make calculations about long-term rates of return on investment.

INDEPENDENCE AND INTERDEPENDENCE

Increased financial integration does not mean, however, that the world economy now runs rampant over national economies. Individual governments (and state institutions such as central banks) can still influence international financial markets through industrial policies, budget deficits, and the manipulation of interest rates. But no single government can now control the system or completely control its own national economy. Even collaboration between governments, such as the G8 meetings (of leaders from the United States, Japan, Germany, United Kingdom, Canada, France, Italy, and Russia), cannot always produce predictable effects on currencies, investments, and trade flows. Nevertheless, collaboration, both formal and informal, and largely under U.S. sponsorship, has become an important feature of the world economy. The high degree of interdependence among states was most strikingly evident in the coordinated response to the global financial and Eurozone crises where bailouts and a range of stimulus programs were simultaneously and aggressively pursued in a number of countries in an effort to stave off a global recession.

MANAGEMENT AND MEDIATION

The crux of the contemporary world economy, then, is the coexistence of national and global structures, and attempts by states to manage the tensions between them. Transnational activities operate within states and under conditions imposed by them. Trade and tariff policies are still of vital importance in regulating trade and investment. For example, even as industrial countries' tariff levels have dropped steadily as a result of the GATT/WTO negotiations of the post-Second World War period, **nontariff barriers** (quotas, voluntary trade restraints, technical barriers, etc.) have often increased, usually as a result of pressure on governments from domestic political coalitions. In a number of countries, especially the United States, there is a growing controversy among economists and politicians over the geographical distribution of the costs and benefits of open as opposed to more "managed" trade. Some commentators argue that certain countries, particularly China and Japan, do not trade "fairly;" rather, they impose too many restrictions on imports while they benefit from the relative openness of other countries.

Financial systems are also important to the geography of the world economy because they mediate between political systems, on the one hand, and the global system of production, on the other. Largely as a result of regulatory policy, national financial systems still differ profoundly in their connection with local industry and in their openness to foreign penetration. The German and Japanese financial systems, for example, are intimately connected with their co-national industrial companies, whereas the U.S. and the British systems are not only independent of but often appear to be at cross-purposes with the needs of the domestic manufacturing industry. This difference has important consequences, particularly in terms of the geography of investment. In Germany and Japan, the fortunes of finance and industry are closely tied. Each has a stake in the other. In Britain, industry has had to compete with a wide range of alternative and often more attractive outlets for bank investment, inside and outside the country. At the same time, financial markets in Britain have been more readily accessible to investment from outside, creating in London a truly global financial center.

In general, national financial systems differ along three dimensions (Zysman, 1983: 69). The first is the process whereby savings are transformed into investments and allocated among competing uses: "Intermediation." The second dimension is the degree to which prices are set in financial markets by dominant financial institutions or by government: "Marketization," or price controls. The third is the amount of government intervention in the financial system: Regulation.

Combinations of the three dimensions differ significantly across countries and so provide distinctive financial environments for internal economic development. The United States and Britain are capital market financial systems in which the intermediation of financial institutions is relatively weak; prices are set in markets; and there is limited state intervention in financial markets. There is little, if any, bias towards local industrial firms or long-run national economic development.

Germany typifies a second model of a financial institution-dominated credit-based financial system in which bank intermediation dominates; prices are set in markets, but the state and financial institutions are closely related. Large banks provide the bulk of the credit to domestic industrial firms and are usually represented on the governing boards of these firms.

Finally, there are government-dominated credit-based financial systems, exemplified by France and Japan. In this case, state entanglement with industry (nationalized industry in France, private industry in Japan) has fundamental limiting effects on the autonomy of markets and financial institutions. Particularly in Japan the financial system has been structured traditionally

Table 3.3 The institutions and ideological basis of the world's dominant capitalisms

Characteristic	US capitalism	Japanese capitalism	European social market	British capitalism
Basic principle				
Dominant factor of production	Capital	Labor	Partnership	Capital
"Public" tradition	Medium	High	High	Low
Centralization	Low	Medium	Medium	High
Reliance on price-mediated markets	High	Low	Medium	High
Supply relations	Arm's length Price driven	Close Enduring	Bureaucracy Planned	Arm's length Price driven
Industrial groups	Partial, defence, etc.	Very high	High	Low
Extent privatized	High	High	Medium	High
Financial system				
Market structure	Anonymous, securitized	Personal, committed	Bureaucracy, committed	Uncommitted, marketized
Banking system	Advanced, marketized, regional	Traditional, regulated concentrated	Traditional, regulated, regional	Advanced, marketized, centralized
Stock market	Very important	Unimportant	Unimportant	Very important
Required returns	High	Low	Medium	High
Labor market				
Job security	Low	High	High	Low
Labor mobility	High	Low	Medium	Medium
Labor/management	Adversarial	Cooperative	Cooperative	Adversarial
Pay differential	Large	Small	Medium	Large
Turnover	High	Low	Medium	Medium
Skills	High	High	High	Poor
Union structure	Sector based	Firm based	Industry wide	Craft
Strength	Low	Low	High	Low
The firm				
Main goal	Profits	Market share, stable jobs	Market share, fulfilment	Profits
Role of top manager	Boss king, autocratic	Consensus	Consensus	Boss king, hierarchy
Social overheads	Low	Low	High	Medium, down

Table 3.3 *Continued*

Characteristic	US capitalism	Japanese capitalism	European social market	British capitalism
Welfare system				
Basic principle	Liberal	Corporatist	Corporatist, social democracy	Mixed
Universal transfers	Low	Medium	High	Medium, down
Means testing	High	Medium	Low	Medium, up
Degree education tiered by class	High	Medium	Medium	High
Private welfare	High	Medium	Low	Medium, up
Government policies				
Role of government	Limited, adversarial	Extensive, cooperative	Encompassing	Strong, adversarial
Openness to trade	Quite open	Least open	Quite open	Open
Industrial policy	Little	High	High	Medium
Top income tax	Low	Low	High	Medium

Source: Updated from Hutton (1995: 282)

to give a long-run orientation to industrial development. For example, the government-run Postal Savings system in Japan channels domestic savings through commercial banks into industrial investments.

For political–institutional reasons, therefore, national economies cannot always be confined solely to creating the optimal conditions for the operation of global industries within them, because national politics reflects conflicts of group and regional interests over tariffs and trade, and national economies have had distinctive trajectories in the development of the financial systems that underpin investment decisions by local firms (see Table 3.3).

State-owned transnational corporations provide another indication of the strength of the relationship between governments and (at least some, selected) industries. These companies may be fully or partially owned by the state, publicly traded entities or not. Of the 30 largest state-owned TNCs ranked by size of foreign assets in 2009, perhaps not surprisingly, only one (General Motors) was a U.S.-based company and none was domiciled in Britain. In contrast, France accounted for six, while Italy and China added another three apiece. Nineteen of the companies were engaged in one of three industries: petroleum exploration, refining, and distribution (7); electric, gas, and water utilities (6); and telecommunications (6) (UNCTAD, 2011).

At the same time, the world economy is developing explicitly to optimize conditions for private business activities, at whatever cost to this or that national economy, including a firm's "own." An increasingly integrated world financial system is one mechanism for this. The "global shift" in production is another. In some LDCs, international economic institutions such as the IMF and the World Bank have become so powerful in setting conditions under which loans will be granted that they have become the *de facto* governments of those countries. In a large number of African states the "internationalization of the state" has gone so far that some commentators have referred to them as quasi-states, unable to direct their own economies without massive external assistance and not offering much in the way of citizenship or economic benefits to their populations. Given the diminished capacity of states to actively direct their economies, what is the significance of states in the contemporary context of globalization?

STATES AND THE GEOGRAPHY OF THE WORLD ECONOMY

The importance of states in the world economy is evident in a number of ways: First, as organizers and mobilizers in the NIEs; second, in the continuing geopolitical rivalry of the DCs; third, in the macroeconomic policies pursued by national governments and central banks to stabilize and reorganize their economies; fourth, in the continued importance of national governments as agents of social and political order within their territories; and, fifth, in the latitude and initiative of lower level governments in attracting and keeping economic activities within their jurisdictions.

1. The NIEs

A remarkable, although perhaps not surprising fact, is that if one examines the larger NIEs in detail (excluding Singapore and the Hong Kong SAR because of their peculiarity as ethnic Chinese city-states), the fastest growing and otherwise best performing countries have had national governments that directly and actively intervened in their economies. In South Korea, for example, successive governments played major roles in fostering economic growth. Adding government savings to deposits in nationalized banks, the South Korean government controlled two-thirds of South Korea's investment resources during the country's period of most rapid growth in the late 1970s. It guided investment in chosen directions through differential interest rates and preferential credit terms. Korean export expansion, the main method of economic growth, was built on an economic base that was stringently protected from foreign imports. In these ways, successive activist governments orchestrated economic growth. "In exchange for subsidies, the state ... imposed performance standards on private firms" (Amsden, 1989: 8). In other countries, subsidies have not always been tied so closely to performance and the result has been lower growth.

The recent experience of South Korea and some other NIEs such as Taiwan, therefore, should not be interpreted as an entirely "market" phenomenon. Economists on the political right and left err when they ignore or systematically devalue the importance of state action in organizing and mobilizing resources for economic growth. Of course, not all states have either the institutional foundations or the resources to "mobilize" for economic growth. South Korea and Taiwan had decisive advantages because of their ethnic homogeneity, their transport infrastructure inherited from Japanese colonialism, a history of land reform, and U.S. investment during the Cold War to help "contain" China and the Soviet Union. In many other cases, ethnic divisions and organized corruption have turned states into barriers rather than facilitators of economic development. Their action or inaction continues to afflict their populations. One thinks of such examples as the Democratic Republic of the Congo, Zimbabwe, and Somalia in Africa (just three examples from a much longer list), Argentina and Venezuela in Latin America, and Sri Lanka and the Philippines in Asia (see Chapters 8 and 10).

2. States and geopolitical rivalry

States at the core of the world economy show few signs of disappearing (see Chapters 6 and 7). Indeed, under the conditions of economic restructuring that have affected most DCs over the past few decades, rivalries have deepened between many countries over trade, monetary, and foreign policy. Successive U.S. governments have used the devaluation and revaluation of the dollar to insulate its economy from the negative impacts of increased foreign competition and import penetration. These actions—similar to the actions taken by China to keep the value of the yuan artificially low to stimulate exports—have been at the expense of other currencies and other countries.

These examples illustrate how some states can use fiscal and monetary policies to stabilize and reorganize their economies at the expense of other national economies. Within their own boundaries, they can also encourage or slow down the processes of restructuring that emanate from changes in the world economy. In the United States in the 1950s and 1960s, defense spending, and housing and transportation policies that stimulated suburban growth helped the development of Fordist firms oriented to national markets (for example, automobile production). In Japan during the same period, the government Ministry of International Trade and Industry (MITI) (reorganized as the Ministry of Economy, Trade and Industry (METI) in 2001) used an industrial lifecycle model to guide investment in new industries as "old" ones achieved "maturity."

Other countries had less direct industrial policies and often operated through government-owned industries (such as electricity or steel) to stimulate new industries and stabilize production. In the United States in the 1980s, defense spending and new financial services (stimulated through the deregulation of banks and other financial institutions) became the focus of government attention. The large deficits in the federal government budget that began to accrue in the early part of the decade further stimulated the financial services industry through the need to attract and reward foreign investment. Regions and localities specializing in military production and finance were primary beneficiaries. Increases in military spending were justified in terms of the security threat posed by the Soviet Union. Not surprisingly, the end of the Cold War to some degree marked a turnaround in priorities with military spending declining from 5.8 percent of GDP in 1988 to only 3 percent of GDP a decade later (although the rise in GDP accounts for a significant portion of this disparity). In the post-9/11 period, fuelled by the wars in Afghanistan and Iraq, military spending by the United States once again increased significantly (see Figure 3.5). In 2011 it accounted for 4.8 percent of GDP and matched the estimated combined investment of the next 14 countries ranked by the size of their military spending (SIPRI, 2013).

3. Macroeconomic policymaking

With the rise of international financial integration and highly speculative financial markets has come a concomitant rise in the political power of central banks. Central banks have the ability to affect conditions in domestic and global markets by manipulating interest rates and the money supply. These banks also function differently in various countries and are subject to distinctive pressures emanating from the conjuncture of external influences and distinctive national institutional policy environments.

For example, where a central bank is politically independent and connections between industry and finance are weak, as in the United States, the central bank will be a "rentier" bank and tend to pursue monetary policies that benefit banks and other financial interests. However, where a central bank is independent but industry and finance are closely linked, as

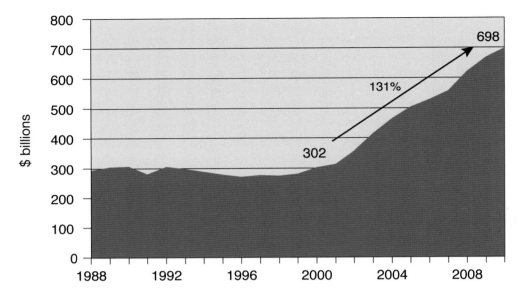

Figure 3.5 United States military expenditure

Source: Based on online World Bank World Development Indicators data at http://data.worldbank.org/data-catalog/world-development-indicators

in many Eurozone countries such as Germany, France, or Italy, the central bank will try to benefit business as a whole, choosing specific policies depending on the relative influence of labor.

Prior to the advent of the euro and creation of the European Central Bank (ECB), the Bundesbank in Germany prioritized the maintenance of price stability, that is, creating a sustainable, low-inflation economic environment. At least since the mid-1970s this strategy seemed to have produced higher aggregate economic growth than would otherwise have been the case. But it also required a general popular fear of inflation (perhaps based on the collective memory of the drastic price inflations of the 1920s and mid-1940s) for acceptance of the tight monetary policies that have been pursued.

Finally, where the central bank is part of the government and lacks "goal independence" (although it may retain "instrument independence" by having independence to select the tools to achieve the prescribed goal such as in New Zealand), industry and finance are linked, and labor is cooperative, for example, in Sweden, central bank policy can be expected to be expansionary and, potentially although not necessarily, inflationary.

4. Governments and peoples

The corporate welfare state that developed under organized capitalism is under threat, however, in many of these countries. The state is certainly still "big" throughout the developed countries and beyond (for example, in India and China), but the Reagan administration in the United States and the Thatcher governments in Britain in the 1980s popularized the view that government spending on social services (Margaret Thatcher's "nanny state") was a drag on national investment and growth because it required taxes to pay for them. Overall, there has been a marked shift away from egalitarian liberalism toward a free market doctrine of **neoliberalism** (see Box 3.1). As the neoliberal paradigm establishes itself globally, this perspective is increasingly spreading beyond the bounds of the English-speaking world.

Box 3.1 Neoliberalism

The failure of Keynesianism (the operational policy framework for egalitarian liberalism) to cope with the economic system shock of the sudden quadrupling of crude oil prices by OPEC in 1973, the consequent over-accumulation crisis, and the subsequent globalization of industrial production, all opened the way for radically different policy perspectives. It also set in motion a shift away from the egalitarian liberalism that had dominated public policy since the 1930s. Just as the idea of market failures had been a powerful notion in the ideological shift from classical liberalism to egalitarian liberalism in the 1930s, so the idea of government failures became a powerful notion in undermining egalitarian liberalism (and especially the Keynesian welfare state) in the mid- to late 1970s. Governments, the argument ran, were inefficient, bloated with bureaucracy, prone to over-regulation that stifles economic development, and committed to social and environmental policies that are an impediment to international competitiveness. As a result, neoliberalism eclipsed egalitarian liberalism, that is, a selective return to the ideas of classical liberalism. Increased taxation (to fund spending on the casualties of deindustrialization), unemployment, and inner city decline led to resentment among more affluent sections of the taxpaying public who were caught up in an ever escalating material culture and wanted more disposable income for private consumption. With pressure on public spending, the quality of public services, public goods, and physical infrastructures inevitably deteriorated, which redoubled pressure from those with money to spend it privately.

The concept of the public good was tarred with the same brush as Keynesianism, as government itself (to paraphrase Ronald Reagan) came to be identified as the problem rather than the solution. Whereas market failures had been the rationale for the ascendance of egalitarian liberalism, government failures became the rationale for neoliberalism. Globalization also played a part: Keynesian economic policies and redistributive programs came to be seen as an impediment to international competitiveness. Labor market "flexibility" became the new conventional wisdom.

As Jamie Peck and Adam Tickell (2002) pointed out, these ideological shifts are part of a continuous process of political–economic change, not simply a set of policy outcomes. They have characterized the process in terms of a combination of "rollback" and "rollout" neoliberalization.

Rollback neoliberalization has meant the deregulation of finance and industry, the demise of public housing programs, the privatization of public space, cutbacks in redistributive welfare programs, shedding many of the traditional roles of central and local governments as mediators and regulators, curbs on the power and influence of labor unions, and a reduction of investment in the physical infrastructure of roads, bridges, and public utilities.

Rollout neoliberalization has meant right-to-work legislation, the establishment of public–private partnerships, the development of workfare requirements, the assertion of private property rights, the creation of free-trade zones, enterprise zones, and other deregulated spaces, the assertion of the principle of "highest and best use" for land-use planning decisions, and the privatization of government services.

The net effect has been a "hollowing out" of the capacity of the national governments while forcing local governments to become increasingly entrepreneurial in pursuit of jobs and revenues, and increasingly pro-business in terms of their expenditures. Indeed, the proponents of neoliberal policies have advocated free markets as the ideal condition not only for economic organization, but also for political and social life. This "ideal" is, of course, more ideal for some than others. Free markets have intensified uneven relationships among places and regions, the inevitable result being an intensification of economic inequality at every scale, from the neighborhood to the nation-state.

Certainly, national governments in the core of the world economy now find themselves in the fiscal crisis that is a "normal" condition in the rest of the world with demands on their resources increasing (for example, aging populations, increasing poverty) as their ability to meet them (for example, loss of higher paying traditional manufacturing jobs and associated revenues) declines. However, governments in the DCs remain the most important agents of social and political order within their territories. They differ only in the degree of "bigness" as measured by laws passed, range and comprehensiveness of programs, scope of government agencies, and number and initiative of employees. The overall involvement of most governments in the lives of their citizens—particularly in Europe as the debt crisis deepened while the mixed-market economies of countries employing the Nordic model enjoyed far greater stability and prosperity—shows few signs of decline.

5. Lower tier governments and economic development

In some countries, local levels of government are able to pursue policies of their own with respect to attracting and keeping economic activities. As some manufacturing and service industries have become more "footloose" following the technological and organizational changes of the recent past (less tied through **agglomeration economies** to specific locations), a variety of factors once marginal to a firm's locational calculus have assumed greater prominence. Some of these can be subsumed under the rubric of local "business climate." In the United States, for example, the northern states of the Manufacturing Belt (the region stretching from Illinois to New York where most manufacturing industry was concentrated from 1880 to the 1960s) tend to have higher personal income tax rates, and greater provision of public goods and services (public education, social services, etc.) than states in the south and west. These conditions provide for less favorable business climates for certain industries than are found in the other regions.

Many states and localities throughout the country, however, have actively pursued businesses in what might be considered a "race to the bottom" by offering tax breaks and subsidies to corporations to relocate. Many U.S. states have offices in Europe and Asia designed to attract foreign investment, but much of the competition for businesses seems to involve attracting established firms and branch plants from other states. This dynamic has undoubtedly contributed to the decentralization of manufacturing industries within the country. The net contribution to national economic welfare is less substantial than might appear.

In contrast, the European Union's Competition Policy attempts to restrict such wasteful zero-sum competition within its territory by placing restrictions on the provision of government incentives such as grants and subsidies to private companies. Increasingly, preparing local labor forces for higher skilled jobs through education and training, and thereby attracting "higher value" industries, is seen as a much more important local economic development strategy (see Chapter 13).

CONCLUSION

In a number of respects, therefore, territorial states and the political regulation of the economy that they provide are essential to understanding the evolution of economic landscapes. But these states, far from being the same everywhere and staying the same over time, are intertwined in complex and contradictory ways with the world economy. States today must operate in a global economic environment in which they have become managers of internal–external transactions rather than, in coordination with national businesses, monopolists over discrete national territories.

Indeed, it has become commonplace to observe that states are both too large and too small for a wide range of social and economic purposes. They are often too large territorially to create full social identities and real national interests as is evident in the proliferation of ethnic and cultural movements around the world. It can also be seen in the difficulties involved in achieving national consensus in many states around national institutions and policies. For example, the so-called welfare state has been under attack in Europe and the United States. But existing states are for many economic purposes also too small geographically. They are increasingly "market sectors" in an intensely competitive, integrated, and interdependent world economy. Two propositions follow from this perspective and inform the rest of this book:

- First, economic power is no longer a simple attribute of states. The growth of world trade, the activity of transnational corporations, the world financial system, global production, and regional trading blocs such as the European Union (EU) and the North American Free Trade Agreement (NAFTA), and the rising power and importance of NIEs point towards a new, highly dynamic, and rapidly evolving world order.
- Second, state, society, and economy are no longer mutually defining. Uneven economic development within and between states has redefined economic interests and political identities across national, regional, and local levels. The economic restructuring associated with globalization has tied many local areas directly into global markets. So local areas are "communities of fate" in a world in which there is less possibility of being shielded from competition within large territorial units. When such places have different orientations to the world economy, because they have different commodities in trade, different trading partners, and different exposures to foreign competition, the possibility for national consensus on trade policy can be significantly reduced. The growing redundancy of national governments as supranational entities (such as the EU) increase in importance, plus the challenge to national regulation from global markets, have conspired to stimulate new and revive old political identities, especially when ethnic and cultural divisions are defined geographically. In this context, therefore, the recent flowering of nationalist and separatist movements around the world is not particularly surprising. New spaces for political regulation at this level may have to coexist with older ones at a larger scale. Institutions at both levels will have to cope with real if also reversible pressures for global interdependence.

3.3 "MARKET ACCESS" AND THE REGIONAL MOTORS OF THE WORLD ECONOMY

Wide acknowledgement that the world economy is undergoing a fundamental reorganization has not meant that there is agreement as to how and why this is happening. Agreement is confined only to the sense that the world economy has entered a phase of **flexible production** in which business operations around the world are increasingly taking the form of core firms (often transnational in scope) connected by formal and informal alliances to *networks* of other organizations—firms, governments, and communities.

The paradox of this trend, and the reason it has generated intense debate, is that while networking allows for an increased spanning of political boundaries by concentrated business organizations, it also creates the opportunity for more decentralized production to sites with competitive advantages. Networks take on different forms with different sectors and in different places. Some have large corporations at their centers with geographically dispersed subcontractors and allied firms (for example, many car manufacturers) whereas others are clusters of firms in high-tech regions (such as California's Silicon Valley) or specialized industrial districts (such as those in Emilia-Romagna and Tuscany, parts of the Third Italy, in which economic growth has been based on clusters of small firms specializing in the same industries such as machine tools, shoes, or woolen textiles).

In all cases, however, sites are never isolated worlds unto themselves. They are linked through social connections and the benefits that come from either spatial divisions of labor (splitting different activities among different locations) or external economies of scale (the benefits that accrue from locating close to similar and complementary producers). The outcome is a world economy in which networks and flows bring together sites widely scattered around the world. Nevertheless, the vast majority of the tightest connections are found in and between Europe, North America, and East Asia. In this respect, the world economy is not yet truly worldwide in its geographical scope.

One account of the source of this shift in the world economy from big, vertically integrated firms organized with reference to national economies to globe-spanning networks of production and finance emphasizes the declining rates of productivity and profits of major corporations in the years between 1965 and 1980. These declines seem tied more to falling rates of productivity than to rising labor costs. Although there have been recoveries in rates of profit at certain times in some economies (such as in the United States since the mid-1980s), these seem fueled in part by suppressing wages and other labor benefits more than by returns to new technologies or new investment. This trend also reflects the "global turn" taken most aggressively by many U.S. firms since the 1970s.

In any case, manufacturing industries in all of the major DCs experienced productivity and profitability crises beginning in the 1970s. From this crisis came the push to rationalize operations, downsize, divest relatively unprofitable activities, and use relocation and diversification strategies to produce higher rates of return for investors. This massive shakeout had a number of consequences including the rapid spread of new technologies, the compression of the "shelf life" of commodities to keep up demand (planned obsolescence), and increased competition to deliver goods and services quickly. Of particular importance, new transportation and communications technologies made it possible for businesses to move physical assets, components, finished products and services, and financial capital ever more rapidly.

In the background lay attempts by the governments of the most powerful states, particularly the United States, to open up the world economy to increased trade and investment across international boundaries. These reflected both the perceived interests of certain businesses in these countries in "going global" to solve their problems and the ideological imperative to build a "free world" economy as an alternative to the closed-off and state-centered economies of the Soviet Union and its satellites. The exhaustion of Fordism also coincided with a number of general changes in the workings of the world economy such as the collapse of the Bretton Woods system for fixing currency exchange rates in 1971, the oil price increases forced on world consumers by OPEC in 1973 and again in 1979, and the world debt crisis following the failure of borrowers (such as semi-peripheral countries including Mexico and Brazil) to pay back the loans made to them from the petrodollars recycled into the world economy by the oil producers.

The emerging character of the world economy can be viewed through two lenses: from the perspective of states and how they fit into the picture as firms reorganize and from the point of view of the firms and how they organize their networks geographically. The former is referred to by some commentators as an emerging "market-access economy" in which states increasingly standardize the rules governing trade and investment in order to situate themselves more advantageously within the evolving international division of labor. The latter can be seen as

Table 3.4 The old and the new pillars of world trade

Old pillars of the free-trade regime	New pillars of the market-access regime
Structure	
1 US model of industrial organization	Hybrid model of industrial organization
2 Separate systems of governance	Internationalization of domestic policies
3 Goods traded and services produced and consumed domestically	Globalization of services; eroding boundaries between goods and services
4 Universal rules are the norm	Sector-specific codes are common
Rules	
5 Free movement of goods; investment conditional	Investment as integrated co-equal with trade
6 National comparative advantage	Regional and global advantage

Source: Based on Cowhey and Aronson (1993: 60, Table 4.1)

territorially based production systems held together through networks and alliances of firms, governments, and communities.

"MARKET-ACCESS" REGIME

Globalization is partly about firms attempting to cash in on the comparative advantage enjoyed in production by other countries and localities, and to gain unimpeded access to their consumer markets. But it is also about governments wanting to attract capital and expertise from beyond their boundaries to increase employment, learn from foreign partners, and generally improve the competitive position of "their" firms. The combination of the two has given rise to a "market-access" regime of world trade and investment. This transformation has eroded the free-trade regime that had increasingly predominated in trade between the main industrial capitalist countries in the post-Second World War period. In its place has emerged a regime in which acceptable rules governing trade and investment have spread from the relatively narrow realm of trade to cover a wide range of areas of firm organization and performance.

Six pillars of this "market-access" system can be identified (see Table 3.4). The first is a move away from the dominance of the U.S. model of industrial organization in international negotiations towards a hybrid model that places less emphasis on keeping governments and industries "at arm's length" and a commitment to encouraging inter-firm collaboration and alliances across as well as within national boundaries. In this model, foreign firms are permitted to contest most segments of national markets, except in cases where clearly demarcated sectors are left for local firms.

A second pillar involves the increased cooperation and acceptance of global rules concerning trade, investment, and money by national bureaucracies as well as an increasingly powerful role played by supranational and international organizations (such as the European Commission for the EU and the World Trade Organization; see Chapter 12). This transformation blurs the lines of regulation between "issue areas" (such as trade and foreign direct investment) and the penetration of "global norms" into the practices of national bureaucracies (for example, the international accounting standards (IAS) proposed by the International Accounting Standards Board Foundation or corporate governance principles drawn up by the OECD).

The third pillar emphasizes the growing level of trade in services beyond national boundaries and the concomitant increased importance of services (banking, insurance, transportation, legal, advertising, etc.) in the world economy. One reason for this is that high-tech products such as computers and commercial aircraft contain high levels of service inputs. Servicing the "software" that such products require has led to an explosion in business of producer services. Another is that producers are demanding services that are of high quality and competitively priced. They can turn to foreign suppliers if appropriate ones are unavailable locally. Banking and telephone industries are two services that have experienced a dramatic increase in internationalization as producers have turned to nontraditional (frequently foreign–offshore) suppliers.

Fourth, international negotiations about trade and investment are now organized much more along sector- and issue-specific lines. One rule no longer fits all. But many of the new rules are essentially *ad hoc* rather than formal.

The final two pillars concern the content of the rules of the market-access regime. One is the increasingly comparable importance of trade and investment due largely to the activities of TNCs in expanding the level of foreign direct investment. Local rules about what portion of a finished product must be made locally (within a particular country) and worries about the competitive fairness of firm alliances, however, have also led to new efforts by governments in the DCs to regulate the flows of foreign investment. "Leveling the playing field" has meant pressure and counter-pressure between governments to ensure at least a degree of similarity in regulation (in, for example, cases of presumed monopoly or antitrust violations).

The final pillar involves the shift on the part of firms from a concern with national or domestic comparative advantage to a concern with global or world–regional competitive advantages. This emphasis reflects the overwhelming attractiveness of "multinationality" to many businesses as a way of diversifying assets, increasing market access, and, at the same time, enjoying the firm economies of scale that come from supplying larger markets.

Production facilities can be located to take advantage of other benefits that come from operating in multiple locations, particularly those offered by foreign sites. Foreign direct investment is often seen as the result of three sets of factors:

- *Ownership advantages*: The advantages that accrue to firms abroad due to their technology and market power relative to competitors.
- *Internalized markets*: The need to ensure returns on research and development (R&D) and other prior investments by controlling production and marketing rather than licensing to foreign firms.
- *Location advantages*: Favorable locational conditions that encourage foreign operations rather than export (market size and needs, production costs, trade barriers).

Business economists tend to emphasize the first two, whereas economic geographers tend to give more weight to the third. In particular, the geographers have tended to use the **product lifecycle** model to explain the trend towards increased relocation of certain production processes in foreign settings. In its original form, this idea did not have any locational significance; it referred entirely to the tendency for products to move from being novelties to mass production to obsolescence. Vernon (1966) gave the product lifecycle a locational component by arguing that as production requirements change as a product "ages" so do locational requirements. In particular, mass production is more labor-intensive than the earlier phases of production. So when a product reaches this phase in its lifecycle, it pays in terms

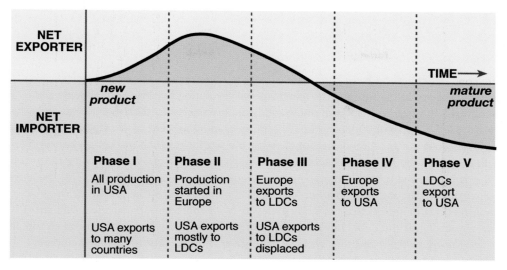

Figure 3.6 The product life cycle model and possible effects on U.S. production and trade (after Vernon, 1966)

Source: Adapted from Wells (1972: 15, Figure 1)

of profitability to move production to where cheaper labor is available. Patterns of imports and exports adjust accordingly (see Figure 3.6).

This model faces a number of criticisms when applied without attention to the specific production requirements of different sectors, such as the overemphasis on labor intensity as a feature of mass production, the importance of automation in much mass production, the implicit assumption that industries and their products "mature" rather than adjust or fail to adjust to changing conditions of production, the lack of attention to customized production in many product categories today, and the neglect of the role of regulatory factors (such as tariffs and other import restrictions) in encouraging the movement of production facilities to countries other than the home one. This last factor, often coupled with the relative strength of a country's currency relative to the currency of a target market, is one of the main reasons why Japanese and European car producers Toyota, Honda, Nissan, Volkswagen, BMW, and Daimler have moved or expanded their production operations in the United States.

The complexity of the world economy means that there is no single model of firm locational behavior. New transportation and communication technologies have created an environment that enables firms to decentralize manufacturing and primary-production activities yet maintain central control. The potential for intensive interaction without geographical proximity is significant. With automation and computerization, an engineer located in a corporate headquarter in Chicago can monitor the production process in a factory in Chakan, India, in real time. Although such advances are reshaping the economic and competitive landscape for businesses, by no means does it signal the end of regional specialization. The evolution of transnational strategies of production does not necessarily imply or require the demise of older ones.

U.S. firms such as Coca-Cola, McDonald's, and Disney have been leaders in the shift to a "borderless world." Implicit in this approach is a sequence of organizational–geographical moves as firms shift from (1) exporting to (2) foreign sales outlets to (3) foreign production to (4) the world as an "investment surface" with production spread over a large number of locations in different countries (see Box 3.2). Firms have learned that they can improve their profitability

Box 3.2 The semiconductor industry and the workings of the market-access regime

Semiconductors are the basic components in electronics technologies. Electronics technologies are at the heart of the informational revolution that has both allowed and encouraged the transformation of the world economy since the 1970s. Semiconductors were a $300 billion global industry in 2012. More than 30 percent of all semiconductors are exported, so it is a good candidate for showing the workings of the market-access regime.

In the 1980s the Japanese makers broke through in world markets, weakening the grip of U.S. producers, and further threatening the already fragile European industry. This challenge was particularly strong in the market for standard memory chips (the DRAM or dynamic random-access memory chips) as innovation was rapid and Japanese manufacturers cut prices steeply. It is the DRAM market that has the largest economies of scale and so can, in the long term, underwrite other types of production. Not surprisingly, semiconductors became a key trade issue. According to IHS iSuppli market research data, in 2011, 11 of the top 25 (and five—Intel, Texas Instruments, Qualcomm, Micron, and Broadcom—of the top ten) semiconductor producers were U.S. companies.

By the end of 2010 the global market share of U.S firms in the semiconductor industry was 48 percent and over 80 percent of these sales occurred in foreign markets. In the U.S., the semiconductor industry is responsible for some 180,000 jobs with capital equipment worth $13 billion and R&D investments of $20 billion (Wilson, 2011). The most lucrative portion of the semiconductor industry has been microprocessors, a segment dominated by Intel (United States), the largest company in the semiconductor field, which has a stranglehold on the production of chips for computers and servers, and ARM (United Kingdom), a company that is a fraction of the size of Intel but has developed an intricate "ecosystem" of relationship with licensees for its low-power, dynamic chips that dominate the mobile phone, mobile computers, and consumer products market (*Economist*, 2012a) and is also used by Microsoft in its Surface tablet designed specifically to challenge the dominance of Apple's iPad.

In Japan, the Ministry of International Trade and Industry (MITI) (now METI) accorded semiconductors the highest priority and worked to establish a major presence in global markets. In practice, Japanese success was the result of the marriage of the *keiretsu* system of big companies at the center of affiliated networks to a national industrial policy. In particular, the Japanese firms enjoyed ready access to capital, and government policy reduced the risks of overextension and fostered domestic competition to achieve the best results in global markets. Out of this Japanese challenge came a whole new approach to global trade and investment in semiconductors. The first element was two U.S.–Japan Semiconductor Agreements (1986 and 1991), which opened up Japan to foreign (U.S.) firms and monitored charges of Japanese "dumping" of semiconductors in the United States at below world prices. The second was the move of Japanese, European, and U.S. producers into a number of international corporate alliances. This shift was particularly critical. To anticipate future costs of R&D, "Firms are turning to ICAs (international corporate alliances) to build common, global infrastructures for the next generation of technologies. Alliances also allow firms to reduce the cost and risk of fielding extensive product lines" (Cowhey and Aronson, 1993: 162).

The net effect is a perfect example of the market-access regime of trade and investment. The new structure of the semiconductor industry is a hybrid of Japanese, U.S., and European models (pillar 1 in Table 3.4). International agreements are the main means by which the industry is regulated (pillar 2). The mix of hardware and software makes it hard to say where the product ends and servicing begins. What is clear is that both must be available on a worldwide basis for a product to be competitive (pillar 3). Since 1986 specialized industry codes have steadily displaced older industry-wide ones (pillar 4). Where chips were once traded relatively freely but without much foreign direct investment, major networks based on international alliances now span national boundaries. Semiconductor firms have become global in their organization as well as in their search for markets (pillar 5). Finally, firms are building global and world–regional advantages in order to give them leverage over home and foreign markets (pillar 6). Product and technology flows are now so globalized that closing markets (national and regional) would doom affected producers to limited market share and to not participating in new rounds of innovation and product development.

if they use their **economies of scope and coordination** (returns to complexity and managerial capability in producing multiple products) to compete effectively for global markets against local producers who may have advantages in **economies of scale and local connections**. Brand names, financial clout, and managerial savvy can overcome these barriers. But this can happen only if foreign markets are opened up to competition, which is the essence of the market access regime.

PRODUCTION NETWORKS AND REGIONAL MOTORS

From the business point of view, the response to the competitive pressures of the market-access regime has been to acquire greater flexibility through technological change, reorganizing labor relations, and establishing links with other firms. One solution has predominated: The creation of networks among producers. This shift has several origins. Perhaps the most significant is the attempt by large firms to reduce the size of their work forces and to outsource less profitable activities to other firms. This process of **vertical disintegration** can be cost saving if a unionized labor force is replaced by a nonunion one, for example, and it also taps into the specialized skills of suppliers and subcontractors.

Another important impetus is organizations' desire to penetrate foreign markets and build a global presence through collaboration (including with erstwhile competitors). So, for example, Toyota entered into a joint venture with General Motors, while Nestlé has a joint venture with General Mills. The focus on production networks highlights the central role of geographical shifts in investment and production as a response to changes in the competitive environment experienced by firms in many economic sectors with the advent of the market-access regime.

Production networks

Four types of network among firms can be distinguished. The first type occurs with *craft-based industries*. These industries are organized around projects more than firms, *per se*. In construction, publishing, film and recording, architecture, and software engineering, high-skilled workforces are employed by firms but share knowledge easily across firm boundaries. Consequently, such industries tend to cluster to take advantage of the external economies implicit in such sharing. External economies include such factors as a labor pool with relevant skills, a broad network of suppliers, excellent educational and training support, and, perhaps most importantly, access to venture capital knowledge about the nature of the business. It also includes intangibles such as a "culture" accepting of innovation, tolerance of failure, local reinvestment, collaboration, promotion on merit, and openness to new enterprises.

The second type of network involves *small firm industrial districts* such as those often associated with the Third Italy. These districts are local integrated networks of producers with different firms specializing in different phases of the production process but competing for work with other local firms when new projects arise. Evidence suggests that they rely in equal measure on external economies of scale in production (collaborative production, local government financing, craft traditions, pools of skilled labor, etc.) and on what can be called non-traded interdependencies—a long history of social collaboration, institutionalized cooperation, and agreement on social conventions governing everyday inter-firm relations.

The third type of network is that of *agglomerated big firm-based production systems* such as in Toyota City in Nagoya, Japan; Boeing and its suppliers in the Tacoma-Seattle region in the United States; and Fiat and its suppliers around Turin in northwest Italy. In some cases, the suppliers pre-existed the emergence of the big firms; in others (as in Japan and South Korea), the dominant company financed the suppliers. Since the 1970s, however, the main process stimulating this kind of network has been the vertical disintegration of the big firms themselves. Whether territorially connected to them or not, large firms can now achieve improved flexibility by using subcontractors to carry out aspects of production that used to be performed within the boundaries of the firm. A high-tech industrial complex such as Silicon Valley is somewhere in between the first or "classic" type of industrial district and the third or agglomerated big firm production system, sharing features of both.

Finally, the fourth type of network is represented by *strategic alliances between firms*. One of the most important innovations of the market-access regime, especially in terms of strategic alliances between international competitors, in this network is that: "Each partner brings to the marriage its own specialty—technology, financial power, access to government regulators or procurement officials—and its own constellation of small firm suppliers" (Harrison, 1994: 138). Each party gains knowledge and connections intrinsic to the other to further their efforts at conquering global markets for their products. One logical consequence of alliances would be merger or acquisition of one partner by the other. But some national laws and customs set limits to these activities (for example, U.S. and Japanese laws restrict foreign acquisition of home-grown firms), and the goal of flexibility is best met by maintaining or recreating alliances rather than engaging in fully fledged mergers.

NEW INTERNATIONAL DIVISION OF LABOR

The first interpretations of the changing character of business organization and the associated changes in the economic geography of production focused on the emergence of a new international division of labor (NIDL). From this point of view, big transnational corporations have created a global economic geography. One of the architects of this viewpoint, Stephen Hymer (1972: 114), imagined that the transnational corporations:

> [W]ould tend to produce a hierarchical division of labor between geographical regions corresponding to the vertical division of labor within the firm. It would tend to centralize high-level decision-making occupations to a few key cities in the advanced countries, surrounded by a number of regional sub-capitals, and confine the rest of the world to lower levels of activity and income.

Table 3.5 "Hymer's stereotype", in which the space–process relationship takes the form
A→B→C

Level of corporate hierarchy	Type of area		
	Major metropolis (e.g., London or Tokyo)	Regional capital (e.g., Brussels or Denver)	Semi-periphery (e.g., Mexico or China)
I Long-term strategic planning	A		
2 Management of divisions	D	B	
3 Production, routine work	F	E	C

Source: Based on Sayer (1985: 13, Table 1).

As a result, the NIDL reflected the hierarchical social division of labor within the big firms themselves. Hymer claimed that a tight **space–process relationship** was coming about in which control and operational activities within firms would be completely separated. As suggested by Table 3.5, the close space–process relationship would take the form A to B to C. But the table reveals other possibilities: D, E, and F.

Logically, a major metropolis (or world city) could dominate all levels in the hierarchy of functions in absolute terms even with the *addition* of foreign operations. The advent of networks was largely responsible for confounding the simple story of the NIDL. Production networks allowed much more complex geographies of production than those predicted by Hymer's simple hierarchy. Big firms have changed their internal structures and external relations in ways that undermine their own internal hierarchies. So the analogy between internal (organizational) and external (geographical) hierarchies now seems overdrawn.

More importantly, much of the explosion of foreign direct investment since 1985 has involved within-core and not core to periphery/semi-periphery flows at a world scale, although in the second half of the first decade of the twenty-first century flows to LDCs increased considerably.

This trend suggests how important market access has also become compared to the search for cheap labor in assembly processes. Increased competition for market shares by U.S. and European companies at home has forced a search for markets elsewhere that cannot be served by an export strategy. Supplier performance, alliances, and service to customers, as well as avoidance of tariffs and other trade barriers have dictated that firms in each region move closer to potential markets in every other region.

GEOGRAPHIES OF PRODUCTION NETWORKS

At the heart of the geography of the world economy under market-access conditions rests an assessment of the tradeoff between the relative benefits to firms of clustering together compared to the relative benefits to firms of conducting economic transactions over space. The former involves the cost-saving role of locating close to suppliers, subcontractors, competitors, and specialized pools of labor. The latter depends on the costs of production involved in overcoming distance in the transactions implicit in production (bringing together inputs, serving markets, etc.).

Table 3.6 illustrates this tradeoff through six possible scenarios.

Table 3.6 Spatial transaction costs versus externalities: six scenarios

	Spatial transaction costs		
	Low	Medium	High
Externalities			
Low	(4)	(5)	(1)
High	(3)	(6)	(2)

Source: Based on Scott (1996)

Resource based or/and resource dependent: In one scenario, producers seek low-cost locations relative to basic inputs and/or markets, and the incentive for firms to cluster is minimal. In addition to resource-based and resource-dependent industries, this scenario often applies to wholesaling and retailing firms in which either transport costs and/or direct access to customers figure prominently in firm locational decisions. The result is locational patterns that conform closely to the distribution of resources and population.

But the situation faced by these firms should not be considered static; rather, it continues to evolve with advances in technology. As Figure 3.7 illustrates, the cost of international transport and communications has declined steadily. New types of ship design, containerization, improved logistics services, and developments in digital technologies have yielded massive improvements in the efficiency of the shipping and land transport industries. As a result, even industries that experience a major "weight loss" during production (with a significant reduction in the amount of output relative to amounts of inputs) the "friction" of distance is far less of a constraint on decision making.

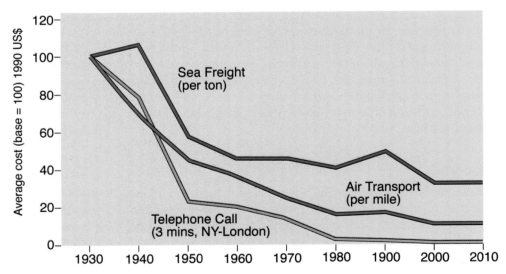

Figure 3.7 Cheaper transport and communications costs on the global highway

Source: Updated from Hargittai and Centeno (2001: 1550, Figure 1)

Agglomeration: A second scenario, one in which external economies are significant but where transaction costs are high, characterizes the situation of industrial districts and high-technology complexes (such as California's Silicon Valley). Firms in this scenario are focused on external economies, require heavy inputs from outside resources, and have an orientation to external consumer markets. Intensive relations between firms encourage agglomeration, but this clustering is constrained by the costs associated with serving outside markets and obtaining outside resources.

Branch-plant industrialization: In this scenario, low transaction costs enable firms to consume external economies at a distance. In this way, external economies can be internalized within a firm (or inter-firm alliance) and then realized through dispersal of functions to locations where they can achieve cost advantages (lower wage bills, etc.).

End of geography: A scenario in which literally anything can be located anywhere as technology enables the radical decentralization of production across all sectors. The absence of spatial limits on access to external economies characterizes this scenario. As yet, it appears to be more fantasy than reality.

Transaction cost determined: Far more likely than the "end of geography" scenario is one in which, although external economies can be obtained at a distance, transaction costs determine attraction to markets or inputs.

Clustering: This final scenario is the most important to the evolving world economy. Spatial transaction costs are assumed to be moderate (on average), but external economies are high. As a result, firms have significant incentive to cluster. Such clusters represent the major novelty in recent times. They are the concentrations of innovative, knowledge-based, and high-value producing industries—the "regional motors" of the world economy (Scott, 1996: 400)—that increasingly drive the world economy. The major reason for this claim is:

> [T]hat contemporary forms of economic production and organization are rife with externality effects, having their roots in the augmenting levels of flexibility, uncertainty, product destandardization, and competitiveness that are some of the hallmarks of contemporary capitalist enterprise.

Despite the increasing returns to agglomeration in many leading sectors (such as high-technology industry, design-intensive consumer goods, and financial and business services), Scott also (1996: 400) argues that many transactions remain intensely sensitive to the effects of distance:

> [W]hile spatial transaction costs have fallen dramatically across a wide front in recent decades, allowing many firms ready access to global markets, there still remain important kinds of transactions that are extremely sensitive to the effects of distance. External economies tend to be well developed in the interaction networks constituted by just such transactions as these and, in order to secure them, producers agglomerate together in geographic space.

The primary geographical implication is that, with increasing world economic integration, the leading production activities will become more concentrated in metropolitan areas and their hinterlands. These locations have long-established competitive advantages acquired typically through trial and error over the years rather than as the result of single overriding locational advantage. This geographical path dependence seems likely to continue proving the wisdom of the old saw that "the rich get richer and the poor get the blame."

Giant metropolitan areas such as Tokyo, São Paulo, New York, Mexico City, Shanghai, Los Angeles, Mumbai, and Seoul, with populations in excess of 10 million in 2012, not to mention at least 60 other urban agglomerations in excess of 5 million apiece, constitute the dynamic centers of the world economy as national boundaries lose some of their grip on channeling the processes governing economic growth. Those who are located in these areas realize a "locational" premium. London reinsurers, Hollywood actors, New York lawyers, and Silicon Valley software engineers extract higher incomes because they embody major specialized activities of specific locales. They have global access in sectors for which there is high global demand.

The flipside of this concentration in the servicing of global markets is the increased marginalization of large parts of the world and their populations (see Chapter 7); a marginalization not necessarily predicated on distance. Indeed, internal to the dominant metropolitan areas are rich and poor districts housing the increasingly polarized income groups that the world economy seems to be bringing in its train. Many of the poor, when they do find employment, find it in providing services for the more affluent, which is increasingly one of the driving forces behind local economic growth. Services for local consumption are responsible for much of the economic growth in large cities. Yet this growth depends on the incomes generated by the goods- and service-producing activities of industries oriented to national and global markets.

Indeed, the growth of smaller cities and areas surrounding major cities is increasingly dependent on the growth of the networked economy. For example, much of the growth of employment and incomes in Britain in recent decades has been concentrated in an arc of "growth areas" extending from Cambridge to Bournemouth, both of which places can be seen as beneficiaries of the growth of the financial service industries in London. Electronic accounting, billing, paperwork, and other back-office functions have decentralized out of the London business district. In this services economy in which both the most lucrative and the poorest paying jobs involve providing services to others (from banking and finance to fast food and check-cashing services), globalization causes London to cast a new shadow over its hinterland.

One theoretical implication is unmistakable: The increased globalization of the world economy is leading not to a spreading-out of economic growth or a homogenization of global space but to heightened differences between regions and localities. Some of the so-called world cities, such as London, Tokyo, and New York, Hong Kong, and Singapore have become centers of (among other things) financial and business services. Other regions, such as the Third Italy, the U.S. Midwest, Taiwan and Shanghai have a focus on manufacturing. Some regions remain primarily agricultural; others are the sites of low-wage assembly or back-office functions; and still others such as Detroit and much of the anthracite coal region in Pennsylvania have become economic wastelands with aging populations, burgeoning informal markets, and limited near-term prospects for economic revival.

No single universal model of business organization can account for all these situations. In an increasingly competitive world economy, they are all the result of adaptive responses to pressures on states and firms to change their ways of doing business.

SUMMARY

In this chapter, we developed the historical–geographical perspective that lies behind the rest of this book. Particular attention has been paid to the specific features of the world economy and its evolution including the following:

- the single world market
- the state system
- the three geographical tiers: core, periphery, and semi-periphery
- temporal patterns and hegemony
- subordination and resistance
- alternative adaptations.

We also explored the changing relationship between states and the contemporary world economy under conditions of globalized capitalism. We emphasized that economic power is no longer best thought of as an attribute of states, and that states and national economies are no longer mutually defining entities. However, we underscored the continuing importance of states as regulators of economic activities. We also emphasized that the world economy still has political divisions that take a geographical form.

We introduced the idea of a "market-access" economy in the third section. We explored how it differs from the previously dominant free-trade regime and the critical role of certain regional "motors" to the world economy. The importance of transnational corporations and strategic alliances between them was emphasized, along with the increasingly complex international division of labor in many economic sectors. The location of economic activities appears to involve a decreased reliance on the costs of assembling the factors of production (that is, spatial transaction costs) and an increased reliance on both traded and nontraded interdependencies (external economies), which encourages a clustering of specialized activities. Rather than encouraging a spreading-out of manufacturing and service industries across the world, therefore, the market-access model of increased global economic interdependence produces a remarkable regional clustering of many activities.

In Parts 2 and 3 of this book, we explore the specific economic–geographical consequences of the evolution of the world economy as described in the first and third parts of this chapter. Part 4 is concerned with some of the manifestations of the globalization/localization nexus at the center of the contemporary world economy that pose challenges to state management raised in the second part: The growth of regional trading blocs and decentralist reactions to the changing world economy.

KEY SOURCES AND SUGGESTED READING

Cowhey, P.F. and Aronson, J.D., 1993. *Managing the World Economy: The consequences of corporate alliances*. New York: Council on Foreign Relations Press.

Eichengreen, B., 1996. *Globalizing Capital: A history of the international monetary system*. Princeton, NJ: Princeton University Press.

Gereffi, G. and Korzeniewicz, M., (eds.), 1993. *Commodity Chains and Global Capitalism*. Westport, CT: Greenwood Press.

Hackworth, J., 2006. *The Neoliberal City. Governance, ideology, and development in American urbanism*. Ithaca, NY: Cornell University Press.

Maddison, A., 2001. *The World Economy: A millennial perspective*. Paris: OECD Development Centre Studies.

Malmberg, A. and Maskell, P., 1997. Towards an Explanation of Regional Specialization and Industry Agglomeration, *European Planning Studies* 5, 25–41.

Massey, D., 1984. *Spatial Divisions of Labor.* London: Methuen.

O'Brien, R., 1992. *Global Financial Integration: The end of geography.* New York: Council on Foreign Relations Press.

Peck, J. and Tickell, A., 2002. Neoliberalizing space, in N. Brenner and N. Theodore (eds.), *Spaces of Neoliberalism. Urban restructuring in North America and western Europe.* Oxford: Blackwell, 33–57.

Scott, A.J., 1996. Regional motors of the global economy, *Futures* 28, 391–411.

Venables, A.J., 2006. Shifts in economic geography and their causes, *Federal Reserve Bank of Kansas City Economic Review* 31, 61–85.

Wallerstein, I., 1979. *The Capitalist World-Economy.* Cambridge: Cambridge University Press.

Part 2

Rise of the core economies

In the next four chapters, we trace the emergence of the world's core economies, following their different paths towards increasing scale and complexity with case histories that illuminate many of the patterns, models, and theories outlined in Part 1. We will detail how the many contours of local, regional, and national economies—no matter how unique or singular they may appear—contribute to a single world economy. In Chapter 4, we explore how the world economy came to be centered on Europe and consolidated through the emergence of merchant capitalism. We also consider how particular kinds of urban and regional change reflect the nature and organization of merchant capitalism. In Chapter 5, we describe the different trajectories that marked the ascent of Europe, North America, and Japan within the world economy, emphasizing the spatial changes consequent on the emergence and evolution of industrial capitalism. In Chapter 6, we detail the globalization of the core economies. In Chapter 7, our focus shifts to the spatial implications of the latest form of

Picture credit:
Linda McCarthy

economic organization of the world's core regions. Although the emphasis throughout this part of the book is on the interactions of dominant forms of economic organization and major dimensions of spatial change, the role of human agency in shaping and differentiating the mosaic of regional landscapes emerges as an important subtheme. What is done, where, and how—under any form of economic organization—reflects human interpretations of how resources should be used. As Ron Johnston (1984: 446) noted, these interpretations:

> [A]re shaped through cultural lenses (which may be locally created, or may be imported); they reflect reactions to both the local physical environment and the international economic situation; they are mediated by local institutional structures; they are influenced by historical context; and they change that context, and hence the environment for future operations.

Chapter 4

Preindustrial foundations

Picture credit: Linda McCarthy

In this chapter, we trace the emergence of an embryonic world economy centered on Europe, and describe the way in which Europeans became, as Robert Reynolds put it (1961: vii), the "leaders, drivers, persuaders, shapers, crushers and builders" of the rest of the world's economies and societies. From these changes, the core areas of Europe forged the template for the economic geography of the contemporary world. It must be recognized at the outset, however, that preindustrial economic development was by no means exclusively a European phenomenon. The early trajectories of other parts of the world often eclipsed that of Europe and were sometimes important in influencing events in Europe itself. We begin, therefore, with a brief review that spans the origins and diffusion of the first, crucial "revolution" in the development of agricultural systems, the rise of ancient empires, the establishment of urban systems, and the spread of feudalism as the dominant form of economic organization. Rather than provide a thumbnail sketch of early economic history, we simply highlight the emergence and spatial implications of certain key socioeconomic forces.

4.1 BEGINNINGS

We start from some basic distinctions provided by the world-systems theory of Immanuel Wallerstein. In his view, at one time all societies were **minisystems**: "A minisystem is an entity that has within it a complete division of labor, and a single cultural framework" (1979a: 17). Such minisystems would include simple hunter–gatherer and some agricultural societies. But as soon as they became tied to empires or the world economy, they ceased to be separate systems. Empires and the world economy are examples of what Wallerstein calls **world-systems**: Units with a single **spatial division of labor** but multiple cultural systems. In the case of a unit with a common political system, there is a world empire. Where no political integration exists, a world economy occurs.

Relatively little is known about the first transitions from primitive hunter–gathering minisystems to larger scale, agriculturally based world empires and world economies. Despite significant advances in the accuracy of archaeological research, we still rely on speculation as much as established facts. It is generally agreed, however, that the transition began in the Proto-Neolithic period (between 9000 and 7000 BCE), when a series of innovations among certain hunter–gatherer peoples established the preconditions for agriculture. These innovations included (1) the use of fire to process food, (2) the use of grindstones, and (3) the improvement of basic tools for catching, killing, and preparing animals, fish, birds, and reptiles. Given these preconditions, the transition to a simple system of **fallow agriculture** (or **shifting cultivation**) was relatively straightforward. It involved sowing or planting familiar species of wild cereals or tubers on scorched land using a slash-and-burn system (cutting down the natural vegetation and burning it to release its nutrients into the soil). This method of cultivation required no special tools and minimized the need for labor-intensive practices such as weeding. When soil fertility in the area declined, the plot was simply abandoned in favor of a new location.

Meanwhile, the domestication of cattle and sheep had begun. By the Neolithic period (7000 to 5500 BCE), stock breeding and seed agriculture had become established techniques of food production; however, the transition from hunting and gathering seems to have occurred slowly and sporadically. Archaeological evidence from a Neolithic village in western Asia, for example, shows that cultivated grains only gradually replaced wild legumes—the major food item in 7500 BCE—over a span of almost 2000 years. Ester Boserup (1981) suggested that there was little incentive to switch to food production until population densities began to increase and/or wild food sources became scarce. From this perspective, then, *demographic conditions as well as technological innovations were a critical precondition for economic change.*

HEARTH AREAS

The weight of available archaeological evidence suggests that the transition to food production took place independently in several agricultural **hearth areas**:

- The earliest hard evidence comes from southwestern Asia, in the foothills of the Zagros Mountains of what are now Iran and Iraq, where radiocarbon analysis has dated the remains of domesticated sheep to around 8500 BCE. Evidence of early Neolithic activity also has been found in other parts of southwestern Asia, particularly around the Dead Sea Valley in Palestine and on the Anatolian Plateau in Turkey.
- A second early Neolithic hearth area was in south Asia, along the floodplains of the Ganges, Brahmaputra, and Irrawaddy rivers.
- Later, from around 5000 BCE, a third hearth area seems to have emerged in China, around the Yuan River valley in western Hunan.
- Finally, evidence suggests independent agricultural organization in four regions of the Americas: The southern Tamaulipas area and the Tehaucán Valley in Central America, coastal Peru, and the North American southwest. In these regions, however, agricultural development not only came later but it was incredibly slow with widespread food production coming to dominate the exploitation of the abundant wild plants and game in those regions only after 1000 CE.

Meanwhile, the agricultural "revolution" had been diffused from southwestern Asia. By 5000 BCE it had begun to spread eastwards, to southern Turkmenia, and westwards, via the Mediterranean and the Danube, into Europe. By 3000 BCE it had reached the Sudan and Kenya (via the Nile), much of India (via Afghanistan and Baluchistan), and had penetrated Europe as far as Britain, Ireland, and southern Scandinavia. By 1500 BCE the last European stronghold of pure hunter–gatherer economies was the zone of tundra and coniferous forest stretching eastwards from the Norwegian coast.

Of course, archaeological evidence is inevitably rather patchy, so the patterns of diffusion from agricultural hearth areas remain a topic of considerable academic debate. More important to us here, however, are the eventual *outcomes* of the transition to food production:

- Most important for the long-term evolution of the world economy were the changes in social organization that resulted from the establishment of settled agriculture. The previous communal social order was steadily replaced by a **kin-ordered system** that laid the basis for a stratified social structure. Kin groups emerged as a "natural" way of assigning rights over resources and organizing the production and storage of food. They also generated new social institutions to deal with the ownership of property and the formal exchange of goods.
- The increased volume and reliability of food supplies allowed much higher population densities and encouraged the proliferation of settled agricultural villages. Together with the social institutions of kin-ordered societies, this transition facilitated the development of nonagricultural crafts such as pottery, weaving, jewelery, and weaponry. Such specializations, in turn, encouraged the beginnings of barter and trade between communities, sometimes over substantial distances.

THE FRAMEWORK OF EARLY URBANIZATION

These outcomes of the agricultural revolution were effectively the preconditions for another "revolutionary" change in the economic and spatial organization of the world: The emergence of cities and city systems. As with the evidence on the agricultural transition, our knowledge of the earliest cities is partly a function of where archaeologists have chosen to dig and partly a function of fortuitous factors like the durability of building materials and artifacts and local climates that preserve or destroy evidence of civilization. It now seems firmly established, however, that urbanization developed independently in different regions, more or less in the wake of the local completion of the agricultural transition. So the first region of independent or "nuclear" urbanism, from around 3000 BCE, was in southwestern Asia, in the Mesopotamian valleys of the Tigris and Euphrates and the Nile Valley (together constituting the Fertile Crescent). By 2500 BCE cities had appeared in the Indus Valley, and 1,500 years later they were established in northern China. Other areas of nuclear urbanism include Central America (from around 1500 CE). Meanwhile, of course, the original southwest Asian urban hearth had generated successive urban world empires, including those of Greece, Rome, and Byzantium.

Explanations of these first transitions to city-based economies have emphasized several factors. Boserup (1981), for instance, stressed the role of local concentrations of population; Jacobs (1969) interpreted the emergence of cities mainly as a function of trade; while the classical archaeological interpretation rests on the availability of an agricultural surplus large enough to facilitate the emergence of specialized, nonagricultural workers.

Another important factor was the emergence of "primitive accumulation" through the exaction of tributes, the control of fixed assets, and/or the control of labor power—usually through some form of religious persuasion or despotic coercion. Once established, a parasitic élite provided the stimulus for urban development by investing its appropriated wealth in displays of power and status. These actions created the kernel of the monumental city but also required an increased degree of specialization in nonagricultural activities—construction, crafts, administration, the priesthood, soldiery and so on—which were organized most effectively in an urban setting.

This kind of expansion, however, could only be sustained in the most fertile agricultural regions where the peasant population could produce enough to support not only the élite but also the growing numbers of nonagricultural workers. In this context, the development of irrigation seems to have been a critical factor. It not only intensified cultivation and increased productivity; it also required the kind of large-scale cooperation that could be organized more effectively in a hierarchical, despotic society. Yet, even in the most fertile and intensively farmed regions, rank-redistributive economies could only expand to a certain point if overall levels of productivity could be increased: Through harder work, improvements in technology, or improvements in agricultural practices. All three of these solutions required more non-agricultural specialists and so reinforced the incipient process of urbanization:

> [A]dministrators and, perhaps, an army to oversee the harder work (their actions may have been accompanied by the élite taking to itself the ownership of land) in the first, craftsmen to create the tools in the second, and also, probably, miners and others to provide the raw materials; and "researchers" to develop the new strains and the new technology (notably irrigation) in the third. Thus the demands for more production are reflected in the urban node as well as in the countryside, and continued growth of the society, to meet the never-satisfied demands of an expanding élite and its associates, leads to self-propelling urban growth.
>
> (Johnston, 1980: 52)

The size of a society's resource base, however, ultimately limited such developments. The logical response to this constraint was enlargement of the resource base through territorial expansion, a process that tended to reinforce and extend the process of urbanization. All these changes involved the creation of city-based jobs. Additionally, whereas small-scale colonial expansion could be organized from one center, expansion beyond the immediate reach of the main settlement (perhaps, a day or two of travel) required establishing secondary settlements. These nodes of the controlled territory acted as intermediate centers in the flow of demands from élite to producers and of goods in return. *As long as growth was a goal, therefore, the empire had to be continually enlarged with an increasing number of urban control centers.* So the expansion of the Greek and Roman Empires laid the foundations of an urban system in western Europe (see Figure 4.1).

Although this transition appears logical and orderly on paper, one should not misinterpret it as a picture of steady growth, expansion, and succession of ancient and classical empires. Urbanized economies were a precarious phenomenon, and many lapsed into ruralism before being revived or recolonized. In a number of cases, this was a result of demographic setbacks associated with war, epidemic, or natural phenomena such as floods or sustained droughts. Such setbacks left too few people to maintain the social and economic infrastructure necessary for urbanization.

An early example of this kind of relapse occurred in the Indus Valley where Aryan pastoralists displaced the urban economy in the middle of the second millennium BCE. Elsewhere, changes in resource/population ratios precipitated the breakdown and decay of urban economies. The demands of repair and upkeep of irrigation systems, for example, coupled with the demands of population growth, sometimes exceeded the available supply of peasant labor. After a while, investments were neglected, armies grew small, and the strength and cohesion of the empire was fatally undermined.

This kind of sequence seems to have resulted in the eventual collapse of the Mesopotamian Empire and may also have contributed to the decay of much of the Mayan Empire more than

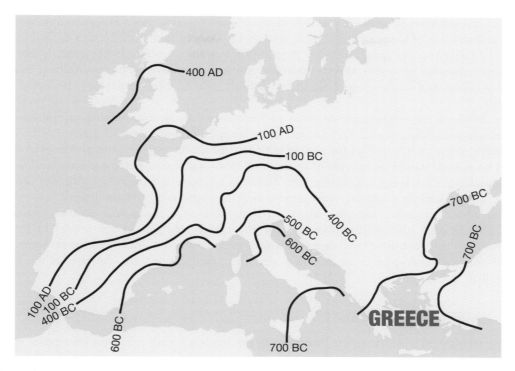

Figure 4.1 The urbanization of the classical world

Source: Based on Carter (1983: 21, Figure 2.2)

500 years before the arrival of the Spanish. Similarly, the population of the Roman Empire began to decline in the second century CE, allowing the infiltration of "barbarian" settlers and traders from the German lands of east-central Europe (ultimately leading to the sacking of Rome in the fifth century CE by the Vandals, an east Germanic tribe).

RURAL CONSOLIDATION

The emergence of urbanization provided an important framework for future development; however, the reorganization and consolidation of rural areas provided the immediate platform for the critical transition to merchant capitalism and the emergence of a European world economy. At the heart of this rural consolidation was the evolution of the elaborate feudal systems of medieval Europe, China, and Japan.

In economic terms, feudal systems were almost wholly agricultural, with 80–90 percent of the workforce engaged in mixed arable and pastoral farming and much of the rest engaged in basic craftwork. Most production filled immediate needs and did not find its way to wider markets. Feudal estates served as the core of the feudal system. Lay or ecclesiastical lords owned the estates and delegated parcels of land to others in return for allegiance and economic obligations, the latter being fulfilled mainly in the form of money dues. The lords, in turn, normally owed allegiance and homage to higher lords from whom they held delegated grants of land. The labor power that ran each estate consisted of a peasant population, most of whom were serfs (descended from slaves and therefore not free) or tenants whose freedom of movement, freedom to marry, freedom to leave property to their heirs, and freedom to buy goods and sell their labor were closely circumscribed by public law. The peasantry was the essential element of the feudal economic system. Peasants provided the human capital (in the form of labor services) and monetary capital (for example, rents in kind, taxes, seigneurial dues, the issuance of money and payments for the use of essential services—milling, baking, olive pressing, and so on—monopolized by the lords) that enabled feudal lords to accumulate wealth.

By 1000 CE the countryside of most of Europe had been consolidated into a series of largely autonomous, feudal agricultural subsystems. Every estate was more or less self-sufficient in the raw materials for simple industrial products. Some of the members of every rural household would be capable of specialized, nonagricultural, part-time activities such as cloth making or basketry; and nearly every community supported a range of specialist artisans and craft workers. In addition, most regions had the capacity to sustain at least some small towns whose existence hinged mainly on their role as ecclesiastical centers, defensive strongholds, and administrative centers for the upper echelons of the feudal hierarchy. Improbably, this economic landscape—inflexible and introverted—nurtured the resurgence of trade and the revival of cities and provided the preconditions for the rise of merchant capitalism in Europe.

4.2 EMERGING IMPERATIVES OF ECONOMIC ORGANIZATION

Before moving on to examine the transition to merchant capitalism and the emergence of the European world-system, we pause briefly to review some of the organizing principles that seem to have been important in delineating the formative stages of preindustrial economic geography:

- Major changes in patterns of economic activity were gradual and incremental, even in hearth areas or core regions.
- Such changes generally preceded the development of critical innovations, particularly in technology and economic organization.
- Such innovations were a necessary but not sufficient condition to bring about radical change; institutional and sociopolitical changes were also necessary in order to exploit them.
- Demographic factors were also critical. Insufficient absolute numbers of potential workers sometimes hindered economic development, while changes in the balance between a population and its local resource base could be important in precipitating either progressive or regressive economic change.
- The law of diminishing returns provided an early impetus for territorial expansion. Colonization was pivotal in the development of hierarchical urban systems and improved transportation. It also stimulated the development of militarism, which induced important changes in spatial organization, for example, elevating the importance of defensive sites for key settlements. Finally, the environmental and social constraints laid bare by the law of diminishing returns were responsible for the emergence of a new geopolitical element—the state.

4.3 EMERGENCE OF THE EUROPEAN WORLD-SYSTEM

In this section, we consider the period that marked the first stirrings of the transition from feudalism to merchant capitalism in the thirteenth century through the creation of the European world-system in the sixteenth and seventeenth centuries. We also examine the proto-industrialization of the early eighteenth century, which served as the foundation for the

Industrial Revolution. We will highlight the emergence, interaction, and spatial implications of the salient aspects of economic change. But, first, we must consider an obvious but often neglected question: Why Europe?

WHY EUROPE?

In the twelfth century, almost half a millennium before Europe embarked on the path of capitalist development that shaped, directly or indirectly, virtually the entire world economy, several well-developed "economic worlds" existed in the Eastern Hemisphere. One was the Mediterranean region, whose principal elements included Byzantium, the Italian city-states, and Muslim North Africa. A second was the Chinese Empire. The central Asian land mass from Russia to Mongolia was a third. The Indian Ocean/Red Sea complex was a fourth. And the Baltic area was on the verge of becoming a fifth.

Why did Europe become the locus of innovatory economic change? Perhaps more importantly, *why not China*? China had approximately the same total population as Europe and, until the fifteenth century, was at least as advanced in science and technology. Chinese ironmasters had developed blast furnaces that enabled the casting of iron as early as 200 BCE. Iron plows were introduced in the sixth century, the compass in the tenth century, and the water clock in the eleventh. The Chinese were also significantly more advanced than the Europeans in medicine, papermaking, and printing, and the production of explosives. They also retained an imperial system with centralized decision making, an extensive state bureaucracy, well-developed internal communications, and a unified financial system, elements ideally suited to economic development and territorial expansion.

China's failure to take off must be attributed in part to its failure to pursue economic opportunities overseas. The Chinese had matched early European exploratory successes by spanning the Indian Ocean from Java to Africa in a series of lucrative and informative voyages; but they simply lacked a comparable interest in further exploration. One explanation for the absence of this colonizing drive is that they saw their own "world" as the only one that mattered. Another is that they were distracted by the growing menace of Mongol nomad barbarians and Japanese pirates. A third explanation is that the centralized power structure of imperial China did not contain enough different interest groups for whom overseas exploration was an attractive proposition.

This last point is seen by some as a facet of a broader set of structural constraints associated with the imperial form. The administration and defense of a large population and land mass perhaps drained the attention, energy, and wealth that might otherwise have been invested in capital development. The imperial system also meant that cultural and social élites tended to be focused on the arts, humanities, and self-promotion *vis-à-vis* the imperial bureaucracy. The centralization of decision making, meanwhile, is seen as having been insensitive to the economic potential of China's estimated 1,700 city-states and principalities. China's imperial framework is also implicated in its failure to develop military technology (after having gained a flying start) in the way that enabled Europeans to turn exploration into domination: Quite simply, the Imperial court suppressed the spread of knowledge of gunnery because it feared internal bandits and domestic uprisings.

Agricultural production served as another important element guiding the trajectories of China and Europe. European agriculture centered on the production of cattle and wheat. In contrast, rice production dominated Chinese agriculture. Because rice production requires relatively little land, China had less need to seek territorial expansion. Conversely, Europe's reliance on wheat and cattle provided a strong impetus for territorial expansion and exploration, while the more extensive use of animal power in Europe meant that "European man possessed in the fifteenth century a motor, more or less five times as powerful as that possessed by Chinese men" (Channu, 1969: 336).

Finally, some writers have emphasized the lack of autonomy of oriental towns compared to their European counterparts. As we will see, the legal and political autonomy of European towns served as a crucial "pull" factor in attracting the rural migrants whose labor and initiative were central to the emergence of merchant capitalism.

CRISIS OF FEUDALISM IN EUROPE

The transition from feudalism to merchant capitalism in Europe remains an issue of considerable debate, largely because we do not know enough about the details or timing of the critical economic and social changes that took place between 1300 and 1450. As a result, a variety of theoretical interpretations have emerged, each emphasizing different elements in the transition. In contrast, scholars generally agree that the overall context for the transition was a phase of economic, demographic, and political crisis brought about by the combination of steady population growth, modest technological improvements, and limited amounts of usable land.

As a result of improvements in plowing techniques, harnesses, and basic equipment in the early feudal period, wheat yields rose significantly and led to a steady rise in population over the twelfth and thirteenth centuries. In response, the feudal economy kept up by reclaiming rough pastureland and woodland. When this began to prove difficult (from around 1250), the population responded by attempting to improve crop rotations and shortening the period the land was permitted to lay fallow. There were limits, however, to such adjustments (Figure 4.2 illustrates the land intensity of a medieval manor in England). The number of cattle that could be kept, for example, was fixed by climatic constraints, which, in turn, limited the quantity of available winter forage; and this, likewise, imposed a limit on the supply of fertilizer for farming. In the absence of further advances in agrarian technology, food shortages were an inevitable outcome. In the wake of shortages, just as inevitably, came epidemics such as the Black Death (bubonic plague) in the 1340s, 1360s, and 1370s. These problems were compounded by climatic fluctuations: The cold winters and late springs of the fourteenth century aggravated the food shortages, while some exceptionally hot summers helped to swell the population of the black rat, host to the rat flea, the most significant vector of the bubonic plague.

Another aggravating factor was the beginning of the Hundred Years War in 1335–1345. The war necessitated a significant increase in taxes, which triggered a downward economic spiral fueled by falling rates of consumption, liquidity problems for noble treasuries, and a shortage of goods, which led to a spike in prices. The shortfall in funds necessitated additional tax increases and provoked a political climate of endemic discontent. The combined result of these pressures was "not only to exhaust the goose that laid the golden eggs for the castle, but to provoke, from sheer desperation, a movement of illegal emigration from the manor" (Dobb, 1963: 21).

The destination of these fugitives was the town, where different laws and tax systems prevailed. The late medieval European town (Cipolla, 1981: 146):

[W]as the "frontier," a new and dynamic world where people felt they could break their ties with an unpleasant past, where people hoped they would find opportunities for economic and social success, where sclerotic traditional institutions and discriminations no longer counted, and where there would be ample reward for initiative, daring and industriousness.

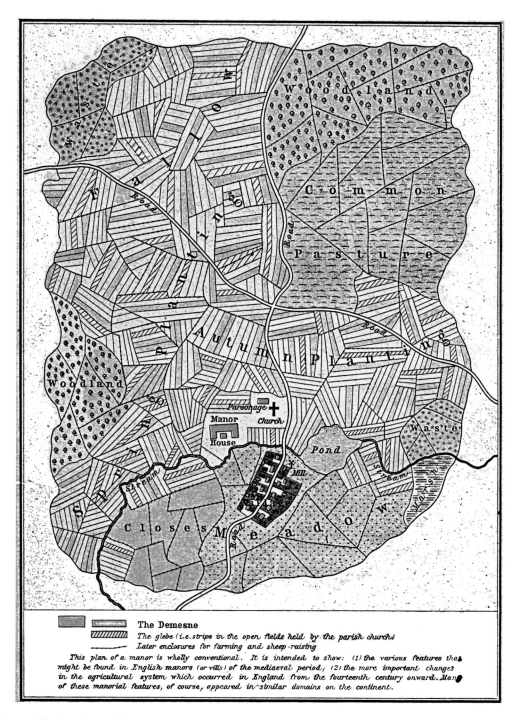

Figure 4.2 Plan of a medieval manor

Source: Based on Shepherd, 1923, *Historical Atlas,* Perry-Castañeda Library Map Collection, The University of Texas at Austin
http://www.lib.utexas.edu/maps/ historical/shepherd/plan_mediaeval_manor.jpg

The attractiveness of towns was not simply a product of the legal status of their inhabitants, however. People had, ironically, begun to prosper at the height of feudal economic development. In order to meet the nobility's more sophisticated and ostentatious requirements, seigneurial incomes had been increasingly realized in the form of cash. This requirement obliged peasants to sell part of their produce on the market in order to pay rents and taxes, and generally sparked trade in commodities. An embryonic pattern of regional trade developed in basic industrial and agricultural produce, and even some long-distance, international trade in luxury goods such as spices, furs, silks, fruit, and wine. As a consequence of this trade the size and vitality of towns increase and a greater number of merchants and craft workers emerged to cope with the demands of the system. This urban vitality served as a major agent in the eventual crisis of feudalism. It underscored the relative inefficiency of the self-sufficient feudal estate and transformed attitudes towards the pursuit of wealth.

RESURGENCE OF TRADE AND EXPANSION OF TOWNS UNDER MERCHANT CAPITALISM

Increased trade and urban growth were both a cause and an effect of the transition from feudalism. They also became hallmarks of the new economic order. As the feudal system faltered and disintegrated, an economy dominated by market exchange replaced it and communities came to specialize in the production of the goods and commodities they could produce most efficiently in comparison with other communities (see Figure 4.3). Merchants who supplied the capital required to initiate the flow of trade became the key group in this system, consequently the label **merchant capitalism**.

In marked contrast to feudalism and earlier rank-redistributive and primitive subsistence economies, merchant capitalism was, at least in theory, a self-propelling growth system, at least to the extent expansion through trade could be realized. Without it, neither merchants nor those dependent on their success—producers, consumers, financiers, etc.—could maintain their position, let alone advance it:

> Mercantile success required the merchants to buy as cheaply as possible, and to sell as expensively as possible; it also demanded that they trade in as large a volume of goods as possible. . . . This created a contradiction, however, for the producers were also consumers (though not of the goods they produced), so that if the prices they received were low, they could not afford to buy large quantities of other goods and thus satisfy the demands of the merchant class as a whole. A consequence of this was a great pressure on producers to increase the volume of goods offered for sale, which meant increasing their productivity, while merchants put pressure on consumers to buy more, even if this meant them borrowing money in order to afford their purchases. Both processes . . . involved producers raising loans which they had to repay with interest; to achieve the latter, they had to produce more (or, if they were employees rather than independent workers, to work harder).
>
> (Johnston, 1980: 33–34)

The regional specializations and trading patterns that provided the foundations for early merchant capitalism were predetermined to a considerable degree by the longstanding patterns developed by the traders of Venice, Pisa, Genoa, Florence, Bruges, Antwerp, and the Hanseatic League (which included Bremen, Hamburg, Lübeck, Rostock, and Danzig; see Figure 4.4) from the twelfth century. As merchant capitalism took hold, centers of trade multiplied in northern France and the lower Rhineland, new routes across Switzerland and southern Germany linked the commerce of Flanders (in Belgium) more closely to that of the Mediterranean, and sea

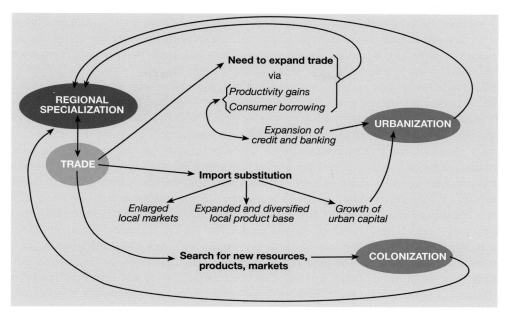

Figure 4.3 The rise of merchant capitalism and the changing space-economy

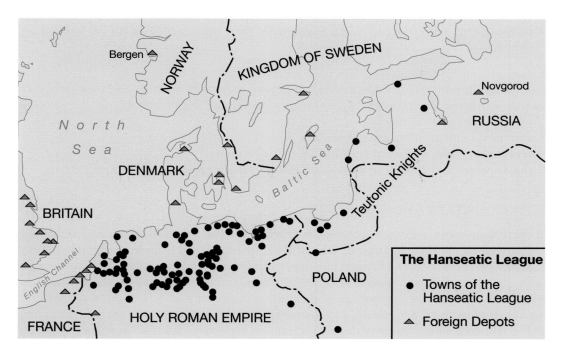

Figure 4.4 Towns and cities of the Hanseatic League
Source: Adapted from Hugill (1993: 50, Figure 2.5)

lanes—across the English Channel, North Sea, and Baltic Sea—began to integrate the economies of Britain, Scandinavia, and the Hansa territories with those of the continental core. Very quickly, a trading system of immense complexity came to span Europe, from Portugal to Poland and from Sweden to Sicily. This trading system was based not on the luxury goods of earlier trade routes but on bulky staples such as grains, wine, salt, wool, cloth, and metals.

The increased volume of trade fostered a great deal of urban development as merchants began to settle at locations that were of particular significance in relation to major trade routes, and as local economies came to focus on market exchange. But, once the dynamics of trade had been initiated, the key to urban growth was a process of import substitution, whereby externally produced goods and services are replaced with locally produced ones. In this way, local economies reinvest their income within their boundaries, which leads to a partial restoration of self-sufficiency and economic autarky. Although some things proved difficult to copy because of the constraints of climate or basic resource endowment, many imported manufactures could be copied by local producers, which increased local employment opportunities, intensifying the use of local resources, and increasing the amount of local investment capital available. As Jane Jacobs argued, cities that replaced imports could then afford new types of goods being produced in other cities. The newly imported innovations, in their turn, might also be replaced with local production, opening up the market for still more innovations from elsewhere. So the cities of Europe:

> [W]ere forever generating new exports for one another—bells, dyes, buckles, parchment, lace, carding combs, needles, painted cabinet work, ceramics, brushes, cutlery, paper, sieves, riddles, sweetmeats, elixirs, files, pitchforks, sextants—and then replacing them with local production, to become customers for still more innovations. They were developing on one another's shoulder.
>
> (Jacobs, 1984: 50)

As a result, patterns of trade and urban growth were very volatile; and long-term local success within the new economic order became increasingly dependent on:

- sustained improvisation and innovation
- repeated episodes of import substitution
- the discovery and control of additional resources and new kinds of resource.

CONSOLIDATION AND EXPANSION

In the fifteenth and sixteenth centuries, a series of innovations in business and technology contributed to the consolidation of merchant capitalism. These included innovations in the organization of business and finance: banking, loan systems, credit transfers, company partnerships, shares in stock, speculation in commodity futures, commercial insurance, courier/news services, and so on. The importance of these innovations lay not only in the way they oiled the wheels of industry, agriculture, and commerce, but also in the way they encouraged savings and facilitated their use for investment. Furthermore, the routinization of complex commercial and financial activity brought with it the codification of civil and criminal legislation relating to property rights (for example, patent laws); a development seen by some as being of critical importance because it provided an incentive for a sufficient number of innovators and entrepreneurs to channel their efforts into the embryonic capitalist economy.

Meanwhile, technological innovations succeeded each other at an accelerated rate. Some of these were adaptations and improvements of oriental discoveries—the windmill, spinning wheels, paper manufacture, gunpowder, and the compass, for example. But Europe also possessed a passion for the mechanization of the productive process as a means of increasing productivity. In addition to improvements based on others' ideas, a welter of independent engineering breakthroughs emerged including the more efficient use of energy in watermills and blast furnaces, the design of reliable clocks and firearms, and the introduction of new methods of processing metals and manufacturing glass.

Innovators jealously guarded these breakthroughs in a hope of monopolizing the advantages they conferred while competitors in other regions went to considerable lengths to acquire new technology at the first opportunity. So, for example, the Venetian government strictly prohibited the emigration of caulkers; and the Grand Duke of Florence gave a reward for the return, dead or alive, of emigrants from key positions in the brocade industry. The French kidnapped skilled iron workers from Sweden; while many governments were happy to provide shelter and handsome rewards for migrant craftsmen who had knowledge of new techniques. These early examples of a "brain drain" were complemented by the practice of temporary migration in the opposite direction in order to acquire new expertise, sometimes legitimately, sometimes covertly. But the most important vector for the diffusion of technological innovations came with the invention of the printing press. Within 20 years of its introduction by Johannes Gutenberg in Mainz around 1450, printing shops had spread throughout Europe, opening up vast new possibilities in the fields of knowledge and education.

Innovations in shipbuilding, navigation, and naval ordnance, however, had the most far reaching consequences for the evolution of the European space economy. By the fourteenth century European shipwrights were building ships skeleton first, as a vast saving of labor in comparison with previous methods. In the course of the fifteenth century, the full-rigged ship was developed, enabling faster voyages in larger and more maneuverable vessels that were less dependent on favorable winds. Meanwhile, the quadrant (1450) and the astrolabe (1480) were invented, and seafarers had acquired a systematic knowledge of Atlantic winds. By the mid-sixteenth century, England, Holland, and Sweden had perfected the technique of casting iron guns, making it possible to replace bronze cannon with larger numbers of more effective guns at lower expense. Together, these advances made it possible for the merchants of Europe to establish the basis of a worldwide economy in under 100 years.

MERCANTILISM AND TERRITORIAL EXPANSION

As we have already seen in relation to China, however, economic strength and technological ability do not necessarily lead to overseas expansion. What, then, translated Europe's economic power and technological superiority to a broader arena?

Figure 4.5 summarizes the most important factors. The large number of impoverished aristocrats produced by European inheritance laws and by expensive crusades and local wars was one important factor. Discouraged from commercial careers by sheer snobbery and encouraged by a culture that romanticized the fighting man, these poverty-stricken gentlemen provided a plentiful supply of adventurers who were willing to die for glory and even more willing to exercise greed and cruelty in the name of god and country, underscoring the importance of the evangelical zeal of the Catholic Church and the political competitiveness of the monarchies during this period.

Above all, however, overseas expansion was impelled by the *logic of merchant capitalism* and the *law of diminishing returns*. As noted, growth could only be sustained as long as output could be increased. After a point, this required food and energy resources that could only be obtained by the conquest—peaceful or otherwise—of new territories. Similarly, merchant

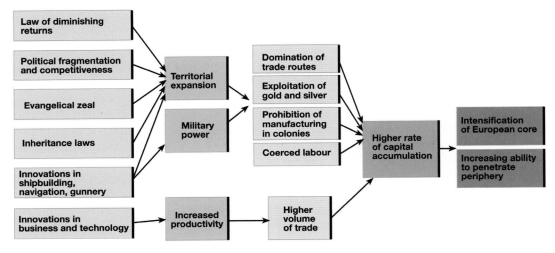

Figure 4.5 The emergence of a European-based world-system

capitalism required new supplies of gold and silver to make up for the leakage through trade with Byzantium, China, India, and Arabia.

Collectively, these motivations found expression in the dogma of mercantilism. Most European countries adhered to this dogma from the sixteenth century to the early eighteenth century. National wealth was measured in terms of the amount of accumulated precious metal (gold or silver), and the fundamental source of economic growth was a persistently favorable balance of trade. This economic "logic" justified not only overseas colonization but also the coercion of plantation labor and the prohibition of manufacturing in the colonies. It also promoted thrift and saving on the domestic front as a means of accumulating capital for overseas investment. It required a high degree of economic regulation, sponsorship, and protection by the government.

There is no need for us to reiterate here the pattern and sequence of European expansion and conquest (though it is worth noting that the overall thrust—overseas from Atlantic Europe rather than inland to the east—reflected the technological superiority of the Europeans on sea relative to land: Asians could counterbalance technological inferiority with weight of numbers until after the mid-seventeenth century when European technology succeeded in developing more mobile and rapid-firing guns). Europeans soon destroyed most of the Muslim shipping trade in the Indian Ocean and captured a large share of the intra-Asian trade. By bringing Japanese copper to China and India, Spice Island cloves to India and China, India cotton textiles to Asia, and Persian carpets to India, European merchants made good profits and with them paid for some of their imports from Asia.

The gold and silver from the Americas, however, provided the first major economic transformation and allowed Europe "to live above its means, to invest beyond its savings" (Braudel, 1972: 268). In effect, the bullion was converted into effective demand for **consumer goods** and **producer goods** of all kinds—textiles, wine, food, furniture, weapons, and ships—which stimulated production throughout the economic system, creating the basis for a "Golden Age" of prosperity for most of the sixteenth century. Meanwhile, overseas expansion made available a variety of new and unusual products—cocoa, beans, maize, potatoes, tomatoes, sugar cane, tobacco, and vanilla from the Americas, tea from the Orient—which opened up large new markets to enterprising merchants.

As European traders came to monopolize intra-oriental trade routes, they literally controlled the flow and patterns of trade between potential rivals. This monopoly enabled European traders to identify foreign articles with a tested profitable market and ship them to Europe where skilled workmen learned to imitate them. Once Europeans begun manufacturing these products, their goods were shipped to the rest of the world:

> For example, Europeans long prized the shawls which were made in the north of India in the Kashmir region; much later Scotchmen [sic] were making imitations of those shawls by the dozens per day; called Paisley shawls, they swept the Kashmir shawls off the general market. Europeans admired the very hard vitrified china of the Chinese, and for a long while bought it to sell to other peoples, taking it from China and distributing it. But then the Europeans began to make it in France and elsewhere, and shortly true Chinese china had become a rare article on the world market while Europe was making and selling enormous amounts of its own "china." For a good while Europeans bought cottons of a very fine quality from India for markets in Africa, Europe, and America, but before too long they had imitated them in England and were shipping cheaper machine-made cottons back to India where they ruined the Indian cotton-weaving industry in its own home.
>
> (Reynolds, 1961: 45–46)

So for Europe, the benefits of overseas expansion extended well beyond the basic acquisition of new lands and resources. In addition to the bullion and the opportunities for import substitution, overseas expansion also stimulated further improvements in technology and business techniques, which added a further dimension to the self-propelling growth of merchant capitalism. New developments were achieved in nautical mapmaking, naval artillery, ship-building, and the use of the sail; and the whole experience of overseas expansion provided a great practical school of entrepreneurship and investment. Most important, perhaps, was the way profits from overseas colonies and trading overflowed into domestic agriculture, mining, and manufacturing. This contributed to an accumulation of capital that was undoubtedly one of the main preconditions for the emergence of industrial capitalism in the eighteenth century.

THE WORLD OUTSIDE EUROPE: TRANSOCEANIC RIM SETTLEMENTS

Outside Europe, the most important features of the economic landscape to emerge as a result of merchant capitalism were the gateway towns and entrepôts established along the coastal rims of the Americas, Africa, and south Asia. These *transoceanic rim settlements* (see Figure 4.6) were of three main kinds:

1. *Trading stations*, such as Canton (now Guangzhou, China), Madras (now Chennai, India) and Goa (India). These locations emerged as the points of contact between Europe and the relatively autonomous economies of the orient. Few Europeans lived in these towns and cities, and only in India was it possible to exercise any secure measure of political control over the large hinterland areas that served as ports.
2. *Entrepôts and colonial headquarters* for tropical plantations, such as Rio de Janeiro (Brazil), Georgetown (British Guiana), Port of Spain (Trinidad), Penang (Malaysia), Lagos (Nigeria), Lourenço Marques (now Maputo, Mozambique) and Zanzibar (Tanzania). In these locales, substantial numbers of European settlers were required for administrative and military purposes, whereas the indigenous population provided field labor and manual labor in the towns. The colonial plantation system made intensive demands on labor, however, and when the indigenous supply was insufficient, the colonizers augmented it by enforced movements of slave labor from other regions, which created distinctive ethnic cleavages among the populations of many colonies.

3. *Gateway ports* for the 13 farm-family colonies on the northeastern seaboard of America (similar settlements were later established in South Africa, Australia, and New Zealand). Although several distinctive groups emerged—the Tidewater Colonies (for example, Jamestown, Baltimore), the New Towns of New England (for example, Boston, Newport), the Middle Colony towns (for example, New York, Philadelphia) and the Colonial Towns of the Carolinas (for example, Charleston, Savannah)—they were essentially a direct extension of the European urban system, peopled by Europeans and oriented much more to their homelands than their hinterlands.

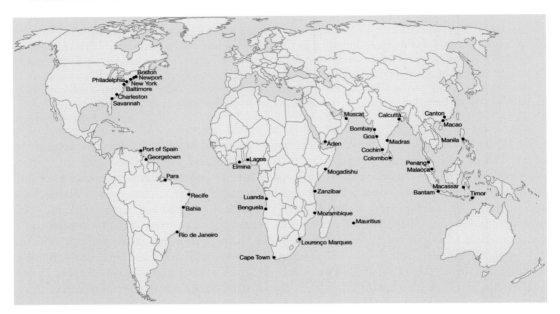

Figure 4.6 Transoceanic rim settlements of the mercantile era

THE SHIFTING LOCUS OF ECONOMIC POWER

The dominant feature of the changing economic geography of Europe in the sixteenth and seventeenth centuries was a dramatic shift in the focus of economic activity from the Mediterranean to the North Sea. At the end of the fifteenth century the Mediterranean was the most highly developed region in the world with central and northern Italy as the hub of economic activity. During the sixteenth century the relative prosperity of the Mediterranean was further enhanced as Spain and Portugal benefited immensely from the influx of treasure from the Americas. By the end of the seventeenth century, however, the Mediterranean had become a backward region in relation to the levels of prosperity generated by the Dutch economy; while England, previously a marginal economy in relative terms, stood poised to threaten the position of the Dutch as world leaders.

Between the extremes of stagnation/regression and dynamic expansion was the experience of France, Scandinavia, Germany, and much of the rest of continental Europe, where a general penetration of economic development and maturing of local economies—a consolidation of merchant capitalism—helped maintain the coherence of the European economy during a period of volatile change in its spatial organization. In detail, therefore, the changing center of gravity of the European economy involved a complex tapestry of overlapping, interlocking, and interacting regional struggles and transformations. The basic processes involved, however, were more general, and they can be illustrated with reference to the decline of Spain and Italy and the rise of Holland and England.

Spain

Spain provides a good example of the importance of import substitution. Quite simply, Spain declined because it had never been fully "developed" to begin with; it had merely been wealthy. The increased demand generated by its acquisition of bullion from the Americas did not stimulate domestic production as much as it might have for two reasons: Bottlenecks in the productive system—the restrictive practices of guilds and the lack of skilled labor, for example—and the complacent attitude of the Spanish élite. In 1675 Alfonso Nuñez de Castro wrote:

> Let London manufacture those fabrics of hers to her heart's content; Holland her chambrays; Florence her cloth; the Indies their beaver and vicuña; Milan her brocades; Italy and Flanders their linens, so long as our capital can enjoy them; the only thing it proves is that all nations train journeymen for Madrid and that Madrid is the queen of Parliaments, for all the world serves her and she serves nobody.
>
> (quoted in Cipolla, 1981: 125)

The treasure of the Americas greatly increased Spain's purchasing power, but this wealth ultimately stimulated the development of England, France, Holland, and the rest of Europe. Additionally, Spain's prosperity induced the government to pursue a persistently warmongering policy that became a serious drain on the treasury. In the course of the seventeenth century, then, as the influx of bullion from Spain's colonies declined (partly through depleted mines), the momentum of the economy evaporated and left insufficient entrepreneurs and artisans to counterbalance an overabundance of bureaucrats, lawyers, and priests and to tackle a mounting national debt.

Italy

Italy's decline was more complex, but its beginning can be dated more accurately: the end of the fifteenth century, when for almost 50 years northern Italy became the battlefield for an international conflict involving Spain, France, and Germany. Famines and epidemics characterized this period as well as severe disruptions to trade that coincided with a blossoming of exchange elsewhere. Buoyed by the international boom in demand during the late sixteenth century, the economy made something of a recovery; but it was a recovery based on traditional methods of organization, which meant, among other things, that competition and innovation were suppressed by the renewed strength of craft guilds.

Between 1610 and 1630 a series of external events led to the collapse of some of the Italians' major markets—the decline of the Spanish economy, disruptive wars in the German states, and political instability within the Turkish Empire. At the same time, many of Italy's competitors had been able to substitute domestic products for Italian imports. At this point, the self-propelling growth of merchant capitalism broke down. Unable or unwilling to respond through innovation and increased productivity, Italian entrepreneurs began to disinvest in manufacturing and shipping. By the end of the seventeenth century, Italy was importing large quantities of manufactures from England, France, and Holland and exporting agricultural goods—oil, wheat, wine, and wool—for which the terms of trade were poor. So foreign trade had been transformed from an "engine of growth" to an "engine of decline."

The Netherlands

The "economic miracle" of the Netherlands in the seventeenth century was launched from a fairly solid platform of trading and manufacturing. Although overshadowed in the early phases of merchant capitalism by the prosperity of nearby Bruges and Antwerp, Holland (and Amsterdam in particular) had steadily developed an entrepôt function for northern Europe (importing flax, hemp, grain, and timber and exporting salt, fish, and wine) which served as the foundation for establishing a manufacturing base. From the stability of this base, the Dutch successfully rebelled against Spanish imperialism and emerged, in 1609, with political independence and religious freedom.

Thereafter, a combination of factors helped the Dutch become leaders of the world economy for more than 150 years. One was the "modernity" of Dutch institutions: relatively few restrictive guilds, a small nobility of landowners, and a relatively weak church after the departure of the Spanish. Another was the vigorous pursuit of mercantilist policies, including not only a strong colonial drive and a significant commitment to merchant shipping but also an uncompromising stance towards competitors. For example, the Dutch blockaded Antwerp's access to the sea from 1585 to 1795, taking over its entrepôt trade and its textile industry. The Dutch were also able to turn their geographical situation to great advantage, developing ocean ports and exploiting the inland waterways that penetrated the heart of continental Europe. They benefitted from a highly developed and very innovative shipbuilding industry whose output completely overshadowed that of the rest of Europe. Finally, the Dutch were the major beneficiaries of the flight of skilled craftsmen, merchants, sailors, financiers, and professionals from the fanaticism and intolerance of the Spanish in Flanders and Wallonia (Belgium).

England

England, at the end of the fifteenth century, was distinctly backward, with a small population (around 5 million, compared to more than 15 million in France, 11 million in Italy, and 7 million in Spain) and a poorly developed economy. The only significant comparative advantage the English held was the manufacture of woolen cloth. The first real break for the English economy came in the first half of the sixteenth century when Italian production and trade collapsed and left the English to capitalize on a sustained increase in woolen exports—a trend that was further enhanced by the progressive deterioration of English currency resulting from Henry VIII's extravagant military expenditures. The boom was halted in the mid-sixteenth century, however, by the recovery of the Italian textile industry and by the war between the Dutch and the Spanish, which disrupted English exports.

By this time, however, English entrepreneurial and expansionist ambitions had become established and were articulated through a strong mercantilist philosophy. Like the Dutch, the English were able to take advantage of their geographical situation, at least in relation to transoceanic trade. They had also developed a strong navy, and gave high priority to establishing a large merchant fleet and to acquiring colonial footholds. Also like the Dutch, they also benefited from the skills of immigrants driven from France and the Low Countries by religious persecution. Innovation, improvisation, and import substitution played their part in ensuring a rapid escape from the mid-century economic crisis and, indeed, in building an economy to challenge that of the Dutch. The development of iron artillery in the 1540s, for example, enabled the English to arm their merchant ships, privateers, and warships more extensively *and* at lower cost. Meanwhile, the exploitation of coal as a substitute for the relatively sparse and rapidly diminishing timber reserves not only helped the English to avoid an energy crisis but also helped to develop new processing techniques. "Concentrating on iron and coal, England set herself on the road that led directly to the Industrial Revolution" (Cipolla, 1981: 290).

SUMMARY

Several organizing principles can be delineated to describe the evolving space economy up to the eve of the Industrial Revolution. First, note that the emerging imperatives of early economic systems described in this chapter appear as recurring elements in subsequent economic epochs, confirming the gradual and incremental nature of major economic change.

We can also confirm the continuing importance of innovations in technology and business organization; although we should note that the innovative process to this point occurred in small steps by way of the gradual accumulation of improvements rather than by distinct bursts of invention which, as we will see, have characterized economic change since the industrial era.

The importance of institutional and sociopolitical factors was also a recurring theme (as, for example, in the constraints of a centralized imperial system on the evolution of the Chinese economy, in the stimulus provided by European laws on property rights, and the role of European governments in implementing mercantilist policies). Similarly, we must acknowledge the continued interaction between demographic change and economic development and, finally, the ongoing impetus for territorial expansion as a product of the law of diminishing returns and, of course, the ego aggrandizement of rulers. In addition, however, we can identify several new dimensions of spatial-economic organization:

- The emergence of a true world economy involving long-distance interaction based on a sophisticated spatial division of labor.
- The progressive elaboration of the world economy in space and across commodities was *uneven*. Some sectors, countries, and regions expanded more quickly than others, and some spheres of opportunity and lines of communication were penetrated more quickly than others, so that its early spread was in a selective, spatially discontinuous fashion.
- The pattern of specialization and the nature of economic interaction within the world economy resulted in the emergence of *core* areas, characterized by such mass-market industries as had emerged (for example, textiles, shipbuilding), international and local commerce in the hands of an indigenous bourgeoisie, and relatively advanced forms of agriculture; *peripheral* areas, characterized by the **monoculture** of **cash crops** by coerced labor on large estates or plantations; and *semi-peripheral* areas, characterized by a process of deindustrialization but retaining a significant share of specialized industrial production and financial control.
- The spatial organization of the European space economy was based around a cluster of core areas in northwestern Europe: southeastern England and Holland together with the Baltic states, the Rhine and Elbe regions of Germany, Flanders (Belgium), and northern France. Peripheral regions included northern Scandinavia, Britain's Celtic fringe (Scotland, Wales and Ireland), east-central Europe, and all of the transoceanic rim settlements and colonies. The semi-periphery consisted of the Christian Mediterranean region, which had been the advanced core area at the beginning of the merchant capitalist era.
- The articulation of the European world economy also produced a distinctive pattern of settlement and urbanization. Merchant capitalism was reflected in the urban landscape by a strengthening of the hierarchical system of settlements and the development of a **central place system**. The overseas territorial expansion associated with merchant capitalism was also reflected in a distinctive urban landscape, as illustrated in Figure 4.7. Johnston (1980: 74) once again provides a succinct description:

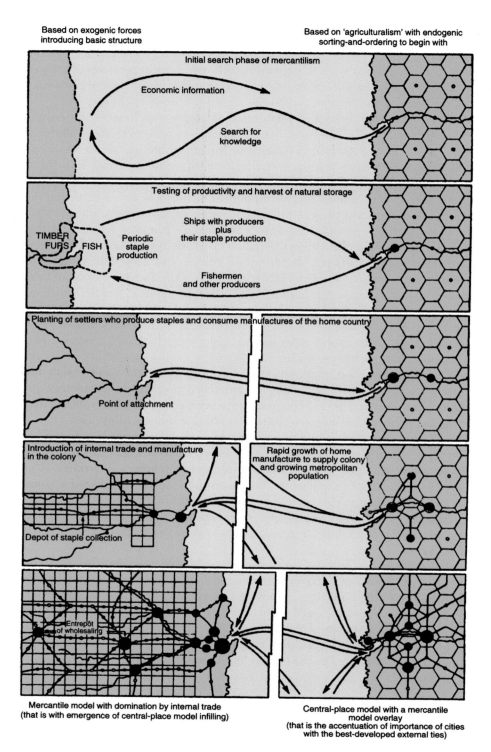

Figure 4.7 Colonialism and urban settlement patterns

Source: Based on Vance (1970: 151, Figure 18)

In the initial stages of mercantile exploration no permanent settlement is established in order to obtain the required products (fish, timber and furs). Then the colony is settled by agriculturalists; the export of their products moves through local articulation points to the colonial port, and thence to the port in the homeland, which grows in size and status relative to its inland competitors. As settlement of the colony expands further inland, so both of the ports increase in size, railways replace rivers as the main traffic arteries within the colony, and internal gateways develop to articulate the trade of areas some distance from the port, while in the homeland places near to the original port benefit from the imports and a new outport is built to handle the larger volume of trade and the bigger vessels.

The emergence of the European world economy brought about a system of internal dynamics that involved three important mechanisms of spatial change:

1. *Strategic investments*: The switching of investment from one area to another by merchants in response to the shifting comparative advantages enjoyed by local producers. These shifts in comparative advantage, in turn, were associated with technological innovations and improvements, institutional changes, currency fluctuations, and so on.
2. *Import substitution*: Communities able to achieve repeated episodes of import substitution, as Jacobs (1984) pointed out, benefit from five aspects of economic development:

    a)   enlarged markets for new imports and innovations
    b)   an expanded and more varied employment base
    c)   new applications of technology to increase rural productivity
    d)   a spillover of employment to rural areas as older, expanding enterprises are crowded out of cities
    e)   growth of city capital.

3. *Militarism and geopolitical change.*

KEY SOURCES AND SUGGESTED READING

Cipolla, C., 1981. *Before the Industrial Revolution. European society and economy, 1000–1700*, 2nd edn. London: Methuen.

Clark, C., 1977. *World Prehistory in New Perspective*. Cambridge: Cambridge University Press.

De Vries, J., 1976. *Economy of Europe in an Age of Crisis, 1600–1750*. Cambridge: Cambridge University Press.

Diamond, J., 1997. *Guns, Germs, and Steel*. New York: W.W. Norton.

Diamond, J., 2005. *Collapse: How societies choose to fail or succeed*. New York: Viking.

Frank, A.G., 1998. *ReORIENT: Global economy in the Asian age*. Berkeley: University of California Press.

Jacobs, J., 1984. Cities and the wealth of nations, *Atlantic Monthly* March, 41–66.

Johnston, R.J., 1984. The World is our Oyster, *Transactions, Institute of British Geographers* 9, 443–459.

Landes, D.S., 1999. *The Wealth and Poverty of Nations*. New York: W.W. Norton.

Reynolds, R., 1961. *Europe Emerges. Transition toward an industrial world-wide society*. Madison: University of Wisconsin Press.

Tilly, C., 1992. *Coercion, Capital, and European States*. Cambridge, MA: Blackwell.

Wallerstein, I., 1980. *The Modern World-System II: Mercantilism and the consolidation of the world-economy 1600–1750*. London: Academic Press.

Chapter 5

Evolution of the core regions

Picture credit: © European Community, 2013

From the second half of the eighteenth century, industrialization rapidly reshaped economic landscapes, introducing new dimensions and shifting the patterns and tempo of the world economy. Today, economic geography within the core of the world economy is dominated by the physical, institutional, and social legacies of industrial capitalism. The economic geography of the peripheral regions has, meanwhile, been shaped by their role in sustaining the industrial expansion of the core economies and more recently the NIEs. In short, few elements of the economic landscape are not a product, directly or indirectly, of the industrial era. In this chapter, we outline the evolution of the economic geography of the industrial core regions, analyzing the major processes involved in the relative ascent and decline of countries and regions within these regions.

5.1 THE INDUSTRIAL REVOLUTION AND SPATIAL CHANGE

The transition during the late eighteenth and early nineteenth centuries from merchant capitalism to industrial capitalism as the dominant form of economic organization is conventionally ascribed to the Industrial Revolution. The Industrial Revolution, in turn, is typically depicted as a revolution in the techniques and organization of manufacturing based on a series of innovations in the technology of production (e.g., Kay's flying shuttle (1733), Hargreaves' spinning Jenny (1765), and Cartwright's machine loom (1787)) and in transport technology and engineering (particularly the development of canal and railway systems). But technological advances were only part of a wider economic, social, and political transition whose origins and preconditions can be found in the Renaissance and Enlightenment. Indeed:

> [P]rior to 1800, living standards in the world economy were roughly constant over the very long run: per capita wage income, output and consumption did not grow. Modern industrial economies, on the other hand, enjoy unprecedented and seemingly endless growth in living standards.
>
> (Hansen and Prescott, 2002: 1202)

The most important context for technological advance was the existence within merchant capitalism of industry organized on capitalist lines by entrepreneurs employing wage labor and producing commodities for sale in regional and national markets. In addition, the capital that had been accumulated through trading provided the means for entrepreneurs to finance investment in the capital-intensive but highly productive technologies introduced during the Industrial Revolution.

From these roots, machine production and the organizational setting of the factory—machinofacture—emerged as the central characteristics of industrialization. While machinery provided the basis for higher levels of productivity, factories enabled this productivity to be exploited to its fullest extent. Gains in productivity were most often achieved through specialization—the assembly-line division of labor—and internal economies of scale. At the same time, the concentration of workers in big industrial units generated urban environments that represented a new dynamic force for economic, social, and political change.

Like merchant capitalism before it, however, industrial capitalism had to confront the twin obstacles of market saturation and the law of diminishing returns. In response, industrialists pursued a variety of strategies. In addition to the constant search for technological advances, industrialists sought:

- new ways of exploiting internal and external economies of scale
- cheaper sources of labor, raw materials, and energy
- greater access to overseas markets
- development of new products, either through new inventions or by the "commodification" of activities previously performed within the household
- formalized relations with labor unions and governments to provide a more stable context (economic, social, and political) in which to operate.

The changes imposed on economic landscapes by the first waves of the Industrial Revolution have been overwritten by a succession of episodes of industrial development, restructuring, and reorganization. These episodes have created significant differences between the major industrial regions; differences that reflect variations in resource endowment, previous patterns of economic development, and, importantly, variations in the relative timing and interaction of these episodes of industrial change.

5.2 MACHINOFACTURE AND THE SPREAD OF INDUSTRIALIZATION IN EUROPE

Although often considered a single, discrete period, the Industrial Revolution included several distinctive transitional phases, each with a unique impact on various regions and countries. As new technologies altered the margins of profitability in different enterprises, so the fortunes of specialized places shifted.

These regional differences, in turn, helped to influence the changing character of capitalism. With the evolution of capitalism came shifts, occasionally of a dramatic nature, in economic, social, political, and cultural relations. These evolving forms of economic organization—interrelated complexes of production, consumption, and income distribution based on the organization of firms—grew in response to the opportunities and constraints created by new production, transportation, and communications technologies. At the same time, this evolution yielded a succession of technology systems that were imprinted differentially across the economic landscape.

Associated with each form of economic organization is a specific regulatory framework—a set of local and historical political arrangements and institutions that emerged to provide appropriate management for the operation of the successive forms of economic organization (e.g., monetary and wage regulation, particular government–business relationships, trading regulation, etc.) and technology systems within the wider national and international context. These regulatory frameworks have four principal functions:

1. regulating the monetary system and financial mechanisms
2. regulating wages and collective bargaining
3. facilitating (or, in some circumstances, constraining) competition, and negotiating the relations between the private sector and public economy
4. establishing the roles of government at various spatial scales.

Three major waves of industrialization in Europe can be identified. Each wave consisted of several phases, and each was highly localized in its impact. The first wave introduced the first technology system of the Industrial Revolution, which included new iron and cotton textile technologies and the use of water power, trunk canals, and turnpike roads. Even within the span of this first wave, however, the imprint was highly differentiated. "Above all," Pollard emphasizes, "the industrial revolution was a regional phenomenon" (1981: 14).

Figure 5.1 clearly highlights that the growth rate of GDP per capita in Europe was close to zero for nearly 1,000 years prior to the first wave of industrialization. In England, the real wage was roughly the same in 1800 as it had been in 1300. Figure 5.1 also demonstrates that population growth was stagnant prior to industrialization, largely reflecting the low pace of technological change.

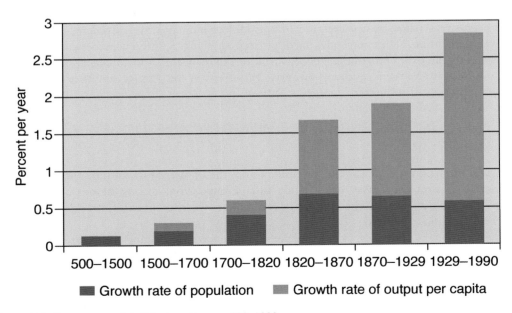

Figure 5.1 Output growth in Western Europe, 500–1990

Source: Based on Galor and Weil (2000: 808, Figure 1)

FIRST-WAVE INDUSTRIALIZATION: BRITAIN

The springboard for the first wave of industrialization, which began in Britain around 1760, consisted of several local hearth areas of "proto-industrialization." These areas had long-standing concentrations of industry based on a wage-labor force using the most advanced of the available industrial processes. The need to locate operations near mineral resources and sources of water power as well as within reasonable distance to local canal systems meant that early industrial activity was highly localized. This localization also reflected the principle of comparative advantage whereby industry had been displaced into the least hospitable locales for agricultural production.

This pattern of proto-industrialization, with its external economies, infrastructural advantages, and well-developed markets, helped to determine the nuclei of industrial development in Great Britain during the first phase of the first wave of industrialization between 1760 and 1790. Although sub-regions such as north Cornwall, south Staffordshire, and north Wales shared the common impetus of certain key resources and innovations, each retained its own distinctive business transitions and industrial style. Much of the required capital was raised locally, labor requirements were drawn (in the first instance) from the immediate hinterland, and industrialists formed regional organizations and operated regional **cartels**.

From the start, then, industrialization was articulated at the regional level. The second phase (of the first wave), between 1790 and 1820, reinforced the position of those embryo industrial regions with a coalfield base and saw the emergence of other regions such as Ulster in Northern Ireland and south Wales as industrialized regions. Meanwhile, the prosperity of the early starters declined markedly as their relative advantages were eclipsed by a combination of three factors:

1. the exhaustion of minerals or the discovery of cheaper alternative supplies
2. the relative inaccessibility of markets due to poor communications or the isolated nature of the locale
3. the lack of size to develop.

The third phase of the "British" wave, between 1820 and 1850, was dominated by the expansion of the railway system. This development did not foster any new industrial regions, but it did widen the market area of the existing industrial regions, drawing more of Britain into the sphere of industrial capitalism.

SECOND-WAVE INDUSTRIALIZATION: A NEW TECHNOLOGY SYSTEM AND NEW FORMS OF ECONOMIC ORGANIZATION

The second wave was characterized by the spread of industrialization to continental Europe. This diffusion did not occur in a straightforward or systematic manner; rather, forms of economic organization and regulatory frameworks emerged to exploit new technologies, primarily leveraging coal, steel, heavy engineering, steam power, and railways. This shift was accompanied by a number of other changes including the development of new labor practices (i.e., the spread of wage-labor norms), the emergence of new corporate structures (e.g., large limited-liability firms with a national rather than local scope), and new relationships between governments and industry (i.e., increased regulation of and investment in key industries by states).

Similar to the British wave of industrialization, the second wave was launched from the proto-industrial regions of continental Europe. Initially, from around 1850, industrialization was concentrated in the Sambre-Meuse region of Belgium and in the valley of the Scheldt in Belgium and France. Subsequent phases saw the spread of industrialization to areas such as the Ruhr in Germany and Alsace, Normandy, and the upper Loire valley in France.

Unlike their counterparts during the first wave of industrialization, would-be competitors on the continent entered a market in which the early movers (i.e., the British) had secured comfortable advantages in technology which translated into a dominant position in the world markets. Britain also had a series of "natural" geographical advantages: A compact territory with a large population, favorable conditions for intensive agricultural production, and a rich variety of minerals including coal.

This **competitive disadvantage** for the industrial regions of continental Europe was compounded by the consequences of the Revolutionary and Napoleonic Wars of the early nineteenth century (as it was in the United States by the Civil War of 1861–1865). Conscription, armed conflict, and military occupation disrupted production and suppressed industrial expansion, allowing British industries to forge further ahead on the basis of the new technology system (and, of course, a constantly evolving and adapting regulatory framework).

But in contrast to their British counterparts, continental entrepreneurs and governments did not have to industrialize by trial and error. By drawing on the British experience—as well as importing British managers, workers, capital, and technology—they minimized missteps and accelerated their pace of modernization. These regions of "inner" Europe differed from one another not only in the mix of industries that gained a foothold, but also by what economic historian Sidney Pollard calls the **differential of contemporaneousness**, whereby new technologies, ideas, and market conditions reached regions simultaneously but affected them in unique ways because they were not comparably equipped to respond to them. Thus, for example:

> Legislation permitting the easy formation of joint-stock companies spread quickly across Europe in the 1850s, and their contribution to overspeculation and wide-spread bankruptcies in the less sophisticated European economies has often been commented on. In banking, the backward economies, using the experience of the pioneers, could bypass some of the difficulties of the latter by enjoying the benefits of more efficient banks, ahead, as it were, of their own stage of economic growth.
>
> (Pollard, 1981: 188–189)

In general terms, however, the cumulative impact of innovations in first- and second-wave industrializers made for convergence: The French Nord (north) began to look and function increasingly like the central belt of the Scottish Lowlands, and the Ruhr began to look and function increasingly like the Sambre-Meuse region. At the same time, areas that had adopted an industrial base and those that had yet to follow suit increasingly diverged in their social and economic complexion. By 1875, the latter still covered a great deal of the map (see Figure 5.2), but many of them were incorporated in the third wave of industrialization between 1870 and 1914.

THIRD WAVE INDUSTRIALIZATION: INTERMEDIATE EUROPE

The third wave of industrialization included "intermediate" Europe—parts of Britain, France, Belgium, and Germany that had not been directly affected by the first two waves, together with most of the Netherlands, southern Scandinavia, northern Italy, eastern Austria, and Catalonia in northeastern Spain. By this time, all European landscapes had begun reorganizing

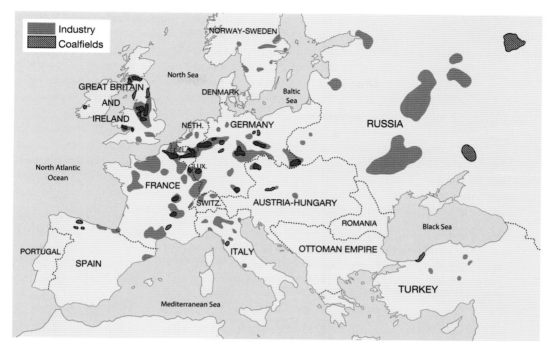

Figure 5.2 Europe in 1875

Source: Adapted from Pollard (1981: xv, Map 2)

in response to the imperatives of a third technology system: One based on steamships, world shipping, the internal combustion engine, heavy chemicals, and heavy engineering.

In the regions of "intermediate" Europe, the imprint of this industrialization was distinctive in several important respects. Prior development played a minor role, and the capital required for industrialization had increased exponentially since the first wave. The combination of these factors and the increasing sophistication of industrial technology led central governments to assume a larger, more proactive role in development. As a result, the economic role of the state among the later industrializers tends to be more pronounced than in the countries of "inner" Europe.

PERIPHERAL EUROPE

The residual territories of western Europe—most of the Iberian peninsula, northern Scandinavia, Ireland, southern Italy, the Balkans, and east-central Europe which Pollard collectively terms the "outer periphery"—remained mainly outside the fold of industrial capitalism and would only be penetrated over the next 50 years to various degrees.

Although a complex interplay of variables contributed to the peripheral status of these regions, an important factor was simply that its entrepreneurs and governments were imitators rather than innovators; they adopted the technologies and forms of organization that had served the pioneer regions well despite the reality of very different economic geographic settings. Railways provide a simple illustration. Rail networks in pioneer regions operated profitably by carrying regular passenger traffic as well as heavy bulk freight like coal, ore, and grain. Extending railway systems to regions that lacked an emerging industrial base and sufficient population density (e.g., in Ireland, southern Italy, Spain, and most of east-central Europe) invited heavy losses. The willingness of states to underwrite such losses reflects the potency of the railways as symbols of political (and economic) virility. What was not foreseen at the time, however, was that integrating national territories did not necessarily result in industrial development; rather, the penetration of the railways to peripheral regions tended to result in their specialization in a subordinate, agricultural role—a special case of Pollard's "differential of contemporaneousness."

Another important reason these regions remained on the periphery can be found in the nature of urban development within the later industrializing regions. In Britain, "inner" Europe, and "intermediate" Europe, a symbiotic relationship between urban and industrial development emerged with cities providing capital, labor, markets, access to transport systems, and a variety of agglomeration economies. In much of peripheral Europe, the "demonstration effect" of these events led to a very different relationship, largely because of the attitudes of the élite:

> Railways were laid to royal palaces, gas or water mains supplied a narrow layer of privileged classes … innovations intended for mass markets were misused for a narrow luxury market and either diverted resources, or led to burdensome capital imports. . . . Above all, the city became the gate of entry to new technology manufactures from abroad, spreading outward from Naples, Madrid, Budapest or St. Petersburg, to kill off native industry as unfashionable.
>
> (Pollard, 1981: 212)

In short, conspicuous consumption precluded import substitution and resulted in cities that inhibited rather than fostered industrial growth.

DISLOCATION AND DEPRESSION

In the first half of the twentieth century, two major wars punctuated the economic development of Europe. The disruption of the First World War was immense. The overall loss of life, including the victims of influenza epidemics and border conflicts which followed the war, amounted to between 50 and 60 million people. About half as many again were permanently disabled. For some countries, this meant a loss of 10–15 percent of the male workforce. In addition, material losses caused a severe dislocation to economic growth: Some estimates suggest that the level of European output achieved in 1929 would have been reached by 1921 had the war not interceded.

Economic dislocation in Europe was further intensified by several indirect consequences of the war. In terms of tracing the evolving economic geography of the core regions of the world, two of these were particularly important:

1. The *relative* decline of Europe as a producer compared with the rest of the world. Europe accounted for 43 percent of the world's production and 59 percent of its trade in 1913, compared with only 34 percent of production and 50 percent of trade in 1923. The main beneficiaries of this decline were manufacturers in the United States and Japan, and Latin America and the British dominions for primary production.
2. The redrawing of the political map of Europe. This transformation created 38 independent economic units instead of 26; 27 currencies instead of 14; and 20,000 extra kilometers of national boundaries. The corollary of these changes was a severe dislocation of economic life, particularly in east-central Europe: Frontiers separated workers from factories, factories from markets, towns from traditional food supplies, and textile looms from spinning sheds and finishing mills; while the transport system found itself only loosely matched to this new political geography.

Just as European economies had adjusted to these dislocations, the **stagflation** crisis of 1929–1935—the Great Depression—created a further phase of economic damage and reorganization throughout Europe. It should be emphasized, however, that the effects of the Depression varied considerably across the various sectors of the economy and from one region to another. The image of the 1930s depends very much on whether one focuses attention on Jarrow or Slough, Bochum or Nice, Glasgow or Geneva.

Meanwhile, the coherence of the European economic world began to disintegrate as individual countries attempted to protect their industries with import quotas and restrictions, currency manipulation, and exclusionary trade agreements. The result was a substantial fall in trade, both in absolute terms and as a proportion of output, with the United States and Japan, once again, reaping the reward.

SECOND WORLD WAR AND RECOVERY

The Second World War resulted in a further round of destruction and dislocation. The total loss of life in Europe was 42 million, two-thirds of whom were civilian casualties. The German

Table 5.1 Growth rates in Europe

	Average annual per capita growth rate of real output	
	1913–1950	1950–1970
Austria	0.2	4.9
Belgium	0.7	3.3
Denmark	1.1	3.3
France	1.0	4.2
West Germany	0.8	5.3
Greece	0.2	5.9
Ireland	0.7	2.8
Italy	0.8	5.0
Netherlands	0.9	3.6
Norway	1.8	3.2
Portugal	0.9	4.8
Spain	−0.3	5.4
Sweden	2.5	3.3
Switzerland	1.6	3.0
United Kingdom	0.8	2.2
Western European average	**1.0**	**4.0**

Source: Adapted from Pollard (1981: 315, Table 9.2)

Box 5.1 Core and periphery in Europe

The cumulative effects of the differential impact of successive waves of industrialization and reorganization have often been interpreted in terms of the core and periphery; the former accumulating capital and economic power, and the latter encountering limitations (natural or imposed) in its quest for economic development. The relative affluence of these core regions is shown in stark fashion in Figure 5.3. Even though the cost of living is notoriously high around London, Paris, and Milan, these regions enjoy a prosperity that is well above the overall level (indexed at 100 for the 27-member European Union). Affluent outliers have also emerged in Southern Ireland, Denmark, northeast Scotland, the Basque country of

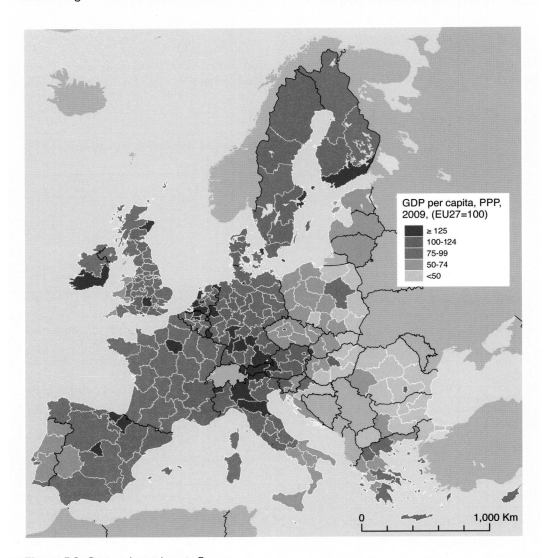

GDP per capita, PPP, 2009, (EU27=100)

≥ 125
100-124
75-99
50-74
<50

0   1,000 Km

Figure 5.3 Core and periphery in Europe

Source: Adapted from Eurostat (2012a: 21, Map 1.2)

Spain, and the capital regions of Finland and Sweden. In contrast, Bulgaria, Romania, Latvia, and much of Poland and the Czech Republic are decidedly peripheral.

The European core has variously been interpreted as the triangular region defined by Lille-Bremen-Strasbourg (the "heavy industrial triangle"), as an axial belt stretching between Boulogne and Amsterdam in the north and Besançon and Munich in the south, as a T-shaped region whose stem extends down the Rhine to Stuttgart. Such definitions can be confusing in their variability, but their main weakness is that they overlook the *interdependence* that exists between core and periphery. It is therefore more satisfactory to think in terms of a core consisting of a number of linkages or *flows* (e.g., capital, migrants, taxes, tourists, consumer goods, etc.) that bind core and periphery together, reinforcing their unequal but symbiotic relationship.

occupation of continental Europe involved ruthless exploitation. By the end of the war, France was below 50 percent of its pre-war level of living and had lost 8 percent of its industrial assets. The United Kingdom had lost 18 percent of its industrial assets (including overseas holdings), and the Soviet Union lost 25 percent. Similarly, Germany lost 13 percent of its assets and ended the war with a level of income *per capita* that was less than 25 percent of the pre-war figure.

After the war, the political cleavage between eastern and western Europe (which resulted from the imposition of what Winston Churchill called the Iron Curtain along the western frontier of Soviet-dominated territory) further eroded the coherence of the European economy and, indeed, of its economic geography. Ironically, this cleavage sparked a surprisingly rapid economic recovery in western Europe: The United States, learning from the mistakes of political leaders in the post-World War I period and believing that poverty and economic chaos would foster communism, embarked on a massive program of aid under the Marshall Plan. This pump-priming action, together with the pent-up backlog of demand in almost every sphere of production, provided the basis for a remarkable recovery. By the early 1950s most of Europe had exceeded pre-war levels of prosperity. By the early 1960s European central banks were in a position to step in, when necessary, to support the U.S. dollar. As Table 5.1 illustrates, growth rates throughout western Europe surged forward to impressive levels.

5.3 FORDISM AND NORTH AMERICAN INDUSTRIALIZATION

The emergence of the United States as a dominant component of the world economy can be attributed to a number of factors including its vast supply of natural resources; a large and rapidly growing market and labor force (thanks, in part, to its open-door policy toward immigration in the late nineteenth and early twentieth centuries); and sufficient size coupled with a capitalist ethos that bred giant corporations with significant research budgets and an incentive to institutionalize the innovation process in a way that has been generally foreign to European industry.

Within the United States the evolving pattern of spatial organization can be interpreted in terms of the interaction of (1) the geography of resources, (2) the introduction of major technological innovations (particularly in transport), and (3) movements of population. Thus:

Major changes in technology have resulted in critically important changes in the evaluation or definition of particular resources on which the growth of certain urban regions had previously been based. Great migrations have sought to exploit resources—ranging from climate or coal to water or zinc—that were either newly appreciated or newly accessible within the national market. Usually, of course, the new appreciation or accessibility had come about, in turn, through some major technological innovation.

(Borchert, 1967: 324)

The history of the development and evolution of the U.S. economy found its clearest geographic expression in changing patterns of urbanization. At the time when Europe was experiencing the first waves of industrialization, the spatial organization of the North American economy was focused on the gateway ports of the Atlantic seaboard, each of which controlled a limited hinterland where the economy was dominated by the production of agricultural staples for export to Europe and the consumption of manufactured goods imported from Europe. From the end of the eighteenth century, however, the North American economy began to break loose from this dependent relationship. Within 100 years, it became the dominant component of the world economy with a closely integrated but highly differentiated urban system.

Several factors related to political independence contributed to this metamorphosis:

1. Political integration under a federal system provided an important stimulus for forging economic links between the component parts of the old colonial system.
2. U.S. capital financed a relatively high proportion of investments which meant that less of the profits "leaked" back to Europe.
3. The federal system also stimulated a proliferation of government employment as every county seat and state capital developed the infrastructure of democracy.
4. Territorial expansion provided a large, rich, and diverse resource base.

Both urban and economic development in this period, however, was constrained by the relatively primitive transportation system of the "sail and wagon" epoch. It was not until the 1840s, when the second technology system of industrial capitalism (i.e., coal, steel, steam, and railways) began to be exploited that U.S. industrialization took off.

FORDISM, TAYLORISM, AND REGIONAL ECONOMIC CHANGE

Although the template of economic geography in the United States had been established largely by 1920, the landscape of industrial capitalism continued to evolve, particularly with the arrival of truck and automobile transportation between 1920 and 1940. Road and air travel, along with improvements in communications, increased the *capacity* and *efficiency* of the economy, and facilitated the functional integration of businesses and regions at an unprecedented pace. The 1920s marked the New Economic Era, a period when consumerism and boosterism edged aside the liberal reactions to industrialism.

The large companies based in major metropolitan centers found themselves ideally situated to capitalize on the increased capacity and efficiency of the economic system. They also exploited new principles of economic organization based on a more intensive division of labor, assembly-line production, and the principles of "scientific" management (also known as **Taylorism** after the mechanical engineer Frederick Winslow Taylor). The resulting increases in efficiency and productivity meant that many goods could be mass produced at low prices for mass markets. The consequent combination of mass production and mass consumption is generally referred to as **Fordism**, after Henry Ford, the automobile manufacturer who led the way in implementing these changes.

Box 5.2 The growth of the U.S. manufacturing belt

The acceleration of industrialization in the United States in the 1840s was in part the result of the diffusion of industrial technology—particularly the wider industrial application of steam and the accompanying changes in the iron industry—and methods of industrial and commercial organization from the hearth area of the Industrial Revolution in Europe. In addition, the demand for foodstuffs and other agricultural staples in North America and abroad stimulated the growth of industrial capitalism as farmers sought to increase productivity through mechanization and the use of improved agricultural implements. Increasing agricultural productivity, in turn, helped to sustain the growing numbers of immigrants from Europe, thus allowing them to be channeled into industrial employment in the mushrooming cities of the United States.

The development of the railway system played a central role in the evolution of this new economic order. Initially, the railways were complementary to the waterways as competitive long-haul carriers of general freight. By the end of the "iron horse" epoch, the railway network had not only realigned the economic system but also extended it to a continental scale. In 1869 the railway network reached the Pacific when, at Promontory, Utah, the Union Pacific railway, building west from Omaha, met the Central Pacific railway, building east from Sacramento. By 1875 intense competition between railroad companies had begun to open up the western prairies as far as Minneapolis-St. Paul and Kansas City. The significance of this transformation was profound:

> Not only did this permit American enterprise to exploit fully the commercial advantages and scale economies of large, diversified natural resources and of the revolutionary technologies evolved in those decades, but it generated rapid, large-scale functional and spatial concentration of finance and management unimpeded by world events, creating a *transcontinental* business mentality. Wide spatial separation of major resources, cities and markets, and adjacency to the easily penetrated Canadian economy all induced mental thresholds for thinking *intercontinental* once imported resources and markets overseas became a necessary ingredient to sustain business activity at home, especially during and after the Second World War.
>
> (Hamilton, 1978: 26; emphasis added)

In short, the railways can be seen as the catalyst that allowed regional economies to develop into a continental economy that stood poised to become the leading component of the world economic system.

Meanwhile, the westward extension of the railways inevitably affected the fortunes of the inland gateway cities. Buffalo and Louisville, for instance, experienced slowed rates of growth and came increasingly to rely on more diversified regional functions. Further west, St. Paul and Kansas City expanded rapidly to become major wholesaling depots. The development of improved transportation networks also led to adjustments in spatial organization within the northeast where fierce competition between the railways and water-borne transport, coupled with equally fierce rivalry between neighboring cities, led to a marked increase in *intra*-regional trade.

In essence, the consolidation of the Manufacturing Belt as the continental economic heartland was the result of initial advantage. With its large markets, well-developed transport networks, and access to nearby coal reserves, it was ideally placed to take advantage of the general upsurge in demand for consumer goods, the increased efficiency of the telegraph system and postal services, the advances in industrial technology, and the increasing logic of economies of scale and external economies that characterized the late nineteenth century. The overall effect was twofold:

1. Individual cities began to specialize as producers geared themselves toward national rather than regional markets:

> Between 1870 and 1890, advances in milling technology and concentration of ownership supported the emergence of Minneapolis as a milling center. Furniture for the mass market centralized in fewer, larger plants using wood-working machinery. . . . The rise of national brewers between 1880 and 1910 is an example of national market firms encroaching on local–regional firms. The brewers in Milwaukee and St. Louis achieved economies of scale in manufacture, used production innovations such as mechanical refrigeration, and capitalized on distribution innovations made possible by the refrigerated rail car and an integrated rail network.
>
> (Meyer, 1983: 160)

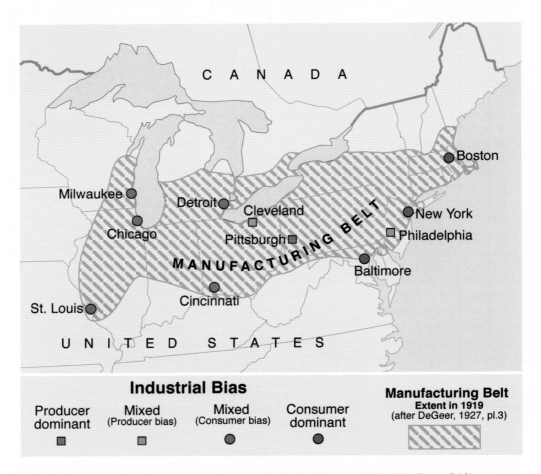

Figure 5.4 The American Manufacturing Belt in 1919 (after Conzen, 1981: 340, Figure 9.13)

Source: Based on Knox *et al.* (1988: 117, Figure 5.1)

Similarly, musical instrument manufacture and men's clothing emerged as specialties in Boston; meat packing, furniture manufacture, and printing and publishing in Chicago; coach-building and furniture manufacture in Cincinnati; textile manufacture in Philadelphia; and so on.

In smaller cities, specialization was often much more pronounced, as in the production of iron, steel, and coach-building in Columbus; furniture in Grand Rapids; agricultural implements in Springfield; and boots and shoes in Worcester. Overall, there emerged a three-part segmentation of the Manufacturing Belt (see Figure 5.4) with a heavy bias towards consumer goods production in the ports of Baltimore, Boston, and New York; a producer goods axis between Philadelphia and Cleveland; and a western cluster of rather less specialized consumer-oriented manufacturing cities.

2. This specialization provided the basis for increasing commodity flows between individual cities, thus binding the Manufacturing Belt together. These linkages, in turn, generated important **multiplier effects** through wholesaling, finance, warehousing, and transportation, adding to the cumulative process of regional industrial growth and increasing the region's comparative advantage. These advantages meant that the Manufacturing Belt was able to attract a large proportion of any new industrial activities with large or national markets, thus stifling the chances of comparable levels of industrialization in late developing regions.

Of course, other regions also industrialized; however, they failed to match the scale and the intensity of the Manufacturing Belt. Instead, these regions supported an array of locally oriented manufacturers as well as some nationally oriented activities based on particular local advantages or raw materials, and they rarely attracted manufacturers of producer goods for the national market.

This new form of economic organization required ever larger companies. A flurry of company mergers soon transformed the business structure of the economy, resulting in a relatively small number of very powerful corporations that stood poised to dominate the economy. By 1920, a mere 1 percent of firms accounted for over 30 percent of all jobs and nearly 50 percent of the country's production:

> The Captains of Industry were clearly in charge. Across the country, territorial communities watched effective control over local production slip out of their grasp. Political power came to focus on the national level of territorial integration which, for the time being, effectively bounded the operation of most businesses.
>
> (Friedmann and Weaver, 1979: 22)

However, the new form of economic organization associated with this new economic era fostered some serious problems. Mechanized agriculture became so "overproductive" that commodity prices plummeted; while the industrial market became unstable as a result of the labyrinth of holding companies that had been created. In October 1929 the stock market collapsed, triggering the Great Depression during which millions of workers lost their jobs. The regional division of labor that had emerged over the previous 50 years meant that some areas suffered particularly acute social and economic problems. The political response took the form of a **New Deal** in which the central government assumed much greater responsibility for overall economic growth and regional economic well-being, a system of public macroeconomic management that came to be known as **Keynesianism** after the doctrine of British economist John Maynard Keynes. In these changes, we can see that a significant shift occurred in the regulatory framework associated with the Fordist form of economic organization. The focus of economic policies shifted from regulation of the supply side to active management of the demand side.

With the outbreak of war in Europe in 1939, the entire North American economy entered a phase of accelerated growth; and in the aftermath of the war the United States and Canada emerged not only with stronger and more efficient industries but also with new technologies and control over new international markets. The United States, in its more outward-looking role as leader of the capitalist world, dictated the pattern of world affairs through the terms of the Marshall Plan, its control of the Organization for European Economic Cooperation (OEEC) (forerunner of the OECD and established to administer U.S. and Canadian postwar reconstruction aid in Europe under the Marshall Plan), and the Bretton Woods Agreement (which established a new framework for international economic relations). By 1960 GNI *per capita* stood at $2,513 in the United States compared with $1,909 in Canada, $1,678 in Sweden, $1,259 in the United Kingdom, $1,200 in West Germany, $1,193 in France, and a paltry $421 in Japan. By this time, however, the economic geography of North America had begun to respond to the imperatives of the new technology systems and forms of economic organization of globalized capitalism, themes we explore in detail in Chapter 7.

5.4 JAPANESE INDUSTRIALIZATION: TWO ECONOMIC MIRACLES

The early rise of the Japanese economy to join the core of the world economy represents a major achievement, and it poses some important questions in relation to the theory and reality of economic organization and spatial change. In particular, how was it that a relatively resource-poor country like Japan was successful in industrializing so soon, when resource-rich regions elsewhere in Asia and in Latin America were not?

Broadly speaking, the answer lies in the fact that the Japanese economy, although organized along feudal lines until well into the nineteenth century, was **autonomous**. It had not been penetrated by the capitalism of the other core regions. Moreover, the transition from feudalism to capitalism took place as a deliberate attempt to preserve national political and economic autonomy.

Even though Japan was "lucky" in not having been politically and economically subordinated, its path forward via industrialization was still obstructed by the other core regions' pre-emption of the technology, infrastructure, and capital necessary for industrial development. How were these obstacles overcome? The answer, again in general terms, lies in the combination of a proto-industrial base and an aggressive expansionist strategy that included flooding overseas markets with cheap products and copying and adapting western technology—a strategy achieved at the expense of an authoritarian government, widespread exploitation, and acute regional disparities.

THE FIRST MIRACLE: FROM FEUDALISM TO INDUSTRIAL CAPITALISM

Japan's transition from feudalism to industrial capitalism can be pinpointed to a specific year—1868—in which the feudal political economy of the Tokugawa regime was toppled by the restoration of imperial power. For over 200 years as the industrial system was developing in the Western Hemisphere, the Tokugawa regime attempted to sustain traditional Japanese society. The patriarchal government excluded missionaries, banned Christianity, prohibited the construction of ships larger than 50 tons, closed Japanese ports to foreign vessels (Nagasaki was the single exception), and deliberately suppressed commercial enterprise. At the top of the feudal hierarchy were the nobility (the *shogunate*), the barons (*daimyos*), and warriors (*samurai*). Farmers and artisans represented the productive base exploited by these ruling classes. Only outcasts and prostitutes ranked lower than merchants.

In terms of spatial organization, the economy was built on a closed hierarchy of castle towns, each representing the administrative base of a local *shogun*. The position of a town within this hierarchy was dependent on the status of the *shogun*, which, in turn, was related to the productivity of the agricultural hinterland. As a result, the largest cities—which became the foundations for subsequent economic growth—emerged among the alluvial plains and the reclaimed lakes and bay heads of southern Honshu. At the top of the hierarchy was Edo (now Tokyo) which the Tokugawa regime had selected as its capital in preference to the traditional imperial capital of Kyoto. Bloated by soldiery, administrators, and the entourages of the nobility in attendance at the Tokugawa court, Tokyo reached a population of around 1 million people by the early nineteenth century. Kyoto and Osaka were the next largest cities with populations of between 300,000 and 500,000 residents. They were followed by Nagoya and Kanazawa with around 100,000 residents each.

With cities of this size, suppressing commerce and preventing the breakdown of the traditional political economy proved to be difficult. As in feudal Europe, peasants fled the countryside in increasing numbers in response to a combination of taxation, technological improvements in agriculture, and the lure of the relative freedom and prosperity of cities. At the same time, cities evolved into important centers of domestic manufacture: Nodes of proto-industrial development that served as the platform for Japan's subsequent development. Meanwhile, prolonged peace had reduced the influence and the affluence of the *samurai*, drawing increasing numbers of them towards commercial and manufacturing activities. So:

> [F]ormer peasants mingled with former warriors in secular occupations coordinated as much by market forces as by feudalistic regulations. A class-based commercial society thus developed despite the efforts of the Tokugawa leaders to maintain the pre-industrial, status-oriented society of old Japan.
>
> (Light, 1983: 158)

By the early nineteenth century, Japan had moved into a period of crises characterized by famines and peasant uprisings and the ineptitude of an introverted and self-serving leadership. In 1853 U.S. Admiral Perry arrived in Edo Bay to "persuade" the *shogunate* to open Japanese ports to trade with the United States and other foreign powers. This neocolonialist threat galvanized feelings of nationalism and **xenophobia** and precipitated a period of civil war among the *shogunate*. The outcome was the restoration of the Meiji imperial dynasty in 1868 by a clique of *samurai* and *daimyo* who considered industrialization a necessity to maintaining national independence.

Under the slogan "National Wealth and Military Strength," the new élite of ex-warriors set out to modernize Japan as quickly as possible. A distinctive feature of the process was the very high degree of state involvement. Successive governments intervened to promote industrial development by fostering capitalist monopolies (*zaibatsu*). In many instances, whole industries were created from public funds and, once established, were sold off to private enterprise at less than cost. Given the prominent concerns over national security, building manufacturing capabilities and capacity in iron and steel, shipbuilding, and armaments played a prominent role in the early phase of Japanese modernization. The state facilitated the importation of industrial technology and equipment, and foreign advisers (chiefly British) supervised the initial stages of development. Meanwhile, the state invested a considerable amount of resources in highways, port facilities, the banking system, and public education in an attempt to "buy" modernization. Similarly, government financed the construction of the railway system (under British direction) before being sold to private enterprise.

The Japanese financed this modernization by harsh taxes on the agricultural sector. As a result, a sharp polarization emerged between the urban and the rural economies; one that was characterized by the impoverishment of large numbers of peasant farmers. Yet the more productive components of the agrarian sector contributed significantly to Japanese economic growth. Improved technology, better seedstock, and the use of fertilizers provided an increase of 2 percent per annum in rice production during the last quarter of the nineteenth century and the first part of the twentieth century, thereby helping to feed the growing nonfarm sector without great dependence on food imports. It is important to note that population growth did not absorb these increases in agricultural productivity as has been the case for most late-comers to industrialization. The Japanese demographic transition arrived after increases in agricultural productivity had helped to finance an emergent industrial sector but in time to provide an expanding labor force and market for industrial products.

Several other factors helped to foster rapid industrialization in Japan in the late nineteenth and early twentieth centuries. One was the cultural order that allowed the Japanese to follow government leadership and accept new ways of life: A recurring theme in modern Japanese economic history. Another was the success of educational reforms: By 1905, 95 percent of all children of school age received an elementary education. Third, Japanese sericulture (silk production) provided the basis for a lucrative export trade that helped finance expenditure on overseas technology, materials, and expertise. It has been estimated that, between 1870 and 1930, the raw silk trade alone was able to finance as much as 40 percent of Japan's entire imports of raw materials and machinery. Finally, and most important, were the benefits deriving from military aggression. Naval victories over China (1894–1895) and Russia (1904–1905), and the annexation of Korea (1910) not only provided expanded markets for Japanese goods in Asia but also provided indemnities for the losers (which paid for the costs of conquest) and stimulated the armaments industry, shipbuilding, and industrial technology and financial organization in general.

By the early 1900s a broad spectrum of industries had been successfully established. Most were geared towards the domestic market in a kind of pre-emptive import-substitution strategy. The textile industry, however, had already begun to establish an export base. Unable to compete with western countries in the production of high-quality textiles, the Japanese concentrated on the production of inexpensive goods, competing initially with western producers for markets in Asia. Their success derived largely from labor-intensive processes in which high productivity and low wages were maintained through a combination of exhortations to personal sacrifice in the cause of national independence and strict government suppression of labor unrest.

JAPAN ADVANCES

During the First World War Japan became a major supplier of textiles, armaments, and industrial equipment on world markets. It nearly doubled its merchant marine tonnage and established a balance of payments surplus. Between 1919 and 1929 this position was consolidated under government sponsorship. Steel manufacturing, engineering, and textiles were further developed, and aircraft and automobile industries were established. Meanwhile,

Japanese innovations began to emerge, weakening its dependence on western technology and providing an important competitive advantage. This pattern of progress was halted, however, by the stagnation of international trade that followed the stock market collapse of 1929. Once again, state intervention provided a critical boost. A massive devaluation of the yen in 1931 allowed Japanese producers to undersell competitors on the world market, while a Bureau of Industrial Rationalization was established to increase efficiency, lower costs, and weed out smaller, less profitable concerns.

Although these interventions helped to sustain Japanese industrialization and improve Japan's overall economic independence, they led directly towards crisis. Western governments—particularly the United States—resisted the purchase of Japanese goods. In Japan, the austerity that resulted from devaluation and rationalization precipitated social and political unrest. In response, the government increased military expenditure and adopted a more aggressive territorial policy. In 1931 the Japanese army advanced into Manchuria and created a puppet state. In 1936 a military faction gained full political power and, declaring a Greater East Asian Co-Prosperity Sphere, engaged in full-scale war with China the following year. In 1939 Japan attacked British colonies in East Asia, and by 1940 the Japanese "had become heavily committed to an industrial empire based on war. In that year, 17 per cent of the entire national output was for war purposes" (Kornhauser, 1989: 119).

By this time, as the rest of the world quickly realized, Japan had attained the status of an advanced industrial economy. The military leaders overplayed their hand, however, by attacking the United States. With defeat in 1945, Japanese industry lay in ruins. In 1946, output was only 30 percent of the pre-war level; and the United States, having weakened the power of the *zaibatsu* and imposed widespread social and political reforms, was set to impose punitive reparations.

THE SECOND MIRACLE: POSTWAR RECONSTRUCTION AND GROWTH

Within five years, the Japanese economy had recovered to its pre-war levels of output. Throughout the 1950s and 1960s the annual rate of growth of the economy held at around 10 percent compared with growth rates of around 2 percent per annum in North America and western Europe. After beginning the postwar period at the bottom of the international manufacturing league table, Japan rose to join the other developed countries by the 1980s (see Figure 5.5). By 1980, for example, Japan had outstripped, even in *absolute* terms, all the major industrial core countries in the production of ships, automobiles, and television sets, and only the Soviet Union produced more steel.

Explanations of this "miracle" have identified a variety of contributory factors. One of the most important was a reversal in United States policy. Cold War strategy, in response to China's pursuit (in 1949) of a communist path to development, dictated that its initial punitive stance toward Japan would be replaced by massive economic aid in order to create a bastion against the spread of communism in East Asia. In return for providing an offshore bastion in the containment of the Soviet Union during the early Cold War, Japan was allowed to re-enter the world economy on terms that were favorable to its economic growth. For example, high Japanese tariffs were tolerated at the same time that the United States sponsored a world economic order based on tariff reduction or removal. The Korean conflict (1950–1953) helped to reinforce this logic and, at the same time, stimulated the Japanese economy through U.S. expenditure on Japanese supplies and military bases.

Once under way, the reconstruction of the Japanese economy drew on some of its previously established advantages: A well-educated, flexible, loyal, and relatively cheap labor force; a large

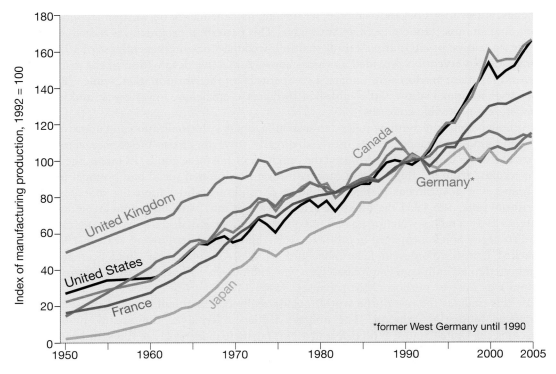

Figure 5.5 Index of manufacturing production for selected countries

Source: Based on online data from U.S. Department of Labor, Bureau of Labor Statistics, Table 3, Output, *International Comparisons of Manufacturing Productivity and Unit Labor Cost Trends, 2012* http://www.bls.gov/fls/#productivity

national market with good internal communications; a good geographical situation for trade within Asia; a high degree of cooperation between industry and government; and a model of industrial organization—derived from the *zaibatsu*—large enough to compete with the transnational corporations of Europe and North America. Several new factors also helped to metamorphose reconstruction into spectacular growth:

1. Exceptionally high levels of personal savings (e.g., 19.5 percent of personal disposable income in 1980, compared with 4.7 percent in the United States), which helped to fund high levels of capital investment.
2. The acquisition of new technology. Between 1950 and 1969 Japan acquired, for around $1.5 billion in royalties and licenses, a body of thoroughly tested U.S. technology that had cost the United States $20 billion *per year* in research and development (R&D). More recently, Japanese investment in domestic R&D has overtaken (in relative terms) that of Europe and North America, providing important advantages in production technology and product design. Overall, Japanese investment in technology in 1980 amounted to 6 percent of her industrial turnover, compared with 1 percent in the United States. Japanese expenditure on R&D as a percentage of GDP has also consistently outpaced investment by Europe and the United States. World Bank data indicate that during the ten-year period ending in 2010, Japan's average annual investment in R&D as a percentage of GDP was 3.3 percent, a level that far exceeded the investment of the United States (2.7 percent), Germany (2.6 percent), France (2.2 percent), and the United Kingdom (1.8 percent).

3. New means of government support. The construction of a rigid and sophisticated system of import protection—tariff and nontariff barriers—shielded domestic markets from overseas competition. A multitude of tax concessions and provisions for financing investments through the Japan Development Bank also fostered the growth of domestic industry. Most important of all, however, was the orchestration of industrial growth by the Ministry of International Trade and Industry (MITI) (reorganized as the Ministry of Economy, Trade and Industry (METI) in 2001). MITI identified key recovery sectors (e.g., steel and shipbuilding) and potential growth sectors (e.g., automobiles, electronics, and computers), and facilitated their development by providing financing, ensuring protection from foreign competition, subsidizing technological development, and arranging corporate mergers. MITI also organized Japanese corporations into business networks (known as *keiretsu*) and established favorable trade, technology, and fiscal policies to help Japanese industry compete successfully in the world economy.

Box 5.3 Regional dimensions of Japanese industrialization

The pace and weight of postwar industrialization have dramatically rearranged the economic landscape that existed at the close of the Tokugawa period. In many ways, the changes wrought on the Japanese landscape parallel those that occurred in response to the industrialization of Europe and North America. Existing urban centers (the castle towns) grew differentially according to their adaptability as regional industrial, commercial, or administrative centers; while new kinds of specialist settlement—ports, mining towns, heavy manufacturing towns and transport centers—emerged and grew rapidly to become major nodes of urbanization. Similarly, the expansion and diversification of the industrial economy imposed a progressive spatial division of labor. The logic of agglomeration economies and economies of scale as well as the strong encouragement of the government through MITI resulted in regional specialization. Within this overall transformation, large company towns emerged as a distinctive feature that reflected the unique role of the *zaibatsu* in Japanese industrialization. The early leaders among the *zaibatsu*—Mitsui, Mitsubishi, and Sumitomo— inevitably came to dominate their host cities; while later established *zaibatsu*, as well as some of the corporate giants spawned by postwar growth, sponsored new company towns in newly industrializing regions (e.g., the city of Hitachi).

Perhaps the most distinctive facet of the geography of industrialization in Japan was the sheer intensity of development crammed into a rather limited amount of suitable land. The importation of raw materials and the manufacturing and export of finished products fueled the economic miracle of the 1960s. As a result, Japan's resurgent industries flourished best in coastal locations near deep-water ports. The megapolitan Pacific Corridor between Tokyo and Kobe is the embodiment of these developments. It developed into the core region of modern Japanese industrialization because it had several deep-water ports, large pools of skilled labor, and a relatively large amount of flat land. The entire region, known as the Tokaido Megalopolis, is comparable in size and scope to the megapolitan area in the United States that stretches from Boston through New York and Philadelphia to Washington. The Tokaido Megalopolis contains more than 50 million people and accounts for more than 80 percent of Japan's total GDP. In 2000 Osaka alone accounted for a GDP that was greater than those of all but eight countries in the world.

Transportation also played a key role in the successful development of the Pacific Corridor. Although port facilities were a precondition for the region's success, inferior connections within the region hampered manufacturers and suppliers that sought to exploit agglomeration economies and restricted the movement of workers and consumers. In response, the Japanese government undertook a massive program of infrastructure investment. The showpiece of this program is the Shinkansen railway system. Opened in 1964 to coincide with the Olympic Games in Tokyo, the bullet trains of the Shinkansen transformed the Pacific Corridor into a daily commuter belt.

Cotton, silk, other textiles, toys, glass, and porcelain dominated the Pacific Corridor's pre-war industrial base. These industries remain in the region but have been dwarfed by the growth of iron and steel, heavy metal products and machinery, shipping and shipbuilding, petrochemicals, paper products, ceramics, automobile and truck manufacturing, cameras, scientific instruments, and electrical and electronic goods of all kinds. Tokyo has grown into a world city with a banking and financial sector that compares favorably to London and New York. The population of Tokyo in October, 2010, was 13.2 million, but the Tokyo-Yokohama urban area has a population in excess of 37 million, making it the largest megacity in the world.

The growth of the Pacific Corridor has inevitably brought problems of crowding, congestion, environmental pollution, and ground subsidence. Additionally, the concentration of economic activity in the Pacific Corridor has resulted in a relative lack of development elsewhere. So Japan is characterized, like Europe and North America, by a center–periphery pattern. In the Japanese case, the periphery consists of northern Hokkaido, Honshu, Kyushu, and Shikoku. Like peripheral regions within older core countries, they have experienced the **backwash effects** of metropolitan development: Selective outmigration, restricted investment (both public and private), and limited employment opportunities. In addition, much of the periphery has a climate that most Japanese find severe, which compounded feelings of deprivation and remoteness.

More recently, the high costs of operating in Japan have begun to weigh on Japanese corporations, straining their allegiance to the nationalist project of economic development. Many larger Japanese corporations have moved production facilities elsewhere in East and Southeast Asia in search of lower production costs and expanding markets. METI attempted to counter the consequent loss of Japanese capital and technology by developing a Technopolis program that sought to establish the infrastructure necessary to lure Japanese capital to domestic high-tech industries. However, METI no longer has direct influence over Japanese corporations, nor do these corporations decide their strategies primarily within the framework of Japan's economic interests.

This decoupling of the systematic interdependence between Japanese government and industry means that places and regions in Japan have become much more interdependent with places and regions elsewhere. It also means that it has become increasingly difficult to sustain the system of lifelong tenure for workers in large corporations, a feature that distinguished Japanese employment practices for decades. Not surprisingly, people's willingness to work long hours and defer consumption in the cause of national economic development has also declined. This unraveling of Japan's successful system of economic development has been reflected in a series of recent economic and political crises. It is also beginning to be reflected in shifting values and lifestyles. Traditional patriarchal values and nationalistic bureaucratic indoctrination have much less meaning for the generation that has grown up in affluence.

This second economic miracle was remarkable not only for its success in terms of economic performance; it reflected a unique path to development, one that combined economic growth with income distribution. Real wages (that is, the effective purchasing power of wages) rose substantially, while income inequality was reduced to one of the lowest levels in the world. Equally remarkable was the interdependence of government and industry, characterized by some as "Japan, Inc." Also important was the exceptional degree of social stability and management–labor cooperation during this period of rapid change. This stability and cooperation, like the interdependence of government and industry, reflected Japanese nationalism and the commitment of the Japanese people to rebuilding the country. The same sense of national identity and purpose fostered people's adherence to traditional values and lifestyles, their willingness to work many more hours than their European and U.S. counterparts, and their willingness to defer consumption, which provided a pool of savings that could be invested in Japanese industry.

5.5 EMERGENCE OF "ORGANIZED" CAPITALISM

The development of the industrial economies of the tri-nodal core brought with it a number of important changes in the nature of economic, social, political, and cultural relations that became woven into the urban and regional landscapes of the industrial core regions. Collectively, these changes characterize what has been called "organized" capitalism. Its principal features include (Lash and Urry, 1987: 3–4):

1. The concentration and centralization of industrial, banking, and commercial capital as markets became increasingly regulated; the increased interconnectedness of finance and industry; the proliferation of cartels.
2. The emergence of extractive and manufacturing industry as the dominant economic sector.
3. The concentration of industrial capitalist relations within relatively few sectors and a small number of countries.
4. The expansion of empires and the control by the core economies of markets and production in overseas settings.
5. The increasing separation of ownership from organizational control and the expansion of complex hierarchies within companies.
6. The rise of a new intelligentsia (managerial, scientific, technological) and of a bureaucratically-employed middle class.
7. The emergence of "modernism," which glorified the sciences and technical rationality, a machine- and future-oriented aesthetic, and a nationalistic frame of reference.
8. The growth of collective organizations in the labor market: Trade unions, employers' associations, nationally organized professions, etc.
9. Regional economic specialization.
10. The dominance of particular regions by large metropolitan areas.
11. Increasing intervention by the state in social conflicts; the development and expansion of welfare policies; and, as a result, greater articulation between corporations, collective organizations, and governments.

Similar to other transitions described in previous chapters, organized capitalism did not emerge in the same way in every location. Three main factors determined the timing of these changes and the extent that a particular economy developed the characteristics of "organized" capitalism:

- *Timing*: Economies that transitioned earliest tended to be relatively less "organized" because late industrializers required greater concentration and centralization of capital to compete effectively with established industrial economies.
- *Institutional transformation*: Whether and to what extent preindustrial institutions survived into the capitalist period:

  > Britain and Germany became more highly organized capitalist societies than France and the United States: this is because the former two countries did not experience a "bourgeois revolution" and as a result, guilds, corporate local government, and merchant, professional, aristocratic, university and church bodies remained relatively intact.
  >
  > (Lash and Urry, 1987: 5)

- *Size*: Smaller countries could compete effectively in the international arena in a relatively small number of sectors. Directing resources toward these activities required greater coordination between the state and industry.

U.S. capitalism was organized fairly early on at the top (e.g., through the concentration of industry; increasing inter-articulation of banks, industry, and the state; and the formation of cartels), but very late and only partially at the bottom (e.g., the development of national trades union organizations, working-class political parties, and the welfare state). British capitalism, in contrast, was organized rather early at the bottom but late at the top; while French capitalism only came to be organized at top and bottom after the Second World War.

So the characteristics of organized capitalism came to be woven into the urban and regional landscapes of the core regions rather unevenly. Meanwhile, they came to represent not just a distinctive set of economic, social, political, and cultural relations, but also the context—that is, the preconditions—for further transformations of capitalism and the new economic landscapes that emerged with the onset of globalized capitalism.

CHANGING ROLE OF THE STATE

It is no accident that the rise of competitive capitalism and the evolution of industrial economies coincided with the emergence of the modern nation-state. Within Europe, the system of nation-states—once established in place of the earlier dynastic kingdoms and empires—fostered the economic, social, and political organization required by the Industrial Revolution. At the same time, competition between nation-states provided a strong incentive for technological innovation. It is important to bear in mind, however, that few nation-states were "natural" entities developed from distinctive cultural or philosophical bases. Rather, they were *constructed* in order to clothe, and enclose, the developing political economy of industrial capitalism. It follows that the process of building nation-states involved the resolution of successive crises that arose from the interaction of territory, economy, culture, and government.

One series of crises arose from the struggle to draw state boundaries that gathered populations with a common identity (or at least the potential for it) into cohesive units. This struggle involved states attempting to build **nations** from a diversity of peoples, and peoples with a common identity attempting to create autonomous **states**. The former has often been perpetrated by powerful regions or groups of neighboring territories. As a result, many **nation-states** came into being with inbuilt core–periphery contrasts and with sociopolitical tensions compounding economic differentials. We examine the reaction to these developments in Chapter 13.

Another series of crises arose from the increasing degree of organization required by capitalism. The evolution of the industrial core regions posed a succession of problems, which resulted in more state intervention in a greater variety of fields. The initial advantage gained by British manufacturers with the advent of the Industrial Revolution soon prompted businessmen elsewhere to realize that the old doctrine of laissez-faire and free trade primarily served the interests of the dominant economy. As a result, governments often assumed the role of protectors—through tariffs and quotas—against low-priced British imports.

Meanwhile, the advent of rail transportation mobilized the state in another sphere—investment in infrastructure. Problems of public health, working conditions, housing, and civil disorder induced further state involvement, as did the need to provide a stable price system for the successful operation of private industry, the need to manage the cyclical fluctuations of the industrial economy, and the need to improve the quality of the workforce and its managers through formal education. Of course, the effectiveness of government intervention depended on economic growth. Nevertheless, the wealthiest economies did not necessarily lead the way in state activity, since public expenditure is mediated through the complex arena of politics.

In detail, then, the development of state functions has been complex; however, two major trends can be identified:

1. The *centralization* of the functions of states, whereby local and regional activities have been rationalized into centralized national bureaucracies as the organization of government has attempted to keep up with the changing scale of economic organization. With increasingly powerful central bureaucracies, the power of politicians at both local and national levels has been constrained and this, in turn, has contributed to crises in the legitimacy of political institutions. One response has been the intensification of demands for the **regional devolution** of power by the representatives of communities in peripheral areas. Another has been the growth of forms of direct action in the shape of grassroots pressure groups. Both phenomena are explored in more detail in Chapter 13.
2. The dramatic expansion of the *public economy* as governments became increasingly drawn into the creation of welfare states. In most core countries, the public economy channels a vast amount of resources into everything from defense, health, education, and income security to transport, infrastructural development, and industrial investment. In 2010, following the global debt crisis, EU statistical office, Eurostat, data indicated that general government expenditure amounted to 50.3 percent of EU GDP with more than half of this amount dedicated to social protection and healthcare.

Indeed, the sheer magnitude of the public economy has come to blur the boundary between the public and the private sectors. Many governments have even been impelled to intervene to prevent the collapse of private corporations (examples have included the U.S. government's efforts to rescue Lockheed and Chrysler in the 1970s, the UK government's efforts in the 1980s to sustain British Leyland, as well as the U.S. government's massive bailout of financial institutions including AIG, Bank of America, and Citigroup in 2008 as well as the auto manufacturers General Motors and, yet again, Chrysler). Governments everywhere have also become the largest single consumers of the goods produced by private-sector enterprise. In short, the public economy is pervasive in its effects on economic well-being.

GEOGRAPHY OF THE PUBLIC ECONOMY

Most public-sector activity has a geographical expression. One of the most obvious examples is the deliberate bias of regional policy and planning, a topic we explore in Chapter 13. In this section, we emphasize the spatial bias—often unintentional—that results from other aspects of the public economy.

The geography of public finance is a complex subject that resists attempts for clear comparisons between countries. It is possible, however, to identify major categories of activity and to illustrate their spatial implications with specific examples:

1. *Wages*: The salaries of government employees including clerks, bureaucrats, and workers in the armed forces, education, public health, nationalized industries, law enforcement, and the courts. Expenditure on these salaries is localized in capital cities. In Washington, DC, for example, almost one-third of all earnings come from federal employment. In addition, government salaries can have an important impact in small or medium-sized cities that have been selected for specific functions—defense installations, for example, or decentralized branches of the bureaucracy (as in the UK government's relocation of social security offices from London to Newcastle-upon-Tyne).

2. *Transfers*: Payments to particular population groups (e.g., the elderly, the unemployed, families with dependent children) and particular industries (e.g., agricultural subsidies and guaranteed prices). These expenditures involve complex flows of monies and are geographically localized only in as much as the "target" populations and industries are localized.

3. *Subcontracting*: Purchasing and outsourcing the delivery of goods and services from businesses in the private sector. This includes a wide range of items—e.g., buildings, roads, dams, power stations, military equipment, office equipment, and publishing—which can have highly localized impacts. Defense expenditure, for example, generally tends to be sufficiently localized (because it is concentrated to a few large corporations such as General Dynamics, McDonnell-Douglas, United Technologies, Boeing, General Electric, Lockheed and Hughes Aircraft in the United States) to create significant multiplier effects. But the resultant spatial bias seems to bear no consistent relationship to core–periphery patterns or to dimensions of economic geography.

4. *Local government expenditure*: In many core countries, spending by local governments has approached the level of expenditure by central governments (although a large portion of local expenditure is always dependent on revenues provided by central governments in the form of grants and revenue-sharing funds). What is most striking about local government expenditure is that, after fulfilling their statutory obligations, local governments vary a great deal in the amount they spend and the categories of their expenditure. These differences reflect a complex relationship between local resources, local needs, and the local political climate.

In order to gauge the net impact of these expenditures, we have to set them against the geography of taxation. Such an exercise is far from straightforward, but we can illustrate the kind of biases that can emerge by reference to federal taxes in the United States. Around 40 percent of all federal revenues are derived from personal income taxes, with another 25 percent coming from taxes on pension trusts and a further 15 percent from taxes on corporate profits. To a large extent, therefore, the geography of federal revenues reflects the geography of income and economic health.

Yet the *structure* of the tax system can have less straightforward geographical implications. The federal tax breaks that are geared to encourage business investment in new equipment and machinery, for example, tend to benefit growth industries in growth areas. By the same token, studies of industrial location have shown that tax differences between states are not a significant factor in inducing industrial relocation.

Although it is difficult to quantify the local effect of these structural characteristics of the tax system, the magnitude of the net flows of monies between the federal government and individual states can be specified. Interstate differentials in such flows have tended to diminish during the postwar period as the federal tax system has become more uniformly progressive and as state variations in per capita incomes have narrowed. That said, the *marginal impact* of the flow of federal funds seems to have been much greater in the South and West. By improving the infrastructure of communications, transportation, sewage facilities, and energy, the federal government helped to establish the *preconditions* for the development of new industries in the Sunbelt. By direct and indirect investment in electronics research, semiconductors, computers, aeronautics, and scientific instruments in the South and West, the federal government enabled the Sunbelt to capture some of the most dynamic activities of the advanced capitalist economy.

INCREASING GLOBAL INTERDEPENDENCE

Although the industrial core regions have been the focus of this chapter, the historical process of industrialization must be viewed from a global perspective. Quite simply, the ascent of the industrial core regions could not have taken place without the foodstuffs, raw materials, and markets provided by the rest of the world. In order to ensure the availability of goods and access to markets, the industrial core countries vigorously pursued a second phase of colonialism and imperialism, creating a series of trading empires.

As soon as the Industrial Revolution gathered momentum, European colonial powers embarked on the inland penetration of mid-continental grassland zones in order to exploit them for grain or stock production—although the detailed pattern and timing of this exploitation was conditioned by innovations such as the railways, barbed wire, and refrigeration—resulting in the settlement of the prairies and the pampas in the Americas, the *veld* in Africa, and the Murray-Darling and Canterbury Plains in Australasia. The emigration that fuelled this colonization was itself a major factor in the economic development of the core regions, siphoning off the "surplus" population generated by demographic transition and swollen by the rationalization of rural economies. Meanwhile, as the demand for tropical plantation products increased, most of the tropical world came under the political control— direct or indirect—of one or another of the industrial core countries.

This expansionism resulted in specialization by the colonies and **client states** in the production of those foodstuffs and raw materials in demand from the industrial core and for which they held a comparative advantage. This specialization, in turn, established a complex pattern of interdependent development that was articulated, above all, in patterns of international trade. From the start, however, this expanded and more closely integrated international system was unevenly balanced. On the one hand, the influence of the core countries on the cultural and institutional organization of the peripheral countries has molded their economic organization to fit core-oriented needs and core-inspired philosophies of "develop- ment" (see Chapters 9 through 11). On the other hand, a variety of barriers and imperfections have blunted the effectiveness of international trade as an "engine of growth." While *some* of the semi-peripheral and peripheral national economies achieved considerable momentum as a result of the stimulus provided by rapidly increasing levels of demand transmitted from the industrial core regions, many did not. In general terms, the effective distance from the industrial core that represented their major market determined the type and profitability of activities in semi-peripheral and peripheral regions. Within the resultant zones of specialization, the beneficial effects of trade were conditioned by a variety of factors including variations in climate, topography, pre-existing systems of agriculture, and population densities. In practice:

> The trade impetus to growth was . . . immensely important for Argentina and Uruguay in Latin America, South Africa and Zimbabwe (formerly Southern Rhodesia) in Africa, Australia and New Zealand, and, to a lesser extent, in Sri Lanka (formerly Ceylon). Elsewhere, there was a significant impact, *but this was inadequate to get sustained development going*, for example on the west coast of Africa. For countries such as India, Pakistan, Bangladesh, Iraq and Iran, *the export trades were too small relative to the total population* to provide much impetus for development, except in very restricted areas.
>
> (Chisholm, 1982: 88; emphasis added)

In short, the benefits of specialization and international trade enabled some regions and countries to ascend within the world-system, but it primarily enhanced the position of those at the top. For the rest, the prospect of economic growth through trade has been diminished. These regions and countries find themselves at a relative disadvantage in accessing capital which, in turn, affects their use of technology and hinders their ability to realize significant increases in productivity. As a result, the amount of labor power required to produce a given quantity of exports from the industrial core will generally be much less than that needed to produce an equivalent value of exports from peripheral countries. From this point of view, the international trade system is characterized by unequal exchange.

Attempts to short-circuit this built-in handicap by borrowing capital with which to purchase new (but not always appropriate) technology have almost always resulted in a debt trap as compounded interest on loans has outpaced increases in the rate of productivity.

In addition, many peripheral countries have been affected by another built-in handicap: the differential elasticities of demand for their products *vis-à-vis* those of their trading partners in the industrial core. In general, the elasticity of demand for primary commodities that are the staples of the periphery is low, that is, large price reductions in overseas markets elicit only a modest rise in demand. Similarly, demand for these products increases only slightly in response to increases in the purchasing power of consumers in the core countries. Conversely, elasticities of demand for manufactured goods are generally high. The net result is that the terms of trade have tended to work to the cumulative advantage of the industrial core.

Most simply, although the world economy has become characterized by interdependent relationships as a result of the spatial division of labor, the periphery has carried the burden of dependency. We explore this theme in detail in Part 3.

5.6 PRINCIPLES OF ECONOMIC GEOGRAPHY: SUMMARIZING LESSONS FROM THE INDUSTRIAL ERA

The case histories in this chapter illustrate how the increasing complexity of economic organization and spatial change make it difficult to generalize about organizing principles or characteristic features. However, the recurrence of certain elements and the emergence of others in the development of the economic landscapes of the industrial core regions during the industrial era can be underscored. Among those elements that carried over from previous eras are the following:

- *Natural resource distribution*—in particular, the importance of iron ore and coal.
- *Demographic change*—the timing of the demographic transition in relation to industrialization and the role of large-scale migrations in relation to changing labor markets.
- *Technological change*—improvements and innovations in transport and communications.
- *Territory*—specifically, colonialism and territorial expansion as responses to the law of diminishing returns.
- *Changes in institutional and sociopolitical settings.*
- *Spatial distribution of investment*—changes in response to the shifting comparative advantages enjoyed by producers in different areas.
- *Import substitution*—as a mechanism of ascent within the world economy.
- *Militarism and geopolitical change.*

In addition, the industrial era saw the emergence of several new dimensions of spatial-economic organization and the increased prominence of others:

- The extension of the world economy to a global scale with a corresponding extension of the spatial division of labor and the consequent intensification of the interdependencies between core, semi-periphery, and periphery.
- The replacement of "liberal" merchant capitalism with a competitive and, later, an increasingly organized form of industrial capitalism characterized by distinct, specialized regional economies organized around growing urban centers.
- The eclipse of the European core of the world economy by the ascent of the United States and Japan.
- The emergence of distinctive core–periphery contrasts within the industrial core territories of the world economy.
- The agglomeration of industrial activity as a result of the logic of economies of scale and the multiplier effect.
- The modification of urban systems by the addition of new kinds of town and city—mining towns, heavy manufacturing centers, power centers, and transport nodes—and the rapid growth of larger preindustrial cities as they benefited disproportionately (because of their established markets, entrepreneurship, trading links, and commercial infrastructure) from the various growth impulses that characterize industrialization.
- The imprint of cyclical fluctuations in the pace and nature of economic activity.
- The "differential of contemporaneousness" in regional economic development—a phenomenon linked to the uneven impacts of the process of technological diffusion and changing technology systems.
- The adaptation of private firms to the changing opportunities and constraints of different technology systems resulting in evolving forms of economic organization from simple manufacturing to machinofacture to Fordism.
- The adaptation of a wider society to these changing forms of economic organization and another evolutionary process—that of changing regulatory frameworks—in which the increasing intervention of governments in economic development was the most important development.
- The emergence of an "organized" form of capitalism founded on the power and authority of independent countries; characterized by a sophisticated interdependence of firms, industries, regions, and governments; and forming the basis for core–periphery relationships at various geographic scales.

SUMMARY

In this chapter, we outlined the evolution of the economic geography of the industrial core regions, analyzing the major processes involved in the relative ascent and decline of countries and regions within the core of the world economy. Important aspects in the evolution of the core economies included the following:

- The Industrial Revolution and how industrialization rapidly reshaped economic landscapes in the DCs such that today economic geography within the core of the world economy is dominated by the physical, institutional, and social legacies of industrial capitalism.
- Machinofacture and the spread of industrialization in Europe associated with machine production and the organizational setting of the factory. While machinery provided the basis for higher levels of productivity, factories enabled this productivity to be exploited to its fullest extent.
- Fordism and associated U.S. industrialization that was partly associated with the emergence of the United States as a dominant component of the world economy.
- Japanese industrialization that drew on established advantages including a high degree of cooperation between industry and government and a model of industrial organization—derived from the *zaibatsu*—large enough to compete with the transnational corporations of Europe and the United States.
- The emergence of "organized" capitalism in the developed countries. A hallmark of this period was the emergence of a workable relationship between business interests and labor unions. Unions had grown in size and strength in the **Progressive Era,** and constituted another increasingly important element of "organization." The role of the state also changed as government expanded the scope of its activities in part to mediate the relationship between organized business and labor.

KEY SOURCES AND SUGGESTED READING

Borchert, J., 1967. American metropolitan evolution, *Geographical Review* 57, 301–332.

Castells, M., 2000. *The Information Age: Economy, society and culture: Volume 1, The rise of the network society*, 2nd edn. Oxford: Blackwell.

Dunford, M., 1990. Theories of regulation, *Society and Space* 8, 297–321.

Galor, O. and Weil, D., 2000. Population, Technology, and Growth: From Malthusian stagnation to the demographic transition and beyond, *American Economic Review* 90(4), 806–828.

Hansen, G. and Prescott, E., 2002. Malthus to Solow, *American Economic Review* 94(2), 1205–1217.

Haywood, J., 1998. *Historical Atlas of the 19th Century World 1783–1914.* New York: Barnes & Noble.

Hugill, P., 1994. *World Trade Since 1431: Geography, technology, and capitalism.* Baltimore, MD: Johns Hopkins University Press.

Hugill, P., 1999. *Global Communications Since 1844: Geography, technology, and capitalism.* Baltimore, MD: Johns Hopkins University Press.

Pollard, S., 1981. *Peaceful Conquest: The industrialization of Europe, 1760–1970.* Oxford: Oxford University Press.

Picture credit: Linda McCarthy

After the Second World War, the economies of the industrial core regions began to enter a substantially different phase in terms of *what* they produced, *how* they produced it and *where* they produced it. This phase is sometimes referred to as advanced or globalized capitalism. It typically involves a combination of ingredients:

> [M]ost especially: the accelerated internationalization of economic processes; a frenetic international financial system; the use of new information technologies; new kinds of production; different modes of state intervention; and the increasing involvement of culture as a factor in and of production.
>
> (Thrift, 2002: 19)

It evolved in response to the increasing inflexibility of the old system of Fordist industrial capitalism. Faced with the saturation of domestic consumer markets, increasing overseas competition, increasing costs of unionized labor and of governmental welfare provision, the industrial corporations of the core economies began to pursue more flexible strategies. They reorganized themselves, redeployed their operations, and revised their relationships with labor unions and governments. The result has been the deindustrialization of the core economies, the industrialization of certain semi-peripheral countries, and the expansion of financial and business services on a global scale.

6.1 TRANSITION TO ADVANCED CAPITALISM

The shift to advanced capitalism has been a result of the cumulative interaction of several processes. As in all previous major economic transitions we have described, the importance of these processes has been revealed only after a period of *crisis* for the old order. In this case, the crisis was thrown into focus by a phase of stagflation and intensified by a sudden increase in the price of oil.

PRELUDE: CRISIS OF FORDISM

The crisis of Fordism emerged abruptly in the early 1970s, throwing into reverse the postwar industrial boom. This reversal is clearly illustrated by the performance of the U.S. economy. In overall terms, the U.S. economy performed exceptionally well from the late 1940s right through to the early 1970s. Real disposable income per capita rose from just over US$2,200 (in constant 1972 dollars) in 1947 to over US$3,800 in 1972. The 1960s were particularly prosperous, with economic growth averaging over 4 percent per year, as a result expanding GNI by 50 percent over the decade. Meanwhile, the average family obtained a real increase of over 30 percent in its disposable income; these were the years of J.K. Galbraith's "affluent society."

After the early 1970s, however, U.S. economic growth averaged only 2.2 percent, while productivity in the private business sector, having increased at around 3.3 percent per year in the 1960s, fell away to 1.3 percent per year in the 1970s. "By 1979, the typical family with a $20,000 income had only 7 per cent more real purchasing power than it had a full decade earlier. The years had brought a mere $25 more per week in purchasing power for the average family" (Bluestone and Harrison, 1982: 4). Unemployment, having remained steady at around 4.5 percent until the early 1970s, almost doubled over the next five years, leveling off at around 10 percent by the mid-1980s. The rate of inflation doubled from around 2.5 percent per year in the 1960s to over 5 percent per year in the mid-1980s. Meanwhile, in 1971, the U.S. economy had moved, for the first time during the twentieth century, into a negative trade balance with the rest of the world; and except for 1973 and 1975, the trade balance has remained negative ever since.

The system shock precipitated by the rise in oil prices in 1973 as a result of the OPEC cartel has been widely cited as a major cause of this downturn (in 1973–1974, petroleum prices quadrupled as a result of the cartel's actions). The evidence is inconclusive, however. Similarly, it has been difficult to establish the influence of other popular scapegoats, such as the role of labor unions in obtaining wage increases in excess of productivity. Rather, the crisis of the 1970s must be seen as the product of a conjunction of trends whose origins can be traced to the 1960s or before. Hamilton (1984) identified several such trends:

1. The stagflation phase of economic cycles. With the downswing of the economic cycle (see Figure 1.2) there was a slowing down of economic growth and a steady fall in profits, particularly in the industrial core countries. This was associated with falling levels of demand for **capital goods**, particularly transport, building, mining and factory equipment (for example, ships, vehicles, machinery, machine tools) and, as a result, steel. Overall, rates of growth in the OECD countries fell from an annual average of 5 to 6 percent between 1963 and 1973 to around 2.5 percent between 1973 and 1978, and less than 1 percent between 1979 and 1982.

   At the same time, rising levels of inflation generally served to reduce profits and limit capital for investment. This resulted in greater dependency on financing investment through the banking sector. This, in turn, meant high interest rates that retarded technological investment and so hindered competitiveness. Inflation also raised labor costs, which increased the urgency of technological investment (particularly automation) at a time when capital was expensive. The result was the widespread depression of both capital-intensive industries (for example, steel, shipbuilding, vehicles, appliances) *and* labor-intensive industries (for example, textiles, clothing, footwear).

2. Increased international monetary instability, which took two major forms:

   - overvaluation of exchange rates as a result of the transition (in the early 1970s) from fixed exchange rates to floating exchange rates. Where currencies were overvalued (for example, the currencies of oil and/or gas producers such as the Netherlands, Norway and the United Kingdom, and, more recently, the USA), the loss of international competitiveness resulted in import penetration and a consequent decline in industrial capacity (see Figure 6.1). In the U.S., import penetration in clothing and textiles increased from 34 percent in 1980 to 55 percent in 1986; import penetration in shoes increased from 50 percent to 81 percent, in computers from 7 to 25 percent; and in automobiles from 35 to 40 percent
   - problems of indebtedness among NIEs and some LDCs following massive borrowing from the petrodollar surpluses created in the OPEC countries (see p. 57). In addition to the international financial instability associated with uncertainty caused by debt rescheduling and fears over national bankruptcies, this created a strong incentive for NIEs and LDCs to increase their exports—of cheap manufactured goods as well as traditional staples—to the core regions in order to obtain the necessary foreign exchange. This, in turn, increased the competitive pressure on the labor-intensive sectors of the core economies.

3. The strengthening, throughout the 1960s, of social values associated with social welfare provision (for example, retirement pensions, healthcare, anti-poverty programs) and environmental protection. Although this created new markets for some products and services, it also raised some industrial costs and contributed to a higher tax burden on both consumers and producers.

4. The introduction of innovations and technological changes in response to escalating energy and labor costs created feedback effects that depressed demand in traditional industrial activities. Energy-saving designs in transport and heating, for example, reduced the demand for steel; while innovations in microelectronics reduced the demand for electromechanical products.

5. A resurgence of political volatility, which reduced the extent of stable business settings and so inhibited several dimensions of world trade, including east–west trade (until 1989 and the fall of communism) and trade involving much of Central America, southwestern Asia and Southeast Asia.

TOWARDS FLEXIBILITY: PRECONDITIONS

Meanwhile, just as in previous major economic transitions, the processes themselves that resulted in the shift to advanced capitalism have drawn on a number of *preconditions* developed during the preceding era. Three main factors are involved:

1. new, enabling technologies in transport and telecommunications
2. changing patterns of demand and consumption
3. corporate restructuring.

1. Enabling technologies and economies of scope

The resolution of the crisis of Fordism and the emergence of advanced capitalism have been made possible, in part, by the availability of permissive technology of two kinds:

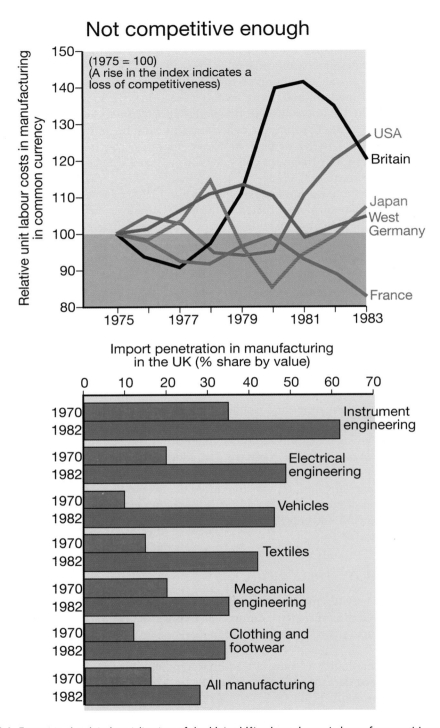

Figure 6.1 Forces in the deindustrialization of the United Kingdom: dramatic loss of competitiveness (1978–83) and consequent import penetration, converting the country from a net exporter to a net importer of manufactures

Source: Based on Hamilton (1984: 352, Figure 2)

- *Circulation*: Improvements in transport and communications technologies (containerization, email, voice-over-Internet protocols (for example, Skype), cloud-computing, communications satellites, etc.), which have reduced the time and costs of circulation, bringing a wider geographic market within the scope of an increasing range of business activities. This global reach has also been advanced through the economic development of peripheral areas and the standardization of products (the latter itself being a function of the development of communications media).
- *Production*: Improvements in production technologies (electronically controlled assembly lines, robotics, nanotechnology, 3D printing, etc.) have allowed for a finer degree of specialization in many production processes, and facilitated a routinization of many operations. This, in turn, has led to the **deskilling** of many production systems while at the same time increasing the *separability* (and therefore the spatial fragmentation) of their constituent parts. This has made it easier for managers to take advantage of new sources of cheaper and less militant labor. Advances in the manufacture and use of synthetic materials have also extended the locational capability of many industries, since raw materials have traditionally been the most restrictive of all factors of production.

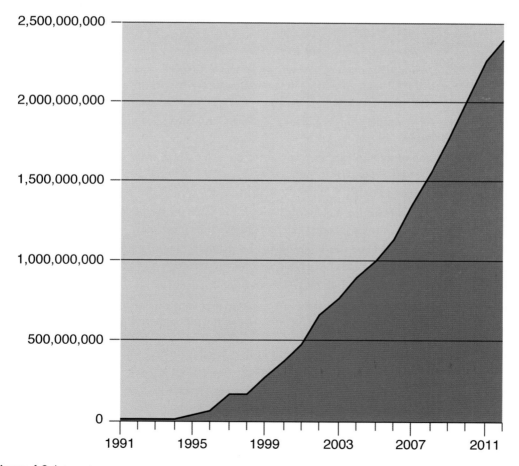

Figure 6.2 Internet users

Source: Based on online data from World Research Institute database http://earthtrends.wri.org/searchable_db/index.php and online ITU statistics http://www.itu.int/ict/statistics

The impact of these enabling technologies is difficult to overstate. During the past few decades the global network of computers, telephones and televisions has increased its information-carrying capacity more than a million times over. Figure 6.2 shows the tremendous explosion that has occurred in the number of Internet users (from 4.4 million in 1991, to almost 400 million in 2000, and over 2.4 billion by 2012) who now access somewhere between an estimated 15–30 billion web pages, to which an estimated 10 million new pages are added every day. In 2010 6.1 trillion SMS text messages were sent, which was equivalent to about 200,000 every second. The emergence of a tri-polar world economy has encouraged the development of round-the-world shipping services that link the three main cargo-generating zones of Europe, the Americas and Asia and the Pacific. So there emerged, by the mid-1980s, the flexible, **intermodal** and global **transportation** services offered by several container lines such as Evergreen (Taiwan), Hapag-Lloyd (Germany) and Maersk (Denmark). In addition, some shipping lines, such as APL and P&O, acquired inland trucking operations, and railway companies became involved in ocean-borne transport (for example, the merger of CSX and Sea-Land in the United States).

Box 6.1 Technological breakthroughs and the information economy

1969:   U.S. Defense Department's Advanced Research Projects Agency (ARPA) launched ARPAnet, an electronic communication network (precursor of the Internet).

1970:   First ATM installed in a U.S. bank; Corning Glass developed industrial grade fiber optic cable.

1971:   Invention of microprocessor (computer on a chip); first email sent on ARPAnet; IBM released floppy disk.

1972:   Atari released Pong.

1973:   PARC labs create the Xerox Alto with many software technologies of later consumer PCs (mouse, ethernet, and graphic user interface).

1974:   TCP/IP invented and this interconnection network protocol ushered in technology allowing different types of networks to be connected.

1975:   Microcomputer invented; Microsoft and Apple launched.

1977:   First successful commercial microcomputer, Apple II, introduced; Microsoft operating systems produced for microcomputers; commercial diffusion of digital switching.

1978:   Two students invented the modem to transfer microcomputer programs to each other via telephone to avoid going out in winter in Chicago.

1981:   IBM released first PC (running Microsoft's MS-DOS); Nintendo's first video game success: Donkey Kong.

1982:   Limited commercial email began; Sony and Philips released the CD (audio and data).

1983:   Motorola introduced first mobile phone (using analog communication).

1984:   Apple launched the Macintosh (first consumer PC with mouse/windows interface); HP launched the laserjet printer.

1985:   Microsoft released Windows 1.0; CD-ROMs (read only) released.

1986:   First commercial mailing list program, LISTSERV, developed.

1987:   PowerPoint 1 released.

1990:   Tim Berners-Lee at CERN created hypertext system that would form the basis of the world wide web; CD-Rs released.

1991:   World wide web released to the public (has 130 sites by 1993); notebook computers introduced.

1992:   First mobile phone with Internet connectivity created.

1996:   First Blackberry launched.

1993:   Netscape orbital satellites allowed GPS to achieve civilian operational capability.

1995:   Amazon and eBay launched; Holiday Inn established the first hotel website with online room registration.

1996:   DVDs introduced (making home VCRs and videos in use since the mid-1970s obsolete).

1998:   Google launched.

2000:   USB flash drives released.

2001:   Wikipedia launched; Apple releases the iPod.

2003:   Intel incorporated Wi-Fi into its Centrino chip; Skype and MySpace.com launched (social networking will become a standard in digital communication); SPAM exceeds legitimate email for the first time.

2004:   Facebook launched.

2005:   YouTube launched.

2006:   Twitter launched; iTunes downloaded its billionth file; commercial cloud-computing milestone reached when Amazon launched Elastic Compute cloud (EC2).

2007:   Apple released iPhone (first smart phone to use multi-touch interface).

2009:   Digital television became broadcast standard in the U.S. and other countries, facilitating web-based TV; Barnes & Noble launched the Nook.

2010:   Apple launched the iPad; YouTube used more bandwidth than the entire Internet in 2000.

2011:   Amazon launched Kindle Fire; smartphone sales exceed PCs sales for the first time.

2012:   Tablets sales exceed laptop sales for the first time.

We should also note that the introduction of new circulation and production technologies has in many cases created a powerful second-order effect: economies of scope, the capacity to provide entirely new products and/or services through the flexible use of the same production or service network. So, for example, the computerized records developed by airlines have lent themselves to an increased scope of business that includes hotel reservations and rental cars. Similarly, the credit records of major retailing firms have provided a base for them to exploit economies of scope. In the United States, Sears exploited access to its retail credit customers by offering insurance services through Allstate, investment services through Dean Witter and real estate services through Coldwell Banker until selling these Sears Financial Network companies in the 1990s.

Another important aspect of enabling technologies concerns their capacity for flexibility during periods of intense competition or changing market conditions. The following example concerns the opportunities for small and medium businesses (SMBs) created by cloud-computing:

> Small businesses (companies with fewer than 100 employees) and midsize players (companies with 100 to 1,000 employees) stand to gain much from the promise of cloud-computing technologies. Cloud computing offers SMBs access to reliable and scalable infrastructure resources (for example, computing and storage), configurable platforms that allow for integration between the business and vendors or customers, and rich application functionalities that can be paid for on an ongoing basis. Consequently, cloud computing offers SMBs the opportunity to enhance or improve IT capabilities in a way they previously could not.
>
> (Diamandi *et al.*, 2011)

Finally, we must recognize the differential of contemporaneousness (see also p. 120) in the uneven spatial impact of these enabling technologies. Following Sachar and Öberg (1990), we can see that new technologies have different implications for different regions within the world economy:

- In the core countries, high tech creates new jobs, particularly in financial and business services. It creates new products, facilitates new production and distribution processes, and new forms of corporate organization, but reduces the need for employment in manufacturing.
- In semi-peripheral countries, high technology brings an increase in manufacturing employment, increases in productivity, and an overall improvement in their competitiveness in the international economy.
- In peripheral countries, new technologies are often too expensive to acquire or deploy. As a result there is a relative decline in both productivity and international competitiveness. To the extent that new technologies are deployed, their main effect is to displace jobs in labor-intensive sectors, as a result adding to a sprawling **informal sector** in urban economies and putting pressure on the public sector to absorb labor in government-sponsored jobs.

2. Changing patterns of demand and consumption

Shifting patterns of consumer demand have also been an important precondition for the evolution of advanced capitalism. Within core countries, Fordism, based as it was on mass production coupled with mass consumption, began to be the victim of its own success as mass markets for many of its staple products—such as cars and refrigerators—came close to being saturated. As the affluent societies of the core countries satisfied more and more of their wants, market saturation could only be avoided by skillful marketing campaigns and continuous modifications in products and packaging. Even so, mass-produced goods were less and less effective in satisfying one of their main roles in affluent societies: that of **positional goods**, possessions that serve as measures of socioeconomic status. As more and more people were able to acquire mass-produced *positional goods*—new cars, TVs, etc. —so more people sought the distinction of custom-made, stylish, high-design and fashionable products. Social distinctions, previously marked by the ownership of a basic set of consumer goods on a sliding scale of size/quality, now had to be established via the symbolism of ensembles of positional goods. The problem for producers was not just that their mass-produced products were rapidly losing their appeal to the most affluent consumers but also that their strategies and processes of production were too inflexible to cater to the many different (and rapidly changing) market niches for positional goods. The result was that many firms began to adopt more flexible forms of production, using new production and circulation technologies in order to exploit a variety of niches within the overall market. One of the classic success stories in this respect is Inditex, the world's largest fashion retailer, and its flagship Zara chain stores (see Box 6.2).

Although the postwar era of steadily increasing affluence in the core economies came to an abrupt end with the stagflation crisis and OPEC oil price increases of the mid-1970s, the shift away from mass markets to niche markets did not slow. Rather, it intensified as a new materialism took root. The main group of people in this change were the baby-boomers whose formative years had been spent in the postwar economic boom. Their reaction to mass consumption was a countercultural movement with a collectivist approach to the exploration of freedom and self-realization. The failure of this countercultural movement (in particular, the failure of the sit-ins, protest marches, general strikes, student–worker alliances and civil disorder of 1968) meant that self-realization slid into self-centered and narcissistic lifestyles: The basis for new market niches. But the real cause of the materialism of the baby-boomers was the shock of emerging onto housing and labor markets just as the economies of the core countries were experiencing a phase of stagflation compounded by the recessionary effects of the OPEC oil price rise. Wages stood still while consumer prices ballooned.

Millions of baby-boomers, raised to take for granted steady improvements in levels of living, found themselves unable to fulfill the American Dream (or the European or East Asian version of it). Their response was to pursue materialism for its own sake. They saved less, borrowed more, deferred parenthood, comforted themselves with affordable luxuries that were marketed as symbols of style and distinctiveness, and generally surrendered to the hedonism of lives infused with extravagant details: Gourmet foods, designer clothes, jewelery, exotic vacations, and high-end electronic and other gadgets. The point here is that all of this represented consumer demand not for mass-produced products but for a rapidly changing array of high-quality products whose value as positional goods had a relatively short life. This meant that producers had to be flexible enough to be able to identify and exploit finely differentiated market niches.

Between core countries, meanwhile, there has been a homogenization of markets. Similar trends in income distribution and consumer tastes have been reinforced by TV and online viewing (including Discovery, MTV and YouTube) and international travel. Together with decreasing relative costs of transport and communications, this has meant that *market niches have merged across national boundaries, making it possible for producers to exploit economies of scale in the production of upmarket products.* To a lesser degree, the same processes have extended from core countries to the more affluent consumers of semi-peripheral and peripheral countries, allowing the marketing of world products (for example, German luxury automobiles, British raincoats, Italian sweaters, Swiss watches, French wines, U.S. soft drinks, Japanese consumer electronics) to global market segments. Barnet and Cavanagh (1994: 15–16, 166) described the emergence of what they call the "global shopping mall":

> The Global Shopping Mall is a planetary supermarket with a dazzling spread of things to eat, drink, wear, and enjoy. . . . [Through] the rise of global advertising, distribution, and marketing . . . dreams of affluent living are communicated to the farthest reaches of the globe. . . . Even in the rural areas of the Philippines any city of over 20,000 will have at least one supermarket, usually a one-room affair about the size of an old New Hampshire general store. In the fishing and rice-farming town of Balanga, Bataan, the San José Supermarket offers . . . Procter & Gamble's Pringle's potato chips, Hormel's Spam, Hershey's Kisses, Nabisco's Chips Ahoy, Del Monte's tomato juice, Planter's Cheez Curls, and Colgate-Palmolive's toothpaste.

Box 6.2 Fast fashion and IT: Zara responds rapidly to changing customer demand

Fast fashion means that a Zara version of an outfit worn by Madonna during the first concert of her 2005 tour could be worn by teenage girls at her final appearance. Zara chain stores are the flagship of parent company, Inditex, the world's largest fashion retailer (overtaking Gap in 2008). Headquartered in Arteixo, Spain, founder Amancio Ortega opened his first store in nearby La Coruña in 1974 selling affordable decently made versions of expensive upscale fashions.

Compared to competitors such as Gap, Zara uses a more vertically integrated business organization. At the company's headquarters, no name designers (copiers?), who are directly in touch with regional sales managers, rotate among teams (in contrast to expensive big name designers such as Stella McCartney or Jimmy Choo at H&M). Allowing greater flexibility and control, most production is in-house in its highly automated company-owned factories or nearby subsidiaries and subcontractors in Spain and Portugal. Relatively little is outsourced to lower cost locations like Bangladesh (and then only for staple garments such as t-shirts). Not needing to sign contracts with Asian subcontractors far in advance, Zara cuts its turnaround time for a garment from concept to store to several weeks and even less for minor style changes to existing garments (compared to conventional twice yearly runway fashion shows of new collections). Zara's huge distribution centers use sorting machines based on overnight parcel companies like FedEx. The company prefers company-owned stores (rather than franchises).

The company's competitive advantage lies in using IT to respond to customer demand as it changes, instead of attempting to predict notoriously fickle fashion trends. Store managers use tablets to enter customer feedback for the designers, linked to point of sale (cash register) processing systems, connected to inventory systems. To maximize the time that store clerks spend helping customers with purchases, and minimize time putting items on display, before shipment all garments are ironed, price and security tagged, and packed on hangers. The staff also note any unsold garments that customers tried on but did not purchase; this allows the designers to plan fashion styles and the staff to adjust orders in response to customer demand based on actual sales and customer feedback. Zara can immediately switch to a new design and cancel a production batch of any garment that does not sell well within a week of arriving in stores. And even amid global fashion trends, each store's inventory is matched to local customers because the clerks have unusual authority to change inventory orders. So in world cities like New York or Tokyo, a 5th Avenue or Ginza store can cater to wealthy tourists in contrast to the younger trendier customers in Soho or Sibuya.

Although fashion advertising is ubiquitous, Zara spends instead on real estate: Handsome or historic buildings in prime locations near the upscale stores with the expensive designer versions of their garments. Advertising is unnecessary because of Z-day when each store receives its twice-weekly delivery of the latest fashions; this keeps customers coming back and purchasing often. Frequent small batch production allows the company to avoid large store inventories that necessitate sales and markdowns. In this way, Zara has changed customer behavior, from waiting months for an outfit to be marked down at Gucci or Prada when space is needed for a new collection, to an impulse purchase; customers buy immediately at full price because the cost is low, the outfit is unlikely to be worn by a co-worker, but will not be in the store on a return trip.

Whether this kind of fast fashion is sustainable is a concern, given that customers are buying many more garments than they really need because the purchase price is so low. Yet there is a high environmental cost in terms of the use of chemicals, water, transportation, and so on. While certainly not the only company outed in Greenpeace's 2012 *Toxic Threads* report, tests indicated that some of Zara's garments contained nonylphenol ethoxylates that break down and form toxic hormone-disrupting chemicals and dyes that can release carcinogenic amine.

3. Logic of corporate restructuring

One of the most important preconditions for advanced capitalism and the globalization of business activity has been the restructuring of the corporate world. In response to changing circumstances, private business has had to develop new strategies in order to survive; strategies that have significantly altered the fortunes of different kinds of cities, regions, and countries.

Two of the most important outcomes overall have been corporate concentration and centralization. **Concentration** involves the elimination of small, weak firms in particular spheres of economic activity, partly through competition, and partly through mergers and acquisitions. **Centralization** involves the merging of the resultant large corporations from different spheres of economic activity to form giant conglomerates with a diversified range of activities. Such corporations are often *transnational* in their operations, having established overseas subsidiaries, taken over foreign competitors, or bought into profitable foreign businesses.

In every industry, there are limits both to the extent to which productivity can be increased and to which consumers can be induced to purchase more. As competition to maintain profit levels becomes more intense, some firms will be driven out of business while others will be taken over by stronger competitors in a process of **horizontal integration**. A successful automobile manufacturer, for example, might buy out other automobile manufacturers.

Meanwhile, the chances of new firms being successful tend to be diminished, since the larger existing corporations are able to draw on economies of scale in order to edge out smaller competitors by price cutting. But even giant corporations cannot prevent market saturation indefinitely; and they are, in any case, always vulnerable to unforeseen shifts in demand. A common corporate strategy has therefore been to engage in **vertical integration** (taking over the firms that provide their inputs and/or those that purchase their output) in an attempt to capture a greater proportion of the final selling price. The automobile manufacturer, for example, might take over companies that make specialized components like engines or car navigation systems; and/or companies that distribute or sell automobiles. The net result is the *concentration of production*, within most industries, in the hands of a diminishing number of increasingly large companies.

Alternatively—or in addition—**diagonal integration** (taking over firms whose activities are completely unrelated to their own) offers the chance of gaining access to more profitable markets and/or less expensive factors of production. Staying with the same example, the automobile manufacturer may buy into energy, advertising, or entertainment companies. The net result in this case is the *centralization* of assets, jobs, production, and decisions about economic life in the hands of an even smaller number of even larger companies.

Table 6.1 Percentage of value of shipments accounted for by the four largest companies in selected manufacturing industries in the USA

Industry	Percentage of value of shipments			
	1992	1997	2002	2007
Cookies and crackers	56	60	67	69
Cigarettes	93	99	95	98
Petrochemicals	—	60	85	80
Semiconductor machinery	41	44	60	64
Turbine and turbine generator set units	79	78	88	68
Telephone apparatus	51	54	56	61
Household vacuum cleaners	59	69	78	71
Household refrigerators and freezers	82	82	85	92
Household laundry equipment	84	90	93	98
Motor vehicles	84	82	81	68
Guided missiles and space vehicles	71	89	96	95

Source: Based U.S. Census Bureau, *Economic Census* data (five yearly; various years)

The extent of these trends can be illustrated in relation to the U.S. economy during the postwar period. Spearheaded by the large corporations that had established themselves through early flurries of horizontal integration in the 1900s (for example, U.S. Steel, International Harvester, American Tobacco, General Electric) and vertical integration in the 1920s (for example, General Foods, B.F. Goodrich, and the major petroleum companies), "big business" began to exert an increasing influence on economic life. Although the incidence of horizontal mergers was greatly reduced by antitrust (antimonopoly) legislation (for example, the Celler Kefauver Act 1950), vertical and diagonal integration proceeded at unprecedented rates, generating around 3,000 mergers per year in the peak years of the late 1960s. Since the early 1970s concentration ratios (the percentage of total sales attributable to the four largest firms) had increased across a broad spectrum of industries. Several major industries, including cigarettes, home refrigerators and freezers, home laundry equipment, and guided missiles and space vehicles, are each almost completely dominated by four (or fewer) firms (see Table 6.1). For example, despite diversification into the growing assortment of competing breakfast options for on-the-go consumers (for example, ready-to-eat breakfast bars, bagels, and muffins or fast-food establishments such as McDonald's), of the top ten ready-to-eat cereal companies in the U.S., Kellogg's (60 percent), General Mills (20 percent), Kraft (9 percent) and Quaker Oats (5 percent) still shared together over 93 percent of the $11 billion in sales in 2012.

Meanwhile, giant conglomerates emerged as a result of diagonal integration. The first of these was Textron Incorporated, established only in 1943 when it sold blankets and other textile products to the U.S. Army. By 1980 it had been involved in buying or selling over 100 different companies in industries as diverse as textiles, aerospace, machinery, watch bracelets, and pens. Textron was by no means an isolated example, however. As early as 1955 the majority of mergers taking place in the U.S. were diagonal, conglomerate mergers. By the early 1990s nine out of every ten mergers involved conglomerate companies.

As a result of all this merger activity, giant conglomerates increasingly influence the world's core economies. For example, between 1950 and 1980 the 50 largest U.S. corporations increased their share of the total value added in *all* manufacturing from less than 20 percent to nearly 30 percent; and the largest 200 increased their share from 30 percent to 50 percent. Similar trends have occurred in the services sector (where the control of variety stores, department stores, car rental firms, motion picture distribution and data processing had become particularly centralized). In the U.S., for example, the ten largest cable operators serve more than 80 percent of all cable subscribers, with the largest two, Comcast with more than 22 million subscribers and TimeWarner Cable with over 12 million subscribers together serving about 47 percent of all cable subscribers. Four companies dominated the mobile phone market in 2012 with the following shares of subscribers: Verizon Wireless (33 percent), AT&T Mobility (30 percent), Sprint Nextel (16 percent) and T-Mobile (9 percent). The top five companies worldwide for tablet shipments (of nearly 130 million in 2012) were Apple (with just over 50 percent of shipments) followed by Samsung, Amazon.com, ASUS (for Google) and Barnes & Noble.

Internet users have even been visiting fewer websites recently (perhaps partly due to dominant websites such as Google not only being a search engine but also offering email and social network services). Online music has also seen a shift to fewer commercial providers as iTunes has become music's biggest retailer; the company also dominates the digital video market with about two-thirds of both TV show and movie sales. Although peer-to-peer (P2P) file sharing has spawned free download companies since the days of Napster, copyright infringement lawsuits forced many to shut down while lawsuits against individuals have made some people think twice about downloading bit torrent files.

Because of their size, the larger elements of U.S. conglomerates have also come to exert an increasing influence in the *international* economy. This is not surprising given that the annual revenue of the very largest business enterprises like Wal-Mart and General Motors are greater than the GDP of countries such as Austria and South Africa. Perhaps more significant is the fact that *the combined overseas output of U.S.-based transnational corporations (TNCs) is now larger than the GDP of every country in the world except the USA itself*. As we will see (Chapter 12), this dimension of the international economy has come to represent a serious economic threat to small countries, and it has prompted the creation of a variety of international and supranational economic and political organizations.

It should be stressed that the United States has by no means been the only core country to generate giant conglomerates. Major TNCs have been bred in Europe, Japan, Canada, and Australasia in response to the same logic that has applied to the U.S.; and some NIEs and peripheral countries now have large home-based transnational corporations. Indeed, these companies have been increasing their share of world markets at the expense of U.S.-based companies; and some of them have extended their operations to the U.S. itself. The South Korean conglomerate Samsung Group, for example (the world's largest maker of TVs, LCD panels, mobile handsets, and computer memory chips, with a global workforce of more than 350,000 and annual revenue of about US$250 billion) has multiple subsidiaries and affiliates including those that produce, market, and sell a wide range of electronic parts and components including smartphones and semiconductors. The company's designers work in design centers in Seoul, London, San Francisco, Tokyo, Shanghai, and Delhi. Overall, the number of transnational corporations increased from just over 7,000 in the early 1970s to more than 80,000 today. Together, these corporations have several hundred thousand foreign affiliates and the largest five hundred account for more than US$15 trillion in worldwide annual sales.

EVOLUTION OF TRANSNATIONAL CORPORATE ACTIVITY

The current importance of transnational corporations in the world economy is the result of an evolutionary process that can be characterized in terms of three distinctive phases (with reference to the experience of U.S.-based TNCs):

- *Phase I*: Beginning in the nineteenth century and extending to 1940, this phase was dominated by investment directed at obtaining raw materials—mainly oil and minerals—for domestic manufacturing operations.
- *Phase II*: After the Second World War, some of the leading corporations began to use foreign direct investment (FDI) in overseas production operations as a means of penetrating foreign consumer markets. Initially, the focus of this investment was western Europe, where the Marshall Plan, NATO rearmament and the U.S. military presence in West Germany provided useful information feedback and points of entry to an expanding consumer market. Meanwhile, the establishment of the U.S. dollar as the world's principal reserve currency at the 1944 Bretton Woods conference (see Chapter 2), had made it much easier for U.S. companies to buy into foreign industries. It was not long before many U.S. firms began to penetrate the expanding markets of parts of the rest of the world, particularly in Latin America. The resulting mergers and acquisitions often led to the restructuring of corporate production processes.

   Bulova Watch is a good example. Bulova originally manufactured watch movements in Switzerland and shipped them to Pago Pago in American Samoa where they were assembled and shipped for sale in the United States. Bluestone and Harrison (1982: 114) reported that Corporation President Harry B. Henshel said about this arrangement: "We are able to beat the foreign competition because we are the foreign competition." Bulova is now headquartered in New York City with operations in Switzerland, the UK, Italy, Canada, China, Japan, and Mexico.

   Between 1957 and 1967, 20 percent of all new U.S. machinery plants, 25 percent of new chemical plants, and over 30 percent of new transport equipment plants were located abroad. By 1970, almost 75 percent of U.S. imports were transactions between the domestic and foreign subsidiaries of transnational corporations. By the end of the 1970s overseas profits accounted for one-third or more of the overall profits of the 100 largest transnational producers and banks.

- *Phase III*: During the 1970s the crisis and destabilization associated with the episode of stagflation brought growing competition from goods produced in the NIEs with cheap labor. In addition, the collapse of the Bretton Woods Agreement in 1971 increased the value of the U.S. dollar, making imported goods cheaper, and making it easier for European and Japanese TNCs to penetrate U.S. markets. In response, U.S. transnational corporations began to restructure their production processes once again, eliminating the duplication of activities between domestic and foreign-based facilities, and reorganizing the division of tasks between them.

Effectively, this third phase has meant:

1. retaining existing facilities that require high inputs of technology and/or skilled labor (for example, headquarter offices in, for example, the United States and Europe)
2. the further redeployment of capital, bringing peripheral countries into the production space of U.S. companies in order to benefit from lower labor costs (in 2011 the costs of hourly compensation for production workers in manufacturing industries in Singapore and South Korea were between about 19 and 23 percent of those for U.S. workers; in Mexico and Brazil, they were between 6 and 12 percent; and in China, they were less than 5 percent)
3. withdrawing from locations where unskilled and semi-skilled labor is more expensive (particularly the United States and Europe—in 2011, for example, the costs of hourly compensation for production workers in manufacturing industries in Germany, Switzerland, and the Scandinavian countries were between about 124 and 181 percent of those for U.S. workers), by using foreign subcontractors.

So, for example, General Electric added 30,000 foreign jobs to its payroll during the 1970s while reducing its U.S. employment by 25,000. Nike is a good example of the consequences of this third phase. This company sold its manufacturing plants in the United States and United Kingdom and now subcontracts out all its production. The geography of Nike's production network has changed over time to reflect changing labor costs across the LDCs. Nike shoe production was initially carried out in Japan, but was later relocated to Taiwan and South Korea. Today, Nike products are manufactured in more than 750 factories by subcontractors employing over 1 million workers across more than 40 countries with low labor costs, including Argentina, Bangladesh, Brazil, Bulgaria, China, India, Indonesia, South Africa, and Vietnam.

6.2 PATTERNS AND PROCESSES OF GLOBALIZATION

Most of the world's population now lives in countries that are either integrated into world markets for goods and finance, or rapidly becoming so. As recently as the late 1970s, only a few peripheral countries had opened their borders to flows of trade and investment capital. About one-third of the world's labor force lived in countries such as the former Soviet Union and China with centrally planned economies, and at least another third lived in countries insulated from international markets by prohibitive trade barriers, and currency controls. With nearly half the world's labor force among them, three giant population blocs—China, Russia and India—have rapidly integrated into the global market.

Globalization, although incorporating more of the world, more completely, into the capitalist world-system, has intensified the differences between the core and the periphery. During the last 50 years, the average per capita income in the wealthiest 20 countries multiplied by 21 times while that in the poorest 20 countries multiplied by 13 times. Some parts of the periphery have really struggled. In Sub-Saharan Africa, economic output fell by one-third during the 1980s. Unlike many other LDCs that managed to restore growth in the early 1990s, Sub-Saharan Africa experienced growth during the 1990s that merely matched the population growth rate (with average annual GDP growth of less than 2.5 percent for 1990–2000) due to "a combination of adverse external developments, structural and institutional bottlenecks and policy errors" (UNCTAD, 2001). Although GDP growth rose to over 4.5 percent during the 2000s, average per capita incomes in Sub-Saharan Africa adjusted for inflation were lower in 1990 and 2000 compared to in 1980, and grew by an average of only

1.5 percent during the 2000s. Indeed, the persistent economic peripherality of Sub-Saharan Africa in the contemporary world economy, exacerbated by the impacts of HIV/AIDS, is considered by some scholars as a more threatening condition than the dependency of the colonial period.

Meanwhile, globalization has resulted in the consolidation of three major world regions comprising North America, Europe and Asia. Most of the world's flows of goods, capital and information are between and within these three world regions. In 2011, for example, WTO data indicate that they accounted for 85 percent of the world's merchandise exports of nearly US$3 trillion (North America 37.7 percent, Asia 31.0 percent and Europe 16.4). They are each other's main export destinations (see Table 6.2). Merchandise trade within these world regions is strong (71 percent of Europe's exports go to European countries, 53 percent of Asia's exports and 48 percent of North America's exports respectively are intra-regional). The United States remains the world's largest trader in merchandise, with 2011 exports and imports totaling US$3,746 billion, followed by China (US$3,641 billion), Germany (US$2,726 billion), Japan (US$1,678 billion), and France (US$1,310 billion).

The United States–European Union (EU) trade relationship is the largest in the world, and it is growing (see Table 6.3). It developed over the centuries since European colonization of North America and has deepened since the Second World War. The United States and European Union are one another's largest merchandise and services trading partners. In 2012 the EU accounted for about 17 percent each of total U.S. exports and imports. Likewise, EU exports to the United States accounted for about 17 percent of its total exports to non-EU countries, while EU imports from the United States accounted for more than 11 percent of its total imports from non-EU countries. Although enjoying trade surpluses with the EU for a

Table 6.2 Inter- and intra-regional merchandise trade, 2011 (US$ billions)

Origin of exports	North America	Europe	Asia	South and Central America	Middle East	Africa	Common-wealth of Independent States
World	2,923	6,881	5,133	749	672	538	530
North America	1,103	382	476	201	63	37	15
Europe	480	4,667	639	119	194	199	234
Asia	906	922	2,926	189	242	152	110
South and Central America	181	138	169	200	18	21	8
Middle East	107	158	660	10	110	38	6
Africa	102	205	146	19	21	77	2
Commonwealth of Independent States	43	409	117	11	24	12	154

Source: Adapted from WTO (2012b:23, Table 1.4)

Table 6.3 United States trade in goods and services with the European Union (US$ billions)

	2002	2007	2012
Exports			
Goods	140.4	242.2	269.7
Services	98.0	182.5	193.8
Imports			
Goods	225.4	356.2	384.3
Services	85.2	145.8	149.7
Balance			
Goods	−85.0	−114.0	−114.6
Services	12.8	36.7	44.1

Source: Adapted from Cooper (2013: 5, Table 2)

number of years, since 1993, the United States has seen growing trade deficits with the EU. Table 6.3 shows that while the United States consistently runs trade surpluses in services with the EU, these are too small to offset its merchandise trade deficits. Merchandise imports into the United States from the EU include cars, computers and components, and machinery such as gas turbines. U.S. service exports to the EU include to travel-, business-, and insurance-related services (Cooper, 2013).

Allen Scott conceptualized this situation as a patchwork of global city-regions, major economic motors, each one being the site of dense networks of specialized but complementary forms of economic activity, together with large and multifaceted labor markets and specialized infrastructures offering powerful agglomeration economies. The central metropolitan area of each regional motor is surrounded by a hinterland occupied by ancillary communities, prosperous agricultural zones, local service centers and the like. The hinterlands of some of these regional motors may coalesce with one another (as in the actual cases of Tokyo/Nagoya/Osaka, Boston/New York/Philadelphia, Los Angeles/San Diego/Tijuana, and more recent LDC examples such as São Paulo/Rio de Janeiro and Hong Kong/Guangzhou). These regional economic motors are linked by intense flows of capital, information, goods and people.

INTERNATIONAL REDEPLOYMENT AND LOCATIONAL HIERARCHIES

The cumulative result of the globalization of economic activity has been the creation of *locational hierarchies* of activities. The aggregate outcome involves:

1. localized concentrations of high-level management in world cities (see Chapter 7)
2. smaller concentrations of mid-level management and administration in large metropolitan areas in core countries, and in the capital cities of NIEs and some peripheral countries
3. clusters of research and development (R&D) activity in high-tech, innovative complexes—*technopoles*—(see Chapter 7) within the core countries
4. regions specializing in advanced, high-tech industrial production, mostly within the core countries
5. decentralized pockets of routinized industrial production in (a) the peripheral regions of core countries, and (b) the metropolitan areas of NIEs and some peripheral countries.

These tendencies, and the fact that they have been influenced so much by the locational strategies of transnational corporations equipped to take a global approach in pursuit of the most profitable redeployment of activities, have contributed to a **new international division of labor (NIDL)** (see Chapter 1).

It should be acknowledged that not all firms or industries are equal in their need or their capability to engage in international redeployment of this kind. It is the largest companies —the transnational corporations—that are in the best position to take advantage of the advances in circulation and production technology. Probably the best-developed example of global production—and the most researched—is provided by the automobile industry, where the clearly defined national markets of the early postwar period have been almost entirely replaced by production and marketing on a global scale. In 1976 Ford introduced the Fiesta, a vehicle designed to sell in Europe, South America, the Asian market, and North America. The Fiesta was assembled in several different locations from components manufactured in an even greater number of locations. The Fiesta became the first of a series of Ford "world cars," which today includes the Focus. Ford's international subsidiaries, which used to operate independently of the parent company, are now functionally integrated, through sophisticated IT systems.

The other automobile companies have organized their own global assembly systems. They employ modular manufacturing for their world cars based on a common underbody platform yet with the flexibility to adapt the interior, trim, body, and ride characteristics to local conditions in different countries: Volkswagen's best-selling European Golf (U.S. Rabbit), for example, GM's Opel Corsa, and Fiat's Palio model deriving from its "Project 178" world car platform. Ford's Focus in North America, for example, has larger front and rear bumpers. Honda produces two distinct versions of the same car from its Accord world car platform: The more powerful, bigger, and more comfortable Accord for U.S. drivers and the smaller sportier Accord aimed at European and Japanese drivers. The top three global automobile corporations accounted for more than one-third of global sales of more than 80 million vehicles in 2012. The top five corporations were, in order of sales: Toyota 9.75 million (with Daihatsu, Hino, Lexus, and Scion), General Motors 9.29 million (which includes Buick, Cadillac, Chevrolet, GMC, Holden, Opel, Vauxhall, and Wuling), Volkswagen 9.07 million (Audi, Bugatti, Bentley, Lamborghini, MAN, Porsche, Scania, SEAT, and Skoda), Ford (incorporating FPV, Lincoln, and Troller) and Hyundai (with Kia).

FLEXIBLE PRODUCTION SYSTEMS

Concurrent with the changing competitive strategies of firms, there have been some significant changes in the organization of production systems in many industries. These are often expressed in terms of a transition from Fordism to flexible production systems. Today, the logic of mass production coupled with mass consumption has been modified by the addition of more flexible production, distribution and marketing systems.

This flexibility is rooted in forms of production that enable manufacturers to shift quickly and efficiently from one level of output to another and, more importantly, from one process and/or product configuration to another. It must be understood as a change that involves flexibility both *within* firms and *between* them. Within firms, a great deal of this flexibility is attributable to the exploitation of new technologies. Computerized machine tools are capable of producing a variety of new products simply by being reprogrammed, often with very little downtime between production runs for different products. Different stages of the production process (sometimes located in different places) can be integrated and coordinated through

Table 6.4 Contrasts in production and labor: Fordism versus flexible production

Fordism	Flexible production
The production process	
– Mass production	– Flexible (small batch) production
– Standardized uniform products	– Product differentiation with a variety of products
– Large buffer inventories of factory stock	– Just-in-time inventories (stock arriving as needed)
– Quality testing ex-post (errors and defective stock detected and rejected late)	– Continuous quality control (defective stock detected and rejected immediately)
– Defective stock unnoticed or stored in buffer inventories	– Immediate rejection of defective stock
– Significant production time lost due to, e.g., long set-up times, defective stock, inventory bottlenecks	– Minimal production time lost as a result of, e.g., shorter set-up times, just-in-time deliveries meeting changing demand
– Resource driven (production dominated by factory stock and output considerations)	– Demand driven (production designed to meet specific consumer and other demands)
– Vertical and horizontal integration	– Vertical disintegration (using subcontractors)
– Cost reduction through, e.g., single tasking where traditional occupational boundaries allow a worker to be very experienced in one specific part of the production process	– Cost reductions through, e.g., functional flexibility where traditional occupational boundaries are replaced by multi-tasking by workers; use of subcontractors with non-union workers
Labor	
– Single task performed by a worker	– Multiple tasks performed by a worker
– Payment per rate (based on job design criteria)	– Individual payment (based on detailed bonus system)
– High degree of job specialization	– No job demarcation
– Little on-the-job training	– Continuous on-the-job training
– No on-the-job learning experience	– On-the-job learning; learning-by-doing integrated into long-term planning
– Emphasis on diminishing worker's responsibility (disciplining of labor force)	– Emphasis on worker's co-responsibility
– Vertical labor organization (e.g., with skilled and unskilled workers in the same industry)	– More horizontal labor organization for key workers (based on skills or trades of workers rather than by industry)
– Greater job security for unionized workers	– High job security for key workers. Increased informal work with no job security and poor labor conditions for temporary workers. No job security for workers employed by subcontractors

Source: Updated from Albrechts and Swyngedouw (1989: 75, Figure 1)

computer-aided design (CAD) and computer-aided manufacturing (CAM) systems. Computer-based information systems can be used to monitor retail sales and track wholesale orders, allowing producers to reduce the costs of raw materials stockpiles, parts inventories and warehousing through sophisticated small-batch, just-in-time production and distribution systems. The combination of computer-based information systems, CAD/CAM systems and computerized machine tools has also helped firms to be flexible enough to exploit specialized niches of consumer demand, rendering geographically scattered upscale markets accessible to economies of scale in production. This kind of flexibility depends on new labor practices as well as new technologies, however (see Table 6.4). There are two main aspects to this. One is the increasingly flexible use of labor within firms, which requires individual workers to perform a wider variety of tasks. Taken to its extreme, this trend has in some instances substituted craftwork for production-line work. The other is the increasingly flexible size and quality of the labor force required at any one plant. This trend has substituted overtime, part-time and temporary employment for permanent, full-time jobs.

Between firms, flexible production can be achieved through the *externalization* of certain functions. One way of doing this has been to restructure permanent and hierarchically structured administrative, managerial, and technical units within large corporations into flatter, leaner, and more flexible forms of organization that can make increased use of outside consultants, specialists and subcontractors. This has led to a degree of vertical disintegration among firms (see Chapter 3). An example is Boeing's 787 Dreamliner that may offer a cautionary tale, however, about production delays and the difficulties of quality control in such an extensive global assembly system. This airplane is sourced with components from Australia, Canada, China, France, Japan, Italy, Russia, South Korea, Sweden, the United Kingdom, and the United States (see Figure 6.3). The worldwide grounding of the Dreamliner followed a January 2013 emergency landing because of an electrical fire associated with the plane's lithium ion batteries. The batteries were manufactured by GS Yuasa Corporation of Kyoto, Japan. Another route to externalization has been to participate in joint ventures, in the licensing or contracting of technology, and in strategic alliances involving design partnerships, collaborative R&D projects and the like. In addition to more traditional (and more expensive) mergers and acquisitions, joint ventures and strategic alliances have become an important contributor to the intensification of economic globalization.

The numerous strategic alliances between the world's largest automakers include parts-sharing agreements and joint ventures in research, as well as in manufacturing. In 2013, for instance, Toyota and BMW formed a strategic alliance to develop hydrogen-based fuel cell technology as well as a new sports vehicle that will run on clean, high-mileage gas. Soon after, Ford, Daimler, and Nissan entered into a strategic alliance to develop hydrogen-based fuel cell technology and vehicles. Other products of strategic alliances include the Nissan Micra/Renault Pulse and the Nissan Sunny/Renault Scala that are manufactured in India in a platform-sharing arrangement.

The Nestlé food company's strategic alliances, for example, include a joint venture with General Mills called Cereal Partners Worldwide (CPW), and one with New Zealand's leading dairy company, Fronterra to supply dairy products throughout the Americas called Dairy Partners Americas. Nestlé also has a joint venture with the Coca-Cola Company, called Beverage Partners Worldwide (BPW), in which the Swiss company cooperates with Coca-Cola in exchanging technologies and in marketing. Nestlé has, for example, licensed its Nestea and Enviga brands of ready-to-drink teas to Coca-Cola in the United States and to BPW for the rest of the world.

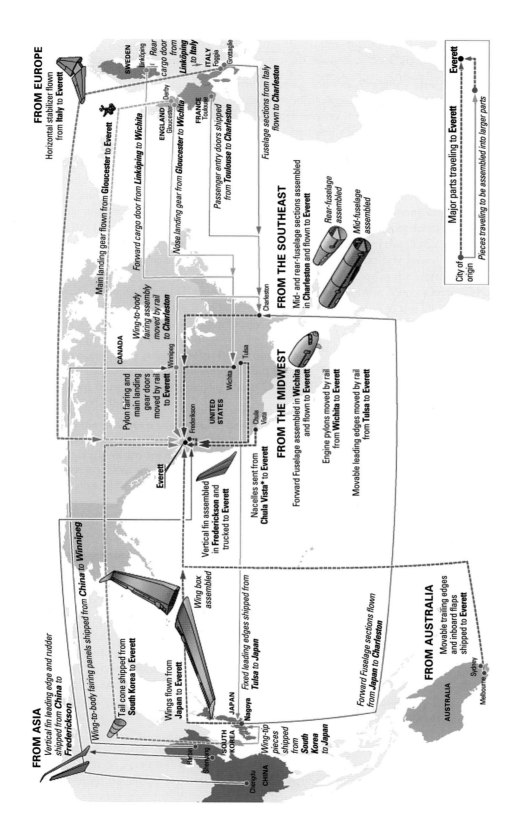

Figure 6.3 Boeing's 787 Dreamliner global production system

Source: Based on *Seattle Times* (2006)

Such alliances have become an important aspect of global economic geography, as transnational corporations seek to reduce their costs and to minimize the risks involved in their multimillion-dollar projects. Strategic alliances serve several functions, they:

- allow transnational corporations to link up with local insiders elsewhere in order to tap into new markets
- provide a quick and inexpensive means of swapping information about technologies that help to improve their products and their productivity
- reduce the costs of product development
- spread the costs of market research.

INTERPRETATIONS OF CORPORATE FLEXIBILITY

The increased flexibility in economic organization can be interpreted in two rather different ways. One interpretation, that of *flexible specialization*, sees the trend toward flexible production systems in a permanent and positive light. In short, new technologies have opened up the possibility for the decline of the large integrated firm and for the growth of a production system organized around clusters of small firms. Much is made of cases such as the Third Italy (see Box 6.4) where such clusters have emerged over the past 40 years. Alfred Marshall's (1920) model of the industrial district is sometimes used to provide a theoretical argument for the clustering of specialized industries in specific localities: This emphasized specialized labor pools, external economies from proximity accruing to firms in the same industry, and the availability of specialized inputs and services. This is, of course, nothing more than a restatement of the main arguments for any kind of agglomeration or **localization economies** (Krugman, 1991: 35–67). It has, however, become popular to restate them in terms of the new institutional economics (Williamson, 1985) as internalizing transaction costs within regions rather than inside firms. In addition, trust, loyalty and partnership between firms are viewed as vital to the establishment of the "new" industrial districts, if not to the old ones of Alfred Marshall's model. Consequently, the social conditions for small-scale production, in combination with (1) a history of artisanal activity, (2) nearby centers of innovation, (3) assistance from local governments, and (4) consensus between labor and management, are all basic requirements for the functioning of industrial districts engaged in flexible production (Harrison, 1992). From this point of view, there is a sociology to these industrial districts that sets limits to its diffusion elsewhere.

On the basis of a series of national case studies (Silicon Valley in the USA, the Île de France technopole in France and the Third Italy), Scott (1988b: 106) described this wave of economic–geographical organization. The basic proposition is that irrespective of the particular industries involved (for example, shoes, clothing, machine tools, computers) there is a major drive toward geographical concentration of industries even for manufacturing so-called mature products (those towards the end of a product lifecycle). Rather than specialization at a regional scale of agglomeration, however, a more localized pattern of specialization is now under way. This is because small firms are its major agents and they prosper best as the providers of goods to rapidly changing markets when they are able to share information, labor traditions and inter-industry links.

Responding to criticisms of the industrial district model (in that it ignores the continuing importance of large firms and exaggerates endogenous (local) conditions relative to world markets and the international division of labor) some authors have provided more synthetic accounts. For example, Scott (1992) later argued that large producers can play an important

Box 6.3 The changing geography of the clothing industry

The clothing industry provides a good example of the way in which local economic geographies are affected by an industry's response to globalization. In the nineteenth century, the clothing industry developed in the metropolitan areas of core countries, with many small firms using cheap migrant or immigrant labor. In the first half of the twentieth century, the industry, like many others, began to modernize. Larger firms emerged, their success based on taking advantage of mass-production techniques for mass markets, and on exploiting principles of spatial organization within national markets. In the United States, for example, the clothing industry went through a major locational shift as a great deal of production moved out of the workshops of New York to big, new factories in smaller towns in the south, where labor was not only much cheaper but less unionized.

Then, as the world economy began to globalize, semi-peripheral and peripheral countries became the least cost locations for mass-produced clothing for global markets. In 1960, less than 7 percent of all apparel purchased in the United States was imported; in 2012, the American Apparel & Footwear Association reported that nearly 98 percent was imported. Leisure wear—jeans, shorts, t-shirts, polo shirts and so on—was an important component of the homogenization of consumer tastes around the world, and it could be produced most profitably by the cheap labor of young women in the peripheral metropolitan areas of the world. The hourly compensation (excluding benefits) of clothing workers in the United States ranges from $8.25 to $14.00; their counterparts in Asia (excluding Japan) average about US$3 an hour, with the labor costs in China and India under US$1, and in Cambodia and Bangladesh less than 25 cents. There are also inequalities in the wages of male and female workers (even in Europe, men in the textile, clothing and footwear industries make 20 to 30 percent more than women). While the retail margin on domestically made garments sold in Europe and the United States is 70 percent or so, the retail margin on clothing made in workshops in countries such as Bangladesh and Vietnam is 100 to 250 percent. A typical example of how the sale price of a $100 garment is divided up would be: $50 to the retailer, $35 to the manufacturer (who spends $22.50 on textiles) and $15 to the contractor, who pays the garment workers $6. The apparel and textile industries together represent the largest industrial employer in the world. Apparel, over half of that industry, employs more than 25 million workers in the garment industry in the semi-periphery and periphery, of whom 75 percent are women. Studies of the industry have shown that some of the female workers are as young as 12-year-old girls from rural villages who have been sent to work as sewing machinists in city workshops, sleeping eight to a room, sewing seven days a week from 8 am to 11 pm. Child labor is also widespread in subcontracting arrangements that make use of homeworkers.

This globalization of production has resulted in a complex set of commodity chains. Many of the largest clothing companies, such as H&M, have most of their products manufactured through arrangements with independent suppliers (about 800 factories in the case of H&M). These manufacturers are scattered throughout the world, making the clothing industry one of the most globalized of all manufacturing activities (see Figure 6.4). H&M has its goods produced in low-cost Asian and European countries. The actual geography of commodity chains in the clothing industry is somewhat volatile, with frequent shifts in production and assembly sites as companies and their suppliers continuously seek out new locations with lower costs.

Although cheap apparel can be produced most effectively through arrangements with multiple suppliers in low-wage regions, higher end apparel for the global marketplace requires a different geography of production. These products—women's fashion, outerwear and lingerie, infants' wear and men's suits—are based on frequent style changes and high-quality finish. This requires short production runs and greater contact between producers and buyers. The most profitable settings for these products are in the metropolitan areas of the core countries—London, Los Angeles, Milan, New York, Paris and Stuttgart—where, once again, migrant and immigrant labor provides a workforce for designer clothing that can be shipped in small batches to upscale stores and shopping malls around the world.

The result is that commodity chains in the clothing industry are quite distinctive in terms of the origins of products destined for different segments of the market. Fashion-oriented retailers in the United States who sell designer products to upmarket customers obtain most of their goods from manufacturers in a small group of high-value-added countries including France, Italy, Japan, the United Kingdom, and the United States. Department stores that emphasize private-label products (that is, store brands, such as Nordstrom) and premium national brands will obtain most of their goods from established manufacturers in semi-peripheral Asian countries. Mass merchandisers who sell lower priced brands buy primarily from a third tier of lower cost, mid-quality manufacturers, while large-volume discount stores such as Wal-Mart import most of their goods from low-cost suppliers in steady growth supplier such as Bangladesh, China and Vietnam (see Figure 6.4).

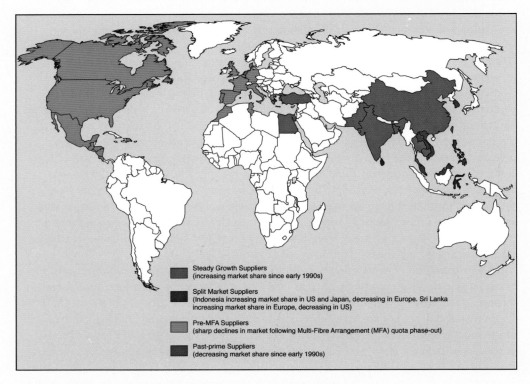

Figure 6.4 The changing global geography of clothing manufacturing

Source: Adapted from *China Sourcing Blog* (2012)

part in inducing and maintaining the growth of (high-technology) industrial districts. He suggests that the usual division of production units between flexible and mass producers is insufficient. He identifies a third type, the *systems house*, flexible producers that benefit from economies of scope flowing from R&D or design synergies, with a variegated internal structure of job specialization and **batch production** (as opposed to mass production) of complex products. These systems houses do not stand alone but are usually connected with nearby flexible producers. They are the hubs for high-technology industrial districts such as those in southern California, Cambridge, Baden-Württemberg, Germany, and Tokyo and Kyoto in Japan. Scott, for one, does not see this model of industrial districts as incompatible with an internationalized world economy. Indeed, he sees this phenomenon as itself "the interlinkage of industrial districts across the globe, . . . as a mosaic of regions consisting of localized networks of transactions (i.e. industrial districts) embedded in global networks of transactions" (Scott, 1992: 274).

A second interpretation of the Fordist/flexible production systems divide sees the methods of flexible production more as a response to a crisis of capital growth in some sectors of industrial production rather than a fully fledged form of economic organization. From this point of view it is the transformation of financial markets and the introduction of flexible production as a means of disciplining the power of labor that attract attention. In the first case, it is argued that through the development in new financial instruments (for example, junk bonds) combined with the spread of sophisticated systems of global financial coordination, the financial system demands increased geographical and temporal flexibility of capital growth (Harvey, 1990). In the second case, the declining rate of profit in the 1970s led firms to a strategy of decentralized production to undermine the power of labor (and reduce wage bills), which had increased under Fordism, yet still maintain centralized control. Many of the often idealized small firms of the Third Italy are indeed subcontractors for larger firms searching for alternatives to their large unionized labor forces.

It is important to note, however, that several criticisms have been directed at these interpretations, most especially the first one of flexible specialization:

- Not all the apparently flexible methods of production are, in fact, that flexible. Simple oppositions such as Fordist and flexible impose a structure on an industrial history that is not all that simple. In particular, the labor processes in different industries, the market and macroeconomic features of different sectors, and the organizational cultures of the firms and areas involved combine to produce a range of industrial geographies along the spectrum between locational fixity and global mobility (Amin and Thrift, 1992: 574).
- Large firms are now adopting many of the methods that were seen as the exclusive province of small firms clustered in industrial districts (as acknowledged by Scott). Some enter into strategic alliances to produce certain items even with firms that are their direct competitors (for example, Toyota with BMW, or Ford with Nissan and Daimler). They can do this even with dispersed production facilities.
- The geographical boundaries of industrial districts are not usually carefully defined; partly this is a result of a lack of consensus about the transactions and flows that must be internalized geographically for a district to exist. Some newer industrial districts sprawl over large areas and overlap with other districts, whereas others are small and exclusive. Does the same logic of production govern both of these types of industrial space?
- Missing from most discussions of both models is attention to specifics (in terms of industries, technologies and limited areas) of flexible production, prior spatial divisions of labor in the affected areas (and elsewhere where a shift to flexible production is not taking place), and

the influence of government policies, especially with respect to technical education, innovation policy, tax incentives and trade barriers.

- More research is needed into how local conditions interact with global competitive conditions to affect the fortunes of industrial districts. What is known suggests that such districts are not immune to the problems of international competition affecting Fordist firms (see Box 6.4 and Chapter 13).

- Scholars who live in places that experienced changes in forms of economic development emphasize the shift to flexible production. The so-called Los Angeles School (Scott is a leading member) used examples of production complexes from southern California, Massachusetts (Route 128), or the Third Italy as if they were drawn from a universal sample or provide a window on the future everywhere. The view of traditional manufacturing workers from the Rustbelt of the USA, for example, would be considerably less sanguine about the break with mass production and the possible universality of flexible production. Many large firms in the northeast of the USA continue to move production abroad or invest in automation. In many manufacturing industries, Fordist principles of production still prevail.

Box 6.4 The myth of the new industrial districts of the Third Italy?

The Third Italy (central and northeast Italy) is an often cited example of one of the flexible production systems organized around clusters of small firms that have emerged during the past 40 years. Small firms are seen as the best providers of goods to rapidly changing markets in a localized pattern of specialization where they are able to share information, labor traditions and inter-industry links. Yet in many cases the tradition and style of the Italian artisanal approach have had to be adapted to modern production methods. Many of the often idealized small firms of the Third Italy are, in fact, subcontractors for larger firms searching for alternatives to their large and more expensive unionized labor forces. Some of the largest Italian shoe manufacturers have even outsourced production to lower-cost east European countries such as Romania and Slovenia, and to China or replaced Italian artisans with cheaper non-EU immigrant workers from countries such as Morocco and China. Even the small upmarket Italian shoe firms and their workers are not immune to the problems of international competition that afflicted Fordist firms. Italian exports have been impacted by increased production in China, India, and Vietnam. Job losses have occurred as some smaller businesses have folded and others have been forced into mergers or acquisitions. International competition has stimulated businesses to strengthen their position at the top end of the market by investing in research and development (R&D), and new technology and design techniques to improve quality and comfort, and to create innovative models and colors. Italy remains the foremost shoe manufacturer in the European Union, and in 2011 was the tenth largest footwear manufacturer and the third largest footwear exporter in the world (after China/Hong Kong and Vietnam). Despite the higher cost of producing shoes in the Third Italy, the small upmarket shoe manufacturers there are betting that they and their quality shoes will survive international competition (Galbraith, 2001).

THE GLOBAL OFFICE AND THE INFORMATION ECONOMY

The globalization of production and the growth of transnational corporations have been associated with another important change in patterns of local economic development: **Producer services** such as financial and business services are now no longer locally oriented ancillary activities but important global industries in their own right. The growing importance of financial and business services was initially a result of the globalization of manufacturing, an increase in the volume of world trade and the growing dominance of transnational corporations. It was helped along by advances in IT. Satellite communications systems and fiber-optic networks made it possible for firms to operate key financial and business services 24 hours a day, around the globe, handling an enormous volume of transactions.

As banking, finance and business services grew into important global activities, however, they were themselves transformed into something quite different from the old, locally oriented ancillary services. The global banking and financial network now handles trillions of dollars every day—no more than 10 percent of which has anything to do with the traditional world economy of trade in goods and services. International movements of money, bonds, securities and other financial instruments have now become an end in themselves because they are a potential source of high profits from speculation and manipulation. Several factors have supported this development:

- The institutionalization of savings (through pension funds and so on) has established a large pool of capital managed by professional investors with few local or regional allegiances or ties.
- Deregulation of banking and financial services, as governments in many countries lifted restrictions and regulations in the hope of capturing more of their growth.
- The quadrupling of crude oil prices in 1973 that generated so much capital for oil-rich countries that their banks opened overseas branches in order to find enough borrowers. In many cases, the borrowers were companies and governments in LDCs that had previously been considered poor investment prospects. The internationalization of financial services soon paid off for the big banks. By the mid-1970s, about 70 percent of Citibank's overall earnings came from its international operations, with Brazil alone accounting for 13 percent of the bank's earnings in 1976.
- A persistent trade deficit of the United States *vis-à-vis* the rest of the world (a result initially of the postwar recovery of Europe and Japan but more recently the growth of NIEs such as China) created a growing pool of dollars outside the United States, known as eurodollars. This supply in turn created a pool of capital that was beyond the direct control of the U.S. authorities.
- "Hot" money (undeclared business income, proceeds of securities fraud, trade in illegal drugs and syndicated crime), easily laundered through international electronic transactions, also found its way into the growing pool of eurodollars. It is estimated that US$1–2 trillion are laundered each year through the global financial system. Even though the margin between these two estimates is enormous, the lower figure of US$1 trillion is still incredibly high.
- The initial response of many governments (including the U.S. government) to balance-of-payments problems was to print more money—a short-term solution that eventually contributed to a significant surge in inflation in the world economy. This inflation, because it promoted rapid change and international differentials in financial markets, provided a further boost to speculative international financial transactions of all kinds.

Together, these factors amounted to a change so important that a deep-seated restructuring of the world economy occurred. Banks and financial corporations with the size and international reach of JP Morgan Chase, HSBC, ICBC (China), Citigroup, BNP Paribas, or Deutsche Bank are able to influence local patterns and processes of economic development throughout the world, just like the major transnational corporations involved in the global assembly system. In addition, key producer services (such as market research, accountancy, advertising, banking, corporate insurance, and legal services) have added an important dimension to the world's economic landscapes. Today, more than four out of five jobs in the USA and UK are in the services sector.

Perhaps most important of all, the combined effect of all this has been to create an information economy. The information economy represents a form of economic production and management in which productivity and competitiveness rely heavily on the generation of new knowledge and on the access to, processing and communication of new information. Some of the most important economic sectors in this information economy are high-technology manufacturing, design-intensive consumer goods, and business and financial services.

BUSINESS SERVICES AND METROPOLITAN GROWTH

A fast growing component of the services sector has been business services. Most of these activities employ personnel either in managerial or information-processing positions. Demand for these services comes from other firms rather than the general public. Consequently, many of these services are located close to their main customers, overwhelmingly in and around large cities. Several factors have acted to reinforce this trend. The first is their internationalization. Agglomeration economies (access to clients and competitors, proximity to technical services, availability of qualified personnel) are so powerful across most of these services that they are disproportionately located in major cities. But ease of communication has made it possible for firms to operate across different cities rather than restrict themselves to one. The spread of manufacturing and conglomerate TNCs has encouraged successful business services firms to follow suit, establishing multicity offices to service their transnational accounts. Second, the deregulation of national markets (especially in banking and finance) also strengthened the relationship between certain large cities with well-established institutions (exchanges and commodity markets) and business services. A relatively small number of centers (London, Tokyo, New York) have benefited disproportionately from this trend. These world cities have become vital control points within a world economy breaking the bounds previously imposed by national restrictions.

BUSINESS SERVICES AND FLEXIBLE ECONOMIES

It can be argued that processes of subcontracting and small firm growth associated with flexible production give rise to new service activities and increase the dependence of manufacturers on the purchase of services from independent vendors. Typically, intermediary functions in economic activity, such as wholesaling, have been regarded as internal to large vertically integrated firms (where all functions are carried out within one firm) or ignored because of an assumption that producers trade directly with one another. However, wholesaling has persisted. Glasmeier (1990) makes a plausible case for the view that high-tech industrial districts, such as Silicon Valley and Baden-Württemberg, Germany, depended from the start on the coexistence of manufacturers and merchant wholesalers. In the case of Austin, Texas, which Glasmeier examines in detail, the wholesalers serve as agents of interregional trade, bringing

Box 6.5 Coming to America?

Concerns about the outsourcing of manufacturing and more recently services have received a great deal of attention in the media and in political circles in developed countries in North America and Europe (see Chapter 11). Although anxiety is high, most service outsourcing still takes place domestically, with much of the rest going within and between the DCs (UNCTAD, 2004). Less than 10 percent of all business process outsourcing (BPO), such as insurance claims processing, and call center credit card services and telemarketing, is done internationally by companies in countries such as India and, more recently, the Philippines.

By some accounts, however, this kind of BPO outsourcing is slowing. While hiring locally can help placate public opinion and offer a response to anti-outsourcing campaigns against Indian BPO companies in the U.S. and Europe, it can also make sense from a business perspective. Most BPO work that can be done internationally in the LDCs is already being outsourced while companies in the U.S. and Europe are increasingly aware of the drawbacks of outsourcing to the LDCs. In addition, studies of job creation in the United States indicate that workers in high-level IT support in the cheaper parts of the country cost only about one-fourth more than those in India. More and more companies in the DCs now prefer their IT and BPO work to be done locally, particularly when that work is strategic and complex (*Economist*, 2013d).

So, recently, the largest Indian BPO companies such as Tata, Infosys, and Wipro have been opening offices in North America and Europe. The largest Indian company, Tata, has about two dozen offices employing about 20,000 BPO workers across the United States, Canada and Mexico:

> Of all the back-office work that has been outsourced, the call-center business is the one that has made the most abrupt exit from India. With information technology, outsourcing firms such as Tata and Wipro are dealing with global companies, but with call centers they are dealing with customers. "We just can't get the accents right," sighs one Mumbai-based outsourcing executive. They tried hard to get workers in Bombay and Bangalore to enunciate their vowels just so. One recent web sketch showed operators imitating Sean Connery, a Scottish actor, for the Scottish market. But many customers had trouble understanding them and were infuriated.
>
> (*Economist*, 2013d)

parts and products from outside the local complex into the local economy. Over time, national and regional wholesalers have displaced local ones in importance to the local manufacturers. This suggests both the importance of merchant wholesalers to the development of industrial districts and the role of exogenous (extra- or nonlocal) agents in local development. The growth of industrial districts cannot be explained just in terms of local social conditions or the nature of manufacturing processes.

Christopherson (1989) argues that the attention given to flexible production in manufacturing has obscured the increasing importance of flexibility in the labor markets of service industries. She points out that by the 1980s in the United States 80 percent of the new jobs were in retail, health and business services, and that perhaps 25 percent of all service jobs were flexible jobs involving part-time work or independent subcontracting. Large firms increasingly dominate the growing service industries but to cut costs they make expanded use of subcontracting and part-time (usually female and minority) employees. In the retail and health sectors, worksites are decentralized and administrative functions separated spatially from the delivery of the services themselves even as large firms become dominant. The services are increasingly standardized from place to place; much like the physical settings such as regional shopping centers, shopping malls, and suburban medical buildings in which they are located.

Spatial homogenization rather than local specialization, therefore, characterizes the spatial pattern of major service industries and the flexible employment on which they rely. This flexibility is more difficult to romanticize than that associated with manufacturing. It involves serious reductions in incomes compared to those paid in the Fordist manufacturing industries. It also reduces the overall bargaining power of the workforce through exploiting gender and ethnic divisions (for example, by paying women less) and spatially dispersed worksites (including services outsourcing, addressed in detail in Chapter 11) to restrict employment security and limit labor organizing.

SUMMARY

In this chapter, we have seen how the crisis of Fordism, coupled with trends associated with corporate restructuring, technological advances, and shifting consumer demand have resulted in the globalization of economic activities that were previously localized within the core economies. Among the salient features of these changes are the following:

- Production hierarchies within the large companies that have come to dominate most industries. These hierarchies have tended to result in separate locational settings for (1) high-level corporate control, (2) production requiring high inputs of skilled labor and new technology; intermediate administration and R&D activities; and (3) routine production.
- The organization of the world economy into three broad international regions:

  1. the highly integrated and very diversified industrial and control centers of the core of North America, Europe, Japan and Australasia
  2. the semi-periphery of resource-exporting countries (such as Saudi Arabia and South Africa), old NIEs (such as Mexico and Brazil), and new NIEs (such as China, India, and Vietnam)
  3. the relatively thinly industrialized periphery that makes up most of the Southern Hemisphere (such as Sub-Saharan Africa), which is highly dependent on the core.

- Persistence within each of these broad regions of nested hierarchies of countries and regions at different levels of economic development. As such the periphery contains core regions and semi-peripheral regions (as, for example, the Lagos/Ibadan region and Abidjan region respectively in West Africa), the semi-periphery contains core regions and peripheral regions (for example, the Calcutta-Hooghly-Howra conurbation and Uttar Pradesh respectively in India), and the core contains regions that are, relatively, semi-peripheral (for example, Greece and east-central Europe respectively in Europe).

KEY SOURCES AND SUGGESTED READING

Christopherson, S. and Clark, J., 2009. *Remarking the Regional Economies: Power, labor and firm strategies in the knowledge economy.* London: Routledge.

Cooper, W.H., 2013. *EU-U.S. Economic Ties: Framework, scope and magnitude.* Washington, DC: Congressional Research Service.

Economist, 2013. The next big thing: Developed countries are beginning to take back service-industry jobs too. *Economist* January 19.

Harrison, B., 2007. Industrial districts: Old wine in new bottles?, *Regional Studies* 41, S107–S121.

Leinbach, T.R. and Brunn, S.D. (eds.), 2001. *Worlds of E-Commerce: Economic, geographical and social dimensions*. Chichester: John Wiley & Sons.

Polenske, K.R. (ed.), 2007. *The Economic Geography of Innovation*. Cambridge: Cambridge University Press.

Scott, A.J., 2006. *Geography and Economy*. Oxford: Clarendon Press.

Scott, A.J. and Storper, M., 2003. Regions, globalization, development, *Regional Studies* 37, 579–593.

Thrift, N.J., 2002. A hyperactive world, in R.J. Johnston, P.J. Taylor, and M. Watts (eds.), *Geographies of Global Change. Remapping the world in the late twentieth century*, 2nd edn. Oxford: Blackwell, 18–35.

UNCTAD, 2011. *World Investment Report 2011: Non-equity modes of international production and developments*. New York and Geneva: United Nations.

UNCTAD, 2012. *World Investment Report 2012: Towards a new generation of investment policies*. New York and Geneva: United Nations.

Part 3

Spatial transformation of core and periphery

In the next five chapters, we examine the spatial transformations of the core and periphery, paying special attention to the changing relationships between core and periphery outlined in Chapters 2 and 3. In Chapter 7, the spatial implications of the latest form of economic organization on the capitalist countries of the world's core regions are examined. In Chapter 8, we examine the spatial transformations in the LDCs that have occurred as a consequence of both an older colonialism and a more recent interdependent global capitalism. Attention is also paid, however, to how the consequences have varied depending on local relations and institutional responses. In Chapters 9, 10, and 11, three major economic activities,

Picture credit: Linda McCarthy

agriculture, manufacturing and services are examined both with respect to their roles in economic development and their changing geographical patterns, particularly from the perspective of the LDCs. The emphasis throughout this section is on the impacts of and responses to the evolving world economy in both the developed and the less developed countries, and to the changing relationships between them.

Chapter 7

Spatial reorganization of the core economies

Picture credit: Linda McCarthy

The evolution of advanced (globalized) capitalism and the emergence of an information economy have led to a significant reorganization of the economic geography of places and regions throughout most of the world. We will examine the nature and implications of these changes for the economic landscapes of the LDCs in Chapter 8. In this chapter, we focus on urban and regional change in the core countries, emphasizing the overall impact of corporate reorganization in creating new industrial spaces and affecting regional economic well-being.

It is important to note at the outset that the globalization of the economy described in the previous chapter has resulted in a relative *increase* in the importance of cities and regions as agents of economic development:

> [I]n a world economy whose productive infrastructure is made up of information flows, cities and regions are increasingly becoming critical agents of economic development. . . . Precisely because the economy is global, national governments suffer from failing powers to act upon the functional processes that shape their economies and societies. But regions and cities are more flexible in adapting to the changing conditions of markets, technology and culture. True, they have less power than national governments, but they have a greater response capacity to generate targeted development projects, negotiate with multinational firms, foster the growth of small and medium endogenous firms, and create conditions that will attract the new sources of wealth, power, and prestige.
>
> (Castells and Hall, 1994: 7)

As at the international level, the major components of urban and regional change have hinged on the redeployment of routine production capacity from high-cost to low-cost locations, and the retention/localization of facilities requiring high inputs of technology and/or skilled labor in key locations with appropriate resources and amenities. As a result, two countervailing trends characterize the economic geographies of Europe, North America, Australasia and Japan: *Decentralization* and *consolidation*. Decentralization has led to an attenuation of regional and interurban gradients in economic well-being; consolidation has contributed to an increased spatial differentiation in terms of the conditions of production and exchange, and the hierarchical structure of control.

At the same time, we have to consider the effects on core countries of wider changes in the world economy and globalized capitalism. We begin, therefore, with a brief outline of the main outcomes of these transformations. In the broadest of terms, three key changes can be identified: An altered relationship between capital and labor; new regional divisions of labor; and new roles for the state. Together, they contributed to the development of a distinctive context for economic development in core countries.

7.1 THE CONTEXT FOR URBAN AND REGIONAL CHANGE

Flexible production has taken hold as companies throughout the developed world have exploited new technologies and new strategies in order to remain competitive in a global economy. In the process, the relationship between capital and labor has been transformed, with corporations recapturing the initiative over wage rates and conditions. New technologies have played a major role in this transformation. The introduction of robotics in factories and information-processing technologies in offices, for example, has made for dramatic increases in productivity but has also created a long-term threat: That of substituting machines for workers that puts them in a weak bargaining position.

Flexible production systems have also created a new interregional and international division of labor, as large corporations have pursued strategies in order to deal with, and exploit, the **time–space compression** introduced by new transport and telecommunications technologies such as intelligent vehicle technologies and cloud computing. Paradoxically, the reduction of spatial barriers has *heightened the importance of geography* and magnified the significance of what local spaces contain because the new flexibility of the business world enables relatively small differences between places to be quickly, if temporarily, exploited. As a result, there has been an acceleration of shifts in the patterning of uneven development based on particular local mixes of skills and resources: *A continuously variable geometry of labor, capital, production, markets, and management.*

Last, but not least, flexible production has required the development of new roles for the state and the public sector: Reduced direct government intervention in the economy and a decreased emphasis on providing for collective consumption (school, hospitals, community services, etc.). The dilemma facing most governments was that deindustrialization accentuated the vulnerability of more and more people while making it increasingly difficult—politically as well as economically—to finance existing programs. As a result, a new conservatism in the orientation of central and local governments emerged. This new conservatism was associated with an ideological stance based on asserting that the welfare state had not only generated unreasonably high levels of taxation, budget deficits, disincentives to work and save, and a bloated class of unproductive workers, but also have fostered "soft" attitudes towards "problem" groups in society. The consequent restructuring of the welfare state was most pronounced in the United Kingdom and the United States, where the Thatcher and Reagan administrations respectively embarked on programs of privatization in health, housing, and education, accompanied by cuts in higher education, in programs for the unemployed, those with disabilities and the elderly, and in regional policy budgets. In the United Kingdom, closer controls on local government expenditure by the national government led to corresponding cuts at the local level, particularly in depressed towns and cities where the need for welfare services is high but local fiscal resources are low.

This last point is central to the emergence of changes in government regulation and intervention that, in turn, is tied in to the emergence of more flexible production. Following Jessop (1992), we can characterize this neoliberal regulation in general terms as involving a commitment to **supply-side** innovation in flexibility, with specific manifestations in several areas:

- a change in the regulation and conduct of labor markets, involving (1) a shift away from centralized collective bargaining towards company- or plant-level negotiations, and (2) an increasing tolerance of insecurity and marginality in the wage relations and employment conditions of unskilled workers
- flatter, leaner and more flexible forms of corporate organization that are suited not only to externalization but also to joint ventures and to public–private partnerships
- more flexible forms of credit: The result of deregulation of financial markets and the internationalization of finance and financial services, which has effectively reduced the degree of control that can be exerted by individual national governments
- displacement of Keynesian welfare states by workfare states:

> The emerging state form will no longer be concerned mainly with securing full employment within a national economy but with guiding and promoting the structural competitiveness of the national economy by intervening on the supply-side to encourage innovation; and it will no longer be concerned to generalise norms of mass consumption but to articulate policies to the need to promote greater flexibility.
>
> (Jessop, 1992: 32)

Table 7.1 illustrates some of the main contrasts between the characteristics of states under the two systems:

- The "hollowing out" of national government as a result of (1) the displacement of national power upwards through international agreements and organizations (see Chapter 12) and downwards to regional and local governments (see Chapter 13) and (2) increasing cooperation among local and regional governments in key fields such as R&D and technology transfer in ways that bypass their respective national states.

Neoliberalism, together with the imprint of flexibility and the globalization of economic activity (as described in this chapter) are described by Lash and Urry (1987) as advanced (disorganized) capitalism; this term distinguishes it from the previous phase that was dominated by a closely regulated and highly organized relationship between labor, capital and government at the level of a country. Advanced capitalism, in contrast, is characterized by:

1. deconcentration of capital within national markets, a growing separation of finance from industry, and the decline of cartels (as a result of the growth of a world market, the increasing scale of industrial, commercial and banking enterprises, and the general decline of tariffs)
2. decline in the absolute and relative size of workers in the traditional industrial sector of core countries as they deindustrialize and the expansion of the services sector with professional, white-collar workers
3. decline in the importance and effectiveness of national-level collective bargaining, and a growth in company and plant-level bargaining (as companies exert their new leverage in order to impose more flexible forms of organization)

Table 7.1 Contrasts in governance: Fordism versus flexible production

Fordism	Flexible production
Policy regulation	Policy deregulation and reregulation
Policy rigidity	Policy flexibility
Collective bargaining	Less unionization
Socialization of welfare with the welfare state providing a social safety net for the needy	Workfare state with reduced and privatized welfare security
International stability through multilateral agreements	International destabilization
National centralization	Decentralization from the national scale and heightened inter-local government competition
Managerial government supporting business with public services	Entrepreneurial government supporting business with corporate subsidies
Indirect intervention in markets through income and price policies	More direct state intervention in markets through procurement and corporate subsidies
Business-financed R&D	More government-financed R&D
Business-led innovation	More government-led innovation

Source: Based on Albrechts and Swyngedouw (1989: 75, Figure 1)

4. increasing independence of large monopolies from direct control and regulation by individual national states
5. decline in average plant size because of shifts in industrial structure, substantial labor-saving capital investment, the hiving off of various subcontracted activities, and the outsourcing of labor-intensive activities to the LDCs and to peripheral parts of core economies
6. decline of metropolitan dominance within core countries—the loss of jobs and population from inner-city areas, and an increase in jobs and population in smaller towns and some rural areas
7. a weakening of the degree to which industries are concentrated in specific countries and regions as a result of the new, variable geometry of the international division of labor
8. decline in the salience and class character of political parties, an increase in cultural fragmentation and pluralism, and individualized consumption.

7.2 SPATIAL REORGANIZATION OF THE CORE ECONOMIES

In this section, we examine two important trends in the economic geography of the core economies, both of which occur within the context of the globalization of economic activity and the shifts within core countries from traditional manufacturing to information economies. The first of these trends is the regional, inter-metropolitan and metropolitan *decentralization* of certain categories of both manufacturing and service employment. The second is the regional and inter-metropolitan *consolidation* of other kinds of activities.

1. SPATIAL DECENTRALIZATION AND EXTERNAL CONTROL

Decentralization operates at regional, metropolitan and inter-metropolitan scales in response to a variety of complex and often crosscutting processes of reorganization and adjustment.

Regional decentralization

Regional decentralization is a product of the migration of some firms and the "births" and "deaths" of others, together with the transfer of productive capacity by plant shutdowns in core, metropolitan regions and the opening of new branch plants (or the expansion of existing ones) in declining or peripheral cities and regions of core countries.

A useful distinction can be made between **diffuse industrialization** and **branch-plant industrialization** (Hudson, 1983). Diffuse industrialization has been directed towards the reserves of unskilled labor in rural regions. The areas of central and northeastern Italy customarily provided classic examples of diffuse industrialization that resulted from the decentralization of companies from the Milan-Turin area in response to the increasing shortage, cost, and militancy of labor there. Diffuse industrialization has typically involved activities in which labor costs were an important part of overall production costs *and* in which there was little scope for reducing labor costs through technological change; so it can be seen as an expression of the product–cycle model of industrial location. Empirical studies have shown that the main attractions of rural locations for such activities have been:

- the availability of relatively low-cost labor
- inexpensive supplies of easily developed land
- lower levels of taxation
- low levels of unionization.

In contrast, branch–plant industrialization has been directed towards the skilled manual labor reserves of declining industrial regions. It has typically involved activities that required significant inputs of technology and of skilled (or at least experienced) labor, and that also required a certain degree of centrality in order to assemble and distribute raw materials and finished products. Good examples are provided by many former textile cities—Dundee in the United Kingdom, for example, and Amiens in France—where branch plants in a variety of light industries (including light engineering and, more recently, IT) have moved in to take advantage of "surplus" labor, cheap factory space and an established infrastructure. It is not only manufacturing activities that are being decentralized, however, some places have attracted white-collar information-processing or wholesaling functions. Of the estimated 5 million customer service agents in the United States, Texas has the most with 450,000 employed in bricks and mortar call centers or using the less expensive cloud-based contact-center technology that allows agents to be "homesourced" through "phonesourcing." In addition to its reliable telecommunications and electric power infrastructure, what attracts companies is the relatively low cost of living and, some have argued, southern hospitality and Texan charm of the call center agents. Kentucky, for example, with its easy access to major population centers via major trucking routes, is home to several of Amazon.com's warehousing and order-fulfillment centers.

The twin processes of diffuse and branch–plant industrialization, combined with the process of mergers and acquisitions, have meant that regional decentralization has come to be

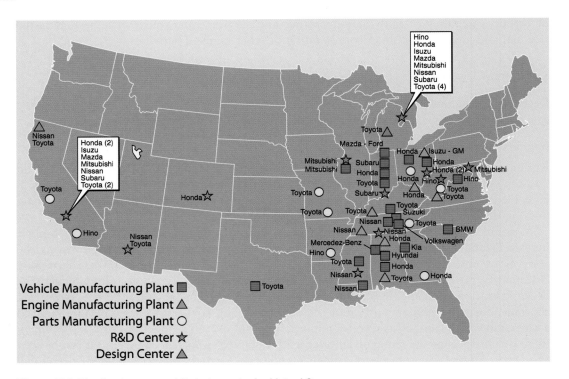

Figure 7.1 The foreign automobile industry in the United States

Source: Based on JAMA (2012: 4) and information on AIADA (American International Automobile Dealers Association) website at http://www.whatisanamericancar.com/ plants/

characterized by increasing levels of **external control**. It has been the large transnational corporations that have been particularly important in influencing the extent and spatial pattern of external control. In the United States, the total number of jobs in foreign-owned firms jumped from 2.0 million in 1980 to 5.6 million by 2012. Many of these jobs have been in manufacturing, and most have been controlled by British, Canadian, Dutch, French, German, Japanese, or Swiss companies. For example, Japanese-owned companies alone account for 12.5 percent of all foreign-owned firm employees. Many foreign automobile companies are clustered in the USA's "automobile alley" of states that include Indiana, Michigan and Ohio, Kentucky, Tennessee and Alabama (Figure 7.1). In high-tech, high-growth industries, about 20 percent of all employment in the United States is in foreign firms.

Because of the degree of external control involved in regional economic decentralization, it has become a moot point as to how much long-term benefit will accrue to the regions involved. On the *positive* side, it can be argued that branch–plant economies benefit by having access to the financial resources, and technological and administrative innovations of the parent firm. Moreover, some locations have attracted "higher order" corporate functions, such as research and development (R&D) (Figure 7.2). The location of foreign-owned industrial R&D facilities in the USA, for example, is concentrated in those areas that also offer specialized expertise in certain university departments: For example, Silicon Valley, around Stanford and Berkeley universities (for computers, semiconductors, and bioengineering), the Research Triangle Park in North Carolina (for biotechnology and telecommunications), and the Boston, Massachusetts region, particularly around the Massachusetts Institute of Technology (MIT) (for computers).

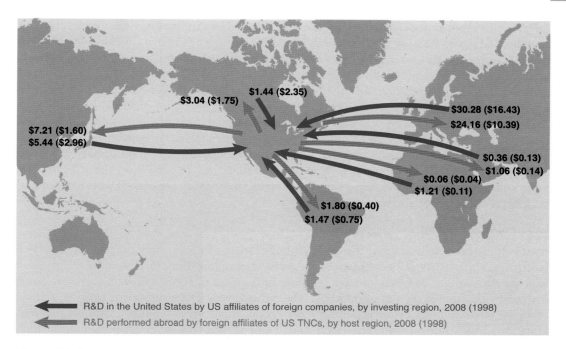

$3.04 ($1.75)

$1.44 ($2.35)

$30.28 ($16.43)
$24.16 ($10.39)

$7.21 ($1.60)
$5.44 ($2.96)

$0.36 ($0.13)
$1.06 ($0.14)

$0.06 ($0.04)
$1.21 ($0.11)

$1.80 ($0.40)
$1.47 ($0.75)

R&D in the United States by US affiliates of foreign companies, by investing region, 2008 (1998)

R&D performed abroad by foreign affiliates of US TNCs, by host region, 2008 (1998)

Figure 7.2 Research and development (R&D) in the USA by U.S. affiliates of foreign companies and R&D performed abroad by foreign affiliates of U.S. TNCs

Source: Based on National Science Board (2012: O-6, Figure O-6)

R&D in some regions is highly specialized in certain industries, such as Detroit, Michigan, for automotive facilities and Richardson, Texas, for telecommunications.

On the *negative* side, it has been suggested that the absence of "higher order" corporate functions in other locations can:

1. limit local employment opportunities, leading to a *deskilling* of the local workforce, to the suppression of entrepreneurial drive and enthusiasm, and to the retardation of technological innovation
2. result in a very open regional economy, so that international economic fluctuations are transmitted into the region relatively quickly. The corollary of this is that because externally controlled plants are poorly integrated with the local economy, their own potential multiplier effects are limited
3. increase the vulnerability of branch–plant economies to the further redeployment of capital—branch–plant economies in the core countries are placed in direct competition with those of the NIEs, which typically have much lower factor costs.

Metropolitan decentralization

Metropolitan decentralization (the relocation of industry and employment from inner-city areas to suburbs) can in fact be traced to the 1930s; but since the early 1970s the process has dominated patterns of urban development in a number of countries. Historically, the major impetus for metropolitan decentralization has been employers' desire to sidestep the increasing militancy of labor in inner-city neighborhoods. Suburban locations have also been attractive to many industries because of the availability of larger tracts of relatively cheap land. Given this basic attractiveness, successive improvements in transport and communications have greatly accelerated the process of decentralization.

Residential suburbanization, meanwhile, has provided labor supplies—including cheap, non-unionized, female labor—that have encouraged the suburbanization of firms. In this way, a mutually reinforcing process was set in motion. At the same time, the intensification of some of the locational disadvantages of inner-city areas—higher taxes, congestion, restricted sites and so on—pushed some firms out. What was most pronounced was the "shakeout" of routine and labor-intensive inner-city areas—some of it destined for relocation in the suburbs, but much more destined for relocation in rural areas, peripheral regions of core economies, peripheral countries or the bankruptcy courts.

Inter-metropolitan decentralization

At the inter-metropolitan level, the most striking aspect of decentralization involved service industries, particularly business services. Corporate reorganization, facilitated by advances in telecommunications, resulted in a general decentralization of routine business services down the urban hierarchies of core countries. The process was geographically selective, with a relatively small number of metropolitan areas (in the northeast and Sunbelt including Dallas, Houston, Los Angeles, San Francisco and Tampa) enjoying a disproportionately high proportion of employment in business services. This phenomenon was surprising to some observers, who had expected that new communications technologies would allow for the dispersion of "electronic offices," and, with it, the decentralization of an important catalyst for local economic development. A good deal of geographic decentralization of offices has occurred, in fact, but it has mainly involved back-office functions that have been relocated from metropolitan and business-district locations to small-town and suburban locations.

Back-office functions are record-keeping and analytical functions that do not require frequent personal contact with clients or business associates. The accountants and financial technicians of high-street banks, for example, are back-office workers. Developments in IT have enabled a large share of back-office work to be relocated to specialized office space in cheaper settings, freeing space in the high-rent locations occupied by the bank's front office. For example, the U.S. Postal Service uses optical character readers (OCRs) to read addresses on mail, which is then barcoded and automatically sorted for delivery. Digitally scanned images of addresses that the OCRs cannot read are transmitted electronically to the remote barcoding system which effectively reads almost all addresses, no matter how poorly written. Mail with addresses that still cannot be read using this automated system is sent to one of two mail recovery centers (formerly dead letter offices) for workers to examine, such as in Atlanta, where wage rates are relatively low.

A prominent example of back-office decentralization from a U.S. metropolitan area was the relocation of back-office jobs in American Express from New York to Salt Lake City (Utah), Fort Lauderdale (Florida), and Phoenix (Arizona).

Internationally, this trend has taken the form of decentralizing back-office functions to offshore locations to save even more in labor costs. For example, U.S. banking giant, Wells Fargo, has its business process outsourcing (BPO) operations in the Philippines while IT corporation, Hewlett Packard, has BPO in Colombia. Similarly, major home improvement retailer, Home Depot, and consumer credit-reporting agency, Equifax, have BPO operations in India. Figure 7.3 is an example of the business model of an Indian BPO provider. This comprises a global network of customer support offices, specialized delivery centers in lower

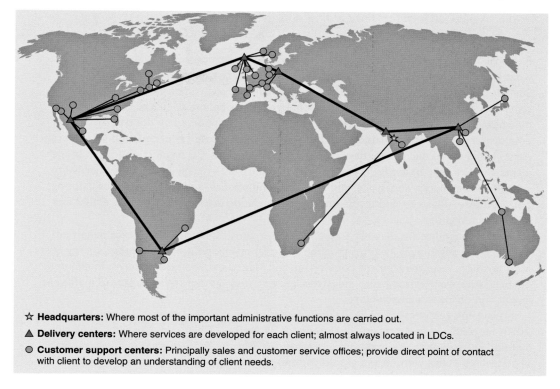

☆ **Headquarters:** Where most of the important administrative functions are carried out.

▲ **Delivery centers:** Where services are developed for each client; almost always located in LDCs.

◎ **Customer support centers:** Principally sales and customer service offices; provide direct point of contact with client to develop an understanding of client needs.

Figure 7.3 Geography of an Indian offshore services provider

Source: Adapted from Gereffi and Fernandez-Stark (2010: 33, Figure 8) http://www.cggc.duke.edu/pdfs/CGGC-CORFO_ The_Offshore_Services_ Global_Value_Chain_March_1_2010.pdf

cost locations and the headquarter offices. This structure allows global BPO companies to be close to their clients and understand what they need, and at the same time to undertake projects that take advantage of variations in employee expertise from different parts of the world (see also Box 6.5). Chapter 11 provides an in depth examination of BPO and the implications for both the LDCs and the DCs.

2. CONSOLIDATION AND AGGLOMERATION

The structural and functional consolidation of certain activities under advanced capitalism has made for countertrends that have strengthened the economic well-being of many of the largest and most central components of the economies of the core countries.

The fundamental reason for the consolidation of certain economic activities in such settings is that:

> Large towns offer larger local markets, with the associated internal economies of scale, plus greater external economies than are available in smaller places, and together these allow production costs which are often significantly lower than those in smaller towns: once transport costs began to fall substantially, so that they were less than the production cost differential between the large-town and the small-town firm, the former could begin the invasion of the latter's market.
>
> (Johnston, 1980: 110–111)

Box 7.1 The Sunbelt

Regional, metropolitan and inter-metropolitan decentralization are individual components in what is ultimately a multidimensional dynamic of spatial change. The growth of the U.S. Sunbelt provides a good example. The Sunbelt phenomenon can be interpreted as the *combined* product of diffuse industrialization, inter-metropolitan decentralization and metropolitan decentralization. Such an interpretation is supported by the types of employment growth that characterized the rise of the Sunbelt: (1) production jobs in branch plants in industries such as textiles, clothing and electronics; (2) production jobs in branch plants and in locally based firms in high-growth industries—mainly in computer hardware, scientific instruments, aerospace, and chemicals and plastics; and (3) service jobs catering both to these industries and to the increased population attracted to the retirement and leisure communities.

In very general terms, Sunbelt states such as Arizona, California, and Texas benefited from relative advantages in terms of labor costs, labor unionization, land costs, energy costs, local taxation, local government boosterism, and federal expenditure patterns. In addition, Sunbelt cities were attractive to industries because they did not have a legacy of inefficient layout and obsolete infrastructure.

Having enjoyed decades of heady growth, the economic downturn that began in the late 2000s hit particularly hard. Sunbelt states such as Arizona, California, Nevada and Florida suffered record job losses and home foreclosures. But the Sunbelt is rebounding, especially in states like Texas with oil and gas drilling and processing.

The sectoral shifts and manufacturing specializations of advanced capitalism have also worked in favor of many large cities and metropolitan regions. Manufacturers of many sophisticated new high-value-added products have been drawn to such locations. The reasons for this are several:

- the complex links that these new products have with established industries
- their dependence on risk capital in the early stages of development
- their need for access to a large, affluent and sophisticated market during the early stages of marketing.

Similarly, large parts of the expanding service sector have been drawn towards metropolitan locations because of the kind of environment and workforce required by information-processing, coordinating, controlling and marketing activities.

Corporate restructuring and new competitive strategies have added to the agglomerative and recentralizing trends of certain economic activities in metropolitan settings. Flexible production requires a new social division of labor with access to a large and fluid labor pool (containing part-time and temporary workers as well as highly skilled workers—attributes that are most readily found in metropolitan settings). Equally important, metropolitan settings are essential to the *externalization* of certain functions and the more extensive use of outside consultants, subcontracting, joint ventures, strategic alliances and collaborative R&D that characterize flexible production systems.

Box 7.2 Agglomeration and the "relational turn" in economic geography: Motorsport Valley

[I]t is precisely the social, institutional, cultural and political embeddedness of local and regional economies that can play a key role in determining the possibilities for or constraints on development; and thus why spatial agglomeration of economic activity occurs in particular places and not others . . . it is not merely a case of recognizing that the mechanisms of economic development, growth and welfare operate unevenly across space, but that those mechanisms are themselves spatially differentiated and in part geographically constituted; that is, determined by locally varying, scale-dependent social, cultural and institutional conditions.

Martin (1999: 75, 83)

Many economic geographers have become increasingly concerned with the ways in which the socio-spatial relations among agents and structures shape the spatial organization of economic activities. In 2003 the *Journal of Economic Geography* devoted a special issue to the "relational turn" in economic geography (see, for example, Boggs and Rantisi, 2003).

Henry and Pinch (2000) offered a case study of the agglomeration of the British motorsport industry in Motorsport Valley to show how this relational economic geography allows economic geographers to conceptualize geographical specialization as the construction of a socially embedded and "relational" economic system. By global standards, the British motorsport industry is a classic example of a leading regional agglomeration. This agglomeration—the Silicon Valley of Motorsport—is located in the vicinity of Oxford (northwest of London) and dominates the world's racing car industry.

Motorsport Valley began as a network of small companies but now includes close links with major international automobile manufacturers, including Ferrari, Ford and Mercedes-Benz. About three-fourths of the world's single-seater racing cars are designed and assembled in this region, including the vast majority of the most competitive Formula One and IndyCar cars. The region is also the base for a large number of rallying teams.

Henry and Pinch argue that Motorsport Valley can best be conceptualized as a "knowledge community" that comprises a socially and spatially embedded economic system facilitating the generation and rapid dissemination of knowledge about the best ways to design and manufacture racing cars. Their case study analysis found that a key characteristic of the industrial organization and labor market of motorsport is a set of processes involving a continual "churning" of people and ideas, in this case, centered on, and within, Motorsport Valley. This "churning" is a process of producing and circulating knowledge within the knowledge community and regional production center of Motorsport Valley. As workers move among companies, for example, they carry with them knowledge and ideas about how things are done in other companies, which helps to raise the level of knowledge throughout the industry and within the region.

Henry and Pinch report on concerns about the geographical mobility of motorsport production within the context of the growth of industry in the NIEs (see Chapter 10). Because the success of Motorsport Valley depends largely on knowledge, it may be easy to shift this expertise to another region, for example, to China where major investments have been made to promote a motorsport industry there. These authors argue, however, that this concern underestimates the (knowledge-laden) production process of the fast moving motorsport industry. "The knowledge of the British motorsport industry is encapsulated in particular people, objects and ways of doing things which are themselves constructed *in a particular place*" (Henry and Pinch, 2000: 140) in this case, Motorsport Valley.

Finally, the national and international redeployment of activities by large conglomerate companies has also contributed to the consolidation of certain activities in the central regions and metropolitan areas of the developed countries. In particular, there has been *a marked localization of two key functions: Headquarters offices and R&D establishments*. Indeed, the distribution of these two functions has come to represent an important dimension of the economic geography of advanced capitalism.

CORPORATE CONTROL CENTERS AND WORLD CITIES

The United States provides a good example of the changing geography of corporate headquarters. Historically, the most striking feature of the geography of corporate headquarters in the USA had been the dominance of the Manufacturing Belt in general and of New York and Chicago in particular. Elsewhere, the pattern of headquarters offices had tended to reflect the geography of urbanization, so that the more important "control centers," in terms of business corporations, had been the major entrepôts and central places that developed under earlier phases of economic development, as points of optimal accessibility to regional economies.

With advanced capitalism the relative importance of the control centers of the Manufacturing Belt has decreased somewhat, with cities in the Midwest, and in the south and the west increasing their share of major company headquarters offices. Atlanta, Dallas, Houston, Minneapolis, and St. Louis have been the major beneficiaries of this shift, although no new control centers have emerged to counter the dominance of New York. One interpretation of this shift is that it is simply a reflection of changes in the central place system: High-order urban areas tend to be higher order business control centers because of their reserves of entrepreneurial talent, the array of support services they can offer, and their accessibility in both a regional and a national context.

In overall terms, "there has been a process of *cumulative and mutual reinforcement* between relatively accessible locations and relatively effective entrepreneurship" (Borchert, 1978: 230; emphasis added). This has made for a high degree of inertia in the geography of economic control centers and this, in turn, has consolidated the economic position of the metropolitan areas of the northeast through the multiplier effects of concentrations of corporate headquarters, whereby the vitality of the corporate administrative sector contributes to the growth and circulation of specialized information concerning business activity, which generates further employment in a relatively well-paid sector and sustains the area's attractiveness for headquarters offices.

The concentration of corporate headquarters offices has contributed to the emergence of a few places within the international urban system as world cities, dominant centers and sub-centers of transnational business, international finance and international business services—what Friedmann (1986) called the "basing points" for global capital. These world cities, it should be stressed, are not necessarily the biggest within the international system of cities in terms of population, employment or output. Rather, they are the "control centers" of the world economy: Places that are critical to the articulation of production and marketing under the contemporary phase of world economic development. Because these properties are difficult to quantify, it is not possible to establish a definitive list or hierarchy of world cities. It is possible to identify world cities on the basis of their role in articulating the functions of the world economy associated with financial markets, major corporate headquarters, international institutions, communications nodes and concentrations of business services (Figure 7.4). On this basis, all three of the dominant world cities—New York, London and Tokyo—are located

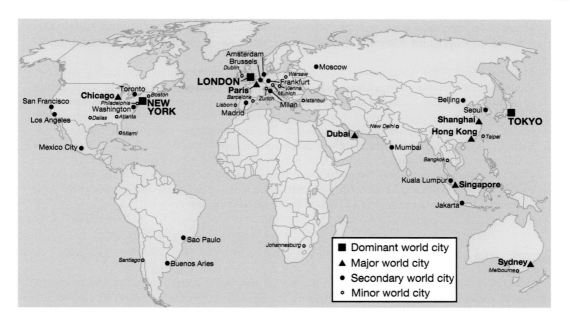

Figure 7.4 The system of world cities

Source: Adapted from Globalization and World Cities (GaWC) 2010 ranking http://www.lboro.ac.uk/gawc/

in core countries, while, importantly, four of the seven major world cities—Hong Kong, Singapore, Shanghai and Dubai—are not. The relative importance of secondary and minor world cities is very much a function of the strength and vitality of the national economies that they articulate.

Despite the limitations of attempting to portray the system of world systems on a map, and the impossibility of capturing the major economic and other interconnections among them, following Friedmann, it is possible to discern a linear character to the world city system that connects, along an east–west axis, three distinct but interrelated subsystems: An Asia-Pacific subsystem centered on the Tokyo-Shanghai-Hong Kong-Singapore-Sydney axis, with Seoul, Beijing, Kuala Lumpur, and Jakarta as important secondary world cities, and quite a few minor world cities such as Taipei, Bangkok, and Melbourne growing in global importance; an American subsystem based on the primary world cities of New York and Chicago, linked to secondary world cities such as Los Angeles and San Francisco in the west, Toronto in the north and Buenos Aires, Mexico City, and São Paulo in the south, which brings Canada and Central and South America into the U.S. orbit; and a European subsystem focused on London and Paris, with linkages across a large number of secondary and minor world cities from Madrid to Moscow and from Brussels to Milan.

Friedmann, and later Sassen (2001), contributed to a better understanding of the global context for how, for example, business services tend to agglomerate in particular corporate control centers, and particularly in world cities. This work, however, tended to result in a focus on the agglomeration within individual world cities such as New York, London, and Tokyo. Following Castells' (2000) notion of a "space of flows," Taylor's (2004) world city network concept additionally stresses the importance of the interrelationships between world cities. For Taylor, the agglomeration of business activities in particular world cities is less important than the global connectivity of these cities. In fact, Taylor's conceptualization of world cities in the world economy fits nicely with Immanuel Wallerstein's world-system theory that forms an important underlying explanatory framework for this book (see Chapter 2):

> Wallerstein's (1979a) description of core processes can be interpreted as city-making processes (both produce spatially clustered high-tech outcomes), and peripheral processes—the development of underdevelopment . . . In the contemporary world-economy, therefore, the core is defined by the processes of new work that are constituting the world city network, and the periphery is the rest of the world beyond the world city network. The semi-periphery is defined in Wallersteinian terms as locales where core and periphery processes are approximately balanced; these are cities in the erstwhile "third world" that are now part of the world city network but are also "mega-cities" (pernicious population "town" growth that is a periphery process). This is what makes cities such as São Paulo, Mexico City, Mumbai, Johannesburg and Bangkok among the most interesting settlements in the first decades of the twenty-first century.
>
> (Taylor, 2007: 296)

AGGLOMERATION AND BUSINESS SERVICES

Business services tend to agglomerate in particular corporate control centers, and particularly in world cities. Barney Warf (2007) argues for the use of actor-network theory to better understand why some forms of service production are concentrated in a small number of urban centers like world cities, while other services are more dispersed globally. He makes a distinction between two kinds of knowledge: Standardized knowledge, which includes forms of information that are easily transmitted from one person to another, such as quantitative data, publicly known rules and standards, and orderly records; and tacit knowledge, which includes information that is not standardized, changes rapidly, and is often not put in writing. Actor-network theory focuses on questions of power, politics and social relations, and highlights the fact that the global service economy is the contingent outcome of different individuals and groups situated in networks. In conjunction with the "cultural turn" in economic geography (see Box 7.5), the use of actor-network theory is a way to help make sense of the emerging geographies of centrality and peripherality unleashed by the globalization not only of manufacturing but also of services:

> High value-added services, using skilled labor and tacit forms of knowledge, are highly agglomerated in the world's global cities. Such functions tend to be deeply embedded territorially and thus the competitive advantages of established centers are difficult to reproduce. In contrast, relatively low value-added service functions, such as back offices, call centers, and offshore banks, are increasingly dispersed to the world's low wage periphery. These operations, relying upon disembedded, standardized knowledge, are footloose and change locations frequently. These two sets of services represent opposite poles of one continuous process that geographically segregates functions on the basis of their value-added and types of skills and knowledges utilized. Both types of services are embodied in people and embedded in local and international contexts, forming complex mixtures of the local and global.
>
> (Warf, 2007: 1)

Research on changing occupational structure and the use of computers in the workplace by Levy and Murnane (2004) supports this assessment. They found that since the 1960s in the United States, the percentage of employees has risen in occupations that emphasize "expert thinking" involving solving problems for which there are no rule-based solutions. Examples include diagnosing an illness of a patient whose symptoms seem strange or repairing an automobile that does not run well but which the computer diagnostics indicate has no problems.

Similarly, the percentage of employees has risen in occupations that emphasize "complex communication" involving interacting with other workers in order to acquire information, to explain it, or to persuade others of its implications for action. Examples include managers motivating people whose work they supervise or an engineer describing why a new design for a Blu-Ray player is an improvement over previous designs. While computers can help, these are not tasks that computers can be programmed to solve, and so are not easily amenable to outsourcing to the LDCs.

In contrast, the percentage of employees in the United States has fallen in occupations that emphasize routine cognitive tasks requiring mental skills that are well described by logical rules. Examples include recording new information provided by insurance customers and evaluating mortgage applications. Because these tasks can be accomplished by following a set of rules, they are prime candidates for computerization, and to outsourcing to the LDCs.

CENTERS OF INNOVATION

As Castells notes (2000: 65), the development of the information technology revolution has contributed to the formation of milieux of innovation where important commercial discoveries interact and are tested in a recurrent process of trial and error. These innovation complexes require "spatial concentration of research centers, higher-education institutions, advanced-technology companies, a network of ancillary suppliers of goods and services, and business networks of venture capital to finance start-ups."

The geography of these complexes has important implications for urban and regional development. Malecki, who has examined the geography of R&D activity in the USA in detail, suggests that the overall pattern can be interpreted in terms of:

1. availability of highly qualified personnel
2. corporate organization.

In relation to the availability of highly qualified workers, he suggests that amenity-rich locations (cities with a wide range of cultural facilities, well-established universities and pleasant environments), which are attractive to highly qualified personnel tend to be favored as locations for R&D activity. Malecki (1991) also noted that existing concentrations of R&D activity tend to be attractive because of the potential for "raiding" other firms.

In relation to corporate organization, it seems that corporate-level or long-range R&D is best performed in or near corporate headquarters in a central laboratory where organizational interaction within a company can be fostered. In firms with independent divisions producing quite different product lines, however, R&D activity tends to be located in separate divisional laboratories. Such a pattern is particularly common for conglomerates that have acquired firms with active R&D programs in existing laboratories. Finally, some industries, whatever the organizational structure of the firms involved, require R&D laboratories to have close links with production facilities, resulting in a relatively dispersed locational pattern corresponding to the pattern of plant location.

The net result of these locational forces is in fact *a marked agglomeration of R&D laboratories in major control centers.* As Malecki points out, most of these major control centers have a significant element of headquarters office activity. Elsewhere, R&D tends to be concentrated in "innovation centers"—university cities with diversified economies, some high-technology activity and a strong federal scientific presence (for example, Austin, Texas; Huntsville, Alabama).

In terms of locational *trends*, Malecki has shown that:

> Although industrial R&D appears to be evolving away from a dependence on some large city
> regions, especially New York, it remains, at the same time, a *very markedly large-city activity*.
> ... The comparative advantage of city size, particularly in centers of corporate headquarters location,
> manufacturing activity and university and government research, shows little sign of reversing.
>
> (Malecki, 1979: 321; emphasis added)

In short, R&D laboratories, like headquarters offices, exhibit a strong tendency for consolidation, accompanied by a certain amount of decentralization. This pattern has important implications for regional economic development because the urban areas in which concentrations of R&D activity exist will in future be able to consolidate their competitive advantage over other areas in generating new products and businesses. They will also benefit from the short-term multiplier effects of employment generation in a particularly well-paid sector. Conversely, cities and regions with little research and development activity will be at a disadvantage in keeping up with the economic content of advanced capitalism.

7.3 OLD INDUSTRIAL SPACES

One of the most striking overall changes within core economies has been the decline in traditional manufacturing employment. Initially, this took the form of a relative decline: Growth in the postwar boom period was much greater in the service sector of most economies than it was in the manufacturing sector. With the globalization of economic activity, however, there has been an *absolute* decline in core manufacturing employment (see, for example, Figures 7.5 and 7.6). Whereas in 1960 manufacturing in the most industrialized countries generated between 25 and 42 percent of the GDP and accounted for similar proportions of their employment, the comparable figures for 2011 were in the range of 15 to 20 percent of GDP and 20 to 30 percent of employment.

The decline has been most pronounced in the early industrializers of northwestern Europe. In the United Kingdom, for example, more than 1 million manufacturing jobs disappeared, in *net* terms, between 1966 and 1976—a fall of 13 percent. This decline affected almost every sector of manufacturing, not just the traditional pillars of manufacturing—shipbuilding (–9.7 percent), metal manufacture (–21.3 percent), mechanical engineering (–14.5 percent) and textiles (–26.6 percent)—but also its former growth sectors—motor vehicles (–10.1 percent) and electrical engineering (–10.5 percent). In the West Midlands, traditionally regarded as a "leading" region in Britain, a net loss of 151,117 manufacturing jobs between 1978 and 1981 helped to redefine the region as part of Britain's Rustbelt. It is within the peripheral regions of the United Kingdom that the problem has been most acute, however. In Lancashire, for example, the textile industry alone shed over half a million jobs.

In short, the decline of the traditional industrial base has been most pronounced in the regions that had come to be most specialized in Fordist industrial production. For some communities in these regions, the consequences of plant shutdowns were disastrous. In Youngstown, which became the symbol of U.S. industrial decline, the closure of the Campbell Steel Works in 1977 eliminated over 10,000 jobs at a stroke. In overall terms, it has been estimated that the mid-Atlantic region (New Jersey, New York, Pennsylvania) experienced a net loss of over 175,000 jobs during the period 1969–1976, whereas the south Atlantic region (Delaware, District of Columbia, Florida, Georgia, Maryland, North Carolina, South Carolina, Virginia, West Virginia) experienced a net gain of over 2 million jobs in the same period. This

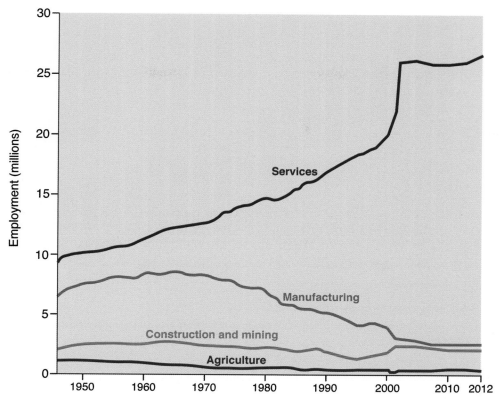

Figure 7.5 Employment by sector in the United Kingdom

Source: Based on online data from U.K. Office for National Statistics, *Workforce Jobs by Industry* http://www.ons.gov.uk/ons/
datasets-and-tables/index.html?pageSize= 50&sortBy=none&sortDirection=none&newquery=JOBS02{lt}Workforce+jobs+
by+industry+%28seasonally+adjusted%29

represents a job loss of 1.5 percent in the mid-Atlantic region and a gain of 24.4 percent in
the south Atlantic region, compared with a net gain of some 15 percent in the U.S. as a whole.

Deindustrialization on this scale brought with it a number of *downward-spiraling* multiplier
effects, including the substantial contraction of major segments of vertically integrated
production chains (for example, ore mining, coalmining, steel production, marine engineering
and shipbuilding) and the disappearance of inefficient, more labor-intensive firms and sections
of production chains (for example, in textiles), leaving the old industrial regions only finishing,
specialized and high-quality product lines (for example, in clothing), which then become
dependent on supply linkages that are "stretched" overseas.

It would be misleading, however, to place too much emphasis on the demise of old
industrial regions. Many cities within such regions have been making the transition to an
advanced economy. Within the Ruhrgebiet of Germany, for example, an economic renaissance
is in evidence at the site of an old steel plant, which was bulldozed to make way for one of
Europe's largest shopping and entertainment centers: A US$1.5-billion megamall with over
200 stores, a 12,500-seat arena, a 1200-seat food court, restaurants, bars, clubs, hotels,
multiplex cinemas, an artificial canal, an amusement park, an aquarium, and 10,500 parking
spaces. The only structure to survive from the steel plant is a huge gas tank, which contains
exhibits and offers a view from the top when using the elevator. CentrO, as the huge shopping

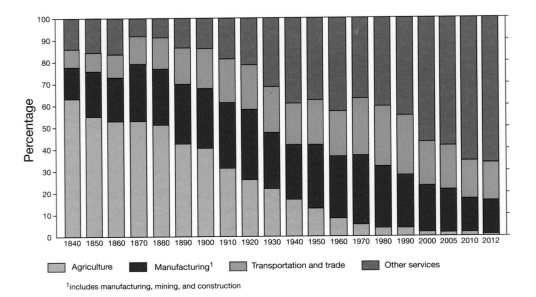

Figure 7.6 Employment shares, by economic sector, USA

Source: Based on online Current Population Survey (CPS) data from U.S. Bureau of Labour Statistics, *Employment and Earnings, 2012* http://www.bls.gov/cps/

center that opened in 1996 near Oberhausen is called, attracts 23 million visitors every year. Such developments show that local efforts can help to counter the regional economic decline associated with deindustrialization. Nevertheless, it is important to note that, in comparison with the jobs lost in traditional industries, employment in retailing, food and leisure provides a much less desirable base: Typically, jobs are less well paid, with less security and fewer benefits. Major new shopping malls such as this one, can also draw shoppers away from existing retail and entertainment establishments in a region.

We should also note that the process of deindustrialization (involving the *relative* decline of manufacturing jobs), while localized within old industrial regions affected every region within core countries. In the United States, for example, even California, an archetypal Sunbelt state, was seriously affected by company shutdowns. In Los Angeles alone, almost 18,000 manufacturing jobs were lost between 1978 and 1982, many of them the result of plant closures by large corporations such as Ford, Pabst Brewing, Max Factor, Uniroyal, and U.S. Steel. In the state as a whole in the single year of 1980, more than 150 large plants closed down, displacing more than 37,000 workers. In short, the overall losses of the Manufacturing Belt concealed a complex and uneven pattern of ebbs and flows.

7.4 NEW INDUSTRIAL SPACES

Advanced capitalism has also been associated with the emergence of *new* industries based on entirely new technologies: Semiconductors and computer software, for example, and more recently, biotechnology, nanotechnology and robotics. So the possibility arises of an entirely new dimension to the economic landscapes of the developed countries, with concentrations of the newest high-tech (sunrise) industries initiating new patterns of urban and regional growth through new production systems with new multipliers of **cumulative causation**.

Studies of high-tech industries in the United States confirm that job creation has been significant. Today nearly 6 million people (just over 4 percent of total U.S. employment) are employed in high-tech industries (comprising about 1.3 million in high-tech manufacturing and more than 4.6 million in the high-tech services industry (communications services with about 1.2 million, software services with almost 1.8 million and engineering and tech services with 1.6 million)).

The growth of some of these industries has certainly been explosive. Employment in software services in the United States, for example, doubled during the 1970s to 250,000, and has now grown to nearly 1.8 million. Some estimates suggest that between now and 2020 employment in robotics will add as many as 3.5 million new jobs worldwide. Of course, jobs in other industries are being *displaced* by the application of robotics. Whether the growth in high-tech industries will be sufficient to cancel out the effects of continued deindustrialization and outsourcing of manufacturing from the core economies is by no means certain, however.

It should also be noted that in terms of occupational structure the expansion of high-tech employment is a microcosm of the trends that have dominated advanced capitalism (in this connection, see also Box 7.3). Studies in California, for example, "suggest that the occupational, ethnic and gender composition of new jobs in high-tech sectors will tend to worsen the current trend toward the 'disappearing middle', that is, toward a labor force bifurcated between high-paid professionals and low-paid service workers" (Markusen, 1983: 19). In relation to corporate structure, high-tech industry is distinctive for its tendency towards the proliferation of small breakaway companies set up by key employees; but at the same time the larger and more established firms have soon been drawn into the process of mergers and acquisitions, either as the dominant element (in horizontal and vertical integration) or as a subsidiary element (in diagonal integration).

TECHNOPOLES

The locational impact of high-tech activities has received a great deal of attention, including technology-oriented complexes, or **technopoles,** and an archetype in Santa Clara county—Silicon Valley—in California. In the 1950s, Santa Clara was a quiet agricultural county with a population of about 300,000. By 1980 it had already been transformed into the world's most intensive complex of high-tech activity, with a population of 1.25 million. Silicon Valley employment in high-tech jobs has risen to nearly 200,000. Each new high-tech job creates at least two or three additional jobs in other sectors—an extremely high multiplier effect.

The initial development of Silicon Valley is generally attributed to the work of Frederick Terman, a professor (and, later, vice-president) of Stanford University at Palo Alto, in the northwestern corner of Santa Clara county. As early as the 1930s Terman began to encourage his graduates in electrical engineering to stay in the area and establish their own companies (one of the first was founded by William Hewlett and David Packard in a garage near the campus; it is now one of the world's largest electronics firms). By the end of the 1950s Terman had persuaded Stanford University to develop a special industrial park for such fledgling high-tech firms, creating a hothouse of innovation and generating significant external economies—including a specialized workforce and a specialized array of business services—which have not only sustained the continued agglomeration of high-tech electronics enterprises but also attracted other high-tech industries. With about 2,500 biomedical companies, California, and the San Francisco Bay area in particular have the largest concentration of biotechnology companies in the country. California's more than 150,000 biotech workers represent more

Box 7.3 The digital divide

The **digital divide** refers to the gap in opportunities between individuals, households, businesses, and areas at different socioeconomic levels to access information technology (IT) for a variety of activities. The digital divide exists both within countries (for example, between urban and rural areas or between richer and poorer neighborhoods) and between countries or groups of countries (for example, between the LDCs and the DCs). Despite the initial promises that IT could benefit everyone everywhere, as Warf (2001: 3, 16) argues:

> [G]eography still matters . . . electronic systems simultaneously reflect and transform existing topographies of class, gender, and ethnicity, creating and recreating hierarchies of places mirrored in the spatial architecture of computer networks. Far from eliminating differences among places, systems such as the Internet allow their differences to be exploited . . . often reinforcing existing relations of wealth and power.

Within DCs such as the USA, for example, while there is increasing use of the Internet by people regardless of race, ethnicity, income, education or gender, Internet users still predominantly belong to urban and suburban, higher income, educated, white households. Globally, the United Nations has reported that the digital divide is closing as the cost of telephone and broadband Internet services fall; this has allowed some LDC governments to expand access to IT. Table 7.2 indicates that the LDCs have very high rates of growth of Internet users. In this connection, Africa is the fastest growing cellphone market in the world. A 2010 report by the International Telecommunications Union reported that more people in Africa use cellphones than anywhere else in the world. Whether making a call, transferring money or checking the market price of crops, personal cell phones, and shared village cellphones are helping to bridge the digital divide in Africa.

At the same time, however, the LDCs continue to have the highest percentages of their populations who are not Internet users (Table 7.2). The primary concern is that lack of access to and development of information, communication and e-commerce technology will prevent many people from benefiting from the knowledge-based economy. In this regard, governments and other organizations operating within the LDCs in particular continue to face enormous challenges as they attempt to promote growth in and access to information technology. Meanwhile, although the digital divide between the LDCs and DCs has been an international topic of debate for a number of years, relatively limited action or funds have been forthcoming from the DCs to help address the issue.

than double the number employed in Massachusetts, the state with the second largest concentration. Stanford University, meanwhile, found itself in receipt of an increasing flood of donations from grateful companies.

This kind of linkage between university research and high-tech activity is seen by many to be the key to maintaining regional competitiveness in the twenty-first century. Not only do the new industries thrive on a symbiotic relationship with one another and university research departments, but key workers also tend to favor technology complexes associated with top-flight universities because they provide abundant social and cultural activities and a job market that allows individuals (and partners) to switch jobs without relocating. Such regions soon acquire a reputation as "the right place to be at," and this often counts for more than cost-

Table 7.2 World Internet users

	Internet users				
	(millions)	(millions)	Percent of nat. pop. (%)	Percent growth (%)	Percent of of world pop. (%)
	2000	2012	2012	2000–12	2012
Asia	114.3	1,076.7	27.5	841.9	44.8
Europe	105.1	518.5	63.2	393.4	21.5
North America	108.1	273.8	78.6	153.3	11.4
Latin America and Caribbean	18.1	255.0	42.9	1,310.8	10.6
Africa	4.5	167.3	15.6	3,606.7	7.0
Middle East	3.3	90.0	40.2	2,639.9	3.7
Australia and Oceania	7.6	24.3	67.6	218.7	1.0
World total	*361.0*	*2,405.6*	*34.3*	*566.4*	*100.0*

Source: Based on Internet World Stats data, available at http://www.internetworldstats.com/stats.htm

of-living or quality-of-life factors. Where, as in Silicon Valley, the "right place to be at" happens to offer the additional bonus of an attractive environment and climate, the result is explosive growth. An important point in this context, as Hall (1981: 536) observed, is that "university systems, even in a country as dynamic as the United States, have a great deal of built-in inertia." Large, top-drawer universities like Harvard, MIT, Berkeley, and Stanford are secure in their status, but few other institutions seem destined to join them. The result is that, outside these potential areas, there are few places in the U.S. where a high-tech industrial base is likely to be developed—apart, perhaps from the Research Triangle (Raleigh-Durham-Chapel Hill) that has already been established in North Carolina around Duke University and the University of North Carolina.

Similarly, there are few environments in other developed countries that are likely to attract a critical mass of high-tech activity, despite the proliferation of technology parks—or, to be more accurate, *designated* technology parks. One exception is France, where an ambitious national program has designated more than 70 "competitiveness clusters" of either world class or national status (Figure 7.7), with special tax breaks and subsidies designed to attract not only high-tech industries such as biotechnology, microelectronics, and photonics but also to develop a supportive infrastructure of universities and R&D labs.

At the same time, technopoles come in a variety of formats:

Most notably, it is clear that in most countries, with the important exceptions of the United States and, to some extent, Germany, the leading technopoles are in fact contained in the leading metropolitan areas: Tokyo, Paris-Sud, London-M4 Corridor, Milan, Seoul-Inchon, Moscow-Zelenograd, and at a considerable distance Nice-Sophia Antipolis, Taipei-Hsinchu, Singapore, Shanghai, São Paulo, Barcelona, and so on.

(Castells, 2000: 421)

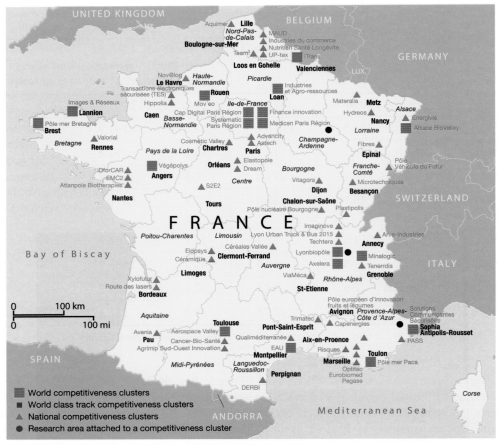

Figure 7.7 French competitiveness clusters

Source: Adapted from Competitiveness Clusters Agency of the French Government (2011: 4) http://competitivite.gouv.fr/

DECENTRALIZATION OF HIGH-TECH EMPLOYMENT

Almost all high-technology complexes are a suburban phenomenon. As Markusen (1983: 26) noted in relation to the early development of the Silicon Valley and Route 128 (Boston) complexes, they were "newly developed, auto-based, suburban areas whose jobs and tax base do not overlay the inner-city poor nor the central city jurisdiction." But, because high-tech firms have tended to be very self-conscious about their "address," these suburban complexes have become crowded and expensive. The outcome has been the familiar combination of corporate functional and spatial reorganization. More routine production tasks and downstream marketing and service functions have been dispersed, while managerial and developmental activities have been retained in order to maximize the external economies of the "right address."

U.S. computer firms have typically kept their R&D and administrative activities in places such as California and Massachusetts while moving their production facilities to southern states to take advantage of lower labor costs. Furthermore, some of the larger corporations in the computer and semiconductor fields have redeployed internationally, also partly in search of cheaper labor, both highly qualified and semi-skilled. There is now an international division of labor in the semiconductor industry, for example. The world's largest manufacturer, Intel, with over 100,000 employees worldwide, has its R&D organized internationally from its head-quarters in Silicon Valley. About half of Intel's wafer fabrication, including microprocessors and chipsets, are produced at the company's facilities in Arizona, New Mexico, Massachusetts and Oregon; the other half is manufactured outside the U.S. in Ireland, Israel and China. Most of Intel's components are subject to assembly and testing at facilities in China, Costa Rica, Malaysia, and Vietnam. As with the local branch–plant economies generated by the decentralization of traditional manufacturing industries, these regional concentrations of decentralized high-tech industry do not always generate many local linkages or multiplier effects.

FLEXIBLE PRODUCTION REGIONS

While the imprint of the high-tech industries of advanced capitalism cannot be said to amount to an entirely new dimension of the economic landscapes of the core countries, their industrial spaces have clearly contributed an additional component to existing landscapes. Meanwhile, however, other industries have been changing, leaving their imprint on the economic geography of the core. The crisis of Fordism, combined with the opportunities afforded by new production-process and circulation technologies, by changing patterns of consumer demand, and by corporate restructuring and new competitive strategies, has led in the last several decades to the emergence of something that *can* be described as a new dimension of the economic landscapes of the core countries. Flexible production regions have emerged in many core economies as a result of the interplay of flexible production systems, existing labor markets and the fixed capital of older industrial spaces.

Flexible production regions, which may contain elements of branch–plant industrialization (see p.183) along with a mixture of other functions and activities in which the emphasis on flexibility results in the externalization of certain functions and the vertical disintegration of organizational structures, which, in turn, lead to locational convergence and spatial agglomeration. Allen Scott, who has contributed most to this interpretation, summed up the central tendency as: "Vertical disintegration encourages agglomeration, and agglomeration encourages vertical disintegration" (1986: 224). The result is a series of regions or production complexes whose dynamics "revolve for the most part around the social division of labor, the formation

Box 7.4 The demise of the Celtic Tiger

Comparing Ireland's incredible economic boom to that of the **Asian Tigers** (Hong Kong, Singapore, South Korea, Taiwan), economist Kevin Gardiner, when working for the U.S. investment bank, Morgan Stanley, coined the term **Celtic Tiger**.

In the mid-1980s Ireland had an unemployment rate of nearly 20 percent, the highest debt per capita in the world, and a GDP per capita of only 63 percent of its nearest neighbor, the United Kingdom. Beginning in the early 1990s, Ireland began to enjoy astonishing growth rates of between 5 and 10 percent of GDP. By the end of the decade, unemployment was down to 4.5 percent, the national debt was down, and the country's GDP per capita had outstripped those of the United Kingdom and even Germany.

The underlying causes of Ireland's economic growth and massive foreign investment, especially from U.S. TNCs, included its openness to international trade and investment, low corporate taxes, low wages, a skilled workforce from decades of government investment in education, a stable national economy, appropriate budget policies, EU membership and adoption of the euro, and regional aid for investment in infrastructure and training from the European Union.

The Celtic Tiger roared until the global economic downturn of 2001 when the economy was impacted by the significant decline in investment in the global IT industry. By the end of 2003, however, the so-called Celtic Tiger Mark 2 was showing signs of recovery with an annual average of about 5 percent in GDP growth. Despite being a small country—with a population of just under 4.5 million—Ireland nonetheless produced one-fourth of all personal computers sold in Europe. TNCs including Intel, Microsoft and Facebook have facilities in Ireland.

Concerns, however, about the future growth of the Celtic Tiger Mark 2 related to the country's rising wages, inflation, infrastructure that has failed to keep pace with the rapid growth, excessive public spending, and the implications for Ireland's continued competitiveness of the accession to the European Union of eight eastern European countries in 2004 and two more in 2007. Additional concerns related to the slowdown in national productivity growth rates, which had been among the highest of the OECD countries. Of greatest concern was the fact that economic growth in Ireland was shifting away from exports and toward domestic housing construction and consumer demand financed by unsustainably high levels of personal borrowing from loosely regulated Irish and foreign banks. When the global financial crisis hit, the Celtic Tiger screeched to a halt as foreign banks stopped lending, exports plunged, and property values plummeted as construction stopped. Having bailed out its banks, Irish public debt, at a mountainous 130 percent of GDP, forced the government to accept a massive bailout of $110 billion from the European Central Bank and the International Monetary Fund. The bailout came with the requirement to implement stringent austerity measures to reduce government debt that resulted in raising taxes, slashing government spending and payrolls, and lowering the minimum wage with social consequences for ordinary citizens including falling living standards and rising unemployment and emigration.

Table 7.3 Propulsive industries and new industrial spaces

Propulsive sector	Typical features	Examples
Design/craft industries		
(a) Labor-intensive e.g., clothing, furniture	Exploitation of "sweatshop" labor; often high level of immigrants; subcontracting	New York, USA Los Angeles, USA Paris, France
(b) Design-intensive e.g. jewellery, watches, ceramics	High-quality products. Extreme social division of labor (but class polarization subdued in some examples)	Jura, Switzerland Emilia-Romagna (Third Italy), Italy, Jutland, Denmark
High-technology industries		
	Segmented local labor markets with skilled managerial cadres and malleable (non-union; temporary) workers	Silicon Valley, CA, USA Route 128, Boston, USA M4 corridor, UK Grenoble, France Montpellier, France Sophia Antipolis, France

Source: Updated from Tickell and Peck (1992: 199, Table 2)

of external economies, the dissolution of labor rigidities, and the reagglomeration of production" (1988a: 181).

The archetypal flexible production region has been the so-called Third Italy (Emilia-Romagna, Tuscany, the Marches, the Abruzzi, and Venetia), where branch–plant industrialization combined with highly skilled local labor markets, well-developed infrastructure, and economies of scale and scope arising from the spatial division of labor between specialized firms to create a regional network of innovative, flexible and high-quality manufacturers whose products include textiles, shoes and ceramics (see Box 6.4). Other examples of flexible production regions based on a similar mixture of design- and labor-intensive industries include Jutland (Denmark) and the Swiss Jura (Table 7.3).

Networks of manufacturers in high-technology industries form the basis of a second group of flexible production regions. Here, the agglomerating tendencies of these industries have resulted in localized growth in relatively particular metropolitan settings: For example Orange County and Silicon Valley in California, Grenoble, Lyon, and Montpelier in France; and the M4 corridor in Britain (Table 7.3).

These examples support Scott's observation that flexible production regions "are almost always some distance—socially or geographically—from the major foci of Fordist industrialization" (1988b: 14). The argument is that the interests of flexibility are best served by avoiding the rigidities (from outdated infrastructure to outdated institutions and labor relations) of Fordist settings. Yet we must recognize that it is quite possible for flexible manufacturers to establish successful enclaves *within* older industrial regions and metropolitan areas. Examples include the design- and labor-intensive clothing industry in New York, clothing, high-tech electronics and furniture in Milan, and clothing and motion pictures in Los Angeles. Equally, it is legitimate to interpret the localized networks of producer services within world cities as a specialized form of flexible production region. We are forced to conclude, therefore,

Box 7.5 Hollywood and the cultural economy of cities

Allen Scott (2005: 1) begins his book, *On Hollywood*, with the following paragraph:

> One of the defining features of contemporary society, at least in the high income countries of the world, is the conspicuous convergence that is occurring between the domain of the economic on the one hand and the domain of the cultural on the other. Vast segments of the modern economy are inscribed with significant cultural content, while culture itself is increasingly being supplied in the form of goods and services produced by private firms for a profit under conditions of market exchange. These trends can be described variously in terms of the aestheticization of the economy and the commodification of culture.

The **cultural economy** can be defined as a group of sectors (cultural-products industries) that produce goods and services—including jewelery, live theatre, music recording, film production—whose symbolic value to consumers is high relative to their practical purpose. Scott's work on Hollywood is of interest to economic geographers because it argues for the vital role of agglomeration economies and localized increasing returns to scale in allowing this "industrial district" to become the largest and most influential cultural-products agglomeration in the world.

As Scott (2010: 193) puts it:

> In one sense, Hollywood is a very specific place in Southern California, and, more to the point, a particular locale-bound nexus of production relationships and local labor market activities. In another sense, Hollywood is everywhere, and in its realization as a disembodied assortment of images and narratives, its presence is felt across the entire globe. These local and global manifestations of Hollywood are linked together by a complex machinery of distribution and marketing. In this manner, Hollywood's existence as a productive agglomeration is sustained, while the images and narratives it creates are dispersed to a far flung and ever expanding circle of consumers.

Scott (2005: 8) offers five thematic arguments that address how the history, geography, economic structure, and cultural energies of the Hollywood motion picture industry combine to hold Hollywood together as a spatial unit, while endowing those producing within this agglomeration with potential long-term competitive advantages.

First, Hollywood emerged as the main center of the U.S. motion picture industry (in competition with New York) because its pioneering model of film production generated an expanding system of agglomeration economies. By the 1920s the "old" classical studio system of production was securely in place and Hollywood had risen to unparalleled dominance nationally and internationally.

Second, following the Second World War, a "new" Hollywood developed as the "old" studio system was transformed into a more diffuse organizational pattern of production. This greatly enhanced the role of Hollywood as a concentrated industrial district, intensifying its place-specific competitive advantages. Important in this process were the significant extension of flexible production systems in Hollywood and the fundamental shift in the role of the major film production companies as they began to act more as sources of financial and coordination services for independent producers in combination with overall marketing and distribution services.

Third, the film production companies in Hollywood, as part of an industrial district, include not only the units of production from which they derive their principal identity, but also the countless other companies in ancillary sectors—including soundstages, set design and construction, prop houses, digital visual effects, agents and talent managers—which directly and indirectly provide the critical physical inputs and services necessary to keep the entire system operational through complex webs of spatial and functional relationships.

Fourth, the Hollywood production system depends on a large number of workers with a diverse range of skills.

Fifth, Hollywood, like other industrial districts, depends for its economic success on an efficient productive base in conjunction with the effective marketing and distribution of its final products. Companies in Hollywood have been remarkably aggressive in marketing, and have established an extensive international distribution system so that their outputs can be exported efficiently. Part of globalization, this diffusion of American culture from Hollywood has had not only important economic but also profound cultural implications globally.

that although flexible production regions represent a new dimension within the economic landscapes of the core, they do not represent an absolute or fundamental break from the old. As with previous transitions, the old order of things does not, and cannot, simply disappear. So:

> As far as the geography of change is concerned, it is necessary to grasp the coexistence and combination of localizing and globalizing, centripetal and centrifugal, forces. The current restructuring process is a matter of a whole repertoire of spatial strategies, dependent on situated contexts and upon balances of power.
>
> (Amin and Robins, 1990: 28)

7.5 REGIONAL INEQUALITY IN CORE ECONOMIES

We have seen that the evolution of advanced capitalism and the emergence of an information economy led to a significant reorganization of the economic geography of places and regions throughout the developed world. This reorganization modified many of the core–periphery patterns of regional development associated with industrial capitalism. Yet core–periphery patterns and regional economic disparities have by no means disappeared. Rather, they have been reconfigured and intensified. Core regions within developed countries have typically been centered on urban–industrial heartlands, modified by the geography of service activities, particularly producer services, and by the imprint of the new spatial divisions of labor associated with globalization, decentralization and agglomeration.

Europe provides a good example of the kind of regional differentiation that is characteristic of contemporary core economies. The overall core–periphery contrasts that were the legacy of industrial capitalism (see Figure 5.3) have carried over into substantial disparities. Figure 7.8 shows the range and variability of regional economic well-being across the European Union, as measured by GDP per capita based on **purchasing power parity** (PPP). While the more affluent regions of the continental core of the European Union (for example, in Germany, the Netherlands, Austria) have per capita incomes that are well over 100 percent of the EU average, some of the least developed EU regions, especially Bulgaria and Romania, have incomes that are under 50 percent of the EU average.

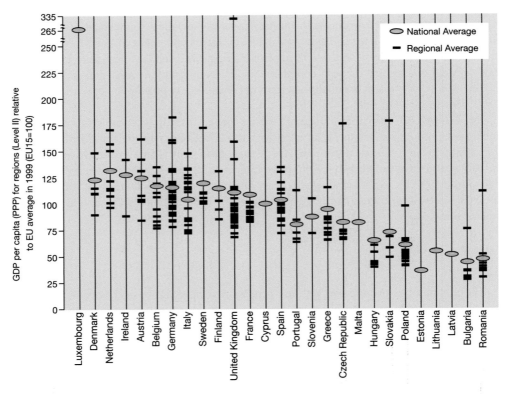

Figure 7.8 Regional inequality across the European Union

Source: Based on data in Eurostat (2012b) http://epp.eurostat.ec.europa.eu/cache/ITY_PUBLIC/1–13032012-AP/EN/1–13032012-AP-EN.PDF

Within this broad pattern of regional inequality there exist sharp local variations. Figure 7.9 shows the geography of inequality among local districts in England, as measured by an index based on a number of deprivation indicators including low income, high unemployment, poor health, and low educational achievement. In general terms, the whole of England to the southeast of a line drawn between the Severn Estuary and Lincolnshire can be considered to be Britain's economic core. At the heart of this region is a prosperous hub centered on Greater London and extending outward for a radius of about 100 kilometers, encompassing the M4 motorway corridor to the west of London, the M3 belt to the southwest and the M11 corridor to the north. In general, high economic performance in Britain is associated with a southeastern location (with the obvious exception of the deprived inner-city neighborhoods of London), with proximity to the motorway system, and with relatively high levels of employment in finance, banking, insurance, and related producer services.

To the north, within the broad periphery of Britain's traditional industrial heartland, the worst performing districts are widely scattered around the former heavy industrial and coalmining districts of Tyneside (Newcastle), Teesside (Middlesbrough), South Yorkshire (Sheffield), Merseyside (Liverpool), as well as in central Scotland and south Wales. The fundamental cleavage reflected in this core–periphery pattern is echoed by a broad spectrum of social and economic data, to the point where it is common to refer to England's political economy in terms of a north–south divide.

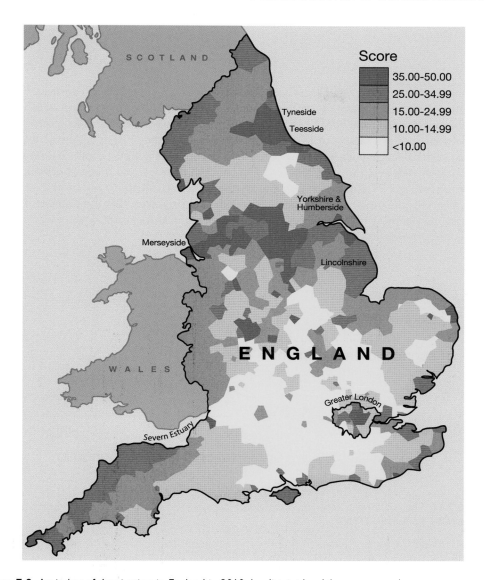

Figure 7.9 An index of deprivation in England in 2010, by district level (average score)

Source: Based on online data from U.K. Department for Communities and Local Government, 2011, English Indices of Deprivation in 2010 https://www.gov.uk/government/publications/english-indices-of-deprivation-2010

INTERPRETATIONS OF REGIONAL ECONOMIC INEQUALITY

A widely known explanation of regional economic inequality is that of Myrdal (1957). This is based on the contention that changes in the location of economic activities in a market economy produce cumulative advantages for one region rather than a straightforward equalization of growth across all regions. **Cumulative causation** refers to the buildup of advantages that occurs in specific geographic settings as a result of the development of agglomeration economies, external economies and localization economies. **Agglomeration economies** are the cost advantages that accrue to individual firms because of their location within such a cluster.

Box 7.6 National economic development and regional inequality

The relationship between overall levels of development and the intensity of regional disparities is central to theory in economic geography. Much of the conventional wisdom on the subject is derived from a major study by Williamson (1965), who examined interregional income disparities in a sample of 24 countries. The results of this analysis suggested that the greatest regional inequalities were associated with countries at a semi-peripheral level of development, with much smaller differences within both the most and the least developed countries (see Figure 7.10). Williamson interpreted these cross-national results as a consequence of the dynamics of economic development, suggesting that the onset of industrial development precipitates sharp increases in regional inequality, which are subsequently reduced as the economy matures.

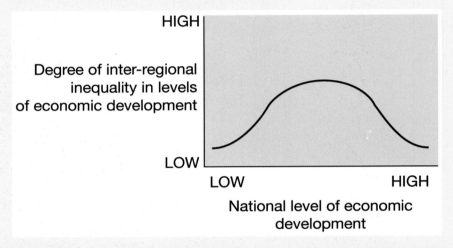

Figure 7.10 The relationship between interregional inequality and levels of economic development, as posited by Williamson (1965)

Both the results and the interpretation of Williamson's work have been questioned, however. Reviewing the large number of empirical studies that followed Williamson's work, Krebs (1982) showed that, even allowing for different measurement techniques, the idea of divergence followed by convergence in regional disparities does not meet with strong support (see, for example, Figures 7.9 and 7.11).

Finally, we must set these trends within the context of longer run economic cycles such that each new phase of economic development, based on new technology systems, and requiring new resources and new markets, initiates a round of "creative destruction" that leaves an indelible imprint on economic landscapes and, therefore, on the pattern and intensity of regional inequality. Furthermore, the related dynamics of the political economy influence the whole question of inequality within countries. So, for example, periods of economic growth and low inflation tend to generate widespread satisfaction and a lack of enthusiasm for altering the status quo at the same time that voters are most likely to feel able to afford to pay for redistributive policies. While we have no adequate single theory or explanation to account for these relationships, it is clear that we must consider regional inequality within the broader context of economic change. Viewed in this way, regional inequality never disappears: It is a perennial consequence of uneven development, of the see-sawing of capital from one set of opportunities to another. We may see regional inequality diminish, over the long term, in one part of the world; but elsewhere, and at different spatial scales, inequality will persist or intensify.

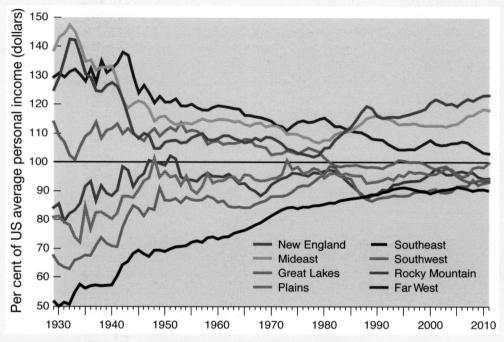

Figure 7.11 Regional trends in per capita incomes in the United States

Source: Based on U.S. Department of Commerce, Bureau of Economic Analysis, online *Regional Economic Accounts* http://www.bea.gov/regional/

These advantages are sometimes known as external economies. **External economies** are cost savings that result from advantages that are derived from circumstances beyond a firm's own organization and methods of production. Where external economies and local economic linkages are limited to firms involved in one particular industry, they are known as **localization economies**. These economies are cost savings that accrue to particular industries as a result of clustering together at a specific location.

Myrdal pointed out that the spiral of local growth involved in cumulative causation would tend to attract people—enterprising young people, usually—and investment funds from other peripheral areas. In some cases, this loss of entrepreneurial talent, labor, and investment capital is sufficient to trigger a cumulative negative spiral of economic disadvantage in these other areas. With less capital, less innovative energy, and depleted pools of labor, industrial growth in peripheral regions tends to be significantly slower and less innovative than in regions with an **initial advantage** and an established process of cumulative causation. This then tends to limit the size of the local tax base, so that local governments find it hard to furnish a competitive infrastructure of roads, schools, and recreational amenities. Myrdal called these disadvantages **backwash effects**: Negative impacts on a region (or regions) of the economic growth of some other region that take the form, for example, of outmigration, outflows of investment capital and the shrinkage of local tax bases. They are important because they help explain why regional economic development is so uneven and why core–periphery contrasts in economic development are so common.

Myrdal recognized that peripheral regions do sometimes emerge as new growth regions, and he provided a partial explanation of them in what he called **spread effects**. Spread effects are the positive impacts on a region of the economic growth of some other region. This growth creates levels of demand for food, consumer goods and other manufactures that are so high that local producers cannot satisfy them. This demand provides the opportunity for investors in peripheral regions to establish a local capacity to meet the demand. Entrepreneurs who attempt this are also able to exploit the advantages of cheaper land and labor in peripheral regions. If these effects are strong enough, they can enable peripheral regions to develop their own spiral of cumulative causation, and change the interregional geography of economic patterns and flows (see Figure 7.12).

Myrdal's model was followed by others based on similar logic. Hirschman's (1958) model assumes "polarization" and "trickling-down effects" but sees early polarization producing a countervailing trend (and, so, interregional equilibrium) rather than the cumulative intensification of initial advantage. Hicks' (1959) model was more like Myrdal's in its emphasis on cumulative causation but it gives greater attention to flows of labor and capital from *growing* to *lagging* regions rather than in the other direction.

Not all industries are equal in the extent to which they stimulate growth. Perroux (1955; 1961) argued that the locations of "propulsive industries" (those that attract other industries and stimulate new ones in the vicinity) serve as the distinguishing characteristic of regions that achieve high rates of economic growth. As a propulsive industry grows, it attracts other linked industries and creates a set of agglomeration economies. A **growth pole** is formed and a growth

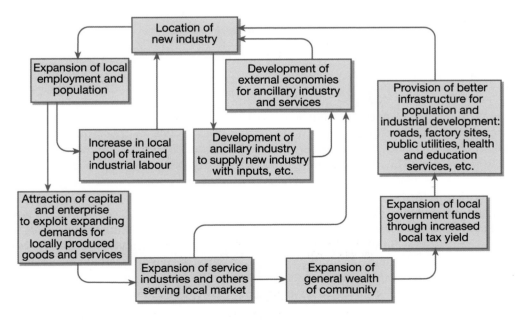

Figure 7.12 Myrdal's model of regional cumulative causation

center develops. In the 1920s shipbuilding was a propulsive industry. In the 1950s and 1960s automobile manufacturing was a propulsive industry, and, in today's information economy high-tech manufacturing, design-intensive consumer goods, and business and financial services are propulsive industries.

Krugman (1991) suggests a more complex and more formal model to account for a core–periphery pattern of economic development and for the possibility of its transformation. He argues that the relationship between demand and production established during an early phase of industrialization locks into place an interregional imbalance through increasing returns to scale in plant operations, transportation costs and demand. As he put it:

> Given sufficiently strong economies of scale, each manufacturer wants to serve the national market from a single location. To minimize transportation costs, she chooses a location with large local demand. But local demand will be large precisely where the majority of manufacturers choose to locate. Thus there is a circularity that tends to keep a manufacturing belt in existence once it is established.
>
> (Krugman, 1991: 15)

However, once the population of a peripheral region reaches critical mass it may serve to stimulate production facilities there. A dramatic shift in regional fortunes may follow. Local "boosterism," faith in a locality's future possibilities, and policies that reflect this can also prove decisive in reversing regional imbalance: "Nothing," therefore, "is forever" (Krugman, 1991: 26). This is an argument for a rapid reversal in regional fortunes, rather than for slow regional balancing. Cores become peripheries and vice versa. Krugman argues that increasing returns to scale, imperfect competition and historical accident conspire to produce geographical concentration. But the identity of the favored region is not set for all time—a new pattern of concentration can break the historic mold.

SUMMARY

In this chapter, we have seen how the crisis of Fordism and the sectoral shifts and changing business structures of advanced capitalism have reshaped the economic landscapes of the industrial core regions. Several aspects of this transition are of special importance:

- The spatial reorganization of core economies has been the product of several interdependent processes, including the globalization of some core-area economic activity, shift from manufacturing towards service industries, development of flexible production processes and competitive strategies, and changes in government regulation and intervention.
- The imprint of the shift towards service employment has had two main dimensions: Deindustrialization and economic decline in regions of traditional heavy industries, and growth in larger metropolitan settings that have attracted higher order producer services.
- This shift overlapped with the globalization of some core-area economic activity, with the result that certain aspects of change have been intensified and others have been introduced. The globalization of economic activity, for example, has intensified the effects of deindustrialization and been associated with the adoption of neoliberal policies.
- Corporate reorganization amid globalization has resulted in an increase in the external control of regional economies.
- Net effects of change have resulted in simultaneous spatial trends involving both decentralization and agglomeration, each highly selective in terms of the regions and economic activities involved.

- The diverse mixes of industry, workers and infrastructure inherited from the industrial era have mediated the broader processes of structural change and reorganization, so that different kinds of regions have evolved in different ways. *There have been four broad trajectories of change.* In the first, restructuring in response to a legacy of declining industry has been the dominant process. Examples include the old manufacturing heartlands of northern England, south Wales, central Scotland and the U.S. Manufacturing Belt. In some rural areas and peripheral regions, in contrast, the dominant process has been one of decentralization of footloose, labor-intensive industries from the metropolitan areas and core regions. Examples include parts of the southern U.S. such as Alabama. A third trajectory is characterized by regions whose industry has developed at or just above the national average, sustained by a consistent supply of new investment. Examples include most of southeast England and, in the U.S., the Boston-New York-Washington, DC-Richmond corridor. Finally, there are some regions whose attributes have made them attractive to new industries and/or to new investments aimed at exploiting new competitive strategies and new production processes. These include the high-tech concentrations of Orange County and Silicon Valley, and flexible production regions such as north-central Italy.
- Overlying these categories there has been a general accentuation of the importance and prosperity of large metropolitan areas, while many smaller towns and, in particular, many of the specialized towns spawned by industrial capitalism (mining towns, heavy manufacturing towns, and so on) have declined.

KEY SOURCES AND SUGGESTED READING

Bryson, J.R., Daniels, P.W., Henry, N., and Pollard, J. (eds.), 2000. *Knowledge, Space, Economy.* London: Routledge.

Castells, M., 2012. *Networks of Outrage and Hope: Social movements in the internet age.* Cambridge: Polity Press.

Cox, K.R., 2005. Global/Local. In P. Cloke and R.J. Johnston (eds.), *Spaces of Geographical Thought.* London: Sage.

Massey, D., 2005. *For Space.* London: Sage.

OECD, 2004. *Regulatory Reform as a Tool for Bridging the Digital Divide.* Paris: OECD.

Scott, A.J., 2010. Creative Cities: The role of culture, *Revue d'Économie Politique* 1, 181–204.

Scott, A.J., 2012. *A World in Emergence: Cities and regions in the 21st century.* Cheltenham: Edward Elgar.

Taylor, P.J., 2004. *World City Network: A global urban analysis.* London: Routledge.

Taylor, P.J., Ni, P., Derudder, B., Hoyler, M., and Witlox, F. (eds.), 2010. *Global Urban Analysis: A survey of cities in globalization.* London: Routledge.

Warf, B., 2013. Contemporary Digital Divides in the United States, *Tijdschrift voor Economische en Sociale Geografie,* 104, 1–17.

Chapter 8

Dynamics of interdependence: Transformation of the periphery

Picture credit: Linda McCarthy

If local isolation were rarely complete and development were rarely totally independent, the evolution of the world economy led to greater and greater interaction between different parts of the world. In this chapter, we focus on the cumulative consequences of this increased interdependence for those regions incorporated into the world economy on terms initially and decisively for many years disadvantageous to them. This is not to say that the terms of interdependence have always remained absolutely disadvantageous, although this is true, for example, in the case of Central America and large parts of Sub-Saharan Africa. Particularly since the 1970s, the major oil-producing countries (for example, Saudi Arabia, Iran, Venezuela, Nigeria, and Indonesia) and the newly industrializing economies (NIEs) (such as Taiwan and South Korea and, most recently, China) have challenged the static picture of a "fixed" industrial core and a "fixed" non-industrial periphery. The world economy now has a vibrant semi-periphery of NIEs and resource-based economies. This chapter begins with a discussion of how existing economies were transformed into colonial ones. A second section identifies the major ways in which these colonial economies were enmeshed and maintained within the world economy. A third section identifies the importance of frameworks of administration introduced by Europeans. A fourth section discusses the cultural mechanisms that facilitated integration into the world economy. The final two sections explore the contexts of change in the nature of interdependence since the 1970s, respectively the global context (the new international division of labor, decolonization, and the Cold War) and several national political–economic strategies or models of development that challenge or, more typically, have adapted to the dominant "liberal–capitalist" one.

8.1 COLONIAL ECONOMIES AND THE TRANSFORMATION OF GLOBAL SPACE

The contemporary world economy began with the global expansion of trade and conquest by European merchants, adventurers and statesmen (see Figure 8.1). But a distinction should be

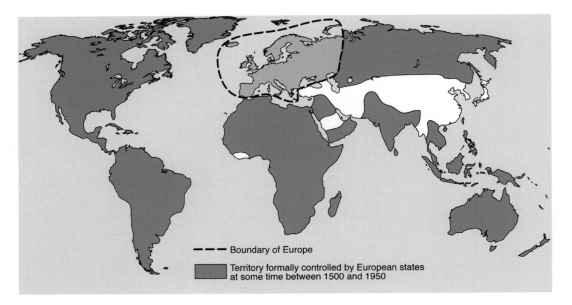

Figure 8.1 Geographical extent of European political control, 1500–1950

Source: Based on Taylor (2000: 106, Figure 3.1)

drawn between pre-capitalist colonial rule, notably that of Spain and Portugal in Latin America, and the new colonialism that was associated with the growth and global expansion of European capitalism, beginning in the sixteenth century and itself undergoing successive shifts in development. The major purpose of pre-capitalist colonialism was the extraction of tribute from subject peoples and its major mechanisms involved political–territorial control. In contrast, the "new colonialism" was associated primarily with economic objectives and mechanisms. Direct political–territorial control, although often advantageous, was not essential. The emphasis initially was on the exploitation of raw materials. After the Industrial Revolution in Britain, however, markets for manufactured goods became an equally important objective. Realizing both of these objectives required a restructuring of the economic landscapes of the colonized societies.

Territorial conquest, with or without the elimination of indigenous peoples, and the planting of either settler enclaves or slave plantations and mining enterprises were the major features of European expansion through the eighteenth century. For much of the nineteenth century, however, many societies that remained or became formally independent were under the economic domination of European and, increasingly, American capitalists. With the "German challenge" to British hegemony in the 1870s, there was a new "scramble" for territorial conquest as rival colonial powers attempted to pre-empt one another, especially in Africa. This coincided with the emergence of capital export as a major stimulus to intervention and domination, as profit rates in the periphery exceeded those in the core.

In both territorial (colonial) and interactional (commercial) forms, capitalist expansion entailed a forcible transformation of pre-capitalist societies whereby their economies were internally disarticulated and integrated externally with the world economy. They were no longer locally oriented but had now to focus on the production of raw materials and foodstuffs for the core economies. Often they became extremely dependent on monoculture in order to confer the blessings of comparative advantage (the purported benefits of specializing in goods that a country can produce at a lower relative cost while importing goods for which its own production costs are relatively higher) on the developed world. At the same time they also provided markets for the manufactured goods exchanged in return. For many parts of the world, this relationship still holds true today.

The dramatic transformation of the existing geography of production that the reorientation towards the core countries entailed is not sufficiently noted. Before the Industrial Revolution and European capitalist expansion, Asian and other countries that are now conventionally characterized as LDCs contained a far larger share of world manufacturing output than did Europe. Bairoch's (1982) calculations of industrial output by world region during the course of the nineteenth century, for example, reveal a picture not simply of a higher rate of industrialization in core countries but also of deindustrialization in the periphery as the cheaper European products forced traditional producers elsewhere out of business.

The division between "developed" and "less developed" economies, therefore, which had been moderate before now grew enormously. This was the main geographical consequence of the evolution of the world economy. The conditions for the development of "national economies" did not exist; neither were they permitted to exist in the "dependent world" that came into existence during the course of the early expansion of the European world economy. The unequal core–periphery structure of the world economy has been in place since the Europeans first ventured out into the world in the sixteenth century. From this point of view, underdevelopment is not an original condition, equivalent to "traditionalism" or "backwardness." On the contrary, it is a condition created by integration into the world economy. At the same time, however, the die is not permanently cast in confining some places and peoples to an underdeveloped condition. Examples such as the United States, Japan, Australia, and the NIEs suggest that upward mobility is possible for some states/regions initially in an underdeveloped state. What is equally clear is that for large parts of the world such mobility is either difficult or next to impossible. To understand the geography of the world economy, we need to understand such cases as well as the "successful" ones.

During the early years of incorporation into the world economy the periphery tended to become specialized in the extraction of raw materials (from gold bullion to furs and spices) and production of plantation crops (such as tobacco, cotton and sugar). As time wore on, plantations and extractive industries were sometimes supplemented by labor-intensive manufacturing that took advantage of cheap colonial labor. By the mid-twentieth century, Latin America, Asia, and Africa were organically linked to and financially dependent on Europe and the United States. The emergence of the United States as a dominant force and the growth of the Soviet bloc, however, undermined the monopoly of political control exercised by the European powers over large parts of the world. A process of decolonization began with the independence of the South Asian countries in 1947. This brought the possibility, however constrained (as the experience of Latin America, politically "independent" since 1820, shows), of more autonomous development.

IMPOSITION OF REGIONAL SPECIALIZATIONS

The nineteenth century was an especially critical period in the creation of colonial economies (see Figure 8.2). Whole regions were made to specialize in the production of a specific raw material (such as gold, spices, or cotton), food crop (such as bananas) or "stimulant" (such as tea or tobacco). Many of these had a prior history, such as the sugar-producing areas of the Caribbean or the cotton-growing regions of the United States, India, and Egypt. But the Long Depression (1883–1896), a major downturn in the world economy—due, among other

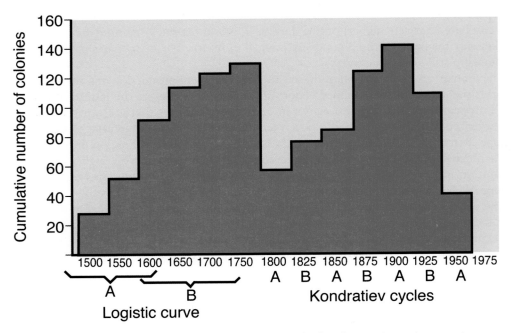

Figure 8.2 Long waves and colonization: the two long waves of colonial expansion and contraction
Source: Based on Taylor (2000: 115, Figure 3.4)

things, to decreased profitability in manufacturing—ushered in a major spurt in the global expansion of capitalism and intensified regional specialization. During this time period more and more resources and labor were drawn into an increasingly differentiated world economy.

Adam Smith and David Ricardo, writing well before this period, had envisaged a global division of labor in which each country would freely choose the commodities it was most suited to produce and freely exchange its optimal commodity for the optimal commodities of others. Unfortunately, this economic vision of comparative advantage, basic to most modern theories of international trade, ignores both the historical–political conditions under which commodities were selected and the costs faced by a specialized commodity economy in terms of vulnerability to the vagaries of "world" demand. Competitive advantage, established through market dominance and political power, makes more sense as the significant determinant of the global map of production.

In the late nineteenth century, "choice" of commodity was often imposed by force or through market domination. Moreover, once embroiled in the global system of regional specialization, an economy had to organize its factors of production in order to foster capital growth, or fall by the wayside. At the same time, other regions, without some initial advantage in raw material, climate, social organization or accessibility, became providers of labor power to the new outposts of global capitalism. Three examples, out of a host of possibilities, illustrate how externally oriented colonial economies were created on the basis of regional specialization: Bananas in Central America, rubber in Malaya, and tea in Sri Lanka.

Bananas are hardly a major food staple, but the creation of banana plantations in the late nineteenth century affected many areas, especially in Central America. Introduced into the Americas by the Spaniards, the banana became a staple crop among the lowland populations of Central America. In the 1870s it became a plantation crop as an American entrepreneur engaged in railroad construction in Costa Rica experimented with commercial banana production to increase the profitability of his railroad. As a result of this initiative was developed the United Fruit Company, incorporated in 1889 as a corporation engaged in the marketing of bananas from Central America in the United States. Over the years, the company produced bananas on plantation-estates in Costa Rica, Panama, Honduras, Colombia, and Ecuador. Geographic dispersal had a number of advantages:

> [It] enabled the Company to offset political pressures in any one host country. Dispersal also allowed it to take advantage of suitable environments in different locations, thus reducing the chance that floods, hurricanes, soil depletion and plant diseases could bring production to a halt in any one of them. To further reduce these risks, the Company acquired a great deal more land than it could use at any one time, to hold as a reserve against the future. In some areas it formed relationships with local cultivators who grew bananas and then sold them to the Company.
>
> (Wolf, 1982: 324)

Much of the labor on the plantations was recruited locally, especially in Colombia and Ecuador, but in parts of Central America workers were brought from the English-speaking islands of the Caribbean. This resulted because of the difficulty the company faced in obtaining laborers from the populated highlands to work in the lowlands, and the firm's preference for a work-force that could be socially isolated and made wholly dependent on the company. The role of these foreign workers gradually decreased as host governments limited immigration and encouraged their native populations to engage in wage labor on the plantations. Bananas are still an important export crop, especially for Panama and Costa Rica.

Wild rubber from Brazil dominated the world market for most of the nineteenth century. In 1876, however, Amazonian rubber seeds were smuggled from Brazil to England where they were prepared for planting in Malaya. Malayan rubber plantations grew from 5,000 acres in 1900 to 1,250,000 acres in 1913. During this expansion, a class of managers for companies operating from London supplanted an original planter class. Laborers were initially imported from southern India but over time many plantations came to employ local Malays. Although plantation production remained dominant, many Malay cultivators tapped their own rubber trees as a source of cash income. Rubber increasingly replaced irrigated rice, a food staple, as the major commodity produced by small-scale proprietors. It remains so to this day.

Finally, among the range of commodities destined for consumption in the industrial world, some were neither foodstuffs (such as bananas) nor industrial crops (such as rubber). Such commodities as sugar, tea, coffee, cocoa, tobacco, and opium were of fundamental importance in the global expansion of the world economy. Explaining the popularity of these "stimulants" is not easy. Some accounts suggest that the work behavior required under industrialization favored the sale of these stimulants (except opium, a special case, as its main initial market was China) because they provided "quick energy" and prolonged work activity. Others suggest that some (for example, sugar and cocoa) provided low-cost substitutes for the traditional and increasingly costly diet of preindustrial Europe. Whatever the basis to demand, by the late nineteenth century the stimulants were of great and increasing importance in world trade (see Mintz, 1985, for an excellent discussion of sugar and its role in world trade). Tea had become "the drink" of English court circles in the late seventeenth century. It came entirely from China. Demand was so great that in the early eighteenth century tea replaced silk as the main item carried by British ships in the Chinese coastal trade. At the time of the American War of Independence, as the Boston Tea Party reminds us, tea was the third largest import, after textiles and iron goods, of the American colonies. Some tea plantations were established in Assam (northeast India) in the 1840s, but, until the opening of the Suez Canal, Indian tea could not compete with the Chinese tea carried by the famous clipper ships around the Cape of Good Hope. With the opening of the canal and decreasing cost of steamship transportation, Indian black teas became commercially competitive with the green teas of China. In the 1870s tea plantations spread with great speed throughout the uplands of what was then Ceylon (now Sri Lanka). This was done by confiscating peasant land through the device of "royal condemnation" and then selling it to planters. By 1903 over 400,000 acres were planted in tea shrubs.

Tea cultivation is extremely labor intensive. To obtain the necessary labor, the Ceylon tea planters imported Tamil laborers from southern India. These Tamils, not to be confused with the long-resident Sri Lankan Tamils of northern and eastern Sri Lanka, now number over 1.5 million people, in a region in which the upland or Kandyan Sinhalese are around about 3 million. As a consequence, an ethnic conflict has been imposed on top of an economic conflict between Sinhalese cultivators and Tamil plantation workers. Tea remains an important export crop in the contemporary Sri Lankan economy.

Many other examples could be added to these three to demonstrate the degree to which regional specialization was the classic motif of the colonial economies as they developed at an accelerating rate in the late nineteenth century. The long depression of that time in Europe and the United States stimulated an unprecedented expansion of the world economy into all parts of the globe as European and American capitalists sought to maintain capital growth in the face of declines in industrial production. Commodity production for a world market was not new but in its late nineteenth-century "explosion" it "incorporated pre-existing networks of exchange and created new itineraries between continents; it fostered regional specialization and initiated worldwide movements of commodities" (Wolf, 1982: 352).

8.2 ECONOMIC MECHANISMS OF ENMESHMENT AND MAINTENANCE IN THE COLONIAL GLOBAL ECONOMY

How was it that the regional specialization that began in the late nineteenth century was possible? And how did it evolve over time? Answering these questions requires us to focus on the means by which an integrated world economy was created: Flows of capital investment, networks of communication and marketing, movements of commodities and people, and transportation—urban networks as channels of diffusion and concentration.

TRADE AND INVESTMENT

The period 1860–1870 inherited from the earlier centuries of colonial expansion two major systems of economic interaction, an Atlantic system built on the "triangular trade" between Europe, Africa, and the Americas (see Figure 8.3) and a Eurasian system built on trade with India, East Asia, and Southeast Asia. In the mid-nineteenth century the Atlantic system in its classic form collapsed. It was replaced by a system based on a mix of competitive colonialism, regional specialization, and investment in infrastructure (especially railways).

Between 1830 and 1876 there was a vast increase in the number of colonies and the number of people under colonial rule (see Figure 8.2). There was also a tremendous expansion of foreign investment by European states and capitalists in the late nineteenth and early twentieth centuries (see Table 8.1). Moreover, there was an important shift in the geographical distribution of both investment and exports. British trade and investment, to use the most important example, shifted away from India, Europe, and the United States, especially with respect to investment in the first two and with respect to exports to the third, from the 1870s on. South

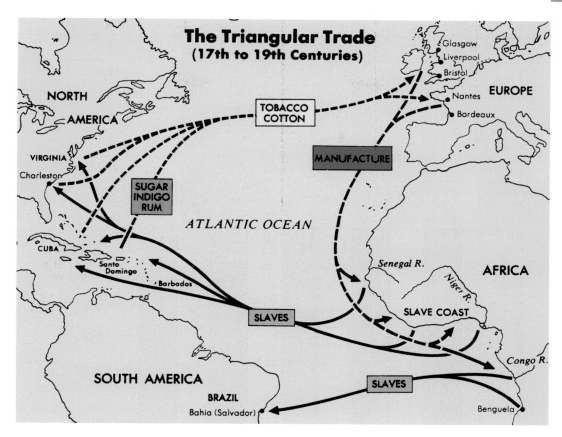

Figure 8.3 The Atlantic system, 1650–1850

Source: Based on Duignan and Gann (1985: 12, Map 1)

America and the British dominions (Australia, Canada, New Zealand, and South Africa) became more important, especially with respect to investment. However, the pattern was to fluctuate considerably over the years as some regions/states increased and others decreased in attractiveness to investors (see Figure 8.4).

The geographical switching of investment, however, was not always obviously economic in motivation. In particular, European incorporation of Africa into the world economy was based largely on competitive colonialism. Local settlers, as in South Africa, sometimes developed their own local "imperialisms," and when challenged by other settlers or hostile natives called in the "Motherland." The relative weakness of many African polities also invited direct intervention. Once one European state was involved, others were tempted to engage in pre-emptive strikes to limit the damage to their "interests." Between 1850 and 1914 Africa was divided into a patchwork of European colonies and protectorates (see Figure 8.5).

TRANSPORT ROUTES

At the global level, the colonial system was bound together by a network of steamship and communication (postal, telegraph, and, later, telephone) routes. These became progressively more dense and interconnected from the 1860s onwards. The Suez and Panama Canals were

Table 8.1 Stock of foreign capital investment held by Europe, 1825–1915 (US$ billion)

	1825	1855	1885	1915
UK	0.5	2.3	7.8	19.5
France	0.1	1.0	3.3	8.6
Germany	0.1	1.0	1.9	6.7
Others	—	—	—	11.4
Total	0.7	4.3	13.0	46.2

Source: Based on Warren (1980: 62, Table 1)

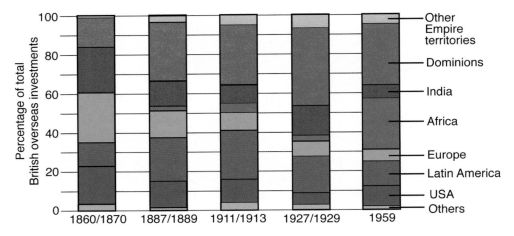

Figure 8.4 Geographical distribution of British foreign investment, 1860–1959

Source: Based on Hobsbawm (1968: 303, Figure 33)

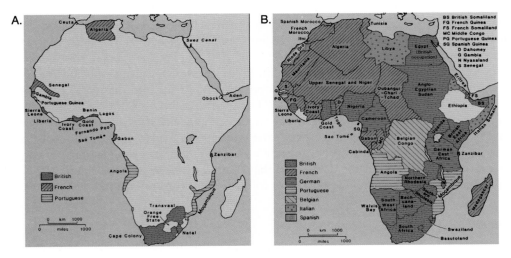

Figure 8.5 Colonization of Africa: (A) 1850; (B) 1914

Source: Based on Christopher (1984: 28–9, Figures 2.1 and 2.2)

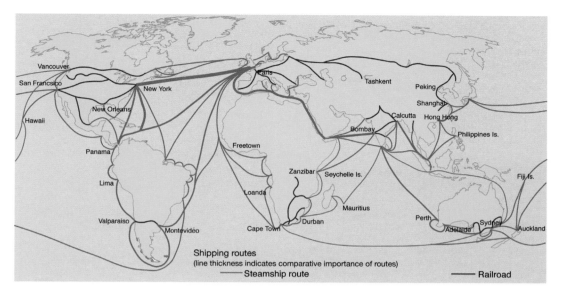

Figure 8.6 World steamship routes, by volume of trade, 1913

Source: Based on Latham (1978: 33, Map 2)

important in providing shorter and less hazardous routes between home ports in Europe and North America and colonial destinations. By 1913 the world economy was effectively integrated by a system of regularly scheduled steamship routes (see Figure 8.6). A world telegraph system enabled orders to be placed and shipments to be embarked for a large number of ports around the world (see Figure 8.7).

Within colonies, railway building was the major mechanism of spatial transformation. In most of Africa and Asia, with the important exceptions of India, South Africa, and north China, railways were not mechanisms for creating integrated colonial economies but, rather, means for moving a basic export commodity for shipment to Europe or North America. However, railways were often an important investment in their own right rather than a burdensome state responsibility. This was especially the case in South America and China, if much less so in India and Africa. In Argentina, for example, although the road and railway networks were oriented towards the River Plate Estuary and the capital city of Buenos Aires, they provided a relatively dense grid for the rich commercial agriculture of the Argentinean pampas (see Figure 8.8). This produced a transport system considerably more interconnected and integrated than the simple linear systems prevalent in Central America, Southeast Asia, and most of Africa. In short, spatial integration into the colonial world economy did not take the same form everywhere.

Whatever the precise nature of the railway networks, there was a tendency for all networks to focus on one or, at most, several coastal ports. These became "privileged" locations, often assuming the role of administrative as well as economic center for the entire colony. Specialization in the export of raw materials and concentration of administrative functions as a result had the effect of stimulating the disproportionate growth of these "links" to the world economy. This was especially marked in India (with Bombay, Calcutta, and Madras) and Africa (for example, Cape Town, Dakar, and Lagos).

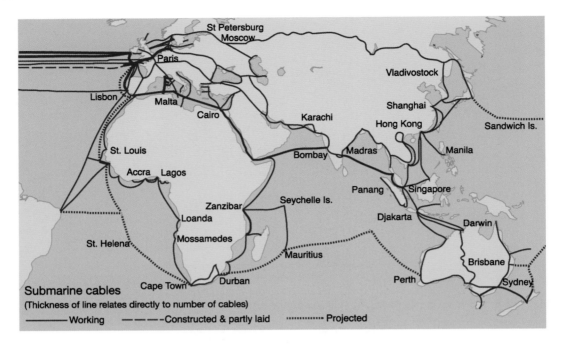

Figure 8.7 Telegraph system in Asia and Africa, 1897

Source: Based on Latham (1978: 36, Map 3)

SETTLEMENT SYSTEMS

But the character of the colonial system also put limits on the growth of the dominant or primate cities. There was only a limited stimulus to the growth of a distinctive urban economy. The orientation of urban networks was towards exploitation of hinterlands rather than an industry- and service-based urban economy. It was only with political independence that a new dynamic for urban growth occurred as the primate cities shifted from being mechanisms for colonial control to their contemporary role "as the corporate representative of the people of the former colony" (Fiala and Kamens, 1986: 28). As Rondinelli (1983: 49) points out:

> [C]olonial activities often stimulated the growth of secondary cities. In some cases they were encouraged to grow as colonial administrative posts or as transfer and processing centers for the exploitation of mineral and agricultural resources in the interior of a country.

Regions without a history of urbanization before colonialism were not surprisingly the most easily and strongly reoriented to the colonial world economy. In Malaysia, for example, cities grew up in the interior where crops were grown for export or where other exportable commodities (tin, especially) were exploited. These cities were connected by railway and road to port cities that grew as processing and transfer points. In regions with a long pre-colonial history of urbanization, such as western Nigeria, roads and railways were often built to bypass traditional centers of trade, such as Benin City, Ife, and Sokoto. New, more effective colonial cities grew up at nodes in the transportation network. As one study notes: "Fortune rode the trains. [Towns] that received terminals grew, but those that did not stagnated or declined, as did many river ports" (Gugler and Flanagan, 1978: 27–28).

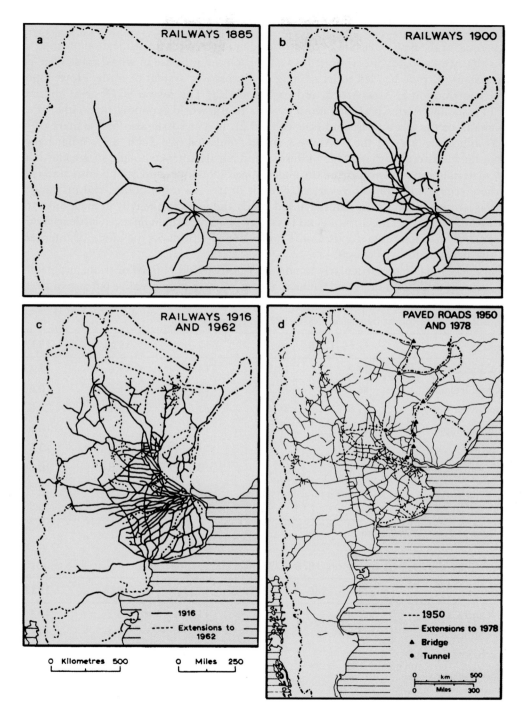

Figure 8.8 Development of roads and railways in the River Plate region of South America, 1885–1978

Source: Based on Crossley (1983: 401, Figure 9.3)

REALIGNMENT OF TRADE AND PRODUCTION

The outcome of the new extension of the world economy and the intensification of trade within regions that were already incorporated was a substantial increase in world trade. The nature of the system of trade has led to its naming as a crossover system of trade. How important was the contribution of Asia, Africa, and Latin America? The answer is: Of great importance. By 1913 Asia and Africa provided more exports to the world economy than either the USA and Canada or the UK and Ireland (see Table 8.2). In 1913 Asia also had a share of world imports almost as large as the USA and Canada combined (see Table 8.3). What happened was that the industrializing countries of Europe and North America bought increasing amounts of raw materials and foodstuffs from the undeveloped economies and ran up large trade deficits with these regions. Britain, however, as a result of its free trade policy, ran up substantial deficits as a result of importing manufactured goods and investing heavily in the industrializing countries (especially the United States and Germany). In turn, Britain financed its deficits through the export of manufactured goods to the undeveloped world. So the circle of international trade and dependence was closed.

India and China were particularly important to this world pattern of trade and payments. It was Britain's trade with India and China that compensated for a negative balance of payments

Table 8.2 World exports by region, 1876–1937, percent of total

Exports	1876–1880	1896–1900	1913	1928	1937
USA and Canada	11.7	14.5	14.8	19.8	17.1
UK and Ireland	16.3	14.2	13.1	11.5	10.6
Northwest Europe	31.9	34.4	33.4	25.1	25.8
Other Europe	16.0	15.2	12.4	11.4	10.6
Oceania	24.1	21.7	2.5	26.3	3.5
Latin America			8.3		10.2
Africa			3.7		5.3
Asia			11.8		16.9

Note: The 1928 values for Oceania, Latin America, Africa, Asia are 2.9, 9.8, 4.0, 15.5 respectively.

Source: Based on Yates (1959: 32, Table 6)

Table 8.3 World imports by region, 1876–1937, percent of total

Imports	1876–1880	1896–1900	1913	1928	1937
USA and Canada	7.4	8.9	11.5	15.2	13.9
UK and Ireland	22.5	20.5	15.2	15.8	17.8
Northwest Europe	31.9	36.5	36.5	27.9	27.8
Other Europe	11.9	11.0	13.4	12.5	10.2
Oceania	26.3	23.0	2.4	23.4	2.8
Latin America			7.0		7.2
Africa			3.6		6.2
Asia			10.4		14.1

Note: The 1928 values for Oceania, Latin America, Africa, Asia are 2.6, 7.6, 4.6, 13.8 respectively.

Source: Based on Yates (1959: 33, Table 7)

with the United States, industrial Europe, Canada, South Africa, and New Zealand. Without the "Asian surplus," Britain would not have been able to subsidize the growth of these other economies. So, far from being "peripheral" to the growth of the world economy, the undeveloped world, especially India and China, was vital.

Between 1918 and 1939 this system of multilateral trade suffered a number of setbacks. One was the overall decline in trade as the world experienced a major depression. But the worst was the decline in Britain's relative position as the linchpin of the colonial world economy. This reflected both successful industrialization in India and China displacing British products (especially cotton textiles) and increased competition from Japan in Britain's "traditional" colonial markets. But another problem was the overproduction of the main export crops and raw materials. As a consequence, commodity prices fell and so did demand for manufactured goods. The successful expansion of plantations and mines, therefore, ultimately undermined the system of capital circulation and growth that their introduction had brought into existence in the nineteenth century.

The onset of the Great Depression of the 1930s effectively ended the expansionist regime of international trade established in the late nineteenth century. The major industrial states reacted to the Depression by raising tariffs and devaluing their currencies. These shifts in economic policy were premised on the assumption that Britain would remain "open" as the linchpin of the colonial world economy. But, as Stein (1984: 375) puts it: "Depression left Britain unable and unwilling to accept an increasingly asymmetric bargain." Not until the 1970s would world trade return to the relative levels that it had achieved in the early 1900s.

Military spending and massive increases in domestic consumption of domestic manufactures provided the keys to economic recovery in western Europe and the United States in the 1940s and 1950s. Although this did lead to increased demand for many of the industrial raw materials and foodstuffs produced in the periphery, there was no longer the crossover system of trading linkages. If anything, the European colonial states and, above all, the United States now came to have direct links to specific sites of exploitation in the periphery without the necessity of the infrastructure and administrative investments that had limited short-run payoffs. This approach favored direct investment and the creation of subsidiaries by transnational corporations rather than portfolio investment and conventional trade. Advantages previously specific to the United States—the cost effectiveness of large plants, economies of process, product and market integration—had become the proprietary rights of large firms. The world was now their oyster, rather than that of the colonial states: "American governments could preach against colonialism while large American [and other] firms colonized the world" (Agnew, 1987: 62).

In the early part of the twentieth century the major share of accumulated foreign direct investment (FDI) was in the less developed countries (LDCs) (62.8 percent in 1914). Total FDI came overwhelmingly from Britain (45.5 percent) and the USA (18.5 percent). Since the Second World War, however, most foreign direct investment has been between the DCs. By the early 1970s only about 30 percent of FDI was directed to the LDCs (which has since fallen to about 20 percent); although after Europe, Latin America was the major recipient region. By the early 1970s also, the USA, with nearly 50 percent, was the major source of total FDI (Dunning, 1983). This has decreased since to about 30 percent with the growth of Japanese, European, and some NIE FDI. The dramatic post-Second World War expansion in FDI, therefore, has not involved all parts of the world on equal terms. In terms of flows of FDI, the LDCs have, on the whole, become less central to the world economy than they were previously. From this point of view at least, the end of colonialism was something of a mixed blessing.

8.3 INFLUENCE OF COLONIAL ADMINISTRATION ON INTERDEPENDENCE

Many of those who colonized the world from Europe, in both the sixteenth and the nineteenth centuries, saw their activities as part of a historic "mission" of western civilization: To bring progress to backward and barbarian peoples. Lord Lugard, the famous British colonial administrator, maintained that Britain stood in a kind of apostolic succession of empire:

> [A]s Roman imperialism . . . led the wild barbarians of these islands [the British Isles] along the path of progress, so in Africa today we are re-paying the debt, and bringing to the dark places of the earth . . . the torch of culture and progress.
>
> (Ranger, 1976: 115–116)

At best the political ideas of the European imperialists were that:

> [P]olitical power tended constantly to deposit itself in the hands of a natural aristocracy, that power so deposited was morally valid, and that it was not to be tamely surrendered before the claims of abstract democratic ideals, but was to be asserted and exercised with justice and mercy.
>
> (Stokes, 1959: 69)

The chief problem was to understand and pacify the indigenous colonized. The Nigerian novelist Chinua Achebe (1975: 5) puts this as follows:

> To the colonialist mind it was always of the utmost importance to be able to say: I know my natives, a claim which implied two things at once: (a) that the native was really quite simple and (b) that understanding him and controlling him went hand in hand—understanding being a precondition for control and control constituting adequate proof of understanding.

This approach provided the ideology for what Hopkins (1973: 189), referring to the British in Africa, has called the "art of light administration," administration without too much long-run investment or explicit (and expensive) violence.

The colonial regimes themselves never amounted to more than a thin veneer of European officials and soldiers on top of complex networks of local collaborators. In India in the 1930s, for example, 4,000 British civil servants, 60,000 soldiers, and 90,000 civilians ruled a country of 300 million people. The British were able to do this:

> [B]y constructing a delicately balanced network through which they gained the support of certain favored economic groups (the Zamindars acting as landed tax collectors in areas such as Bengal, for example), different traditional power holders (especially after the Great Mutiny of 1857, the native princes), warrior tribes (such as the Sikhs of the [sic] Punjab), and aroused minority groups such as the Muslims.
>
> (Smith, 1981: 52)

This kind of brokerage system was to be found in every colonial territory without a large European settler population. Sometimes a foreign economic presence was crucial (the Chinese in Southeast Asia; the Lebanese in West Africa; European settlers in Algeria and Kenya). Often there were alliances with new or traditional ruling groups (the princely states in Malaya; the Ottoman bureaucracies in Tunisia and Morocco; the Hashemite family in Mesopotamia and Syria). Above all, local rivalries were exploited to advantage, as in Madagascar, India, and China. Even in the face of nominal local political independence, as in China or Latin America, colonial imperatives and administrative models had considerable influence through imported school curricula and business practices (for example, British influence was strong in Argentina and Venezuela; German influence was strong in Chile and Brazil).

Alliances and administrative structures were far from static and differed from colony to colony and between colonial powers. But one change *was* permanent. The new colonies, often vastly bigger than the territorial units they superseded, created markets of unprecedented size. Internal tolls and other restraints on trade disappeared. Sumptuary laws that prevented people of low status from acquiring luxury goods were abolished. All forms of servitude that interfered with the wage economy were outlawed. The great tribal migrations of eastern and southern Africa were brought to a close. New judicial methods were introduced and old ones eliminated. Schools and hierarchical systems of local administration were established.

The colonial powers operated in different ways. In Africa, for example, the British administration was more civilian and decentralized than the French and Belgian administrations. Its officials:

> [P]rided themselves on being gentlemen and amateurs, rather than on being military, legal or administrative specialists. The British pioneers set up an administrative hybrid based partly on British metropolitan models and partly on models derived from colonial India and Ireland.
>
> (Gann and Duignan, 1978: 355)

In particular, there was a dispersal of administrative power.

Any description of the particularities of administration in the various territories would require much more space than is available here (see Gann and Duignan, 1978; Gifford and Louis, 1971, for some of the details). One example must suffice. In Nigeria, Britain's most populous colony in Africa, the coastal (Lagos) and northern (Kaduna) regions were administered in completely different ways. The coastal region had a longstanding commercial base and export trade, tied to Liverpool and England's northern (for example, Lancashire) industries. Consequently:

> Lagos governors . . . tried to please north country British businessmen by emphasizing the needs of trade, communications and public health, by avoiding wars and punitive expeditions, by their reluctance to impose direct taxation, and by their determination to maintain a policy aimed at "peaceful penetration" and commercial development.
>
> (Gann and Duignan, 1978: 209)

The northern region was a borderland and its international trade was limited:

> In this region the tone of administration was military; the British ruling group was linked to London and the Home Counties [the southeast of England] rather than to Lancashire . . . Government emphasized prestige instead of profit, hierarchy in place of diversity.
>
> (Gann and Duignan, 1978: 209)

In all colonies, however, priority was given to communication, transport, and medical care. Railways were built both to promote agricultural and mineral exports and to facilitate the movement of police and army detachments. Post offices, telegraphs, and telephones gradually tied together the local administrative units. Indicative of the centrality of transportation networks to colonial administration was the fact that public works departments were often the first government units established in a territory. As Gann and Duignan (1978: 271) emphasize: "By 1914 all the British African dependencies possessed a basic infrastructure of specialized services, the most important of which was the creation of a modern transportation network."

An important cultural import into the colonies, therefore, was the assumption that the state should both encourage development and provide social services—education, agricultural instruction, etc.:

> The very notion of the state as a territorial entity independent of ethnic or kinship ties, operating through impersonal rules, was one of the most revolutionary concepts bequeathed by colonialism to post-colonial precedent . . . All of them have taken over, in some form or other, both the boundaries and the administrative institutions of their erstwhile Western overlords.
>
> (Gann and Duignan, 1978: 347)

8.4 MECHANISMS OF CULTURAL INTEGRATION

The imposition of colonial rule and, more generally, western penetration of societies outside Europe, involved a great deal of violence and war. But, once established, "law and order" involved the imposition of western values as much as terminating local conflicts and suppressing practices (witchcraft, infanticide, bride burning) that Europeans regarded as "barbaric."

The social effects of European values were paradoxical. On the one hand, old values were destroyed, as missionaries and schoolteachers attacked animistic creeds, polygamy, and other customs. Families often broke up as some members "converted" to Christianity and others did not. On the other hand, some people used the new ways to establish new bases of authority. In particular, western-educated natives became indispensable to European rule and influence. Interpreters, clerks, foremen, and police sergeants were cultural pioneers; they represented the new order and profited from it.

The economic effects were also double-edged. As restraints on trade disappeared, commercial agriculture and trade spread in extent and intensity. Yields increased as agricultural techniques improved, and trade proved more profitable as new communications linked previously isolated interiors with coastal entrepôts. Yet, as a consequence, the certainties and rhythms of local life broke down and traditional skills were devalued. Above all, new types of consumption, while adding to the comforts of life of those with sufficient disposable income, led to the destruction of many local industries and the growth of dependence on manufactured imports from the colonial "Motherland." In this context, obligations to community and chief began to weaken. Money became the major metric for assessing social status. This was in part because money could help purchase an education:

> Education, in turn, brought power and influence. These new opportunities profoundly affected life in the village, and the village ceased to be an almost self-contained unit, absorbing all the interests of its people. Instead, cash-cropping and wage labor for limited periods gradually came to occupy a much more central position in the cultivator's life.
>
> (Gann and Duignan, 1978: 367)

The growth of "free" labor was a process that was "always uneven and idiosyncratic" (Marks and Rathbone, 1982: 13). It depended on spatial variation in the extent of competition for labor; conflicts of interest between firms and the colonial states; and the availability of alternatives to wage labor. But once colony-wide labor and other markets were effectively created, the prospering of commercial enterprises (both foreign and indigenous), such as mines and plantations, depended on an increasingly efficient and productive labor force to operate new equipment and machinery. This required measures to both increase labor force stability (housing, minimum wages) and attempts to upgrade the health, literacy, and skills of employees. Of course, employee organizations also played a role in pressuring for these changes. Consumption demands, and so demands for higher incomes, tended to increase in concert with the increase in permanent wage employment.

It is evident that in many colonies there were dramatic improvements in education and health. In Africa in the period 1910–1960 the number of children attending school grew much more quickly than had school enrolments in Europe in the boom years of 1840–1880. There were also significant increases in life expectancy (for example, in Ghana in 1921 it was 28 years; in 1980 it was 45; and by 2000 it was 57) and reductions in infant mortality rates. To the British colonial authorities education was an important means of inculcating both "modern" work habits and a commitment to the class structure of colonialism. At the center of colonial education, was:

> [T]he idea of work—taught to those who lacked property—emphasizing regularity, the organization of time and human energy around the work routine, and the necessity of discipline. It was a moral and cultural concept . . . Prohibitions against drinking and dancing . . . were as much a part of changing concepts of labor as forced recruitment, vagrancy laws, and the insistence that workers put in regular hours.
>
> (Cooper, 1980: 69–70)

But, above all, it was necessary "to get workers to internalize cultural values and behavior patterns that would define their role in the economy and society" (Cooper, 1980: 70).

This conception of education was particularly characteristic of the British colonies. Other colonies either failed to develop the capitalist labor markets to which it was a reaction (for example, sections of the French colonies on the southern margins of the Sahara Desert) or were severely underfunded for public activities (for example, the Belgian Congo and Portuguese colonies). But British policy brought a price. It was precisely the educated élite that "formed the vanguard of the nationalist movements, and the more 'disciplined' African workers became, the more effective were their trade union organizations in pursuing not only economic, but political anti-colonial objectives" (Sender and Smith, 1986: 66). As taught in colonial schools, western concepts of "democracy" and "justice" served to undermine the legitimacy of the western empires that had introduced them.

The growth of wage labor incorporated women as well as men into the colonial world economy, although at significantly lower levels of participation. More importantly, however, the spread of wage employment disrupted existing sexual divisions of labor, often to the detriment of women. Women remained powerfully constrained by family, marriage, religion and so-called "domestic duties," even while engaged in wage employment. But norms, power, and traditional patterns of authority based on gender, as well as age, caste, and lineage, were subject to radical challenge because of wage labor, education, migration, and urbanization. Lonsdale (1985: 730) quotes from one Zulu chief in southern Africa who in 1905 expressed his opposition to the cultural changes of the time as follows:

> Our sons elbow us away from the boiled mealies in the pot when we reach for a handful to eat, saying, "we bought these, father," and when remonstrated with, our wives dare to raise their eyes and glare at us. It used not to be thus. If we chide or beat our wives and children for misconduct, they run off to the police and the magistrate fines us.

Cultural change was not always one way, however. Some pre-capitalist and pre-colonial social and religious institutions were strengthened. More accurately, perhaps, "new" syncretic traditions were invented out of elements of past traditions. Mission-educated élites often invented mythic histories of ancient empires and new nationalist traditions both to legitimize their quests for political power and to protect themselves in the new labor markets. Adaptation was at least as common as straightforward assimilation in many colonial settings.

Ultimately, however, cultural incorporation, whether by adaptation or assimilation, undermined the sociopolitical relation of dominance/subordination that colonialism existed to reproduce. Even when the idiom of dominance was *improvement* rather than *order,* native groups were not always persuaded of the natural superiority of colonial ways. Langley's (1983: 223) conclusion concerning American colonial adventures in the Caribbean is a fitting epitaph to the cultural contradictions of the entire colonial enterprise:

> Striving to teach by example, they found it necessary to denigrate the cultural values of those whom they had come to save … Their presence, even when it meant a peaceful society and material advancement, stripped Caribbean peoples of their dignity and constituted an unspoken American judgment of Caribbean inferiority. Little wonder, then, that the occupied were so "ungrateful" for what Americans considered years of benign tutelage. But, then, Americans do not have in their epigrammatic repertory that old Spanish proverb that Mexicans long ago adopted: "The wine is bitter, but it's our wine."

8.5 CHANGING GLOBAL CONTEXT OF INTERDEPENDENCE

The colonial world economy began to disintegrate after the Second World War. The crossover trading system effectively ended in the 1930s. The Second World War's buildup of industrial capacity for military purposes in the United States and the Soviet Union produced two superpowers without overseas colonial empires and European decolonization further undermined the colonial world order. Respectively, they and their allies defined the so-called First and Second Worlds. In the 1950s and 1960s the then so-called Third World (of politically independent but often politically nonaligned and always less prosperous countries) was born. Large parts of Asia and Africa now joined Latin America as a largely nonindustrial and ex-colonial but still "dependent world" (see Figure 8.9). The initial tendency was to attempt to achieve economic self-reliance through national strategies of industrialization and diversification of trading partners away from the particular colonial power (France, Britain, etc.). This was the first change in relation to the global context of interdependence.

END OF COLONIALISM AND NATIONAL STRATEGIES OF DEVELOPMENT

As new states came into existence, so did attempts to stimulate industrial development and economic growth. From the Depression of the 1930s on and especially during the Second World War, stagnation and shipping blockades had encouraged some import substitution in the colonies. Increasingly, traders and merchants in the richer peripheral countries looked to manufacturing industry as a source of capital growth. Leading politicians also saw in industrial development both national and personal advantage (see Chapter 10 for more details on the preference for industrialization).

Most importantly, however, the barter terms of trade (the ratio between the prices of exports and the prices of imports) for many of the basic commodities exported to the core countries deteriorated for many years only recovering somewhat recently in the face of dramatic increases in world prices for foodstuffs and some other primary commodities since 2007 (see Table 8.4). Basic commodity prices (largely those for raw materials) also declined relative to

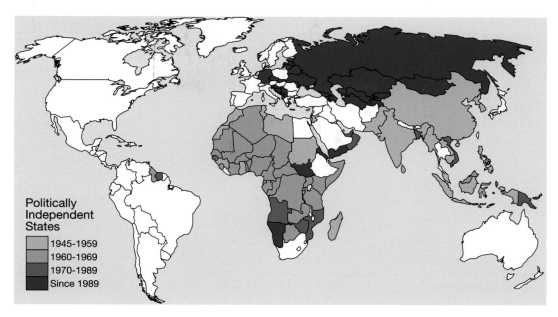

Figure 8.9 New states of the world since the Second World War

Source: Updated from Edwards (1985: 209, Figure 8.1)

those for manufactures, recovering less dramatically compared to price indices for commodity groups in recent years (see Table 8.4). One consequence of the long-term trend was a widening gap between what peripheral countries received for their exports and what they had to pay for their imports. The resulting deterioration in the general barter terms of trade between richer and poorer countries was widely viewed as inevitable—given the low elasticity of demand for LDCs' primary commodities (as incomes rose in developed countries there was not a parallel increase in the consumption of goods from LDCs), the persisting low wages in the primary production sector in the LDCs (lowering local consumption power) and the high protective barriers for competing primary production (mainly temperate agricultural products, such as sugar beet and cereals, and largely erected in the 1930s) in the developed countries.

In fact, while real prices for almost all primary commodities did decline since the early twentieth century until recently, the rate and volatility of decline has varied considerably across primary commodities and relative to manufactures. It was this volatility of real prices received by primary commodity exporters, combined with a *perception* of a deteriorating barter terms of trade, that inspired the first post-independence leaders to embark on policies that led away from specialization in primary commodity production. Trends until very recently did not offer much encouragement as prices for the principal nonfuel primary commodities in world trade stagnated, were volatile, or went down. The stagnation of prices for primaries against manufactures was especially marked until the 2000s for the least developed or poorest countries (see Table 8.4).

Pessimism about the future prospects for primary commodities was reinforced by the view that the primary sector was inherently backward compared to the manufacturing sector: Because of the latter's multiplier effects and, allegedly, greater economies of scale. Consequently, the pursuit of industrialization could be justified "theoretically" as well as materially (see Chapter 10).

Table 8.4 Barter terms of trade and world prices for primary product commodity groups

	1970–1972	1980–1982	1990–1992	2000–2002	2007	2011
Index of barter terms of trade: primaries against manufactures (1990–1992 = 100)						
Least developed countries (e.g., Haiti, Mali, Afghanistan)						
Agricultural products	121	120	100	86	102	129
Manufactured products	175	165	100	76	100	95
Other less developed countries (e.g., Argentina, Egypt, Pakistan)						
Agricultural products	125	125	100	97	115	115
Manufactured products	169	164	100	90	98	96
Price indices (2000 = 100)						
Tropical beverages	62	182	98	89	148	212
Vegetable oilseeds	74	150	114	103	226	307
Agricultural raw materials	40	111	125	97	164	223
Food	58	152	121	121	169	269
Minerals	49	100	116	92	313	339

Source: Based on online UNCTAD Annual Terms of Trade Indices at http://unctadstat.unctad.org/ReportFolders/reportFolders.aspx; online UNCTAD Annual Commodity Price Indices at http://unctadstat.unctad.org/ReportFolders/reportFolders.aspx?sRF_ActivePath=P,8,37&sRF_Expanded=,P,8,37

The promotion of industry came from either protection—from reserving domestic markets for domestic industry—or, later, establishing export enclaves and attracting foreign and local capital to finance branch plants. Once established, and/or protected through their early, vulnerable years, new industries would be able to compete globally. A predilection for protection, in many cases, reflected both a positive interpretation of the past practices of such countries as the United States, Japan and Germany, which had in their day protected **infant industries** and the conception of an "activist" state common to many ex-colonial territories.

During the 1960s exports of manufactures from LDCs grew quickly, from around US$3 billion in 1960 to over US$9 billion in 1970. As a percentage of total world trade in manufactures this was an increase of from under 4 percent to 5 percent. In the 1970s growth was even more rapid. By 1980 LDC-manufactured exports were more than US$80 billion or over 9 percent of the world total. This growth is part of the new international division of labor (NIDL; see also pp. 45–49). One of the most notable features of the period 1960–1980, however, and perhaps even more notable since then, has been the polarization of performance and prospects between different regional groupings of LDCs. Almost 75 percent of the total of LDC-manufactured exports comes from 11 NIEs (Brazil, China, Hong Kong, India, Indonesia, Malaysia, Mexico, Singapore, South Korea, Taiwan, and Thailand). In most of these cases—especially Hong Kong, Malaysia, Mexico, Singapore, South Korea and, Taiwan—industrial growth has been export led (export enclave) rather than import substitution (protection). Since 1980 this pattern has become institutionalized, with the bulk of the foreign direct investment, bank lending and trade relating to manufacturing production outside of the DCs concentrated in East Asia, parts of Latin America, and Eastern Europe.

Just over 66 percent of all LDC merchandise exports (manufactures, food, agricultural raw materials, fuels, ores and metals) are now sold to DCs. Both high levels of protection in the DCs and the risk of increased protection in the future have limited this market. Since the early 1970s the level and uncertainty of protective trade barriers to LDC manufactured exports have increased tremendously, particularly so-called hard-core nontariff barriers such as quotas, voluntary export restraints, and the Multifiber Agreement (MFA) (involving quotas by the DCs on textile and clothing imports from the LDCs between 1974 and 2005). This is especially the case for relatively more finished products. Between 1966 and 1986, the share of imports affected by all nontariff measures increased by more than 20 percent for the USA, around 40 percent for Japan and 160 percent for the EC. By 1986 21 percent of LDC exports to the DCs were covered by these barriers even as global average tariff rates and coverage declined. The Uruguay round of the General Agreement on Tariffs and Trade (GATT), however, incorporated some significant steps towards trade liberalization. Beginning in 1994, voluntary export restraints were abolished and the highly protectionist quota regime for textiles and clothing began to be phased out.

Manufactured goods can be classified into two main product groups: producer goods—capital goods (for example, machinery and equipment, including transport equipment) and intermediate goods (for example, raw materials, and semi-finished items)—and consumer goods, of which textiles and clothing are the largest single category in international trade. Capital goods account for about half of all manufactured goods traded in the world economy (up from 46 percent in 1990). But different world regions account for different shares of the two product categories. For example, DCs supply 65 percent of world exports of machinery and transport equipment and 35 percent of textiles and clothing. LDCs are important mainly as suppliers of textiles and clothing and certain light industrial products, mainly consumer goods (with the important exception of electronics components and automobile components, in the case of countries such as China, Brazil, Mexico, and South Korea). The specialization of trade flows between LDCs and DCs, therefore, extends today beyond the distinction between primary commodities and other goods to apply *within* the category of manufactured goods.

ROLE OF TRANSNATIONAL CORPORATIONS

The growth of trade in manufactures in the 1970s, after a 40-year period in which manufacturing production was intensively concentrated in the DCs and there was more limited DC–DC as well as DC–LDC trade in manufactures, was influenced by the growing significance of transnational corporations and of contractual cooperation between firms in different countries. Transnational corporations (TNCs) have long been active in manufacturing in LDCs. As we saw in Chapter 6 and discuss further in Chapter 10, they tended at first to duplicate plants around the world in order to gain access to protected markets or to make use of local raw materials. The production by TNCs of cars (for example, in Brazil), agricultural engineering products (for example, in South Africa and Mexico) and pharmaceuticals (for example, in India) across a range of LDCs are examples. This kind of manufacturing production still exists, especially in countries with large internal markets. In Brazil, for example, in the mid-1970s to take an extreme case, almost 50 percent of industrial output was produced by TNCs and more than 90 percent of TNC production was sold locally.

Since the 1970s, however, much TNC involvement in LDCs has also involved what is known as **global sourcing**. As a result of technical change, especially reductions in transportation costs, and the appeal of cheap (often female) labor in certain countries, production activities that once were adjacent spatially can now be dispersed widely. Many so-called light industrial processes are especially suited to the separation of various stages of production. In particular, labor-intensive stages (as in the product lifecycle model) can be located to take advantage of both the enormous international spread in wage levels and the exchange rate fluctuations between currencies that have been a feature of the world economy since the early 1970s. With respect to wage levels, the footwear industry faces wage costs of about US$15 per hour in the United States but under US$1 in Bangladesh, the Philippines, and Trinidad and Tobago; Chinese textile workers earn less than US$3.50 per hour compared to a U.S. rate of about US$19 (the higher skills and productivity of workers in the textile industry translate into higher wages compared to the clothing and footwear industries).

One industry that has engaged in global sourcing on a massive scale (and, perhaps, for this reason, is somewhat exceptional) is the consumer electronics components and products industry. This industry has two characteristics that have encouraged the shift to global sourcing: Discrete production segments, of which some are extremely labor intensive and require a "flexible response" because of short product cycles that make automation uneconomic; and compact products (parts and components) that can be shipped relatively cheaply. East Asian locations with cheap, reliable, literate and tractable (largely female) labor forces have been especially attractive to this industry (and some others such as textiles and clothing). Governments have often facilitated the process of establishing component and assembly plants through the provision of export processing zones (EPZs), subsidies and tax advantages, and the enforcement of the "political stability" highly valued by TNCs and their local subcontractors.

CHANGES IN MARKETS FOR PRIMARY COMMODITIES

For many LDCs, however, there is still a heavy dependence on trade in primary commodities (see Chapter 9). But the primary commodity sector has become extremely heterogeneous with respect to trading conditions since the Second World War. Four major categories stand out in this regard: Fuels (mainly petroleum), nonfuel minerals, grains, and other agricultural products. These product groups have experienced very different price movements and, to some extent, quantity fluctuations over the past 50 years. The nonfood commodity prices have been especially volatile. Generally, manufactures have increased in price to the disadvantage of primary commodity exporters. But there have been two periods, 1949–1952 and 1973–1980, when demand for primary nonfood commodities was extremely strong and commodity prices surged. In particular, in the years 1973–1980, inflation and uncertain economic conditions in the DCs boosted the prices of agricultural raw materials. Since 1980, however, at least until the 2000s, the relative price strength in the fuels and agricultural products groups largely disappeared and the value of commodity export earnings in these sectors sank precipitously in relation to the prices of manufactures and minerals. Perhaps the most negative price movement from the perspective of most LDCs has been in the price of grains. There was a long-term decline in world grain prices until around 2007. This reflects tremendous increases in production the world over, but especially in the United States and other DCs. Normal yields per hectare are now more than twice what they were in 1950. The real price of wheat, however, is now about half what it was 100 years ago. Climate change and huge increases in demand, however, are now seeing this trend threatened as at no time in the previous 50 years.

Across all primary commodities, commodity agreements between producing and consuming countries (cocoa, tin, sugar, and natural rubber) and producer cartels (most famously, OPEC for petroleum) have failed to reduce volatility and consistently raise the prices of primary commodities relative to those of manufactures because of fundamental differences of interest between producers and consumers and among producers. Even OPEC, after successfully raising the price of oil between 1973 and 1979, has been riven by conflict and the failure to attract some major oil producers (such as Britain, Mexico, and Norway) to its ranks. This failure has encouraged further attempts at industrialization as the major strategy of economic development. However, price volatility acts to reduce the industrial potential of "mineral economies" (Auty, 1991). This is one of the Catch-22s of the contemporary world economy. Dependence on oil is said to sometimes lead to what is called the **resource curse**. This suggests that rather than a blessing, reliance on a primary commodity for which there is tremendous world demand can give rise to a range of negative effects such as encouragement of corruption (particularly when state-owned companies have a monopoly), a rise in the exchange rate between the country's currency and others that can then raise inflation and squeeze out investment in agriculture and manufacturing, and pressure to share revenues by investing in prestige projects and providing subsidies that do not stimulate long-term economic growth.

Some countries, however, especially those in Sub-Saharan Africa, could probably benefit from increased attention to primary commodities. The macroeconomic policies of many African governments have worked to undermine their region's shares of world export markets across a range of primary commodities. Since the early 1970s real export earnings have remained stagnant or declined significantly in 25 out of 33 countries in Sub-Saharan Africa for which data are available. The whole of this region, with around 700 million people, has export revenues that are less than those of Singapore, a city-state of 5 million people.

Inelasticity of demand in the DCs (expressed in deteriorating barter terms of trade) cannot completely explain the magnitude of these declines; neither can the absence of commodity diversification, since those countries with a relatively diversified structure of agricultural exports—such as Tanzania—have not experienced more favorable trends in export earnings than more specialized ones. Sender and Smith (1986: 127) explain the absolute decline of Sub-Saharan Africa's contributions to world commodity markets in terms of "the continued dominance of anti-trade ideologies and export pessimism" that are:

> [P]robably explained by the political hegemony of nationalism. It remains expedient for the national bourgeoisie, or for those determining the form and nature of state intervention, to deflect criticism by resort to anti-imperialist rhetoric and to blame foreign scapegoats for economic failure.

However, this is probably too narrow a perspective. Political agendas and social problems of a more general nature have also played important roles. In the immediate post-independence period, considerable political energy was expended in diversifying import and export markets rather than building larger ones. This was a direct result of trying to slay the "colonial dragon" as the newly independent countries tried to become less dependent on their former colonial powers. Governments have also been faced with major ethnic divisions and rivalries, fragile political institutions, and "superpower" infiltration and manipulation. The Nigerian Civil War in the early 1970s, frequent military coups d'état and American or Soviet covert operations in most African countries are symptomatic examples of the diversions from economic policymaking that have faced political élites in Sub-Saharan Africa (and to a lesser extent also in Latin America and Asia) since the 1940s.

DISPARITIES WITHIN THE PERIPHERY

Disparities among LDCs have increased substantially since the 1970s. Several groups have emerged and can be distinguished. First, there are those NIEs that have grown rapidly and are important exporters of manufactured goods. These include the "old" NIEs such as South

Korea, Taiwan, and Hong Kong, and newer NIEs, predominantly in Asia (for example, Malaysia, Thailand), but including Brazil and Mexico. In 1960 the East Asian NIEs only accounted for 5 percent of total LDC exports and in 1980 for 10 percent but, by 2005, this had risen closer to 50 percent.

Then there is a group of countries that experienced reasonable growth until the late 1970s but because of high debt loads struggled during the 1980s in particular. Examples would include Argentina, Brazil, and the Philippines. A couple of countries in this group with less serious debt problems, Costa Rica and Colombia, managed to continue their economic diversification away from primary commodities. A third group remains very dependent on raw material exports but export demand has held up to some extent. Examples would include oil exporters such as Nigeria, Ecuador, and Cameroon.

Finally, there are two groups of low-income countries mainly in Sub-Saharan Africa and Asia. The Asian group—Afghanistan, Bangladesh, China, India, Nepal, Pakistan, and Sri Lanka—is populous and, until recently, its countries isolated themselves from the world economy through protectionist policies. China has been the most aggressive in opening up to trade and foreign investment and the impact of this is now most apparent in China's coastal areas, especially around Hong Kong and in the vicinity of Shanghai (see Chapter 10). Sub-Saharan Africa has had the poorest record of economic growth over the past 20 years. Most countries in the region have experienced declining or stagnant export earnings in the 1980s and 1990s. They are heavily dependent on foreign aid and investment by multilateral institutions such as the World Bank (but on the dangers of over-aggregating the African case, see Grant and Agnew, 1996).

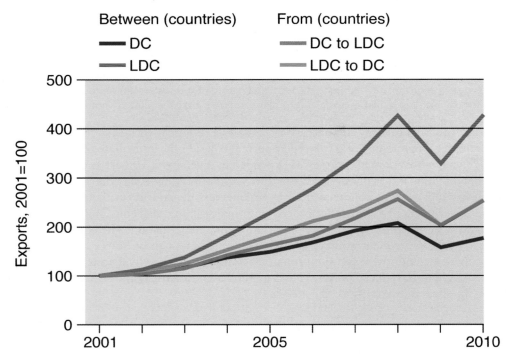

Figure 8.10 Exports between and from developed and less developed countries, 2001–2010

Source: Based on *Economist* (2013a)

An important consequence of the growth of the NIEs, on the one hand, and the disparities between countries outside the historic core countries, on the other, has been increasing trade between countries within the less developed category relative to that between the LDCs and the DCs. Increasingly, countries such as China are major sources of manufactured exports to the various groups of countries identified here. At the same time, such countries are becoming major importers of raw materials and foodstuffs from those countries still concentrated in the production of primary commodities. As a result, trade between LDCs as a category has increased at a faster rate than has trade between developed countries and from LDC to DC and vice versa (see Figure 8.10). This suggests that the core–periphery structure of the world economy is undergoing a fundamental reordering in terms of the identity of the countries whose economies are driving world trade with China and Southeast Asia joining the United States, Europe, and Japan.

EFFECTS OF THE COLD WAR

The end of colonialism did not usher in an era of equivalently "sustainable" national develop-ment everywhere in the former colonial world; that much should be clear from the preceding discussion. The factor initially most responsible for this was the Cold War between the United States and the former Soviet Union, which, while encouraging "aid" programs of one kind or other, also encouraged militarization and political instability. After the Second World War the world was effectively divided into two spheres of influence with large parts of the new so-called Third World of former colonies as a zone of superpower competition (see Figure 8.11). In certain cases, such as, for example, South Korea and Taiwan, "superpower" aid (U.S. in these cases) contributed to economic growth. In many African countries, aid helped achieve major improvements in physical and social infrastructure, although much of the most productive aid did not come from the superpowers, which have specialized in military aid and technical assistance (intelligence gathering), rather than direct economic assistance. International agencies (the UN, World Bank, etc.), and some European countries (particularly the Scandinavian countries and the Netherlands) have provided much of the more economically "useful" aid (on international aid, see Chapter 2).

In other cases, models of development were imported from either the United States (free enterprise) or the Soviet Union (central planning) and then supported/undermined from outside by each of the superpowers. This often led to increased militarization both of govern-ments and national budgets as internal opponents were repressed and external patrons satisfied. Between 1960 and the collapse of the Soviet Union in 1991, more than 11,700,000 people were killed in 143 major wars and episodes of political violence (those with more than 1,000 deaths attributable to them). Most of these were in the LDCs.

GLOBALIZATION OF CAPITAL

A second change in the global context of interdependence since the demise of colonialism and the rise of national development strategies, and increasingly important since the 1970s, has been the increased pace and internationalization of the world economy as noted in Chapters 3 and 6. Capital has become much more mobile, both in time and space. For example, before 1972, currency exchange rates changed once every four years on average, interest rates moved twice a year and companies made price and investment decisions no more than once or twice a year (see Figure 3.3). All this has changed. There is now an almost constant review of prices and investment decisions, a constant instability and disorder. This places an even greater

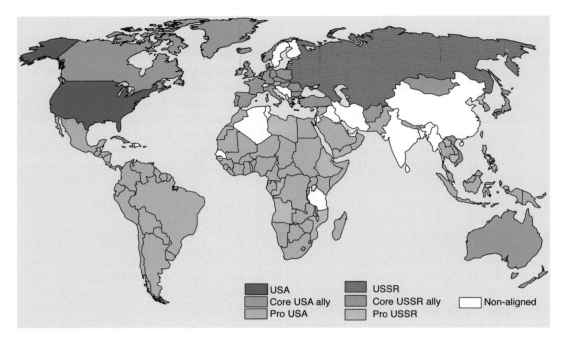

Figure 8.11 Soviet and U.S. spheres of influence, 1982

Source: Based on O'Loughlin (1989: 302, Figure 11.1)

premium on the ability of firms and governments in LDCs to react within some coherent national framework to changes in the global political–economic environment.

Nowhere is this clearer than with respect to the world financial system. As we have seen, the growing integration of the world economy and the loosening of government control over exchange rates (when the major industrial governments, led by the United States, abandoned the Bretton Woods system of semi-fixed currency exchange rates in 1971) stimulated the growth of a massive *private* international monetary system. This system was organized around eurodollars, a term that originally meant U.S. dollar deposits in banks in Europe but now refers to dollars that circulate outside the United States, which are used for world trade and are not regulated by the U.S. government. This global currency mushroomed between the mid-1970s and the early 1980s as a result of the enormous dollar surpluses earned by the OPEC countries from oil sales. The large banks that received these funds sought borrowers who:

> [C]ould be charged enough interest to enable the banks to earn a profit. The banks first moved Eurodollars into Third World countries, saddling them with a US$1 trillion debt burden by the end of 1986, compared to less than US$100 million in 1973. Instead of supporting new productive investment, however, a large portion of this debt went into luxury consumption. Over US$200 billion disappeared through capital flight from the Third World back to the industrial countries.
>
> (Wachtel, 1987: 786)

This phenomenon was particularly deleterious for poor oil importers (who had to finance their oil imports) and countries with ambitious development plans (into which outside capital could be pumped). Many of them are still locked into debt repayment schedules that require them to devote the lion's share of their export earnings to servicing their debts. In Latin America, for example, since the early 1980s there has been a persistent net outflow of capital "reflecting a 'scissors movement' of declining grant aid and mounting debt repayments" (Cook and Kirkpatrick, 1997: 63).

After 1982, when Mexico and certain other major Latin American borrowers effectively defaulted on loan repayments, the eurodollars shifted into funding the U.S. budget deficit, into firm mergers and acquisitions in the United States and Europe and into the world's stock markets. But the burden of debt, in both the form of repayment and the inability to borrow fresh capital with outstanding debt, emerged as a major new barrier to economic development. In the 1980s net financial flows to LDCs initially decreased as commercial bank lending dried up. Since 1986 government and multilateral agency (World Bank, etc.) aid and foreign direct investment by TNCs have increased substantially. But only in 1989 did total flows equal the 1982 figure and, within this, aid increased from 38 to 55 percent of the total. Since then private investment has increased somewhat, including both portfolio and foreign direct investment, but debt service still soaks up significant proportions of all capital inflows. This is borrowing to pay back what was previously borrowed rather than productive investment in economic activities that improve the lot of the LDCs and their populations. In the years between 1990 and the Asian financial crisis in 1997 in particular, most new bank lending from the DCs went to Asia, whereas the other regions had to rely more on foreign direct investment (as in Latin America) or on aid and multilateral loans from international organizations (for example, the International Monetary Fund in Sub-Saharan Africa).

DEBT AND INSTABILITY

If the 1970s are remembered for two major oil price shocks (in 1973 and 1979) and a persisting downturn in the world economy, the 1980s were marked by three global phenomena. One was the international debt problem, which effectively undermined growth in many LDCs for most of the decade. Although a variety of international agreements renegotiating interest payments reduced the overall severity of the international debt crisis in the 1980s, debt burdens remain at historic highs and effectively undermine the possibility of future credit-led economic development (Corbridge, 1993, offers a thorough overview of this phenomenon). The absolute debt load of LDCs peaked during the mid- to late 1990s at around 39 percent of their combined GDP after declining in a cyclical fashion to 33 percent in 1990 from a previous high of around 37 percent in 1986. So debt is a problem that still haunts many LDCs' economies. Its impact varies widely, however, and is not a direct function of the absolute size of the debt. For example, as of 1999 Mexico had a huge debt of US$166.9 billion but its debt–service ratio (interest and principal payments as a share of all exports) was "only" 25.1 percent, whereas, at the other extreme, Colombia with a debt of US$34.5 billion had a debt–service ratio of 43.5 percent. Of course, many developed economies also now have huge government debt loads, in large part because of bailing out failing banks in the context of the global financial crisis of 2008. A second was the volatility of exchange rates, illustrated most vividly by the steep appreciation of the U.S. dollar until 1985, followed by a dramatic descent until the late 1990s, at which point it began to appreciate again so that by the early 2000s it had reached levels not seen since the mid-1980s. These shifts have had important effects on the imports and exports of countries whose currencies are pegged in value to global currencies such as the U.S. dollar. Commodities become more or less expensive depending on the shifts in the value of the global currency. The net impact has been negative for nearly all LDCs. Finally, the U.S. federal government and national trade deficits, emerging spectacularly after 1983, produced growth in the USA at the expense of stagnation elsewhere. Investment that could have gone to the LDCs under other circumstances was diverted to the USA to finance the deficits. This also

Box 8.1 The Asian financial crisis

The early 1990s in East Asia was a time of rapid economic growth. Doubts about the East Asian Miracle surfaced in 1996, however, as observers pointed to a slowdown in exports, excess capacity in many industries and declining earnings. In 1997 the currency and financial crisis that erupted in Thailand in July spread to South Korea, Malaysia, and Indonesia; Hong Kong, the Philippines, and Singapore were less hard hit; China and Taiwan were least affected although they still experienced a slowdown in growth.

The crisis in East Asia deepened in late 1997 and had spread to the rest of the world by 1998. Speculative attacks in other semi-peripheral countries—notably Brazil and the Russian Federation—caused economic difficulties and capital flight. On Wall Street, the Dow Jones Industrial Average recorded its second and third biggest losses in August. Efforts by the individual East Asian countries themselves, combined with assistance from the IMF and World Bank among others, helped bring the crisis under control to the extent that, by March 1999, the Dow closed above 10,000 for the first time in its history. The East Asian economies rebounded and enjoyed a growth rate of 4.1 percent in 1999 and close to 6 percent by 2000.

Some of the factors that precipitated this crisis in East Asia, which have been targeted with macroeconomic and structural policies, include national financial weakness (overvaluation of currencies pegged to the U.S. dollar, large volumes of short-term capital inflows and exposure to short-term debt), inadequate regulatory oversight of financial and other businesses, high corporate indebtedness, failed management and overcapacity in key manufacturing sectors. Governments in the countries that were most affected learned their lessons. Since 1999 they have tended to manage their currencies more actively, restrict capital movements, and shift from what Studwell (2013, 223) calls an "economics of efficiency"—focused entirely on global competitiveness—to an "economics of development"—oriented to education, nurturing of new industries, *and* competition.

raised interest rates on outstanding loans such as those held by heavily indebted LDCs. Consequently, high interest rates and low commodity prices rather than the debt loads incurred in the 1970s were probably the major barriers to growth in the LDCs in the 1980s. On balance, the 1980s was not a good decade for the LDCs as a whole. Likewise, the period since 1990 has hardly been stellar given the generally declining terms of trade for primary commodities, the debt loads incurred but not yet paid off and the economic woes of a number of the NIEs precipitated by the Asian financial crisis of 1997. With the global financial crisis since 2007, many of the world's poorer countries, with the exception of the NIEs this time around, also find themselves crowded out of world credit markets and unable to refinance previous debt loads.

CONTINUED CORE–PERIPHERY POLARIZATION?

From one point of view, the net effect of the changes in the global context for interdependence has been an increased division between the LDCs, on the one hand, and the DCs, on the other. National income and purchasing power statistics support this interpretation. But, from another point of view, the periphery has, in fact, developed rapidly. This interpretation is supported by data on output, health and education. One way of reconciling these discordant interpretations is to argue that the post-colonial world economy has come to rest on increasingly diffused global production but has lacked a similar attainment of a global spread of consumption. The relatively low incomes available in LDC factories and plantations have put a cap on local purchasing power even as local labor forces were made more efficient (through improved health and education) and increased their output.

The problem with this reconciliation and the interpretations on which it is based is that they are geographically over-aggregated. The experience of different groups of LDCs has been different, as suggested previously by the "grouping" of countries. On the one hand, some of the Latin American countries, for example, are relatively large and have relatively high levels of per capita income (for example, Brazil, Argentina). Some of them did achieve considerable income growth in the 1960s and 1970s on the basis of industrialization to satisfy local markets (import substitution). Most of them, however, were heavy importers of oil (Mexico and Venezuela were exceptional) and their industrial sectors were generally uncompetitive in world markets. They were hit in the 1970s by the combination of oil price rises and their failure to switch to export-oriented manufacturing in the boom years of the late 1960s. They had to borrow to ease the oil shock adjustment instead of paying for it with export earnings. They are now caught in a **debt trap** of accumulated loans compounded by the high interest rates of the early 1980s, and the growth of trade barriers to the manufactures they export to the United States and Europe.

The Sub-Saharan African countries, on the other hand, are much poorer per capita on the average than Latin American or East Asian countries and their low level of output in all sectors is undermined by their even faster rates of population increase. They are economies with small industrial sectors and a heavy dependence on primary commodities. As commodity prices dropped in the 1980s (see Table 8.4), the Sub-Saharan African countries were forced to borrow to maintain minimal levels of consumption. Their debt burden is similar to that of the most indebted Latin American countries. The consequence, as in Latin America, is a general reduction in the standard of living but in contexts where it is already desperately low.

Finally, Asian countries have, on average, managed the best over the past 30 years. They have been more successful in maintaining economic growth as a secular trend and adjusting to short-run cyclical downturns such as the world recessions of 1974–1975, 1979–1982, 1989–1993, 2000–2002 and 2007 to the present day. The East Asian NIEs, the most dynamic economies in the region, have been able to expand their export of manufactures. They now account for almost 60 percent of LDC total exports of manufactures. Although, like the Latin American countries, they borrowed heavily in 1974–1975 to adjust to the oil price increases, their export performance has allowed them to keep relatively good borrowing terms and adjust more easily to the massive interest rate increases of 1980–1981. Yet, they are not without their own difficulties. Even before the Asian financial crisis, growth rates were slowing throughout East Asia in the mid-1990s indicating that there may be limits to an export-based strategy of economic development when established markets stagnate or decline and growth is more reliant on cheap labor and high savings than on technological and organizational innovation. The rise of the East Asian NIEs, however, has altered the global development picture in fundamental ways. For one thing, their high rates of economic growth have translated into declining national poverty rates. Because of the numbers of people involved this can be made to seem as if this is a trend across the entire periphery. In fact, poverty reduction has been much less evident elsewhere. This means that there are increasingly significant differences *within* the periphery and between semi-peripheral and truly peripheral economies with respect to poverty and overall quality of life.

8.6 ALTERNATIVE MODELS OF DEVELOPMENT

In the face of the failure of many LDCs to maintain, let alone increase, their production output and the consumption levels of their populations, the models of development on which national development efforts have been based have been called into question. This coincides with the growing questioning of the American model of competitive capitalism (especially the lack of effective national trade and industrial policies) in the USA and the collapse of the Soviet command economy model in the (former) Soviet Union and Eastern Europe. Neither of these models can any longer be said to offer a simple way out of the "development impasse." The spread of more liberal and open trading policies in the 1980s, 1990s, and 2000s did produce benefits for some countries, particularly the NIEs, but these have been strictly limited geographically. The open economy/liberal model is undoubtedly still dominant at the moment; sponsored by the main international economic institutions (such as the IMF, WTO, and World Bank) as well as by the TNCs and major global banks. But after the global financial crisis of 2007 onwards, it too is now in question within the developed core as much as elsewhere.

It is in this context that new models have appeared to replace previous ones. Perhaps the three most important ones today are based on (1) a synthesis of liberal reforms and social democracy as practiced in parts of Europe, particularly in Sweden, (2) the Chinese experiment in globalization since 1978, and, to a much more limited extent, (3) Islamic economic practices. Each of these alternative models is noted briefly.

The European experience is seen as relevant because it combines both a focus on economic growth with an emphasis on reducing income inequalities and social exclusion. The problem is that historical experience suggests that both rarely take place at the same time. It has been only with considerable struggle that subordinated groups have been able to wrest various programs and social rights from their national governments often in the face of resistance from dominant groups such as local and foreign capitalists. The great Chinese experiment since 1978 is also unique in that it combines a government regulated macro-economy with considerable decentralization of power over industrial and financial affairs to a range of other groups including domestic capitalists, local governments, and TNCs. Finally, practices and beliefs drawn from the Islamic religion have become important in southwest Asia, North Africa, and other parts of Asia (for example, Indonesia). The prohibition of usury or "excessive" interest charged on monetary loans is one of the more concrete and obviously appealing features of Islamic economics. But, as yet, no system of political economy based on Islamic principles has been established in any country (including Iran). The conclusion of Katouzian (1983: 164), one of the leading authorities on Islamic economics, seems appropriate:

> While one may empathize with the desire to construct an indigenous ideology that can be identified with the Islamic beliefs and practices of its advocates particularly in view of the havoc caused by selective application of Western ideas under the late Shah [of Iran], it is no more to be expected that Islam can provide a comprehensive economic system than that the latter could be based on Christianity, Judaism, or any other traditional religio-political system.

In practice, what is happening is more by way of different adaptations to a dominant liberal capitalism than the adoption of full-blooded alternatives to it. If some places have seen a full-scale adoption of a market-access capitalism in which barriers to the flow of goods and capital have been radically reduced, others have seen the continued or renewed imposition of state regulation. Some places with heavy state regulation have been able to successfully incorporate themselves into global trade and capital circuits primarily through the export of manufactured goods and services (see Table 8.5). These are usually populous states such as India and China

Table 8.5 Adaptations to global capitalism

	State-regulated	Market-access
Export of manufactures/services	China, India	Taiwan, South Korea
Export of oil and gas	Russia, Venezuela	Nigeria, Saudi Arabia

with distinctive competitive advantages in different sectors. Smaller countries emphasizing exports typically must make themselves much more open to the potential turbulence of the world economy. Resource-oriented economies likewise divide into two main groups according to the degree of state regulation and direction. Here, scale of production also seems to matter but the patterns are more unstable in that the role of different strong political leaders, such as Putin in Russia and Chavez in Venezuela, and their ability to exploit upswings in global demand for their countries' oil and natural gas, seem to be more determining of policy choices than the structural characteristics of their economies *per se*.

SUMMARY

In this chapter, we surveyed the dynamics of interdependence between the core and the periphery of the world economy from the colonial period to the present day. We have identified the following points as being of critical importance:

1. Existing economies were transformed into colonial ones through regional specialization in primary commodity production.
2. In the late nineteenth and early twentieth centuries, a "crossover" multilateral system of trade, with Britain as its linchpin, integrated the world economy.
3. The "crossover" system was progressively displaced by foreign direct investment (FDI) from transnational corporations (TNCs). American firms were especially important.
4. Colonialism created the conditions for wage labor and gave priority to improving communications, transportation and medical care. The European-style territorial state became accepted as the basic political unit for regulating economic activity.
5. Western values had paradoxical effects. On the one hand, values of work discipline and private property were disseminated. On the other, new syncretic traditions were invented.
6. With decolonization, new states came into existence that attempted to encourage industrialization.
7. For many years much manufacturing in the LDCs was import substitution. Since the 1960s, however, TNCs have engaged in global sourcing: Dispersing some production functions to appropriate sites in LDCs and exporting components/products back to the USA, Europe or Japan. Some NIEs have developed their own export-oriented industries.
8. Many LDCs are still heavily dependent on the export of primary commodities, the prices of which are highly volatile and have tended to decline against those of manufactured goods over time.
9. Cartels and production agreements have largely failed to stabilize the production or prices of most primary commodities. The success of OPEC in relation to petroleum beginning in the 1970s is the one exception.
10. The Cold War between the United States and the former Soviet Union and the increased pace and internationalization of the world economy placed serious constraints on

development efforts. The global debt crisis of the early 1980s has been another especially important constraint.

11. The integration of production within the world economy has not been matched by an integration of consumption. However, different world regions of the periphery have had different experiences in this regard: The Asian countries (especially the East Asian NIEs) have been most successful, the countries of Sub-Saharan Africa least so.

12. Alternative models of development, from Europe, China and Islamic traditions, have arisen to challenge the dominant U.S./Soviet ones because of the failure of the dominant ones to manage social inequalities or generate sustainable economic development. But more typical in practice have been adaptations to the dominant liberal capitalism of the globalizing world economy involving greater or lesser degrees of state intervention versus more whole-hearted acceptance of market-access policies.

The next three chapters take off from this general perspective on the transformation of the periphery and semi-periphery to examine contemporary patterns of agriculture, industry and services, paying special attention to the changing relationships between core and periphery outlined in Chapters 2 and 3.

KEY SOURCES AND SUGGESTED READING

Amsden, A., 2007. *Escape from Empire: The developing world's journey through heaven and hell.* Cambridge, MA: MIT Press.

Banerjee, A.V., Bénabou, R., and Mookherjee, D. (eds.), 2007. *Understanding Poverty.* Oxford: Oxford University Press.

Cook, P. and Kirkpatrick, C., 1997. Globalization, Regionalization and Third World Development, *Regional Studies* 31, 55–66.

Corbridge, S., 1993. *Debt and Development.* Oxford: Blackwell.

Diakosavvas, D. and Scandizzo, P., 1991. Trends in the Terms of Trade of Primary Commodities, 1900–1982: The controversy and its origins, *Economic Development and Cultural Change* 39, 231–264.

Fiala, R. and Kamens, D., 1986. Urban Growth and World Polity in the Nineteenth and Twentieth Centuries: A research agenda, *Studies in Comparative International Development* 21, 23–35.

Grant, R.J. and Agnew, J.A., 1996. Representing Africa: The geography of Africa in world trade, 1960–1992, *Annals of the Association of American Geographers* 86, 729–744.

Powell, A., 1991. Commodity and Developing Country Terms of Trade: What does the long run show?, *Economic Journal* 101, 1485–1496.

Sender, J. and Smith, S., 1986. *The Development of Capitalism in Africa.* London: Methuen.

Studwell, J., 2013. *How Asia Works: Success and failure in the world's most dynamic region.* London: Profile Books.

Wachtel, H.M., 1987. Currency without a Country: The global funny money game, *The Nation* 245, 26 December, 784–790.

Weinthal, E. and Luong, P.J., 2006. Combating the Resource Curse: An alternative solution to managing mineral wealth, *Perspectives on Politics* 4(1), 35–53.

Chapter 9

Agriculture: The primary concern?

Picture credit: Linda McCarthy

"Development" is often equated with the structural transformation of an economy whereby agriculture's share of the national product and of the labor force declines in relative importance. Agriculture has often been viewed as a "black box from which people, and food to feed them, and perhaps capital could be released" (Little, 1982: 105). This perspective, long dominant among planners and politicians and common to both the U.S. and the former Soviet models of development, reflected the low elasticity of demand for food (demand increases very little with higher incomes), the secular global trend towards higher labor productivity in agriculture (the same output can be produced by fewer workers because of technology, fertilizers, etc.), the limited multiplier effect of agriculture on other economic activities and the secular tendency for the barter terms of trade to turn against countries that export primary commodities and import manufactured goods.

However, it is almost certain that the world's population will rise to over 9 billion by the middle of this century. It is equally certain that most of the growth in population between now and then will take place in the LDCs. Consequently, these countries in particular will need to increase their food production to supply the additional people and to increase their standard of living. At the same time they face two major constraints: Much land is unsuitable for agricultural purposes (see Figure 2.12) and their involvement with the world economy often reduces their food self-reliance without sufficient compensation in other sectors.

The purpose of this chapter is to describe the contemporary state of agriculture in the periphery of the world economy. To this end, the chapter is organized as follows: A first section establishes the importance of agriculture as an economic sector and stresses the dual trends of increased agricultural production for the world market and decreased food self-reliance; a second section discusses the general relationships between land, labor, and capital in the periphery with special attention to efforts at rural land reform; third, the capitalization of agriculture in the periphery by transnational corporations is described; fourth, and last, the role of science and technology in agriculture in the periphery, especially in the form of the so-called Green Revolution, is assessed.

9.1 AGRICULTURE IN THE PERIPHERY

The countries of the periphery have all been significantly involved with modern commercial farming since the beginning of western colonization in the sixteenth century. But subsistence and production for local markets have remained of great, if decreasing, importance. Malassis (1975) identifies four types of agricultural system in the periphery: (1) the "customary" farm involving common ownership of land for both cultivation and grazing; (2) the "feudal or semi-feudal" estate, hacienda and latifundia; (3) "peasant agriculture," including minifundia (small, subsistence farms), commercial farms, and share cropping; and (4) capitalist plantation or mechanized agriculture based on wage labor. These four types of farm organization produce three types of commodity: (a) commercial foods, primarily cereals for the domestic market; (b) subsistence foods, primarily for personal use; and (c) export crops, where the major market is overseas. The historical trend in agriculture in most countries of the periphery has been from (1) and (2) to (3) and, especially, (4) in farm organization and from (a) and (b) to (c) in types of agricultural commodity.

However, the three continents of the periphery—Africa, Asia and Latin America—differ in terms of agricultural organization and performance. Above all, Sub-Saharan Africa is, or has been until recently, abundant in land and sparse in population; Asia is largely short of land relative to population; and Latin America contains both areas with large populations and areas with few inhabitants. Agriculture is also of much greater relative importance in Sub-Saharan Africa and Asia than in Latin America, both in terms of employment and contribution to national product.

WOMEN'S WORK

It is also important to recognize that in agriculture in the LDCs it is the women rather than the men who are overwhelmingly more important as the source of workers. Indeed, the gender dimension is not a secondary consequence of variations in agricultural organization but "a fundamental organizing principle of labor use" (Joekes, 1987: 63). Regional differences are apparent, however, indicating the contingencies of resource endowment and carrying capacity. More women are involved in agriculture in Africa, relatively speaking, than elsewhere. In 2000 the UN FAO estimated that at least 75 percent of all women in the labor force in Sub-Saharan Africa were involved in agriculture, compared to 68 percent in India, 70 percent in China, 62 percent in other low-income Asian countries (such as Bangladesh and Cambodia) and 35 percent in middle-income Asian countries (such as Malaysia and South Korea). In Latin America, the comparable figure is a very low 10 percent. This reflects the greater degree of mechanization (and export crop orientation) in Latin American agriculture and higher levels of female rural to urban migration compared to other regions. Official figures may capture the female day laborers on larger commercial farms but certainly miss many of the subsistence farming activities carried out predominantly by women. Labor force participation data usually involve very narrow definitions of agricultural activity focused on land cultivation and large-scale livestock keeping. While many of the men migrate to find work, they leave behind the women, whose largely unrecorded role in agriculture includes tending to the fields and the animals. Women also do most of the domestic work: Processing food crops, preparing meals, fetching water, collecting fuel wood, and caring for the children, elderly and sick (an increasing burden in the face of the HIV/AIDS pandemic):

However, although women have the primary responsibility for managing resources, they usually do not have control. National law or local customs often deny women the right to secure title or inherit land, which means they have no collateral to raise credit and improve their conditions.

(UNFPA, 2001: 7)

In a few communities, government agencies and nonprofit organizations that have recognized this untapped economic potential have begun to provide women with information, education and access to credit.

FORMS OF AGRICULTURAL ORGANIZATION

Forms of agricultural employment and organization also tend to differ among the regions of the world. Mechanized agriculture and export crops have become of greatest importance in Latin America. Green Revolution agriculture has become most widespread in producing wage and peasant foods in lowland Asia with pockets in Latin America and North Africa. "Resource-poor" agriculture, producing a range of crops, predominates in Sub-Saharan Africa and areas of poor soils and drainage elsewhere. Production differences reflect these organizational and endowment differences.

While per capita food production in the periphery has not matched that of the core, and in many cases has not kept up with population increases, spectacular growth in the production of specific crops for export to the core was characteristic of the 1970s and 1980s in particular. In Latin America by the late 1970s, commercial agriculture, centered primarily in the large farm sector, was estimated to account for half of all agricultural production, nearly one-third of the cultivated area and one-fifth of the entire workforce. For example, sugar production increased by over 200 percent in El Salvador, Guatemala, and Honduras between 1965 and 1977. The production of sugar in these three countries has continued to increase since then, by over 100 percent. Beef production in the Dominican Republic grew at 7.6 percent per annum between 1970 and 1979. It continued to grow at 4.6 percent annually throughout the 1990s. Soybean production, relatively unimportant in Brazil before 1970 at 1.5 million metric tons, rose to almost 26 million metric tons per annum by 2010. The expansion of export production and regional specialization has been most characteristic of agriculture in Latin America. In Sub-Saharan Africa, however, export crops have failed to maintain global market shares even as total agricultural production increased. This reflects both declining productivity in the export sector and government attempts to direct investment into industrialization rather than agricultural commodities. Food production has been dismal, particularly in the context of rapid population increase. In Asia, both productivity and production have increased enormously because of fertilizers and the application of new technologies, but most growth has been in cereals (especially rice and wheat) production rather than "special" export crops such as those of growing importance in Latin America (for example, fruits and beef). The problems for the Asian countries are their high land–population ratios and the competition they face from agriculture in the United States and Europe in the crops (such as wheat and rice) in which their growth has been concentrated. U.S., EU and Japanese subsidies and market protection for agricultural production deprive Asian (and other LDC producers such as Argentina) of both higher prices and international markets. Lower production of cereal crops in the core of the world economy would produce higher world prices (through a decrease in the amount produced) and greater access of LDC producers to DC markets.

Box 9.1 The coffee commodity chain

Coffee is the world's second most valuable traded commodity, behind oil. There are about 25 million farmers and coffee workers in over 50 countries producing coffee. Coffee was historically developed as a cash crop in colonial economies, planted by peasants or wage laborers on large plantations for sale in the core countries. It is currently the largest food import of the United States. The coffee **commodity chain** today involves a string of producers, middlemen, exporters, importers, roasters and retailers before reaching the consumer (see Chapter 1). Global consumption has increased tremendously over the past 20 years, owing much to the Starbucks® phenomenon: The spread of coffee shop franchises of this or other similar brands all over the world. Around 70 percent of world production is of Arabica beans, used for higher grade and specialty coffees, with 80 percent coming from Latin America. The rest is Robusta coffee, grown mainly in Africa and Asia. Typically, coffee farmers and workers receive extremely low wages relative to the final retail price of coffee. This has encouraged the development of the **fair trade** in coffee movement to try to improve working conditions and wages for producers (see Box 9.2). Most small farmers sell to middlemen, while large estate owners usually process and sell their crop directly overseas at prices fixed by the New York or other international coffee exchanges. Most importers purchase green coffee from established exporters and estate owners in producing countries such as Brazil and Colombia. They then hold the stocks selling gradually to roasters to both maintain the price and to control supply. Importers, therefore, are the key agents in the coffee commodity chain. Roasters, of whom there are around 1,100 in the United States today, usually have a set of recipes and sell to large retailers under such brands as Maxwell House (Kraft) and Sanka (Philip Morris), Folgers and Millstone (Procter & Gamble), and Nestlé. Although these large roasters account for 60 percent of green coffee volume in the USA, some roasters produce as few as 500 bags a year for the specialty coffee market of high-end coffee shops. With the highest profit margins in the value chain, roasters are a key link, therefore, on the road from producers to consumers. Retailers sometimes now roast their own beans but, by and large, they sell either beans or coffee to the general public. Supermarkets and other shops account for about 60 percent of retail sales with coffee shops making up the rest. Other foodstuffs and industrial raw materials follow similar commodity chains geographically linking together producers and consumers living and working in different places.

PROBLEMS

Each of the three major regions of the periphery/semi-periphery faces distinctive problems with respect to its agriculture. For Latin America, it is the expansion of export crops at the expense of local food crops. As a consequence, food imports are often necessary. For Sub-Saharan Africa, it is the total deterioration of agriculture in the face of population pressure on marginal land, low productivity, government bias against investment in agriculture and fluctuations in export earnings. Food imports are now an absolute necessity. For Asia, production of cereals has increased greatly but prices have been low because of global gluts. So increased agricultural production has not generated the capital necessary for investment in other sectors, such as industry. When prices increase, local populations must pay the increase or substitute other cereals that are imported, more often than not, from Europe or the United States. Between

Table 9.1 Food production per capita for selected countries (2004–2006 = 100)

	1980	1985	1990	1995	2000	2005	2010
China	36.7	46.3	55.4	71.2	86.4	100.5	116.7
Malaysia	51.8	61.8	76.7	84.4	86.8	100.3	107.3
Indonesia	58.7	67.4	76.8	88.6	83.4	98.1	114.4
Philippines	100.9	88.3	94.5	92.1	93.0	100.5	101.6
Sri Lanka	114.8	109.3	98.0	105.9	100.8	102.2	117.9
Mexico	84.7	87.4	83.3	90.6	94.1	98.5	100.4
Ghana	61.1	60.6	55.0	82.6	89.4	99.9	111.5
India	77.8	86.1	91.7	96.0	99.2	100.1	114.2
Nigeria	54.8	56.5	69.3	84.8	91.5	99.7	89.5
Bangladesh	89.0	85.6	85.8	80.0	97.1	102.8	122.6
Côte d'Ivoire	98.5	92.9	96.3	99.7	104.2	98.3	96.1
Haiti	162.2	156.1	126.0	101.0	108.1	101.6	100.7
Zimbabwe	122.2	136.9	112.7	75.2	111.9	90.5	94.9
United States	89.4	95.9	91.2	93.5	100.9	99.4	104.6

Source: Based on online data at FAO (FAOSTAT), available at http://faostat.fao.org

1980 and 2010 food production per capita increased substantially and consistently only in China, Malaysia, and Indonesia among all LDCs (see Table 9.1). The situation did improve more recently in India and, alone among African countries, in Ghana. At the other extreme, some Sub-Saharan African countries such as Côte d'Ivoire have seen stagnation and others such as Zimbabwe significant declines in food production.

Although the world as a whole produces sufficient food for everybody, 850 million people in the LDCs, one in five of the population is chronically undernourished. As many as 2 billion people fill themselves daily with adequate food calories but lack a diet balanced in needed nutrients. Hunger and inadequate diets are especially serious in Africa, where 36 percent of the population is chronically undernourished. Comparable figures are 9 percent in Latin America and 14 percent in Asia. In these world regions conditions have improved since 1970 when 19 percent and 40 percent, respectively, were chronically underfed. In Africa, there has been little or no improvement (35 percent in 1970). The remarkable improvement in Asia owes much to improved rural healthcare, which protects people from falling sick and losing income or work and subsequently disrupting family food supply, and increased crop yields. Another way of putting the food problem would be to compare food production per capita in the three regions. In this perspective, Asia has seen an impressive 78.3 percent increase from 1961 to 2010 and Latin America has experienced a 54.8 percent increase. In Africa, food production per capita has *dropped* by 12.2 percent over the same period.

In large parts of the periphery today, agriculture is a vulnerable sector: either oriented externally or subject to the vagaries of world market prices without the protection and subsidies

Box 9.2 From free trade to fair trade?

Developed countries such as the United States and those in the European Union (EU) continue to provide billions of dollars in agricultural subsidies to their own growers while limiting access to their markets by poor farmers in the LDCs. Oxfam developed an index of these double standards using ten measures of trade policies in the DCs, including average tariffs, the size of tariffs on agricultural imports, and restrictions on imports from the LDCs. The index scores for the EU and the United States were at the top of the list of DCs that call for free trade but limit access by the LDCs to their markets. Compounding this situation has been the volatility in world prices during the last few decades for the sale of primary products such as tropical beverages and food that farmers in LDCs depend on for income.

Believing that aid alone is not the answer to poverty in the LDCs, Oxfam and other aid organizations have worked to establish a fair trade model. Seeking to set up alternative trading links between producers and consumers, people such as Michael Barratt Brown founded organizations such as Twin Trading. Influenced by ideas of sustainable development, unequal exchange and dependency, Barratt Brown, in his seminal book, *Fair Trade*, pointed to the deteriorating trading position of many LDCs due to the decline in basic commodity prices for their exported food and raw materials relative to the prices of their imported manufactured goods.

The fair trade model recognizes the weak bargaining position of many small producers at the beginning of the commodity chains that underpin the world economy. This model is an attempt directly to connect consumers in the DCs with producers in the LDCs through a network that includes features such as long-term trading contracts that offer price stability for farmers. Fair Trade Labeling Organizations International (FLO), an umbrella organization of three producer networks, 19 labeling groups, and three marketing organizations, maintains the standards for the fair trade label and certifies cooperatives that meet these standards. Goods displaying the fair trade label are sold with a guaranteed minimum price that includes a social premium paid by the consumer to the democratically organized cooperatives to be spent on infrastructure investments—processing facilities, schools and hospitals—for the benefit of members. Fair trade goods meet environmental sustainability standards and ILO conventions covering labor practices.

The fair trade label applies to a variety of goods, including coffee, tea, cocoa, bananas, and honey. The first fair trade coffee was imported in 1973 into the Netherlands from a small-farmer cooperative in Guatemala. In 2003 the United States overtook the Netherlands as the largest destination for fair trade coffee. Today, over 250 coffee cooperatives representing more than 700,000 farmers across more than two dozen countries in Latin America, Africa, and Asia are certified by the FLO. Annual sales of all fair trade products are growing at double-digit rates and are estimated to be worth well over US$1 billion. At the same time, however, the total value of fair trade products—accounting for 0.5–5.0 percent of all sales in their product categories in Europe, the United States, and Canada—is minor in relation to the overall flows of international trade.

Gavin Fridell identified three perspectives on the fair trade network. The "decommodification" perspective incorporates notions of "ethical trade" and depicts fair trade as a challenge to the commodification of goods under global capitalism. Jeff Popke sees the ethical trade movement as the reincarnation of the traditional Marxist concept of "defetishizing" the commodity in order to expose its underlying unequal social relations of production. Although challenging the core values of global capitalism—competition, growth and profit maximization—this approach has limited potential to change the global trading system because it depends completely on revealing the conditions of global inequity to consumers in the DCs.

The "alternative" perspective offers a rights-based approach to development that "makes trade fair" as a replacement for free trade and the consequent dominance of DCs and TNCs. This perspective not only highlights the plight of farmers in the LDCs, but also confronts organizations such as the WTO about the structural causes of poverty, such as the trade barriers and subsidies of the DCs. The main criticism of this approach, however, is that its successes have only been possible because it has remained part of the dominant paradigm, and as such is limited by the constraints of consumer demand, limits on the price of fair trade goods imposed by the market, and the growth potential of fair trade niche markets.

As a result, the "shaped-advantage" perspective is seen as most accurately reflecting the overall impact of fair trade so far. This more moderate approach seeks to help poor farmers to improve their position in the existing global market through the help of nongovernmental organizations (NGOs). The more market-oriented approach of this "microeconomic tinkering" is seen as accounting for its recent success in assisting certain groups of poor farmers to enter the global market on better terms. Of concern, however, are the dangers of "mainstreaming" that require the farmers to deal with TNCs such as Starbucks® in order to get fair trade products more widely into the hands of DC consumers and the lack of an explicit component that directly confronts the structural causes of poverty.

enjoyed by agriculture in the core. Yet it is absolutely vital. Vast numbers of people are still employed in or are immediately dependent on agriculture. And, whatever the model of economic development adopted, any hope of improving living standards in general depends on increasing agricultural production.

9.2 LAND, LABOR, AND CAPITAL

Agriculture in the contemporary periphery rests on a foundation of agrarian history and recent changes can only be understood in this context. Central to agrarian history the world over has been the impact of market forces on landholding patterns and the structure of rural social relationships. Although rural areas are often characterized as static and traditional, the historical record shows frequent changes in agricultural practices and labor relationships in response to global and domestic political–economic conditions. But some features of land-holding systems and rural life have persisted from the period of incorporation into the world economy. In this section the mix of "old" and "new" in the agricultural organization of different parts of the contemporary periphery (Latin America, Sub-Saharan Africa, and Asia) will be examined.

LATIN AMERICA

In Latin America, conquest and colonial domination created patterns of subsistence and commercial agriculture based on large landholdings. After independence, this characteristic and its corollary, an exploited and powerless peasantry, became firmly entrenched as the region was firmly tied into the world market as a producer of primary commodities. Between the 1850s and 1930s the various countries of Latin America came to depend on the export of one or two primary export commodities to the industrial countries—first Britain and later the United States. The older hacienda system, albeit complex and varied in its particulars from place to place, went into decline to be replaced by a plantation system that already had a considerable history in the sugar plantations of northeast Brazil and the Caribbean (see Table 9.2).

The growth of export-oriented agrarian capitalism was associated with the emergence of a politically powerful landed élite linked to foreign investors and commercial agents dealing in primary commodities. Agriculture for domestic consumption was largely ignored and through control over governments the agricultural élite was able to increase its hold over land, labor, and capital.

Table 9.2 Land, labor, capital, and markets: haciendas and plantations

	Haciendas	Plantation
Markets	Relatively small and unreliable; regional, with inelasticity of demand; attempt to limit production to keep prices high	Relatively large and reliable; European, with elasticity of demand; attempt to increase production to maximize profits
Profits	Relatively low; highly concentrated in small group	Relatively high; highly concentrated in small group
Capital and technology	Little access to capital, especially foreign; operating capital often from Church; technology simple, often same as that of peasant cultivators	Availability of foreign capital for equipment and labor; foreign direct investment late in nineteenth century; relatively advanced technology, with expensive machinery for processing
Land	Size determined by passive acceptance of indigenous groups; attempt to monopolize land to limit alternative sources of income to labor force; unclear boundaries	Size determined by availability of labor; relatively valuable with carefully fixed boundaries; much unused land; relatively cheap
Labor	Large labor force required seasonally; generally indigenous; informally bound by debt, provision of subsistence plot, social ties, payment in provisions	Large labor force required seasonally; generally imported; slavery common; also wage labor
Organization	Limited need for supervision; generally hired administrators/managers, absentee landlord	Need for continual supervision and managerial skills; generally resident owner/manager

Source: Based on Grindle (1986: 30, Table 3.1)

The concentration of landholding and the marginalization of peasant agriculture did not occur without resistance. Agrarian uprisings and social banditry were widespread. In Mexico the 1910 uprising was a major impetus to the revolution; strikes were extremely common in the corporate plantations of coastal Peru in the period 1912–1928; in Colombia, rural violence by agrarian tenant syndicates directed against commercial coffee producers lasted well into the 1930s. The 1930s also was a period of rural unrest in the Brazilian northeast and in El Salvador among dispossessed peasants and unemployed plantation workers.

When the world economy collapsed in the Great Depression of the 1930s so too did export-oriented agriculture. This spurred the emergence of active nationalists, often in the military, who wanted to increase industrialization and diminish reliance on the export of primary commodities. Between 1930 and 1934 there were 12 forcible takeovers of power—from Argentina to Peru to El Salvador. Argentina, Brazil, Chile, Colombia, Mexico, and Uruguay all instituted import substitution industrial strategies. These led to a massive movement of people off the land. For the region as a whole, in 1920, only 14 percent of the population lived in urban areas, but by 1940 the proportion had risen to 20 percent. In Argentina, Chile, and Uruguay, urban percentages reached 35–45 percent of the population. One major consequence of this was a decline in the grip of the landholding élite over national politics in some countries as urban professional and working classes grew in size and influence.

This change, however, can be exaggerated. Many countries continued to rely on the export of one or few primary commodities—the Central American and Caribbean countries, but also Argentina, Colombia, and Chile—and rural land remained concentrated in the hands of the landed élite. What was different was the emergence of nationalist and populist movements committed to industrialization rather than export agriculture.

Pursuing policies of import substitution had important effects on agriculture. For one, manufacturing surpassed agriculture in its contribution to gross domestic product in a number of countries (Argentina, Brazil, Chile, Mexico, Uruguay, and Venezuela) in the 1940s. Much of the new capacity was concentrated in or near the capital cities of the states that were its major sponsors (Buenos Aires, Rio de Janeiro, Santiago, etc.).

Industrialization required a "draining" of agriculture for resources (cheap food, raw materials) and capital (foreign exchange, taxation). As a consequence, a premium was placed on efficiency in agricultural production. This was thought to require large holdings, the spread of technological innovation and capitalization (heavy capital investment). Between 1940 and 1960 there was a massive migration of people from the countryside to the cities as a consequence of mechanization and the expansion of large landholdings at the expense of small tenants and proprietors.

In the 1960s import substitution became increasingly expensive as the "easy phase" emphasizing light consumer goods was played out and the prodigious expense of moving into heavier capital goods became apparent. In a process that accelerated during the 1970s, a development model based on export promotion slowly displaced import substitution. According to this model, agriculture had been neglected and, although no substitute for industrialization, more efficient production of domestic food crops and increased agricultural exports were important in both maintaining political stability and obtaining foreign exchange. After 1965 public investment in rural areas and agriculture increased in a large number of Latin American countries.

Government policies have discriminated heavily in favor of the larger landowners. The geographical distribution of official credit, research and extension, infrastructure, mechanization and Green Revolution inputs reflects the geography of landholding. In Peru, for example, about half the credit supplied by the Agricultural Development Bank between 1940 and 1965 went to cotton growers, who were among the wealthiest coastal agricultural exporters. Food crop producers—largely peasants—were mainly ignored by the bank. In Mexico in 1970, mechanization was used on 25.7 percent of the crop area of farms of more than five hectares but was used on only 4.3 percent of the crop area of farms that were under five hectares in size. In Brazil, all government policies have tended to reinforce the emphasis on commercial agriculture in the south and east regions at the expense of the northeast and small-scale producers everywhere.

This is not to say that large-scale capitalist agriculture has completely displaced peasant production. Far from it. A large section of the agricultural labor force is still "part-peasant" in that it supplements its wage earnings with the produce of its often less-than-subsistence plots. This serves to sustain capitalist agriculture through reducing the costs of reproducing a labor force. In many parts of Latin America, therefore, large-scale capitalist agriculture and small-scale peasant production still coexist uneasily. The past is still present.

SUB-SAHARAN AFRICA

In Sub-Saharan Africa, unlike Latin America (or Europe), access to labor not land was always the basis of economic and political power. From 1830 to 1930 agriculture in Sub-Saharan Africa underwent an incredible expansion in the form of small-scale commercial farming. Some commercial farming had existed prior to this period, for example in the Hausa-Fulani and Mandinka states of northwest Africa, but the introduction of new crops and the expansion of existing ones into previously uncultivated areas increased the scale and geographical distribution of commercial agriculture. Of special importance were such crops as cocoa, cotton, coffee, groundnuts, and oil palm, which were grown mainly for export markets. They spread along with European traders, the introduction of foreign capital, the shifting objectives of native farmers and traders, and, finally, colonial rule. This was the "cash crop revolution" (Tosh, 1980) that brought Africa into the world economy and capitalism into Africa.

Colonial rule involved massive intervention in existing agriculture through forced labor and taxation. Taxation, in particular, provided a fresh stimulus to cash cropping. In some parts of Africa, especially the east and south, taxation also encouraged labor migration to mines, plantations, and industries established by European settlers. In West Africa, however, labor migration pre-dated colonial rule. It was of a seasonal nature and involved the integration of farming in the interior with migration to more fertile but labor-deficient coastal areas. In West Africa, cash cropping by small-scale farmers and long-distance labor migration at harvest time were indigenous phenomena that increased in intensity after the onset of colonial rule. Elsewhere, cash cropping and labor migration were relatively novel and related much more to either European settlement (as in South Africa, Zimbabwe, or Kenya) or European initiatives in mining and plantation agriculture (as in Zambia and Zaire).

Another distinctive feature of West Africa as compared, for example, to Kenya was that the production of food and cash crops was complementary rather than competitive. Even today food crops such as plantains, cocoyams, and peppers are grown to provide shade for young cocoa trees. Moreover, the period of peak labor demand for cocoa harvesting (November–February) complements the peak labor demand periods for the cereal-growing areas to the north (May–July and February–March). Cocoa farms, therefore, have rarely faced a maximum price for labor and the commercial cocoa industry can coexist with the market for labor in food crop production.

In Kenya, however, the European settlers specialized in the production of food crops and their production cycle matched that of subsistence producers. They consequently had to compete for labor with the subsistence sector. In addition, the establishment of estates or plantations in Kenya involved the confiscation of land from subsistence producers and the subsidy of commercial production at the expense of the subsistence sector.

Throughout Africa, the rate of agricultural production slowed markedly during the 1930s and the Second World War. It was only in the 1950s, when world prices for many export crops increased as the industrial countries entered into their long boom of the 1950s and early 1960s, that there was a rapid expansion in export crop production. But the increase in demand for Africa's export crops was short lived, peaking as early as 1956. Since then cash cropping and commercialized livestock farming have been concentrated in the districts where they were dominant 50 years ago. With the exception of sugar, most new planting (of cocoa, coffee, or tea) has taken place within the areas that were already the major producers in the early 1950s.

In those districts in which agricultural production has intensified or expanded, it has involved different types of farming. For example, in Côte d'Ivoire, plantations have been the major agent of growth, whereas in Ghana, Kenya, and Sudan it has been small-scale peasant cash crop production that has been responsible for most growth. Indeed, in Kenya the small-scale farming sector has largely replaced the plantation sector as the most dynamic in terms of commercial production.

Total agricultural production (cash crops and food staples) has increased substantially in Sub-Saharan Africa since 1980 (see Table 9.3). However, the rate of population increase over the region as a whole has meant that there has been less increase overall in per capita terms, with some countries experiencing a decrease in per capita agricultural production (Table 9.3). Most African governments have adopted policies that seek to depress food prices to feed their burgeoning populations. This often leads them to set higher prices for large-scale producers because of presumed efficiencies (and political influence?). Penalizing the food production sector is meant both to stimulate export crop production and feed increasingly large urban

Table 9.3 Index of total (and per capita) gross agricultural production: selected African countries (2004–2006 = 100)

	1980	1990	2000	2010
Côte d'Ivoire	49.6 (105.2)	71.1 (102.3)	101.8 (110.7)	105.1 (96.0)
Ghana	31.1 (61.5)	37.7 (55.1)	79.3 (89.6)	125.6 (111.4)
Kenya	45.5 (99.8)	67.9 (103.3)	75.5 (86.2)	124.9 (110.0)
Malawi	44.1 (90.8)	52.2 (71.5)	98.2 (112.3)	155.0 (133.6)
Mozambique	61.5 (105.1)	59.9 (91.9)	84.5 (96.4)	111.9 (99.4)
Nigeria	28.5 (52.8)	48.5 (69.5)	80.2 (90.7)	100.2 (88.5)
Senegal	55.6 (111.5)	79.1 (118.7)	107.4 (122.8)	147.9 (129.2)
Sierra Leone	57.7 (94.1)	68.8 (89.2)	52.9 (65.9)	123.4 (108.5)
Zambia	48.3 (95.9)	67.2 (98.1)	78.8 (88.6)	142.0 (124.5)
Zimbabwe	80.4 (138.6)	100.5 (120.6)	132.3 (132.9)	95.7 (95.7)
Sub-Saharan Africa	74.9 (124.6)	82.9 (108.2)	96.0 (102.5)	115.1 (109.4)

Source: Based on online data at FAO (FAOSTAT), available at http://faostat.fao.org/

populations. In fact, it has discouraged farmers, especially the mass of small-scale farmers, from increasing their production through investment in increased productivity.

The trade policies of DCs and the advice their experts offer have also contributed to the problems of African agriculture. North America, the European Union (EU) and Japan may practice fairly free trade in the manufactured goods and services in which they may have comparative advantages but they are relentlessly protectionist about foodstuffs; precisely the sector in which African countries can offer competitive products. For example, U.S. government subsidies to its sugar, tobacco, and groundnut farmers lead to lower prices for U.S.-produced crops than would be the case without the subsidies. This deprives African producers of potential markets. With respect to advice about cropping decisions, Africans have received some of the worst advice ever offered by people from one part of the world to another. The litany of disasters resulting from advice offered by foreign experts is much too long to provide here. Two examples must suffice. In Burkina Faso and elsewhere in the dry Sahel region of north-central Africa, the UN Food and Agricultural Organization (FAO) encouraged local farmers to grow potatoes. A bumper crop resulted, which then rotted unsold in local markets where potatoes were seen as an exotic crop without any history in local diets. By Lake Turkana in East Africa, Norwegian experts persuaded Turkana cattle herdsmen to give up their cattle and take up fishing only to find out that the cost of chilling the fish exceeded what they could bring in city markets. Not only was the fishing equipment a wasted investment but the Turkana were now also without their cattle. They ended up on food aid provided by the surpluses bought up by the U.S. and other governments as a result of overproduction brought about by their subsidy programs to cereal producers and dairy producers.

But countries differ in the relative extent to which farmers must bear the brunt of tax- and price-setting policies. It all depends on the political base of governing élites and the origins of marketing organizations. In Ghana and Zambia in the 1980s, for example, urban-based politicians put the burden on small-scale farmers to a much greater extent than the rural-based politicians of Kenya. In Ghana, the Cocoa Board is a patronage organization, whereas in Kenya, producers control the marketing organizations. Interestingly, the increase in total agricultural production in the 1980s was higher in Kenya than in Ghana and Zambia (see Table 9.3). In Kenya, this benefited both food production for domestic consumption and increases in sales of export crops. Subsequent liberalization and privatization in Ghana, however, have contributed to significantly increased production in the 1990s (Table 9.3).

Three trends have nevertheless been fairly general over the past 30 years. One has been the increased importance of wage labor, especially with respect to export crops. This has further monetized the rural economy and reduced the degree of reliance on domestic groups (families) as sources of farm labor. This in turn has reinforced the role of long-distance migration in agricultural labor and given some districts the specialized role of "migrant labor reserve" for other districts in which export agriculture is important. For example, even with restrictions on international migration, Togo and Benin in West Africa have been a major source of temporary and permanent migrants to Côte d'Ivoire and Ghana (see Figure 9.1).

A second trend has been the changing role of women in African agriculture. Women have become central to the production of food crops on small-scale farms such as those that dominate throughout Sub-Saharan Africa. As Swindell (1985: 179) puts it:

As men have become more involved in commercial cropping and non-farm occupations, so women have become increasingly responsible for the cultivation of food staples. This is especially true in those areas where the out-migration of men is persistent, and it could be argued that the expansion of commercial cropping and the industrial labor force has been built on the backs of women farmers.

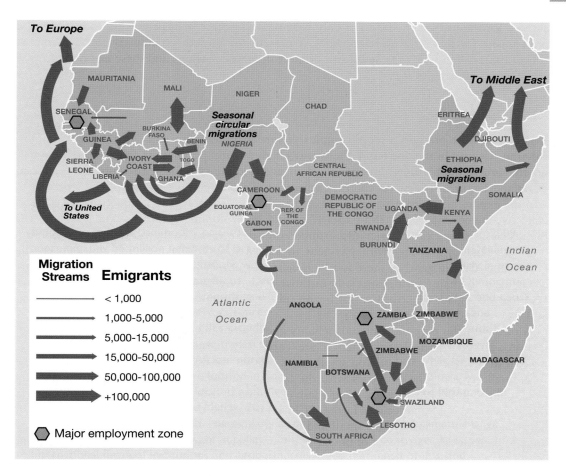

Figure 9.1 External migration flows in Sub-Saharan Africa

Source: Based on Aryeetey-Attoh (1997: 136, Figure 5.4) and Marston *et al.* (2008: 248, Figure 5.32)

The third trend has been growth in agricultural production through extending areas under cropping or grazing rather than through raising yields. Green Revolution technologies (high-yield varieties, fertilizers, etc.), mainly addressed to cereal production, have been either inappropriate or not widely adopted in Sub-Saharan Africa. Whatever the cause, however, commercial agriculture has become extensive rather than intensive. This has led to farming on poor soils in areas with unreliable rainfall and the displacement of subsistence agriculture onto ever more marginal terrain. Sen (1981a) implicates this trend as a major factor in the famines that have afflicted many parts of Africa over the past 20 years. Civil wars, poor food distribution networks, and the degradation of soils through lack of crop rotation have also played some part.

Although much of African agriculture has become increasingly commercialized, it remains largely small scale and still involves domestic groups or families. The level of agricultural production, however, has not kept up with the world's highest rates of population increase. In many countries, there are now major national food deficits. At the same time government policies in many countries have had the effect of discouraging agricultural production both for food staples and export crops. But in most countries, farming must remain the dominant activity for the foreseeable future if only because an increase in agricultural productivity is a prerequisite for industrial development. At present the growth of industry through import substitution is limited by the small size of most domestic markets and these can only grow if the incomes of farmers rise.

ASIA

Asian agriculture presents a more complex picture than agriculture in Latin America or Sub-Saharan Africa. On the one hand, the world's highest rural population densities are here but, on the other, populations are organized in agricultural systems with quite different and distinctive features. The major contrast, at least until recently, was between China, where there was no export agriculture to speak of and the rural economy has been organized around collective ownership (from 1954 to 1979), and those countries such as Malaysia and the Philippines, where export agriculture (rubber and sugar, respectively) is important and share-cropping tenancy (renting with payment in kind to landlords) predominates outside the plantations. But, in general, there is a high incidence of tenancy in Asian countries and share cropping is its major form, especially in those areas in which rural population densities are very high (Bangladesh, Java, Central Luzon (in the Philippines), the West Zone of Sri Lanka, and eastern and southern India).

Along with the preponderance of tenants goes a concentration of landholding, although less on average than in Latin America (see Table 9.4). It has been estimated that about 87 percent of the world's 500 million small farms (under two hectares in size) are located in the Asia-Pacific region. The countries with the largest numbers of these small farms are China (193 million), India (93 million), Bangladesh (17 million), Indonesia (17 million), and Vietnam (10 million). About 95 percent of farms in China are smaller than two hectares. In India, about 81 percent of the farms are this small and they account for 44 percent of the total cultivated area. In Bangladesh, 96 percent of the farms are under two hectares and account for 69 percent of the area of cultivated agriculture (Thapa and Gaiha, 2011).

Table 9.4 Average farm size, 2010

Region	Average size (ha)	% < 2 ha
East Asia	1.0	79
South Asia	1.4	78
Southeast Asia	1.8	57
Sub-Saharan Africa	2.4	69
West Asia and North Africa	4.9	65
Central America	10.7	63
Europe	32.3	30
South America	111.7	36
United States	178.4	4

Note: 1 hectare = 2.477 acres

Source: Based on Deininger and Byerlee (2011: 28, Table 1.3)

The evidence from many Asian countries, including Bangladesh, China, the Philippines, Thailand, Pakistan, and India, indicates that average farm size has decreased over time. Other evidence from India suggests two types of change in historical patterns of rural social structure: The growth in some areas of the class of self-employed cultivators or rich peasants, favored by 1950s' land reform (for example, Gujarat) and the transformation of large landowners into capitalist farmers employing migrant laborers (for example, Punjab). Both changes are signs of increasing commercialization of agriculture even as sharecropping tenancy persists in marginal areas to provide labor reserves for seasonal and cyclical purposes at little or no cost to the commercial sector.

In the colonial period, governments concerned themselves either with plantation agriculture or with raising taxes from other forms of agriculture. In India, the British created a class of landed aristocrats called *zamindar*s as revenue collectors for the government. The *zamindar*s, however, did not have any real interest in improving agriculture. Over time they and other intermediaries became an immense burden on actual cultivators whose rents included not only revenue for the government but also income for the various intermediaries. After independence, India, Pakistan, and other countries in South and Southeast Asia where this system prevailed, abolished intermediary tenures. However, many of the old intermediaries continued to cultivate their holdings through tenants and sharecroppers on the same exploitative terms as before. Only in China, South Korea, and Taiwan did land redistribution lead to an effective abolition of the power of large landlords. This was arguably crucial in providing the basis for later spectacular economic development as small household farms gained incentives to dramatically increase production. In China, this was spoiled for a generation by large collective farms that drained rather than invigorated the agricultural sector. The recent return to smaller, equal plots in China has again provided the basis for maximizing agricultural output; what Studwell (2013, xiii) calls "highly labor-intensive household farming."

Since independence, however, total agricultural production has increased at rates at least commensurate with population growth in most Asian countries. Unfortunately, much of the growth has been concentrated in export crops or cereals (wheat, rice) rather than across the board. Moreover, the unequal social structure of most rural areas has ensured an upward drift of the benefits of increased production. Rural poverty has increased as agricultural production has increased.

A major source of increased production of cereals (especially wheat) since the 1960s has been the Green Revolution. This had its most significant impact in both Punjab regions (since Partition, there is one in India and one in Pakistan) and the Indian state of Haryana where irrigation facilities could be utilized. Benefits have accrued disproportionately to large farmers and the technologies involved (new seed varieties, heavy applications of chemical fertilizers) cannot be applied in areas without irrigation facilities: 80 percent of the cultivated area in India, 90 percent in Bangladesh.

In general, over the past 40 years most Asian governments have not favored agriculture. Many have pursued pricing and credit policies similar to those noted earlier for Sub-Saharan Africa. This seems also to be true at least for considerable periods in the case of China. Indian development plans until the late 1970s were systematically biased against the agricultural sector. Yet there is a direct relationship between agricultural yields and a price structure that favors the agricultural sector. The countries with the highest ratios of product prices (for example, rice) to input costs (for example, fertilizer cost) are also where yields are highest. The three countries with the highest rice yields per hectare in Asia, Japan, South Korea, and Taiwan also have perhaps some of the poorest soils in Asia. Government policies (especially subsidies for inputs such as fertilizers) and egalitarian rural social structures (all farmers are rewarded) are the most plausible causes of the differences in crop yields. One negative effect of this, however, is a high level of water pollution produced by the heavy use of subsidized fertilizers.

According to the United Nations, three-fourths of the world's "absolute poor" (those unable to maintain a minimum nutritional standard) live in Asia and more than four-fifths of them live in rural areas. The most common feature of the rural poor in the region is landlessness or limited access to land. Poor rural households tend to have larger families, lower educational attainment and higher underemployment. While the Green Revolution helped to achieve the most rapid and widespread decline in poverty, hunger and premature death in history during the 1970s and 1980s, progress has since stalled. Appropriate technologies remain to be developed for the agriculturally marginal areas that are home today to about 40 percent of the rural poor. In China, for example, nearly 65 million officially recognized poor people live in remote, mountainous areas: "In the 1990s, poverty reduction fell to less than one third of the rate needed to meet the United Nations' commitment to halve extreme poverty by 2015" (IFAD, 2002: 2).

9.3 RURAL LAND REFORM

In Latin America and Asia, the landholding and tenurial systems have been periodically "reformed" as a result of pressure from peasant movements, government attempts to make agriculture more efficient and productive, and external pressures from TNCs and international development agencies. Certain models have sometimes been followed depending on whether efficiency or equity has been the overriding goal. In the former case, the Taiwanese and South Korean experiences are emphasized; in the latter, the Chinese experience is often the model. However, in practice, agricultural reform, especially land reform, is overwhelmingly a socio-political process rather than a technical one of choosing a model and then following it.

At one time or another, but especially between 1960 and the early 1970s, virtually every country in Latin America and Asia passed land reform laws. A wide range of arguments have been proposed to justify a role for land reform in agricultural development. There are perhaps four justifications that have been most common and they have appealed differentially to different social groups. The first of these is a "conservative" argument: Land reform is a minimal concession for political stabilization. The second is a "liberal" argument: Land reform is needed to create a class of capitalist farmers and expand the domestic market. Third, there is the "populist" argument: Small farms are more efficient (and equitable) than large ones. Fourth, the "radical" argument: Peasants are rapidly being dispossessed of their status as independent producers and are prisoners of cheap food policies and agro-export policies, consequently land reform towards collective production (collective farms, state farms) is necessary, if insufficient, for economic development.

Most actual land reform policies have been of the "liberal" type, concerned with creating a reform sector. About three dozen land reforms are classified in Table 9.5, including those in the same country when a land reform program was later redefined (for example, Chile). All the diagonal reforms on this table are redistributive ones in the sense that they either increase the size of the reform sector without changing the non-reform sector (1, 7 and 13 in Table 9.5) or involve expansion of the reform sector (25). Reforms 2, 3 and 4 are oriented towards eliminating "feudal" (or other pre-capitalist) remnants from agriculture rather than redistributing land. In each case, the transition to capitalism is dominated by (2) a landed élite, (3) farmers or (4) peasants.

The only possible reforms, as opposed to drastic changes, once a capitalist agriculture has been established are either shifts in the type of agrarian structure (8, 9, 14) or distributive

Table 9.5 A typology of land reforms

		Post-land reform				
	Mode of production in whole society	Semi-feudal	Capitalist	Capitalist	Capitalist	Socialist
Pre-land reform	Mode of production in agriculture — Land tenure	Semi-feudal estates and reform sector	Capitalist estates and reform sector	Capitalist farms and reform sector	Peasant farms	Socialist farms
Semi-feudal	Semi-feudal estates	(1) Mexico 1917–34 Taiwan 1949–51 Colombia 1961–67 Chile 1962–67	(2) Bolivia 1952– Venezuela 1959– Philippines 1963–72 Ecuador 1964– Peru 1964–69 Colombia 1968–	(3) Mexico 1934–40 India 1950– Guatemala 1952–54 Egypt 1952–66 Iran 1962–67 Chile 1967–73	(4) South Korea 1950– Taiwan 1951–63 Iraq 1958–	(5) China 1949–52
Capitalist	Capitalist estates	(6)	(7) Costa Rica 1962–76	(8) Peru 1969–75 Philippines 1972–79	(9)	(10) Cuba 1959–63 Algeria 1961–71
Capitalist	Capitalist farms	(11) Guatemala 1954–	(12) Chile 1973–	(13) Mexico 1940– Dominican Rep. 1963– Egypt 1961– South Africa 1994–	(14) Zimbabwe 1980–	(15)
Capitalist	Peasant farms	(16)	(17)	(18)	(19)	(20)
Socialist	Socialist farms	(21)	(22)	(23) Russia 1991–	(24)	(25) Cuba 1963– China 1952–78 China 1979– Algeria 1971–77

reforms within a given type (7, 13, 19, 25). All reforms can give way to counter-reforms: Chile essentially switched to (12) from (3) after the 1973 military coup, Guatemala returned to (11) from (3) after the military coup of 1954. The Chinese, Cuban, and Algerian cases are ones of land reform involving the collectivization of agriculture that were part of more "radical" programs of sociopolitical change. But since the early 1980s China has shifted to a mixed system of collective ownership but private use. The recent redistribution of land from large capitalist farms to small peasant holdings in Zimbabwe and some other African countries has destroyed the existing agricultural system without affecting either improvements in production or in the incomes and status of the new farmers.

The most widespread and successful (in the sense of lasting) land reforms have been those facilitating the creation of a capitalist agriculture (1–5 in Table 9.5). In Latin America, the combination of antifeudal land reforms with more spontaneous development of capitalism has both removed most feudal remnants and put an end to reform efforts. A similar conclusion can be drawn for Asia. Except for China, reform efforts generally ended in the early 1970s. By and large, they cannot be said to have lived up to their promise for the needs of the bulk of the rural population irrespective of the nature of the reform undertaken. However, in some cases, such as Taiwan and South Korea after the Second World War, and China's land privatization since 1978, rural land reform appears to have served as a prerequisite for later industrialization by increasing crop yields and through increased rural earnings providing capital for industrial investment.

9.4 CAPITALIZATION OF AGRICULTURE

Spontaneous change, therefore, has now become much more important than reform in agricultural development. Over the past 20 years there has been a substantial increase in direct and indirect investment by transnational corporations (TNCs) in the agriculture of a number of peripheral countries. In many countries, TNCs, attracted by cheap land and labor, appropriate physical conditions, improved infrastructure and a decline in the relative profitability of other sorts of investment, have increased their involvement in export-oriented agriculture and the production and distribution of seeds, pesticides and fertilizers. Thailand, for example, which exported no pineapples in the early 1970s, had, by 1979, become the major world exporter after Hawaii because the U.S. company Castle and Cooke had moved a major part of its pineapple operations out of Hawaii. Similarly, the Philippines, which exported no bananas in the 1960s, had become one of the world's major exporters by the mid-1970s. This was again due almost entirely to new TNC investment. So, just as TNCs that specialize in manufacturing use global sourcing, agricultural TNCs have turned to multiple sites of production to lower labor costs, gain year-round supplies for seasonal crops (for example, strawberries in January in Europe from Chile) and to avoid labor and environmental regulations. Over the past 30 years the global food industry has been one of the world's fastest growing industries.

Of great importance, however, was the prior emergence in Europe, Japan, and North America of a highly capital-intensive agriculture serving a food system in which consumers increasingly demanded high-value products (such as lean beef, chicken products, and fresh fruit and vegetables) at the same time as marketing and distribution were concentrating in the hands of large-scale wholesalers and supermarket chains. Economies of scale could be realized within large vertically integrated firms that supplied the new wholesalers and direct retailers. Global sourcing is an extension into the periphery of a shift towards industrialized agriculture that was well under way by the 1950s in the United States and Europe, with beef cattle lodged in lots for fattening and chickens stacked on top of one another in battery houses. The recent demand for organic produce, very fresh fruit and vegetables and worries about contamination of the beef food chain—prompted by the outbreak of so-called mad cow disease (bovine spongiform encephalopathy (BSE)) in Britain—however, may signal the limits of the globalization of food production when consumer tastes and demands in urban and export markets resist the imposition of mass-produced items. Different food products now have different food systems associated with them. Only some are amenable to global sourcing.

It is in Latin America that the capitalization of agriculture by TNCs has been both most extensive and intensive. Of the six countries usually identified as the "new agricultural

Box 9.3 Agribusiness and the developed countries

Direct corporate involvement in agriculture—agribusiness—has been an inevitable outcome of the logic of specialization and economies of scale. With greater specialization, farms become less autonomous and self-contained as productive units, making for the penetration of an integrated, corporate system of food production, processing and distribution: Agriculture has become increasingly drawn into a food-producing complex whose limits lie:

> [Well] beyond farming itself, a complex of agro-chemical, engineering, processing, marketing and distribution industries which are involved both in the supply of farming inputs and in the forward marketing of farm produce.
>
> (Newby, 1980: 61)

It is in the actions of food-processing conglomerates like Archer Daniels Midland (ADM), Cargill, ConAgra, Monsanto and Nestlé, Newby suggests, "that the shape of agriculture and ultimately of rural society in virtually all advanced industrial societies is decided" (1980: 62). The most common form of corporate involvement in agriculture has to do with the forward contracting of produce at a fixed price. This not only weakens the independence of farmers, but also tends to transfer income from farmers and rural communities to the processing industry. Forward-contracting arrangements also reinforce the overall structural changes affecting agriculture:

> They encourage both fewer, larger holdings and increased specialization so that the size of individual enterprises can be enlarged to fully achieve the prevailing scale economies. This trend . . . is likely to lead to both a reduction in the numbers employed in agriculture, and a decline in the managerial role of those farmers remaining . . . leaving them caretaker functions.
>
> (Metcalf, 1969: 104)

Rural landscapes have also been affected as the logic of industrial production and centralization has been applied to agriculture. In northwestern Europe, for example, field systems have been rationalized, hedgerows and dykes removed, and mechanization has virtually eliminated the need for a large labor force, leaving the fields of most farms devoid of human life for most of the year. Factory farming has brought poultry and pigs indoors permanently, while many cattle spend their winter months indoors, and there are now zero-grazing techniques that may see them inside year round. Only the sheep steadfastly refuse to acknowledge the laws of industrial production, stubbornly refusing to prosper in regimented and sanitized conditions.

countries" in which agricultural investment has been concentrated, four are in Latin America: Argentina, Brazil, Chile, and Mexico. The other two are Hungary and Thailand, where the governments have promoted agricultural investment for urban and export markets, focusing on such high-value food products as meats, fruits, and vegetables. Sometimes control is exercised directly by purchase of land and involvement in production. For example, between 1964 and 1970 U.S.-based TNCs purchased 35 million hectares of agricultural land in Brazil alone. Increasingly, however, TNCs and international development agencies (the World Bank, the U.S. Agency for International Development (AID), etc.) are encouraging traditional rural élites to become commercial élites, practicing mechanized farming of export crops that are processed and marketed by the TNCs or by contracting out to peasant producers. These strategies reflect both fear of the revolutionary potential of peasant movements in traditional agrarian social structures and the need for TNCs to keep a low profile lest they become the targets of nationalization drives.

AGRIBUSINESS

The impact of agribusiness investment in the agriculture of the periphery, therefore, is not restricted to the development of export enclaves or plantation enclaves as was characteristic of an earlier phase in the development of the world economy. Rather, its most important effect is probably the way in which it channels capital to a class of rural capitalists and so consolidates TNC control over entire national agricultural systems. The penetration of peripheral agriculture by international agribusiness is, in effect, just another aspect of the new international division of labor.

Between 1966 and 1978, for example, U.S. investment in Latin American agriculture expanded from US$365 million to US$1.04 billion, growing from 15 percent to 21 percent of total U.S. foreign direct investment (FDI) in Latin America. This investment was heavily concentrated in Argentina, Brazil, Mexico, and Venezuela where the growing urban middle and upper classes provided a domestic supplement to U.S. demand for so-called luxury foodstuffs (meat, fruits, and vegetables). As demand grew for the fertilizers, pesticides, herbicides, improved seeds, and agricultural machinery needed by the "new" agriculture, TNCs such as Du Pont, W.R. Grace, Monsanto, Exxon, and Allied Chemical were increasingly involved in local production.

TNCs and foreign portfolio investment capital have been involved in a variety of ways. In the state of Sinaloa in northern Mexico, for example, 20–40 percent of the credit for agricultural production in the 1970s came from north of the border. In Argentina, the amount of foreign capital in beef production decreased, while it increased in the packing and processing industries. In Mexico and Central America, contract production linked national producers with TNCs. Foreign banks have become major agricultural lenders. For example, the San Francisco-based Bank of America became heavily involved in Guatemala in the 1970s, lending for major development projects such as converting forest to pasture for beef production, and providing speculative export loans.

Since the 1990s foreign companies and state-owned corporations (particularly from China and South Korea) have been active in directly buying up large tracts of land in Africa and, to a degree, in Latin America. So-called land grabs, these land purchases are directed at guaranteeing access to agricultural land (and raw materials) in the future for countries and agribusinesses whose land supplies (and raw materials) closer to home are running out. As subsidies encourage farmers in North America and Europe to convert more and more crops into biofuels (in 2012, 40 percent of the U.S. corn/maize crop ended up as ethanol), the entire global food supply chain is extending globally. It is not simply the capture of agricultural land from local users and production that is problematic. More seriously, the land grabs define captive supply chains with the land locked into corporate and national chains that leave less production (particularly of food crops) available for open exchange. Gigantic commodity trading houses such as Glencore, Cargill, Archer Daniels Midland, and Bunge have been particularly active in organizing this global agricultural system. But they are not alone. In Southeast Asia, for example, Vietnamese companies have ruthlessly carved out large rubber plantations in Laos

Box 9.4 The great land grab?

By Bart Yavorosky

Before the spike in commodity prices in 2007–2008, FDI rarely targeted land. But in 2009, approximately 56 million hectares of large-scale farmland deals were announced (Deininger and Byerlee, 2011). These acquisitions were concentrated in the LDCs, particularly in parts of Africa (Sudan, Ethiopia, Mozambique, Tanzania, Madagascar, Zambia, and the Democratic Republic of the Congo), Latin America, and Southeast Asia (Philippines, Indonesia, and Laos). In contrast, the majority of investors represented DCs such as the United States and European countries, as well as NIEs including Brazil, China, India, South Africa, South Korea, and the Gulf States. Many of these transactions reflected regional or colonial patterns. For example, 96 percent of South African purchases were located in Africa, while 87 percent of South Korean acquisitions were located in Southeast Asia and Melanesia. Similarly, Portuguese investments were located exclusively in the former colonies of Mozambique, Sierra Leone, and Angola (Land Matrix, 2012a).

The potential for a land grab—particularly as population growth and climate change impact arable land and fresh water supplies—raises a number of controversial issues. International financial institutions and agribusinesses argue that much of this land, particularly in Sub-Saharan Africa, has low yields (often one-third or less of its potential). Raising yields could help feed the estimated 1 billion hungry people in the world including the 240 million in Sub-Saharan Africa. In theory, modernizing agricultural production could create jobs and generate revenues in places where most people are engaged in subsistence farming. The acquisition of land by foreign investors also has the potential to positively transform rural economies (but only if it is coupled with transparency, good governance, strict environmental regulation, and investments in infrastructure that benefit local populations).

In contrast, many NGOs and scholars contend that land purchases often dispossess the most vulnerable local people who are removed from land they had cultivated or grazed for generations, create few jobs due to the mechanization inherent to industrial–agricultural practices, and do not increase the amount of food available to local populations because production is geared toward exports.

Large-scale farming also raises concerns about water supplies. Nearly 65 percent of the acquired land that has been put into production has been used to raise water-intensive crops such as jatropha, soybeans, and corn (Land Matrix, 2012b). At the current acquisition rate in Africa, if the acquired land is cultivated with water-intensive biofuel and flex crops (crops that can be used for food or nonfood purposes), the water required for irrigation would exceed the available supply by 2019 (Oakland Institute, 2011).

using loans from Deutsche Bank (Germany's largest bank). So, major financial institutions also figure prominently alongside resource companies and agribusiness in this latest reorientation of world agriculture.

Local impacts of agribusiness

The consequences have been manifold. At a global level there has been a marked reorientation of Latin American export agriculture from Europe to the United States. Before the Second World War exports were strongly oriented to Europe. At a national level there has been an extraordinary expansion of some crops at the expense of others, especially traditional food staples. Some crops that were not widely produced in the 1960s have grown at a rapid if volatile pace reflecting climatic trends and shifts in external demand: Rice in Brazil, Venezuela, and Colombia; soybeans in Paraguay, Argentina, and Brazil; and palm oil in Ecuador (see Table 9.6).

Profitable products destined for affluent urban and foreign markets have replaced the food staples. In Chile fruits and livestock replaced wheat and sugar beet; sorghum replaced corn in Mexico and Brazil; livestock replaced the basic crops throughout the region as indicated by statistics showing the vast expansion of permanent pasture lands at the same time croplands either decreased in area (as in Mexico and Venezuela) or increased only moderately (as in Costa Rica, Colombia, Panama and Honduras). In some places, increased livestock production also stimulated the expansion of feed grain production, often on land that formerly produced the food staples of middle- and low-income groups.

Shortfalls in food staple production have necessitated the increased import of basic food items. Until 1993 agricultural exports grew steadily even if they did not keep pace with imports. Since then, however, weaker economic conditions in the United States and Europe have reduced overall demand for Latin America's agricultural exports (such as January strawberries) at the same time as the cost of imported food (and other products such as fertilizers and machinery) increased appreciably. The increased preference of affluent consumers in North America and Europe for local produce has also eaten into the potential demand for foodstuffs imported over long distances. So agricultural production tends to recapitulate both the volatility of

Table 9.6 Selected export crops in some Latin American countries (metric tons)

	1970	1980	1990	2000	2010
Rice					
Brazil	94,968	1,442	1,427	466,960	473,426
Colombia	5,160	41,330	54,764	37	102
Venezuela	60,056	17,088	0	60,242	4,413
Soybeans					
Paraguay	41,293	537,300	1,794,618	2,980,060	3,922,310
Argentina	26,800	3,500,000	10,700,000	20,200,000	13,616,013
Brazil	1,508,540	15,155,804	19,897,804	32,734,958	25,860,785
Palm oil					
Ecuador	150,000	244,930	835,697	1,339,400	145,781

Source: Based on online data at FAO (FAOSTAT), available at http://faostat.fao.org

demand and the declining terms of trade associated with the historic production of agricultural commodities in peripheral economies.

Nevertheless, the penetration of foreign agribusiness has had important effects on rural populations. One effect has been the increased concentration of land holdings in the hands of capitalist farmers and TNCs such that:

> Throughout the region, tenants and sharecroppers were replaced by agricultural workers, and permanent workers were displaced by part-time laborers. Given these changes, landowners could minimize the costs of maintaining a labor force through periods when it was not needed and expand cropping or live stocking areas by taking over lands that had been assigned to resident laborers, tenants, and sharecroppers. Labor costs were thus reduced for the entrepreneur, and the available pool of laborers, forced to provide for their own maintenance during inactive periods, was enlarged.
>
> (Grindle, 1986: 98)

Another effect has been to increase the need to borrow, and so indebtedness of surviving peasant farmers and part-time laborers. Debt is nothing new for peasant farmers. As the meaning of subsistence changed in a monetized economy to include "urban goods" and processed foods, so too did the importance of money. In the past, money was obtained through the sale of labor for cash wages or sale of market crops. Debt arose because of the need to store and transport crops and pay for inputs before cash was available. Often yields and cash wages were so low that more debt was incurred merely to survive. Today debt is also incurred by the necessity of competing against the capitalist export sector for land, inputs, and water resources.

THE CYCLE OF INDEBTEDNESS

In order to manage the higher debt load, peasants must farm their land more intensively. This only exacerbates the problem. Traditional farming methods such as crop rotation and fallow agriculture are replaced by monoculture to grow the most remunerative crop. This process leaches and depletes the soil, leading to poor harvests and soil erosion. As a consequence, more fertilizers and new seeds are required, which deepens the cycle of indebtedness. Warman (1980: 238) described the cycle of indebtedness that has followed the increased capitalization of agriculture in central Mexico:

> The peasant has to combine several sources of credit, on occasion all of them, in order to bring off the miracle of continuing to produce without dying of starvation. He does it through a set of elaborate and sometimes convoluted strategies. Some people plant peanuts only in order to finance the fertilizers for the corn crop. Others use official credit to finance planting a cornfield or for buying corn for consumption in the months of scarcity, while they resort to the local bourgeoisie or the big monopolists in order to finance a field of tomatoes or onions. Many turn to usurers [money lenders] to cover the costs of an illness or a fiesta . . . Given what they produce in a year, what is left after paying the debts does not go far enough even for food during the dry season, much less for starting a crop on their own. For them, obtaining a new loan is a precondition for continuing cultivation, one that must be combined with the sale of labor if they are to hold out to the next harvest. Each year the effort necessary to maintain the precarious equilibrium increases.

Peasants, then, are survivors as much as victims. Increasingly, wage labor has come to provide a major portion of family income even for peasants who own land. Often this has involved temporary long-distance migration. In Guatemala, for example, the coffee, cotton, and sugar harvests in December and January involve the seasonal migration of an estimated 1 million highland Indians. Temporary wage labor on nearby plantations and capitalist farms, however, is perhaps the major form of adaptation.

DRUG CROPS

In some areas, peasants have also supplemented their incomes by switching to the cultivation of drug crops. The market for these crops in the United States and Europe has grown exponentially since 1970 and the crops can be grown in remote areas on low-grade soils. Given the illegality of drug crops in world trade, remoteness becomes a virtue rather than the liability it is in more legitimate trade. Afghanistan and Myanmar (Burma) are important sources of heroin destined for U.S. and European markets. In three Latin American countries, Peru, Colombia, and Bolivia, the value of cocaine exports is estimated to be US$800 million per annum. Of course, much of the proceeds goes to drug barons, public officials, and intermediaries. But for many peasants, the drug traffic is one of the only ways they have of paying their debts in an effort to respond to the disruptions resulting from the capitalization of agriculture by TNCs and foreign investment. Profits from drugs also fuel the insurgencies of ethnic and political opponents of existing governments. The main routes of surreptitious export change frequently in response to both new alliances between producers and intermediaries and successful efforts by police forces at intercepting the drugs before they hit the streets of U.S. and European cities. Some commentators see the laundering of profits from the international drug business as a major activity in some offshore financial centers. The drugs business is not new. It has ancient roots. In the nineteenth century opening up China to the export of opium from India was one of the main causes of the war between Britain and China that was, as a result, called the Opium War. Illicit though it may now be, the global trade in drugs fits into the long history of the trade in stimulants—tea, coffee, opium, etc.—as an important part of the growth of the world economy (see Chapter 8).

THE CASE OF THE BEEF BOOM IN CENTRAL AMERICA

An interesting case study in the capitalization of Latin American agriculture is the so-called beef boom in Central America in the 1970s and early 1980s. This led to the emergence of Central America as a major supplier of beef to the United States when it had been previously relatively insignificant. It resulted from the tremendous increase in demand for beef in the United States as a result of the emergence of fast-food franchises such as McDonald's and Burger King. The new franchises were not particularly demanding of high-quality beef; what they wanted was quantity that could be formed into patties of equal size and weight by sufficient grinding and tenderizing. But the quantity needed was so huge that the fast-food chains (and frozen-meals makers) needed to look beyond the USA for their supply. Sources such as Australia, New Zealand, and Canada were subject to severe quota limitations that were part of intensive tit-for-tat trade negotiations on the part of the U.S. government in the GATT. South America was excluded because of the prevalence of foot and mouth disease. Central America was favored by U.S. government policy to help "friendly" governments diversify their exports in the face of the perceived "geostrategic threat" from Cuba and the Soviet Union in the region. By 1979 Central America had acquired 93 percent of the share of the U.S. beef quota available to LDCs.

A number of TNCs and individuals found it profitable to respond to the demand for beef from Central America. Some very large U.S. companies became involved through subsidiaries and joint ventures. For example, R.J. Reynolds owned huge grazing ranches in Guatemala and Costa Rica at that time through its then subsidiary, Del Monte, and directly processed and marketed its beef through a variety of outlets: Ortega beef tacos, Chun King beef chow mein, and Delmonte Mexican foods. It also sold beef through Zantigo Mexican outlets (Kentucky Fried Chicken). One of the largest firms in the Central American beef business was Agrodinamica Holding Company, formed in 1971 with 60 percent of the stock owned by wealthy Latin Americans and 40 percent of the stock owned by the ADELA Investment Company (a private investment company with offices in Washington, DC, Luxembourg, Zürich, and Lima that operated entirely in Latin America). This operation controlled thousands of acres of pasture in Central America, owned numerous packing plants, and ran a Miami (Florida) beef import house and wholesale distributor.

Other TNCs became involved in supplying the beef business with inputs (grass seed, barbed wire, fertilizers, feed grains, and veterinary supplies). Pulp and paper companies, such as Crown Zellerbach and Weyerhauser, invested in cardboard box factories to supply packinghouses with containers for shipping the beef. Finally, fruit companies with access to large blocks of land turned them into moneymaking properties.

TNCs, however, were not the only beneficiaries. Wealthy families with access to large amounts of marginal and forestland turned them into pasture. Some urban-based professionals (lawyers, bankers, etc.) also became involved as "weekend ranchers" of peripheral areas previously untouched by commercial agriculture.

The massive displacement of peasants by ranchers and cattle, however, met with tremendous resistance. As Williams (1986: 151) put it: "The receding edge of the tropical forest became the setting of a conflict between two incompatible systems of land use, one driven by the logic of the world market, the other driven by the logic of survival." The violence and civil war throughout much of Central America in the 1970s and 1980s bore no small relationship to the expansion of the beef export business.

The Central American beef boom ended in the 1980s, however, due to declining international beef prices and reduced U.S. demand as real incomes stagnated and consumers became more health conscious. The U.S. Congress also passed a more restrictive meat import act in 1979, which significantly reduced Central America's access to the U.S. market. Beef exports from Costa Rica, Guatemala, and Honduras were prohibited on several occasions as the USA enforced laws prohibiting the import of substandard beef and beef with pesticide residues. During the 1980s and early 1990s the U.S. government also prohibited meat imports from Nicaragua and Panama for political reasons. To make matters worse, cattle ranchers' costs (inputs and taxes) were rising. A major blow was the decision in 1987 by Burger King, which at one time bought 70 percent of Costa Rica's beef exports, to stop buying Latin American beef because of criticism of the "hamburger connection" (Kaimowitz, 1995).

This criticism relates to other important consequences associated with the capitalization of agriculture: deforestation and environmental degradation. Much of the loss of forest in Central America, the Amazon Basin of Brazil, and in Southeast Asia has been due to the extension of ranching as well as timber extraction and the burning of timber as fuel wood. While the rate of Deforestation in Central America is estimated to have declined from 4,000,000 hectares per year in the 1970s to 300,000 hectares in the 1990s, at this current rate, Central America will lose its remaining forest in fewer than 60 years. A related stimulus to the incredible pace at which tropical forests have been disappearing since the 1970s has been the need to pay off the debts incurred in expensive industrialization campaigns. The opening of forestland to capital-intensive agriculture has been one strategy for swapping natural resources for income to repay debts. Five of the world's "mega-debtors"—Brazil, India, Indonesia, Mexico, and Nigeria—all rank among the top ten deforesters. The conversion from forest to pasture or cultivation can also increase soil compaction, soil erosion and nutrient depletion—particular problems in marginal locations with less fertile soils. During the rainy season, erosion in places with steeper slopes can contribute to devastating flooding and mudslides in the flatland areas. The capitalization of agriculture in the periphery, therefore, has had correlates other than increased productivity, the establishment of comparative advantage in export crops and the increased import of food crops.

9.5 SCIENCE AND TECHNOLOGY IN AGRICULTURE

The beef export boom in Central America would not have been possible without the importation of techniques of "scientific agriculture." In this context, this involved creating "new" breeds of cattle by combining "beefier" attributes with high resistance to pests and tropical heat, transforming pasture management by sowing higher yield grasses and fertilizers, enhancing water supplies by digging new wells and ponds, and providing better veterinary care to cattle herds.

The past 40 years have witnessed an intensive drive on the part of international development agencies (such as the Food and Agriculture Organization (FAO) of the UN and the World Bank), some national governments, and agribusiness to introduce scientific farming into agriculture in LDCs. The results have been controversial. From one point of view, yields have been increased and, especially in parts of Asia but also to a degree elsewhere, agricultural productivity and production have been significantly increased. Of particular importance have been the new wheat, maize, and rice varieties associated with the so-called Green Revolution. It is generally acknowledged that the gains from these new varieties (and the fertilizers and irrigation they require) have been concentrated in certain districts of India, Pakistan, and Sri Lanka, the central Philippines, Java in Indonesia, peninsular Malaysia, northern Turkey, and northern Colombia. In addition to increased yields, the new techniques can involve an increase in demand for labor in land preparation, fertilizer application, and harvesting, and increases in the wages of agricultural laborers (as in Indian Punjab). Doubts are sometimes expressed, however, about the sustainability of these trends in yields and labor use.

From another point of view, scientific agriculture is largely an instrument of commercialization and capitalization rather than a mechanism for improving agricultural productivity and production *per se*. This is not to say that new seed varieties, fertilizers, etc., are always inappropriate; rather, that it all depends on the sociopolitical context in which they are applied. In particular, research efforts in scientific agriculture have been heavily biased towards certain commodities that are either most important in the DCs or significant in world trade. The very small amount of research on important food staples such as cassava, coconuts, sweet potatoes, groundnuts, and chickpeas is especially noteworthy. The "research system" gives high priority to export crops such as cattle, cotton, and sorghum and to those such as rice and wheat that have "wide adaptability": The ability to transfer a new variety from one region to others. Wide adaptability can be criticized, however, for its potential in reducing genetic variety and making crops more vulnerable to disease.

A more frequent criticism of scientific agriculture, particularly in its manifestation as the Green Revolution, is that it primarily benefits larger, more prosperous farmers who have readier access to the necessary inputs and credit sources. At the same time it encourages the debt cycle among poorer peasants and part-time laborers discussed earlier. Farmers must take out loans to pay for the increasingly expensive inputs that only with increasing prices relative to production can they possibly repay. Moreover, the new varieties require increased dependence

Box 9.5 Science and rice

For half of the world's population, overwhelmingly in Asia, the lifecycle revolves around rice. In Vietnam, a child's first solid food is rice gruel. In Taiwan, chopsticks stuck in a mound of cooked rice symbolize death. Getting a good job in Singapore is an "iron rice bowl," and unemployment is a "broken rice bowl." The characters for Toyota and Honda, the great Japanese car companies, mean in the Japanese language, "bountiful rice field" and "main rice field," respectively. For people in places in which rice has long been the main staple of everyday diet, rice means just about everything that is important: Birth, death, power, wealth, virility, fertility, vitality, and so on. The oldest recorded cultivation of rice occurred in what is today Thailand in 4000 BCE, although the crop is thought to have originated in Africa. Its cultivation spread widely but rice became the staple crop in Southeast and East Asia. Elsewhere, wheat and other cereal grains tended to be more important. The great advantage of rice lies in its yields that, on average, are twice as large as those of wheat. Today, rice feeds more people than any other crop. Although more wheat is harvested annually than rice, over 20 percent of that harvest goes to feed animals. Virtually the entire annual rice harvest (598.2 million metric tons in 2000) goes to feed people, mostly in Asia, where more than 60 percent of the world's population lives. In Bangladesh, Cambodia, Indonesia, Laos, Myanmar, Thailand, and Vietnam, 56 to 80 percent of daily calories come from rice. Rice has what botanists call "developmental plasticity": It can grow in a wide variety of circumstances. It flourishes best, however, in the humid tropics. The three largest producers, China, India, and Indonesia, produce and consume about 60 percent of the world's rice. With only 4 percent of the world's rice in world trade, a stable local supply is crucial to the food supply of most Asian countries. All the world's exports, about 24 million tons, would not meet demand from India for more than two months. From the 1930s to the 1950s rice yields in Asia stagnated, while improved healthcare led to a doubling of the population. The application of chemical fertilizers did little to improve the situation. The established types of rice grew, but they grew too tall, fell under their own weight and rotted in the flooded fields in which they were cultivated. A new strategy came in the early 1960s as a result of research on new hybrid varieties of rice carried out at the International Rice Research Institute (IRRI) in Los Banos, Philippines. IR8, one of the first new varieties, was spectacularly successful in raising yields. It grew faster—maturing in 130 rather than the usual 180 days—and allowed farmers to harvest two or even three crops a year from the same land. It also produced twice as much rice as either of the parent varieties. This variety and subsequent ones were so successful in doubling the world's rice crop that they were called "miracle varieties." They and new wheat varieties led to the declaration of a **Green Revolution** in which the war on hunger and famine was said to have been won. This was premature. By the 1980s the IRRI had engineered 250 new varieties of rice that are planted in 106 countries; but, at the same time, world rice production has flattened out and the population has kept on growing. A simple answer might be just to plant more land in rice. In Asia, however, little or no land is left for expansion. So the pressure is on to increase yields even further through more varieties better fitted to specific ecological conditions and, due to genetic modification, pest resistant. Insects and diseases destroy nearly 25 percent of rice crops. The question of the moment is whether or not yields can be increased indefinitely even with genetically modified varieties. The leveling-out of production in recent years might suggest that the limits to scientific agriculture in rice production have now been reached.

on the acquisition of energy-intensive inputs (such as fertilizers and agricultural machinery), largely controlled by TNCs.

The substitution of feed crops—crops for feeding animals rather than direct human use—and export food crops for local food crops has been one important recent impact of scientific agriculture. Some observers refer to this as the "second green revolution," meaning that it has produced a new wave of crops, whereas the earlier trend produced greater yields of staple crops. This is not only biased in favor of farmers with capital, it also can lead to the neglect of food crops fundamental to local diets. As a result, while exporting increasing quantities of meat and fruits, some countries find themselves having to import beans, wheat, and maize to feed their rural populations.

Evidence from such diverse settings as Mexico, India, and Bangladesh suggests that where capital-intensive agriculture is introduced into areas with an uneven distribution of resources it exacerbates the condition of the rural poor by marginalizing subsistence systems, such as share cropping, and encourages the polarization of land control between a class of capitalist farmers, on the one hand, and the mass of the rural population, on the other. The impact of scientific agriculture, therefore, cannot be separated from issues of social structure.

SUMMARY

Since the early 1960s until 2010 GDP growth rates have been generally faster in the less developed countries than in the developed countries (5.2 percent per annum compared to 2.2 percent). In addition, despite large rates of population growth, the per capita incomes of the periphery taken as a whole have grown at about 3.2 percent per annum. Agricultural production has also increased, in contrast to the stagnation of the colonial period in many Asian and some African countries. Food production per capita in Latin America and Asia grew by 5 to 10 percent from 1960 to 1970 and from 1970 to 1980. In Asia, these rates accelerated to over 20 percent from 1980 to 1990 and from 1990 to 2000. In Latin America, although the rates slipped somewhat from 1980 to 1990, the growth of food production rebounded to a more than 15 percent increase from 1990 to 2000. Only in countries with birth rates of 3 percent or more, as in parts of Sub-Saharan Africa, or where there were major social upheavals, such as Central America, Bangladesh, Cambodia, and Vietnam, is this picture particularly misleading. Throughout the periphery, the incidence of chronic hunger and malnourishment has declined since 1970, however, despite civil strife, wars, and natural disasters such as droughts, floods, and earthquakes.

At the same time, however, the incidence of rural indebtedness and poverty and the loss of land for food production to meet local demand have increased enormously. This is because increased agricultural production in the context of the world economy is no guarantee that the people involved in achieving it will see its fruits. This chapter has attempted to show how this can be the case by detailing the effects of progressive commercialization and capitalization. When export crops displace subsistence uses and food staple production, increased agricultural production does not necessarily benefit rural populations. Far from it: They often find themselves ensnared in webs of poverty and indebtedness that are the direct product of modern scientific agriculture in contexts where there are few alternatives to agricultural employment. In reaching this conclusion, the argument of this chapter has involved making the following major points:

1. Agriculture is often given a subsidiary role in models of development followed by governments even when it is a vital source of sustenance and employment. For a variety of

reasons, national government pricing and credit policies have tended to drain agriculture in favor of the industrial–urban sector.

2. The three continents of the periphery—Africa, Asia and Latin America—differ significantly in terms of agricultural organization and performance.

3. It is also important to recognize that in agriculture in the LDCs, it is the women rather than the men who are overwhelmingly more important as the source of workers, especially in Sub-Saharan Africa.

4. There is a long history of commercial agriculture in the periphery. Until recently, however, it was a plantation or export enclave sector surrounded by a largely subsistence sector.

5. Agriculture, particularly when based in small labor-intensive household farms and leading to increases in food and crop production of benefit to local communities, has a vital role to play in laying the foundation for subsequent economic development across other sectors.

6. Rural land reform has tended to encourage the development of capitalist agriculture rather than benefit the interests of peasant farmers.

7. Rural land reform, and the recent activities of governments and transnational corporations have produced a much more widespread commercialization and capitalization (increasingly capital-intensive type) of agriculture. This has been most marked in Latin America but can also be seen elsewhere.

8. "Scientific" agriculture has tended to reflect and reinforce the capitalization of agriculture even as it has increased yields for a limited number of agricultural products, mainly a few staples such as rice and wheat and others in the export trade.

KEY SOURCES AND SUGGESTED READING

De Janvry, A., 1984. The Role of Land Reform in Economic Development: Policies and politics, in C.K. Eicher and J.M. Staatz (eds.), *Agricultural Development in the Third World*. Baltimore, MD: Johns Hopkins University Press.

Fine, B., 1994. Towards a Political Economy of Food, *Review of International Political Economy* 3, 519–545.

Goodman, D., 2003. The Quality "Turn" and Alternative Food Practices: Reflections and agenda, *Journal of Rural Studies* 19, 1–7.

Hayami, Y., 1984. Assessment of the Green Revolution, in C.K. Eicher and J.M. Staatz (eds.), *Agricultural Development in the Third World*. Baltimore, MD: Johns Hopkins University Press.

Hesse, M., Schmitt, J., and Wagner, W. (eds.), 2013. Deutsche Bank backs Ruthless "Rubber Lords," *Der Spiegel* (English Edition), May 15.

Joekes, S.P., 1987. *Women in the World Economy*. New York: Oxford University Press.

McMahon, P., 2013. *Feeding Frenzy: The politics of food*. London: Profile Books.

Rosenthal, E., 2013. As Biofuel Demand grows, so do Guatemala's Hunger Pangs. *New York Times*, January 5.

Rosset, P., Patel, R., and Courville, M. (eds.), 2007. *Promised Land: Competing visions of agrarian reform*. San Francisco: Food First.

Studwell, J., 2013. *How Asia Works: Success and failure in the world's most dynamic region*. London: Profile Books.

Swindell, K., 1985. *Farm Labor*. Cambridge: Cambridge University Press.

Warman, A., 1980. *"We Come to Protest": The peasants of Morelos and the national state*. Baltimore, MD: Johns Hopkins University Press.

Chapter 10

Industrialization: The path to progress?

Picture credit: Wikipedia

In the 1950s and 1960s the development strategies of many less developed countries placed considerable emphasis on manufacturing industry, which was considered to be the leading sector of economic development. More recently, as the DCs have "lost" some branches of manufacturing to locations in the periphery and some LDCs have embarked on aggressive export-oriented development strategies, it seems that efforts at industrialization can pay off. But what exactly has been the result of several decades of industrialization in the LDCs?

In the LDCs over the past 40 years value added in manufacturing (MVA) has risen at a rapid pace. This is in a historical context in which manufacturing has declined as a proportion of total world value added, going from 35 percent in 1985 to 27 percent in 2008. Services grew from 59 to 70 percent in contribution to total world value added over the same period. The increase in MVA in the LDCs is relative, however. The LDCs supplied 8.2 percent of world MVA in 1960, 14.4 percent in 1980 and still only 30.8 percent by 2008. Moreover, manufacturing industry has been highly concentrated. From 1966 to 1975, four countries, representing 11 percent of the population of the LDCs, accounted for over half the increase of the LDCs' MVA. Eight countries (Argentina, Brazil, India, Indonesia, Iran, Mexico, South Korea, and Turkey) with 17 percent of the total population produced about two-thirds of the increase. From 1975 to 2005, however, Argentina dropped out and China, Malaysia, Singapore, Taiwan, and Thailand were added to the list. The presence of China and India on this list raises the proportion of the population of the LDCs in the high-growth category to 60 percent.

At the same time, growth rates of MVA have been lowest in the poorest countries. In Sub-Saharan Africa, the growth of manufacturing has been particularly slow. In 1975 manufactured production represented 5 percent of the GDP of the Sub-Saharan African LDCs as opposed to 16 percent for the Asian ones and 25 percent of those in Latin America and the Caribbean. By 2008 the comparable figures were approximately 16 percent for the Sub-Saharan African LDCs, 32 percent for Asia (35 percent in East Asia, 18 percent in South Asia), and 21 percent for Latin America and the Caribbean.

In all LDCs, irrespective of their growth rates, industrial production has been characterized by a particular expansion of heavy industries: Iron and steel, machinery, and chemicals. Over the entire period 1950–2000 this expansion was more rapid than the growth of food processing or textile, clothing, and shoe industries. This point needs emphasizing because of the tendency to assume, because of increasing exports of goods such as clothing and shoes to Europe and the United States, that light and consumer goods industries have grown the most. However, since the 1970s, the industrial mix of many LDCs has undergone significant change. After the Second World War, industrialization, even if involving foreign investment by TNCs, was largely concerned with import substitution. Since the early 1960s the possibilities of substitution have dwindled in the face of the mounting costs of establishing heavy industries and as subcontracting and global sourcing by TNCs have replaced the older strategy of direct establishment of subsidiaries. In this changed global context, a fundamental reorientation has taken place in the most industrialized LDCs in East Asia and Latin America. In these countries, an increasing proportion of industry is oriented towards exporting manufactured goods, mostly to the developed countries.

The shift towards an export orientation was facilitated by rapid economic growth (and increasing consumer incomes) in the developed countries (especially in Europe and Japan) and the liberalization of world trade beginning in the 1960s. Above all, however, it reflects a change in national industrialization strategy. The role of the state remains central. Like import substitution, export-oriented development strategies involve a strong managerial role for the state in adjusting to changing global pressures. The states that have been most successful in doing so, such as China, South Korea and Taiwan, are now among the leading industrializers. But the ideological and institutional context has changed fundamentally. The focus has moved from self-sufficiency in a world of national economies to gaining competitive advantage in a world economy organized around principles of market access.

In this chapter, the progress of manufacturing industrialization in the LDCs is examined in four complementary ways. First, the national and global stimuli to industrialization are described. Particular attention is directed towards the role of industrialization in national ideologies of modernization, the practical basis to the demand for industrialization and the global context for the shift from import substitution to export-oriented export strategies. Second, the problems facing industrialization in the LDCs are reviewed, focusing on the limits to industrialization posed by certain national-level and global constraints. Third, the geographical pattern of industrialization is surveyed at global, regional and urban scales. Finally, industrialization in Russia and China is profiled. China, in particular, is examined for its experience of industrialization over the recent past because it calls into question the sustainability of the core–periphery structure of the world economy as it has previously been organized. Along with such growing industrial powers as Brazil, Russia, and India, China's rapid rise within the global manufacturing system signals a shift towards a very different world economy. The BRIC (Brazil, Russia, India, and China) economies are seen by some commentators as the beneficiaries of increasingly export-oriented globalization.

10.1 NATIONAL AND GLOBAL STIMULI TO INDUSTRIALIZATION

The central attention given by many LDCs to industry is partly a result of the prestige of this sector, which is widely considered the hallmark of development. Although the notion of "industrialization-in-general" can be criticized on grounds of vagueness and lack of attention to the specific mix of industries and their relation to the needs of the mass of the population, industrialization figures prominently in most national ideologies of modernization. Perhaps

China for part of the 1960s was something of an exception, at least in theory; but even there the lure of industrialization as a development strategy proved stronger than ideological commitment to rural–agricultural development.

Interestingly, in the core of the world economy, particularly in Europe, industrialization has always given rise to various "discontents." Contempt for the production of worldly goods shows up even in the writings of the classical economist, Adam Smith. Later concerns have been more with the nature of industrialization, in particular the "balance" between heavy and consumer goods industries. These ideas have had their most vocal expression among those East Europeans who decried the "overemphasis" of the now defunct communist governments on heavy industry and Latin American complaints about the "lack" of heavy industry. However, the association between industrialization and progress is now strongly established. Only environmental activists, a rare breed in most LDCs, question the unrelenting priority given to industrial development.

THREE MYTHS

Three ideas of mythic proportions are at the center of the claim that national industrialization is the path to progress, even though they are of questionable empirical validity. Interestingly, they all involve negative views of agriculture as much as positive endorsements of industry and they all imply a simple sectoral logic of development as movement from agriculture to industry. The word "myth" does not imply falsehood so much as an unexamined idea that comes to guide thought and action.

First, agriculture is viewed as having more limited stimulus effects on other economic activities than industry. In other words, industry is seen as providing multiplier effects (new knowledge, new products, stimulative effects on the service sector, etc.) that agriculture cannot provide. The best refutation of this particular idea is the key role that agriculture played in the early industrialization of Europe and the continued importance of agriculture in the economies of the developed world. Of course, in each case investment was required to develop forward linkages to consumer industry (food processing, etc.) and backward linkages to input providers (fertilizers, etc.). In each case, farmers were also important as consumers of industrial products, when not penalized by low prices for their products, and as significant financiers of industrial investment, through savings and taxation. Adelman (1984) has proposed that precisely these stimulative features of agriculture can be used to substitute "agriculture-demand-led-industrialization" (ADLI) for import substitution and export-oriented development models.

Second, farmers have a reputation for conservatism whereas industrialists (and workers) are viewed as agents of modernization. Imprisoned in ancient and traditional cultures, farmers, especially peasant farmers, are without dynamism and rationality. Yet, again, this idea is easily refuted by evidence from all over the world. For example, as shown in Chapter 9, there is a strong link between producer prices and yields of rice in different Asian countries. Corn (maize) production in Thailand, bean production in the Sudan, and wheat production in India and Pakistan have all increased as prices have increased and decreased when prices have declined. These are hardly indications of conservatism and lack of responsiveness to commercial incentives.

Third, for many governments industry is seen as the only productive sector. Only in industry, the argument goes, are there increasing marginal returns through economies of scale in production. Moreover, the average productivity of workers in industry is higher than that of those in agriculture. However, the productivity of other factors, capital in particular, is probably higher in agriculture. In most of the countries for which data exist, the gross marginal capital–output ratio is lower and so productivity is higher in agriculture than in other sectors. In the United States, the only country with a sufficiently long statistical series, the total productivity of all factors has increased faster in agriculture than in other sectors.

PRACTICAL RATIONALES FOR INDUSTRIALIZATION

Whatever the empirical merit of the three ideas, they have become firmly entrenched and associated with modernization through industrialization. Manufacturing industry is widely viewed as the path to progress and it figures prominently in most national development ideologies and plans. These ideologies, whatever precise roles they reserve for "private" business and state direction, have been reinforced by certain practical problems facing most governments. There are perhaps three that appear most important. One of them concerns the terms of trade in exchanging primary commodities for manufactured goods. As suggested in Chapter 9, there are good grounds for pessimism about the growth potential in general of primary production (raw materials and agricultural exports) because of the long-term trend of deteriorating barter terms of trade with manufactured goods. However, for specific primary commodities and specific countries, investment in primary production can be preferable. Since 2000 prices of many basic commodities have trended upwards and the barter terms of trade between primaries and manufactured goods has become less disadvantageous. Nevertheless, by and large, governments have not been persuaded of the virtues of agricultural (or resource) specialization. They can even point to the case of the OPEC cartel—the most successful attempt in the history of the world economy to bolster the price of a primary commodity—to illustrate the limitations of primary production. From its dominant position in 1973–1974, OPEC has become less and less able to govern the world price of oil. This reflects both adjustment strategies in consumer countries (energy conservation, shifts to non-OPEC suppliers), and the emergence of political conflicts and different production strategies among member countries (the Iran/Iraq war of the 1980s; long-term market share and conservation of oil reserves versus rapid production, for example, Saudi Arabia versus Nigeria). It is easy to infer from the experience of OPEC the long-run limitations of a development strategy based on primary production for world markets.

Second, many LDCs have massive unemployed and underemployed populations concentrated increasingly in urban areas. Deteriorating living conditions in the countryside (see Chapter 9) and the availability of better public services in urban areas have encouraged large-scale rural-to-urban migration, often in the absence of industrialization. To survive in the cities, people engage in a wide range of "informal sector" economic activities as street vendors, shoeshine boys, stall keepers, public letter writers, auto mechanics, taxi drivers, subcontractors, tailors, drug dealers, and prostitutes. Sometimes these activities can be linked to industrialization of a formal variety through subcontracting, but often they cannot. International migration, both temporary and permanent, is sometimes an alternative for those with better education and greatest initiative. But, among other things, this produces a "brain drain" that poor countries can ill afford. It is in this context that expansion of employment in manufacturing industry can often become an important national imperative.

A third incentive for industrialization comes from the state-building activities of national élites. National industrialization can be a "prestige" goal around which national populations can be mobilized. All governments are also under pressure to industrialize in order to compete with other countries. Pressure comes from both domestic élites, especially the military, and from foreign allies and patrons. Some of the emphasis on heavy industries undoubtedly derives from this pressure. The significant growth of military industries in the LDCs is directly related to it. Industry is also an important instrument of political favoritism and patronage. Governments can reward "loyal" social and ethnic groups and punish "disloyal" ones, by directing industrial activities towards some places and away from others. Industrialization, therefore, often involves political stimulation, of both "noble" and "ignoble" varieties.

The national industrialization drives that took place in the aftermath of decolonization had limited effects until the late 1960s, except in those countries with large domestic markets and long-sustained import substitution policies (for example, Brazil and Mexico). The spread and intensification of industrialization since the late 1960s coincides to a certain extent with the declining rate of profit in the DCs (see Chapter 3) and the consequent shift in strategy by TNCs from high-wage/high-consumption forms of production in the DCs (Fordism) towards spatially decentralized forms of production in which low-wage labor forces are important in certain phases (see Table 10.1). This suggests that the changing global context has been fundamentally important in stimulating the recent expansion of manufacturing industry in some LDCs. In other words, the **new international division of labor (NIDL)** in manufacturing and its associated spatial decentralization of many production activities (largely related to the international product lifecycle model) are closely related to the "crisis" of capital growth in the DCs. A changing geography of manufacturing employment has been the result (see Table 10.2).

Table 10.1 Labor conditions in global manufacturing

	Average hourly compensation in manufacturing in 2011 (US$)	Strikes and lockouts in manufacturing 1990–2008 (average annual)
United States	35.53	8
Canada	36.56	7
United Kingdom	30.77	36
Germany	47.38	Not available
Portugal	12.91	100
Australia	46.29	129
Japan	35.71	33
Mexico	6.48	16
Poland	8.83	10
South Korea	18.91	77
Singapore	22.60	0
Taiwan	9.34	Not available
Philippines	2.01	27

Source: Based on United States Department of Labor data, International Comparisons of Hourly Compensation Costs in Manufacturing, available at http://www.bls.gov/web/ichcc.supp.toc.htm; International Labor Office (ILO) LABORSTA Internet data, available at http://laborsta.ilo.org/

Table 10.2 Changing geography of manufacturing employment: paid employment in manufacturing (millions of people)

Region/country	1980	1995	2005	2010	% change
United States and Canada	21.4	20.3	18.5	15.8	−26.0
Japan	12.1	11.9	11.4	10.4	−13.2
Western Europe[1]	28.2	25.9	23.7	23.8	−15.6
Core countries	*61.7*	*58.1*	*53.6*	*50.0*	*−18.9*
South Asia[2]	60.5	61.3	62.7	63.1	+4.0
Southeast and East Asia[3]	68.6	115.3	96.7	126.7	+84.7
Latin America[4]	8.7	9.1	16.2	22.5	+158.0
Semi-peripheral and peripheral countries	*137.8*	*185.7*	*175.6*	*212.3*	*+54.0*

Notes:
1 Austria, Belgium, France, Germany, Italy, Netherlands, Portugal, Spain, Sweden, United Kingdom
2 Bangladesh, India, Sri Lanka
3 China, Hong Kong, Malaysia, Philippines, Singapore, South Korea, Taiwan, Thailand
4 Brazil, Mexico, Venezuela

Source: Based on Peet (1987), p. 781; UNIDO (Reference Information: http://www.unido.org/); United Nations (Monthly Bulletin of Statistics On-Line: http://esa.un.org/unsd/mbsdemo/mbssearch.asp); International Labor Office (ILO) (LABORSTA Internet data: http://laborsta.ilo.org/)

The extent to which cost advantages for TNCs (particularly low-cost labor) have entirely spurred the growth of the NIEs can be exaggerated. In the first place, TNCs do not dominate production in the NIEs. Even in Brazil, a country that for many years had had a high proportion of TNC investment relative to local sources, foreign direct investment (FDI) declined from the mid-1970s, until this trend was reversed during the late 1990s as a result of currency and trade reforms. Second, governments in the NIEs have controlled foreign investment, preferring investment by banks and other financial interests to FDI: "Productive capital has largely remained under the control of local corporations; most foreign capital has entered countries either as bilateral or multilateral aid or as financial capital" (Webber and Rigby, 1996: 453). Nevertheless, export-based industrialization, as opposed to import substitution, does allow firms to pay lower wages than they would have to if the local economy had to absorb all of the supply. This has broken the local geographical bonds between production and consumption on which organized capitalism was based. The expansion of industry in the LDCs, therefore, has some basis in cost advantages, even if these are not ones that accrue largely to TNCs and it has had the consequence of increasing the competitiveness of world markets across a wide range of manufacturing industries.

But perhaps the best and most simple evidence that costs alone cannot explain the spread of industrialization to the LDCs is that export-oriented industrialization did not concentrate where wage rates were lowest. It has been concentrated mainly in certain countries with other characteristics. The most successful ones, the East Asian NIEs of China, South Korea, Taiwan, Singapore, and Hong Kong, were ones in which the state and local industrial capital were closely interlocked and receptive to massive foreign investment. Moreover, they are the ones in which much of the foreign investment was in the form of private bank credit or strategic alliances between TNCs and local partners rather than direct investment by TNCs. In this way, they maintained more local control over investment decisions. In addition, the privileged geopolitical arrangements of South Korea and Taiwan with the United States were important in opening U.S. markets and making available U.S. aid and investment in return for the East Asian countries' forward role in "containment" of mainland China and the Soviet Union. Since the late 1980s there has been an increased level of FDI in East Asia, much of it flowing between East Asian countries rather than emanating from outside the region. The East Asian financial crisis of 1997 reinforced this trend as some FDI that would otherwise have gone into Southeast Asian countries (for example, Thailand, the Philippines, and, especially, Indonesia), was diverted to East Asian countries (for example, South Korea, Taiwan, as well as China and Hong Kong). Indeed, between 1991 and 2000 private financial flows to the LDCs grew rapidly after stagnating in the 1980s but "official" (governments, World Bank, etc.) investment tailed off having been more important than private flows throughout the 1980s. This has led some commentators to see a foundation being laid for a future boom in peripheral industrialization. However, though the global context may well trigger the *possibility* for industrialization, especially of the export-oriented type, it never guarantees its realization. Various contingencies —political, economic, social—intervene to determine *where* it takes place.

10.2 LIMITS TO INDUSTRIALIZATION IN THE PERIPHERY

There are a number of constraints that will probably limit the spread and intensification of the export-oriented industrialization that has lain behind the impressive growth of the NIEs in particular and the semi-peripheral and peripheral countries in general since the 1970s.

PROFIT CYCLES

The declining rates of profit in the DCs—between the 1960s and the mid-1980s—may have been cyclical rather than secular. There is evidence that corporate profitability in the USA, for example, began to rebound in the late 1980s after a 15-year downturn. The more critical analyses of this trend argue that the return to profitability has been fueled by stagnant wage rates. This may mean that locating production facilities or engaging in subcontracting in the LDCs can be less attractive to U.S. TNCs as labor costs go down in the core. However, the figures (as reported by the U.S. Department of Commerce, for example, 4.57 percent average after-tax return on capital in manufacturing in the USA in 1988 compared to 2.87 percent in 1985, 3.99 percent in 1975 and 4.80 percent in 1965) may reflect the falling dollar after 1985 helping U.S. manufacturers in foreign markets and the increased return on foreign investments rather than improving domestic profitability (for example, the European operations of General Motors lifted what would otherwise have been an even more dismal rate of profitability in the late 1980s). Certainly the picture is a complex one, with not only countries but also industrial sectors (electronics, automobiles, etc.) having different trajectories in profit rates over time.

In addition, some of the major U.S. TNCs, such as IBM and GE, that pioneered the shift in production to the new international division of labor (NIDL) through the creation of foreign assembly operations in the LDCs have become lackluster performers, challenged in the most profitable production lines by more innovative smaller firms with currently more localized patterns of production in industrial districts because of their reliance on immediate response from independent subcontractors. Overall, however, many companies with their origins in different countries, not just those that faced profit crises in the 1970s, have developed and become committed to supply chains that extend across multiple locations in different countries.

PROTECTIONISM

There are inherent limits to the generalization to other LDCs of the export successes of the NIEs. To begin with, the period 1950–1973 was one of unusually high growth in world trade. Whereas global trade expanded only at 1 percent per annum during 1910–1940, the period 1953–1973 saw an increase in total trade of 8 percent per annum, and of 11 percent per annum for manufactures. The consequence was an increased interdependence in the world economy as trade barriers were lowered. Since the late 1970s, however, protectionism has remained a response in the DCs in times of declining rates of economic growth and increased unemployment, domestic inflation and balance of payments deficits. It is the LDCs that have become the major advocates of global trade liberalization. Cline (1982) estimated that if all the LDCs had the same export intensity as Hong Kong, Singapore, South Korea, and Taiwan, there would be a shift from 16.4 to 60.4 percent of aggregate DC manufactured imports originating in the LDCs. Given the reliance of the NIEs on DC markets, the likelihood of this expansion without protectionist responses seems extremely unlikely. As Cline (1982: 89) concludes:

> It is seriously misleading to hold up the East Asian G4 [Gang of Four, i.e., Hong Kong, Singapore, South Korea, and Taiwan] as a model of development because that model almost certainly cannot be generalized without provoking [a] protectionist response ruling out its implementation.

Of course, Cline's generalization from the current NIEs to all LDCs represents an extreme scenario. There is still considerable scope for building up export markets in both DCs and other LDCs. In particular, there is evidence that the established NIEs are now transferring some of their more labor intensive industries to countries that have a short-run comparative advantage. For example, Taiwanese firms have subsidiaries in China and Malaysia, and Hong Kong firms have subsidiaries in China and the Philippines. Among the LDCs, consequently, adjacent countries are not necessarily condemned to see their neighbors who enjoy an initial advantage permanently monopolize the positions they have won in world markets, even though they must of necessity start out as subservient to them.

ACCESS TO TECHNOLOGY

The TNCs that have been involved in setting up subsidiaries or engaging subcontractors in the NIEs must now respond to protectionist pressures in the DCs in which their final markets are concentrated. This will involve substituting radical new automation technologies for low-wage labor. Even if the technology diffuses evenly at a global scale, production costs will decline more steeply in the DCs than in the NIEs. In addition, economies of scale associated with batch production for customized as opposed to mass markets and external economies associated with a close geographical integration of component suppliers and final assembly in **just-in-time** (JIT) and zero inventory production systems can reduce the attractiveness of global sourcing. There is already evidence that some assembly of electronic circuits is being brought back to DCs from LDCs, and the introduction of new technologies and new production systems may further reduce the need for cheap, unskilled labor.

It is important to note again, however, that the growth of the NIEs is not uniquely due to the activities of foreign TNCs. Domestic firms, including an increasing number of TNCs from the NIEs themselves play important roles, particularly in such export sectors as electronics, automobiles, textiles, clothing, and shoes. By 2000, for example, foreign sales accounted for more than 25 percent of total sales for the ten largest East Asian TNCs (as ranked by foreign assets, for example, Hutchinson Whampoa from Hong Kong, Samsung from South Korea). Only in Latin America and, to a certain degree, in China have the expansion of foreign TNCs and the growth of exports been closely related; but the control of foreign TNCs over most of the new automation and information technologies can limit their transfer when these TNCs are not locally dominant.

DEBT AND CREDITWORTHINESS

The success of export-oriented industrialization has been tied to the growth of enormous debt loads underwritten by FDI, official aid agencies and international banks. Among the most problematic features of the world economy by the late 1990s was the absence of any new groups willing to finance industrialization in the LDCs and the appearance of a global credit shortage because of the emergence of Eastern Europe (especially the former East Germany) and the former Soviet Union (particularly Russia) as competitors in world financial markets. The institutions that did play a central role in the past are largely unwilling to do so now. Perhaps the only institution that has substantially increased its lending and encouraged others to do likewise is the International Monetary Fund (IMF). But the IMF still lends relatively little compared to the size of LDCs' current account deficits and attaches conditions to loans that many countries with strong state direction find undesirable and damaging to long-run development.

Even in the 1970s lending, especially private bank lending, was concentrated in the LDCs with relatively high per capita incomes and those with already impressive growth records. The four largest borrowers (Mexico, Brazil, South Korea, and the Philippines) accounted for over 60 percent of total accumulated non-OPEC LDCs' debt to international banks in December 1982. Much of the money lent came from recycled petrodollars in the context of declines in credit demand from traditional clients in the DCs in the wake of the 1974–1975 recession. This situation is not likely to repeat itself even though in the 1990s private lending from banks and stock markets to East Asia and Latin America did increase after declining precipitously in the 1980s.

RESEARCH AND DEVELOPMENT (R&D) CAPACITY

The leading sectors of industrial growth in the world economy are the information technologies. The USA, Europe, and Japan lead the way in world demand for these technologies and their products. Consequently, success in the global information processing and electronics industries requires a research and development (R&D), production, marketing and political lobbying base inside all three regions of demand. Given the importance of politics—especially in the form of lobbying against protectionist threats—established TNCs and countries with large public sector investments in information technologies are at a distinct advantage. Those companies and countries struggling to gain entry beyond the labor intensive, assembly level will be faced with formidable barriers, not least the absence of local protection in the face of U.S.-led attempts in such international institutions as the GATT over the past ten years at deregulation of LDC domestic markets.

POLARIZATION AND INSTABILITY

The non-city-state NIEs (the experience of Singapore and Hong Kong is not immediately relevant to most LDCs) with the highest growth rates—particularly China, South Korea, and Taiwan—have enjoyed a unique set of circumstances that are not generalizable to other peripheral settings. In particular, the last two inherited the transport and education infrastructure imposed by Japanese colonialism. Later, massive U.S. aid in the 1950s—for geopolitical purposes—stabilized their economies and tight controls on imports plus government allocation of foreign exchange and capital promoted export-led growth. In addition to land reform (see Chapter 9) in South Korea and Taiwan (and latterly in China), state enterprises were a key part of national growth strategies in the 1970s, accounting for 25 percent to 35 percent of total fixed investment. In South Korea—following the practice of Japan—interest subsidies and other incentives that rewarded performance were employed to induce private companies to develop major industries and focus on exports.

Crucially, relatively equal income distribution in South Korea—where the richest 20 percent receive 5.3 times the income of the poorest 20 percent—has allowed the government to adopt policies pursuing efficiency and growth with limited social and political unrest. By contrast, Latin American countries such as Brazil (until its fiscal reforms, beginning in 1994), where the comparable income differential exceeds a factor of 26, have run larger government budget deficits through financing redistributive programs to the poor that help prevent political unrest while yielding to the demands of the rich to limit taxation. In such circumstances the extreme inequality of incomes limits government fiscal flexibility (see Table 10.3). Inevitably, politics in the face of great income inequalities tends to favor the status quo and works against the innovation needed for manufacturing industries to prosper.

WEIGHT OF POPULATION GROWTH

Finally, it seems that export-oriented industries can make, and have already made, an important contribution to the creation of new employment in certain small countries. However, it would probably be a mistake to believe that these industries can make anything other than a marginal contribution to employment in the periphery as a whole. As of 1980 the World Bank estimated the total number of direct jobs created by export industries in the poorest LDCs as between 2 and 3 million, about 10 percent of total industrial employment in these countries, or, in other terms, less than 0.5 percent of the total labor force of some 850 million people. Even considering multiplier effects, the total number of direct and indirect jobs created came to 5–10 million, around 1 percent of the total labor force. As this labor force of over 1 billion today is growing at an annual rate of 2.0 percent (down from 2.3 percent in the 1980s and 1990s), the number of workers added every year to the total labor force is about twice as large as the entire labor force employed in jobs created by export-oriented manufacturing. Across all LDCs, therefore, export industries cannot possibly absorb the additional labor force arriving each year or even reabsorb those unemployed due to cyclical shifts in demand for export products. Even restricting attention to seven NIEs (Brazil, Egypt, India, Mexico, the Philippines, South Korea, and Taiwan), total jobs created during the 1960s by the export of manufactured goods represented only 3 percent of total employment. However, for small countries with high levels of industrial exports the picture is somewhat different. So in Taiwan in 1969 one job in six was created by manufactured exports and in South Korea in 1970 one job in ten was created by exports of all kinds. These numbers doubled in the 1980s. Moreover, for the city-states such as Hong Kong and Singapore, with their dynamic producer service

Table 10.3 International variations in the concentration of incomes (higher figures indicate greater inequality)

	Year	Ratio of income of lowest 5th : income of highest 5th of the pop.
Low Human Development Index		
Zambia	2004	15.3
Burkina Faso	2003	6.9
Côte d'Ivoire	2002	9.7
Ethiopia	1999	4.3
Medium Human Development Index		
Brazil	2004	21.8
Colombia	2003	25.3
South Africa	2000	17.9
China	2004	12.2
Philippines	2003	9.3
Thailand	2002	7.7
Tunisia	2000	7.9
Indonesia	2002	5.2
Sri Lanka	2002	6.9
India	2005	5.6
Russian Federation	2002	7.6
High Human Development Index		
United States	2000	8.4
United Kingdom	1999	7.2
New Zealand	1997	6.8
Canada	2000	5.5
Germany	2000	4.3
Norway	2000	3.9

Source: Based on UNDP (2007: 281–284, Table 15)

(finance, organization, research) sectors as well as large concentrations of export-oriented manufacturing industries, the figures are even higher. By 2000 in Taiwan, one of the most consistently successful NIEs over the past 20 years, the 22 million population had a GDP per capita (US$17,400) about one-third larger than that of Portugal (US$11,060) or Greece (US$11,960), universal schooling and healthcare, and the world's fourth largest gross international reserves (holdings of monetary gold, special drawing rights, reserves of IMF members held by the IMF, and holdings of foreign exchange under the control of monetary authorities) after Japan, China, and the USA. Taiwanese business is a major investor throughout Asia, contributing to the industrialization of coastal China even though "officially" still on a war footing against the mainland (see later section on China).

It is not, however, only the absence of large dependent peasant and urban populations that gives the smaller states of East Asia distinct advantages in relative job creation from export-oriented manufacturing. These countries have concentrated until recently on labor-intensive export industries, such as textiles, clothing, and electrical and mechanical assemblage. These industries use a great deal of labor per unit of production and this labor is unskilled. They also employ disproportionately large numbers of women whose social position can be exploited to suppress wages and limit union organization. As Joekes (1987: 90) puts it: "Women's lesser education and their expectation (born of past experience) of receiving little training make them apparently suited to unskilled occupations and, most importantly, prepared to stay at such unskilled jobs, however monotonous they may be."

Moreover, these countries initially managed to limit capital-intensive production in order to maximize the return on their higher labor/capital ratios in export-oriented manufacturing. This sets them apart from other NIEs that have much lower capital/labor ratios in export industries and higher levels of capital-intensive production for domestic markets (for example, Brazil). It also leads to an improvement in income distribution as well as employment creation, since the heavy reliance on unskilled labor implies that relatively more of the incomes generated by export industries will go to the poorest segments of the population and less to those classes rich in capital or technical skills. The peculiar historical development of the East Asian NIEs in job creation by export industries, therefore, is not readily duplicated by larger countries—or some smaller ones, for example, Puerto Rico—in which incomes and wealth are divided unequally and in which powerful sociopolitical classes have a vested interest in maintaining the status quo.

10.3 GEOGRAPHY OF INDUSTRIALIZATION IN THE PERIPHERY

As stated earlier, beginning in the late 1960s and 1970s a number of LDCs underwent a rapid process of industrialization, financed in part by the export of investment capital from the developed world. These countries, the NIEs, were the ones where growth was fastest. But growth was not limited to them. What is most important is to grasp the dynamics of industrialization underway since the 1970s. If the 1980s, the 1990s, and the period since 2000 generally have not seen the same overall dynamism, with very important exceptions such as China and India, it is the process of industrialization that demands our attention.

Before examining this, the actual dimensions of the huge transformation in the location of world manufacturing need describing at the outset. Since the 1970s the center of gravity of world manufacturing has begun to shift away from the DC heartlands it had come to occupy in the nineteenth and twentieth centuries. The result by 2010 was a very different ordering of countries in terms of manufacturing as a percentage of GDP (see Figure 10.1). The rapid ascension of countries in East and Southeast Asia to the top spots in the ranking is the most important feature. Of course, it is incorrect to see manufacturing as the sole driver of economic growth anywhere. So, the relatively modest positioning of the United States (and some other countries) in Figure 10.1 should not be interpreted as signaling economic decline or subordination. This is merely a statement of where manufacturing is now located rather than how it is financed or who controls it or its markets.

TRAJECTORIES OF INDUSTRIALIZATION

It is how the shifts in the relative location of manufacturing have taken place or the nature of the trajectories followed in different places that concern us. Figure 10.2 illustrates one way

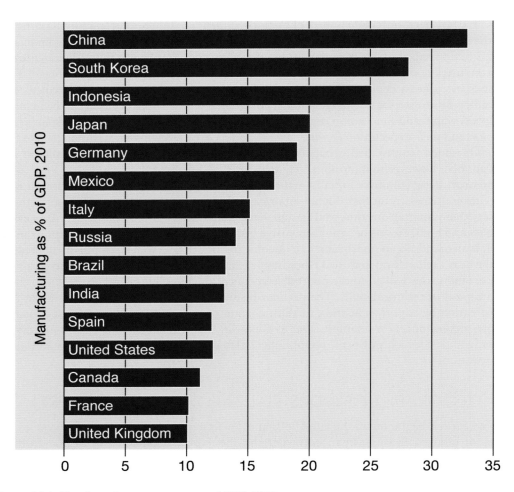

Figure 10.1 Manufacturing as a percentage of GDP, 2010

Source: Based on *Economist* (2012b)

of doing this. This shows the groupings of countries resulting from applying Sutcliffe's (1971) three "tests" of industrialization:

- Test One, at least 25 percent of GDP in industry.
- Test Two, at least 60 percent of industrial output in manufacturing.
- Test Three, at least 10 percent of the total population employed in industry.

This last test measures the impact of the industrial sector on the population as a whole. Seven groupings and two paths to industrialization result from applying the three tests. The groupings are as follows:

A   Fully industrialized countries that pass all three tests.
B   Countries that pass the first two tests but with limited "penetration" into the population or the economy as a whole. These are semi-industrialized countries.
A/B Borderline cases (for example, Greece).

C   Countries that pass the first and third tests. A large industrial sector (in mining or oil) affects the population widely, but manufacturing is weak.

D   Countries that pass the second test only. A small industrial sector dominated by manufacturing.

E   Countries that pass the first test only. A substantial industrial, but non-manufacturing, sector has limited impacts on the population.

O   Other—non-industrialized—countries, failing all three tests.

There are two possible paths to industrialization given this categorization. They are E → C → A, where, for example, a mining enclave expands to involve the total population and manufacturing develops later; or D → B → A, where manufacturing leads to increases in industrial output and later towards a more industrial labor force.

If data for 2004 are used as well as data for 1975, a number of shifts are discernible. A number of D → B → A moves are clearly visible; most of the NIEs (see later) fit this model (see Figure 10.1). The E → C → A path, characteristic in the past of the USA, Australia, and South Africa, and the expected route, perhaps, of the members of OPEC and other mineral-rich economies, is also of importance, but as yet at an early stage. It is clear, however, that, although there are some important examples of industrialization in the periphery, most LDCs are not industrializing, are industrializing quite slowly or are deindustrializing (losing manufacturing industry), a new phenomenon since 1975.

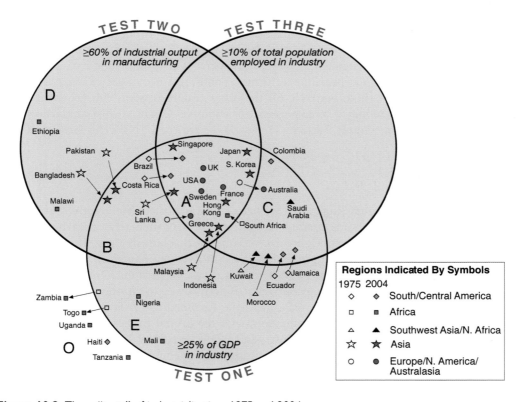

Figure 10.2 Three "tests" of industrialization, 1975 and 2004

Source: Based on online data from World Bank, World Development Indicators http://data.worldbank.org/data-catalog/world-development-indicators

One problem with this type of analysis is that it provides minimum thresholds for defining industrialization but misses the point that after achieving these levels, economies typically begin to lose jobs in manufacturing and experience an expansion of their producer and financial service industries. Manufacturing in itself is then no longer the driving force behind economic growth. This is what has happened in all of the DCs, where services are now a much more important component of their economies than is manufacturing. For example, in the USA by 2000, services accounted for 72 percent of GDP, manufacturing had fallen to 18 percent, and only 15 percent of jobs were in manufacturing.

The conclusion about the general lack of industrialization across the periphery as a whole is reinforced by focusing more directly on rates of growth in manufacturing output. Figure 10.3 shows that the high rates of growth in the period 1995–2005, the years immediately preceding the global economic crisis beginning in 2007 that has probably limited subsequent development, were relatively concentrated geographically. There is no "exclusive" list of NIEs, as different indicators lead to the inclusion of different countries. Also, there has been some volatility over time. The 1980s and 1990s were not particularly good decades for the NIEs as a whole. Latin American NIEs such as Brazil and Mexico fared worse in the 1980s than in the 1990s. The Asian financial crisis impaired the economic performance of many Asian NIEs in the late 1990s. Back in 1979 the OECD recognized ten countries as NIEs (Brazil, Hong Kong, Greece, Mexico, Portugal, Singapore, South Korea, Spain, Taiwan, and the then Yugoslavia) on the grounds of "fast growth of the level and share of industrial employment, an enlargement of export market shares in manufactures, and a rapid relative reduction in the per capita income gap separating them from the advanced industrial countries" (OECD, 1979). In Figure 10.3, only two of these NIEs had manufacturing output growth rates between 1995 and 2005 exceeding 5 percent per year (South Korea (7.6) and Singapore (5.3)), two had 3.6 percent (Spain and Portugal), one had 3.1 percent (Mexico), one had 2.5 percent (Greece), one had 1.3 percent (Brazil), the former Yugoslavia no longer exists and growth rates are not available for the other two. Of the other countries with high growth rates (5 percent or more), some are major oil exporters (for example, Oman (10.6)) or starting from tiny industrial sectors (for example, Vietnam (11.0)). Some are "new" NIEs (for example, China (10.1) and India (5.5)).

EXPORT PROCESSING ZONES

The NIEs such as South Korea, Singapore, Mexico, Taiwan and, predominantly, China, have been especially active in export-oriented industrialization particularly in early stages of developing export-oriented sectors (Figure 10.4). Some of this industry was attracted to and concentrated geographically in **export processing zones (EPZs)** (see Chapter 2). These are limited areas in which special advantages accrue to investors. These include duty-free entry of goods for assembly, limited restrictions on profit repatriation, lower taxation, reduced pollution controls and constraints on labor organization (strikes banned, etc.). The intention of EPZs was to attract local and international companies with the capital to set up factories oriented to export production; and much of the industry in EPZs is owned by TNCs. According to the World Bank (2002), the number of these EPZs mushroomed from only a handful in the 1970s to more than 500 in 73 countries by the late 1990s. Most of the EPZs have clustered in NIEs around major markets: North America, Europe, and Japan. The share of employment in the EPZs is substantial in some LDCs, especially when taking into account that the agricultural and informal sectors still employ a significant percentage of the population. While Mauritius is particularly high at 17 percent, EPZ employment shares are about 5 percent in the Dominican

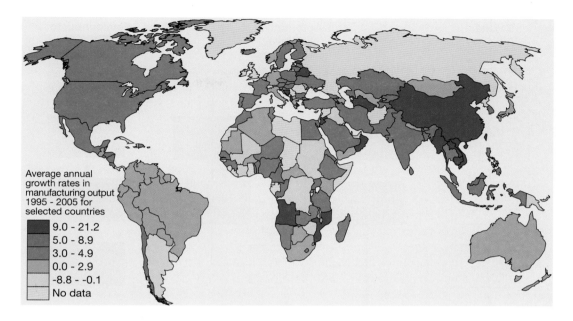

Figure 10.3 The geography of growth in manufacturing output, 1995–2005

Source: Based on World Bank, *Country at a Glance Tables,* available at http://www.worldbank.org/

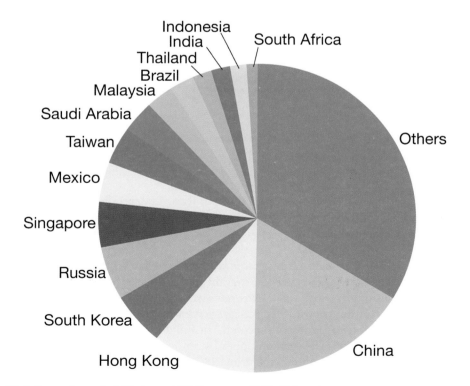

Figure 10.4 Share of certain NIEs in total LDC exports, 2005–06 by value

Source: Based on online data from *CIA World Factbook 2007,* available at http://www.odci.gov/cia/publications/factbook/

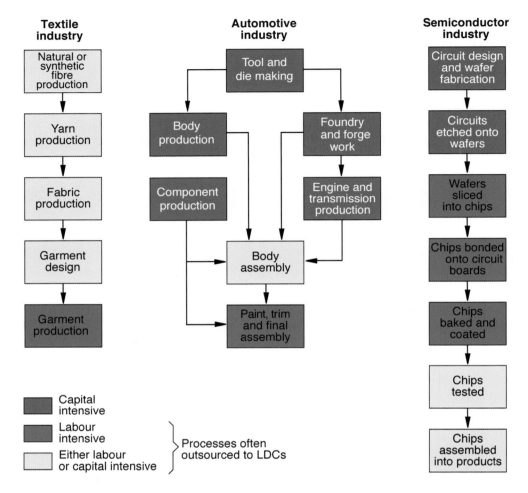

Figure 10.5 Labour processes in three manufacturing industries

Source: Adapted from Crow and Thomas (1985: 49)

Republic and 2 to 4 percent in Honduras and Costa Rica. EPZ production is concentrated in textiles and clothing, microelectronic assembly, and the assembly of cars and bicycles. More specifically, it is the labor-intensive stages of production that are characteristic of the EPZs (see Figure 10.5).

In Mexico by 1978 60 percent of the *maquiladoras*—the assembly plants located in EPZs along the United States border since the program of tax breaks for U.S. business began in 1965—were engaged in electrical and electronic assembly and 30 percent were involved in textiles and clothing. In the Asian EPZs, over 50 percent of the employment is in electronics, with the clothing and footwear industries second. The buildup in electronics has been especially marked since the early 1970s. The total amount of employment in EPZs, however, is relatively small. In Mexico, *maquiladora* employment in 1980 was only around 110,000, although, by 1992, the year before Mexico joined NAFTA, it had grown to 500,000, and by 2001 it exceeded 1 million. An estimated 2.5 million workers are directly employed in EPZs in Asia. Most of the workers are unmarried women between the ages of 17 and 23.

Such young women, earning around US$3.00 per hour, accounted for between 70 and 90 percent of employment in Mexican EPZs in 2008. Within Asia, young women account for 94 percent of EPZ employment in Vietnam, 90 percent in Sri Lanka and Thailand, 87 percent in Malaysia, 77 percent in Taiwan, and 74 percent in the Philippines. Young women are preferred as workers because their wages tend to be lower than men's and because they are considered "nimble fingered" and "more able to cope with repetitive work" (Armstrong and McGee, 1986). Certainly, very little training is required and wages are very low. Often an exploitative trainee system is used in which "trainees" are paid only 60 percent of the local minimum wage and are repeatedly fired and rehired so as to obtain a permanent 40 percent reduction in the wage bill.

Many of the EPZs also do not appear to have generated major multiplier effects. Links to local economies are generally limited especially in Latin America. In the Mexican case the North American Free Trade Agreement (NAFTA) with the United States and Canada has removed the need for the special border EPZs now that the whole of Mexico has become a giant EPZ. However, Scott (1987) has shown that in some Asian cases, the activities of TNCs led to the growth of both "diffusion facilities" owned by local firms engaged in higher level (not solely assembly) operations and locally owned subcontract assembly houses. The former are concentrated in South Korea, Taiwan, and Hong Kong, the latter are found in Thailand, Malaysia, Singapore, the Philippines, Hong Kong, Taiwan, and South Korea. In Malaysia, for example, the presence of foreign TNCs in the electronics industry since the 1970s has contributed to the development of local suppliers (see Figure 10.6). Many of these local companies have been established by former employees who, after acquiring technical and marketing expertise from TNCs, left to set up their own spin-off companies.

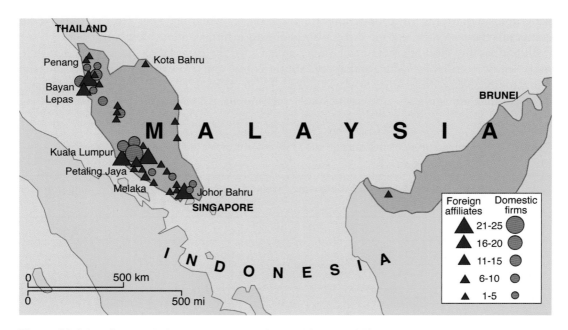

Figure 10.6 Locally owned electronics industry plants in Malaysia, 1999

Source: Adapted from UNCTAD (2001: 74, Box Figure II.4.1)

TRANSNATIONAL INVESTMENT PATTERNS

Foreign direct investment by U.S. TNCs has shifted dramatically over time. In 1969 most investment in electronic assembly was in Hong Kong, South Korea, Taiwan, and Singapore. By 1983 the Philippines and Malaysia had become relatively more important. By 2002 China had become the prime location for U.S. chip plant construction. This kind of foreign direct investment, therefore, is relatively footloose and sensitive, not only to marginal shifts in wage rates, local fiscal conditions, and political instability, but also, in the case of China, to potential markets in populous LDCs, especially after they become members of the World Trade Organization (WTO) which opens them up to common trading rules and provides mechanisms for managing trade disputes. To spread their risks, however, many of the top U.S. TNCs that are engaged in electronic manufacturing operations in Asia have plants or subcontractors in a number of different locations in different countries (see Table 10.4). Intel, for example, has three different manufacturing facilities including one in China, Texas Instruments also has a plant in China plus two in Japan, while one of Micron Technology's two plants is in Singapore. Both Qualcomm and Broadcom outsource all of their manufacturing to Asian subcontractors in locations including China and Taiwan.

The overall global trend of TNCS moving to offshore phases of production to multiple locations to take advantage of labor-cost, infrastructure, tax and other benefits shows no sign of slowing. Even in the aftermath of the 2007 global financial crisis the net trend is still upward for U.S. and European companies (see Figure 10.7). The supply chains of major companies—from Apple and Intel to Benetton and Zara—are now networked globally to take advantage of small differences in costs in order to suppress prices in their increasingly competitive major consumer markets. In those sectors—such as clothing—most susceptible to shifts in labor costs, because they are so relatively labor intensive and substitute technologies are not available, geographical patterns of production are increasingly volatile. This reflects the overall competitiveness of such sectors. But it also relates to macroeconomic considerations.

In textiles and clothing there has been a tendency for the most labor-intensive activities to move away from Hong Kong, Singapore, South Korea, and Taiwan to other parts of Asia. Bangladesh, China, Indonesia, Sri Lanka, and Thailand have become especially important: "The clothing industry uses little capital and is very mobile. All you need is a shed, some sewing

Table 10.4 Manufacturing plants outside the USA owned or leased by top US electronics corporations, 2013

US rank	Company	China	Germany	Ireland	Israel	Japan	Singapore
1	Intel	1		1	1		
2	Qualcomm	*					
3	Texas	1	1			2	
4	Broadcom	*					
5	Micron Technology				1		1

Note: * Fabless company that outsources all manufacturing to Asian subcontractors

Source: company websites

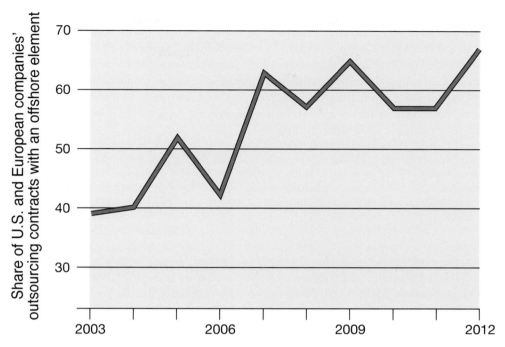

Figure 10.7 Share of U.S. and European companies' outsourcing contracts with an offshore element, 2003–12

Source: Adapted from *Economist* (2013b)

machines, and lots of cheap nimble fingers" (*Economist*, 1987: 67). Partly this has been a product lifecycle effect, as wage rates for unskilled labor have increased in the older NIEs. Perhaps of greatest importance in this geographical shift, however, has been the imposition since 1974 of strict quotas (quantity restrictions) on textiles and clothing by the USA and European countries on imports from established producers in many LDCs. These Multifiber Arrangement (MFA) quotas set businesspeople from Hong Kong and South Korea, the major figures in this industry, searching for countries with higher quotas and low production. Hong Kong has also redirected its production to non-quota high-quality specialized fashion items and used China to fill its own quotas on basic clothing products. China has been allowed a large expansion simply because the DCs are keen to expand their exports of capital goods (for example, steel, weapons) to such a large market and do not want to invite retaliatory action on their exports in the face of a clothing quota decrease.

The Agreement on Textiles and Clothing (ATC) is bringing trade in textiles and clothing in line with the rules of the WTO. In 2005, the MFA—intended as a temporary measure to give the DCs time to adjust to competition from the LDCs—began to be phased out. In the meantime, some DCs have shown a willingness to remove quotas for individual countries that open their markets in return. In 2001, for example, the EU lifted all quotas for textile and clothing imports from Sri Lanka in exchange for tariff reductions and commitments on nontariff barriers to trade. During the 1980s Japanese sources displaced U.S. and European ones as the major investors in the countries of Asia. Even as Japanese foreign direct investment in the USA and Europe followed an incredible growth rate during the late 1980s, Japan's stock of investment in Asia, already higher than that of the USA or Europe by 1985, grew faster than either of the other two. Despite its economic woes during the 1990s, Japan's investment in Asia remained higher than that of the EU or the USA (in 1998, for example, Japan's FDI in East Asia was US$72 billion, compared to US$65 billion from the EU and US$45 billion from the USA). This has led to talk of an East Asian economic bloc forming under Japanese dominance. Certainly, Japan's investment in manufacturing in neighboring Asian countries has involved Japanese companies moving production abroad in a sector in which Japan has been losing comparative advantage. Moreover, increasing Japanese FDI in Asia has not been restricted to investment in manufacturing:

> More recently, increases in FDI have been associated with the rapid growth of spending on research and development by Japanese firms. Some Japanese FDI has also been triggered by policy actions that make it advisable to move production of certain goods to other countries. Trade frictions . . . may have encouraged Japanese firms to invest in Asia so as to build "export platforms" for the U.S. market.
>
> (Bayoumi and Lipworth, 1997: 12)

In fact, the evidence suggests that the region is moving, but only slowly, towards an intra-regional bias in trade and direct investment flows such as is evident in Europe. East Asian trade and investment barriers are being reduced or eliminated as part of voluntary efforts by Asia Pacific Economic Cooperation group (APEC) members (see Chapter 12) and in conjunction with WTO agreements. As yet, however, East Asia in general and Southeast Asia in particular, including such countries as Malaysia, Indonesia, Thailand and the Philippines (despite their membership in ASEAN; see Chapter 12), do not form an incipient trading bloc diverting trade through some deliberate strategy of economic regionalism.

However, there is undoubtedly a degree to which regional proximity to a historically dominant trading and investment partner underpins the success of the old NIEs and the emerging NIEs close by. Most of the successful export-oriented NIEs are adjacent to other ones and at least one important already industrialized economy. But although this is necessary, it is not a *sufficient* condition. Other factors such as underlying productivity conditions, political regime strength and stability, adaptability (rather than factor endowments or resources), and a mix of import substitution and export-oriented industrialization are among the most vital ingredients. The economies of East Asia, including the emerging NIEs of Southeast Asia, have major advantages with respect to all of these factors. It is interesting to see how the formerly centrally planned economies of Eastern Europe have gained competitive advantage in like manner from their proximity to Western Europe and reasonably well-developed infrastructures and well-trained labor forces (as well as membership of the European Union, of course).

REGIONAL LINKAGES AND INDUSTRIAL EVOLUTION

It should be emphasized, however, that although the EPZs and the spatial division of labor within East Asia are symptomatic of the importance of offshore production for final markets in the developed countries, it would be mistaken to see peripheral industrialization solely in these terms. For one thing, local markets for electronic components and semiconductor devices have grown rapidly in East Asia over the past few decades. U.S. factories now ship about 27 percent of their production to consumers in East Asia. This has encouraged the establishment of marketing, sales and after-sales service facilities in the region, especially in Hong Kong and Singapore. These two centers now function as nodes in a global system of producer services located in major world cities. They have also become important global banking and financial centers as a result of their coordinating roles in East Asian manufacturing industry. One factor facilitating this has been the international networks between ethnic Chinese groups in East Asia and North America.

Also of great importance, however, in that local firms have developed a wide range of industries oriented towards producing final products for export. Beginning in the early 1960s, for example, South Korea pursued an aggressive export-oriented industrial policy. In the early 1960s familiarity with manufacturing acquired during the earlier import substitution period was used to develop a number of export **infant industries** by obtaining manufacturing licenses, loan capital and imported business know-how (see Table 10.5). These were mainly labor-intensive activities. But even with low wages it was not until the late 1960s that productivity and quality were high enough to make these industries, textiles, clothing, and footwear, internationally competitive. The South Korean government subsidized this process by using currency depreciations to boost exports, making tax concessions and providing cheap loans.

During a second stage, 1967–1971, other **infant industries** were encouraged as the initial group achieved international competitiveness. These were more technologically advanced industries such as electronic assembly and shipbuilding. In shipbuilding, production went from 25,000 gross tons in 1970 to 996,000 gross tons in 1975; the ships built were also increasingly large and simple (such as supertankers and bulk carriers) with greater potential for automated production. By 1976 the Korean shipbuilding industry was globally competitive.

In the early 1970s a third "wave" of industries was in the process of creation including motor vehicle assembly and consumer electronics. For instance, the automobile industry, which began in South Korea in 1967, produced 83,000 units by 1977 (and 3.5 million units by 1985). By the early 1980s, therefore, South Korea had acquired a wide range of internationally competitive and self-sustaining industries. And, despite the increasing technological sophistication of each wave of innovation, considerable emphasis was still placed on the original labor-intensive industries; although some of these were increasingly decentralized to other Asian locations. The South Korean policy of widening its manufacturing base is the most successful model of the kind of development policy being pursued by all the NIEs. The country's technology-intensive industry drive since the early 1980s has been toward more high value industries, including semiconductors, bioengineering, aerospace, and even robotics (Table 10.5). Peripheral industrialization, therefore, is not just the offshore processing or assembly work for TNCs that a simple focus on EPZs would imply.

AGGLOMERATION AND NEW INDUSTRIAL COMPLEXES

Peripheral industrialization is organized at the intra-national and urban levels as well. Coastal and metropolitan areas have been favored locations, because of infrastructural advantages and ease of external access. Even in countries with little export-oriented industry this pattern is evident. In Nigeria, for example, and despite the transfer of the federal capital in 1991 to the more centrally located Abuja, well over 50 percent of the country's industrial employment remains concentrated in Lagos and five other coastal states. In East Asia and Latin America much of the new manufacturing industry of the past 30 years is found in the major metropolitan areas. Both foreign and indigenous investment tends to be attracted by the amenities, basic infrastructure and political access characteristic of the larger urban areas (see the section headed "Production networks and regional motors" in Chapter 3 for why this is the case). When urban growth is already concentrated in a primate city, such as Bangkok in Thailand, recent growth has tended to reinforce its primacy. This seems to be especially true of foreign direct investment.

Table 10.5 Stages in South Korea's export-oriented industrial development

	1962–1971: Industrial takeoff		1972–1981: Heavy/chemical industry drive		1982–present: Technology-intensive industry drive	
	1962–1966	1967–1971	1972–1976	1977–1981	1982–1991	1992–present
Infant industries	Textiles Clothing Footwear	Electronic assemblies Shipbuilding Steel Chemicals (fertilizers)	Motor vehicle assembly Consumer electronics Special steels Precision goods (watches, cameras) Turnkey plant building Metal products	Automotive components Machine tools Machinery assembly Simple instruments Heavy electrical machinery assembly	Semiconductors Precision machinery	Microelectronics Bioengineering Aerospace Optics Chemicals (fine) New materials Robotics
Industries becoming competitive		Textiles Clothing Footwear	Electronic assemblies Shipbuilding Steel Chemical (fertilizers)	Motor vehicle assembly Consumer electronics Special steels Precision goods Turnkey plant building Metal products	Automotive components Machine tools Machinery assembly Simple instruments Heavy electrical machinery assembly	Semiconductors Precision machinery
Self-sustaining industries			Textiles Clothing Footwear	Electronic assemblies Shipbuilding Steel Chemicals (fertilizers) Textiles Clothing Footwear	Motor vehicle assembly Consumer electronics Special steels Precision goods Turnkey plant building Metal products Electronic assemblies Shipbuilding Steel Chemicals (fertilizers) Textiles Clothing Footwear	Automotive components Machine tools Machinery assembly Simple instruments Heavy electrical machinery assembly Motor vehicle assembly Consumer electronics Special steels Precision goods Turnkey plant building Metal products Electronic assemblies Shipbuilding Steel Chemicals (fertilizers) Textiles Clothing Footwear

Source: Updated from Linge and Hamilton (1981: 33, Table 1.9); Kai-Sun *et al.* (2001, Chapter 3)

The case of Japanese direct investment in East Asia is illustrative. It has been highly concentrated in the national capitals and their immediate vicinity. This is the case whether investment is measured by number of firms, employment or capital, although capital concentration is most pronounced. The Japanese electronics industry has had three major manufacturing agglomerations in East Asia (see Figure 10.8): (1) Singapore and vicinity, including nearby Johor Bahru in Malaysia and Batam Island in Indonesia; there is also significant Japanese investment along the Singapore-Malaysia corridor, which links agglomerations in Malaysia from Kuala Lumpur as far north as Penang; (2) Taipei and vicinity, including Hsinchu Industrial Park and western corridor cities running from Taichung to Kaohsiung; and (3) Hong Kong and vicinity, including major concentrations in Guangdong Province such as Shenzhen, stretching northwards along the east coast of China to South Korea (Aoyama, 2000). In the recent past, China has been the overwhelming beneficiary of Japanese companies outsourcing production from relative higher cost to lower cost locations. More recently, however, Chinese companies themselves have been engaging in outsourcing (see Table 10.6).

In a number of urban areas there are also incipient industrial complexes redolent of Silicon Valley in California and other high-technology complexes in developed countries (also see Chapters 2, 3, and 7). They are made up of both foreign-owned (U.S., European, and Japanese) and locally owned assembly plants, and a surrounding constellation of linked activities. Hong Kong, Manila, Seoul, Singapore, Penang (see Figure 10.5), and Taipei all have such complexes. Scott (1987) used the Manila case to illustrate how a complex can arise in an initial context of low general economic development. The Manila complex originally consisted of a core of nine major U.S.-owned semiconductor branch plants. These were served by 14 locally owned subcontract assembly houses and three specialized capital-intensive "test and burn" facilities; there were also a number of specialized tool and die, and metal shops. Some of these were

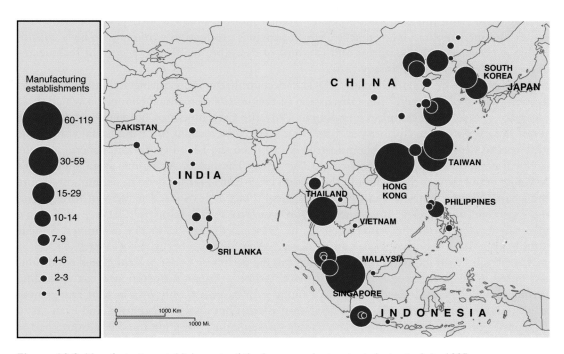

Figure 10.8 Manufacturing establishments of the Japanese electronics industry in Asia, 1995
Source: Based on Aoyama (2000: 228, Figure 2)

Table 10.6 China Mobile Limited's cellphone global production system

- Telecommunications service provider determines system architecture (China Mobile)

- Suppliers of cellphones and components (China, South Korea, Taiwan)

- Suppliers of design platform (integrated device manufacturer (IDM): EU, South Korea, USA; design houses: China, Taiwan, USA)

- IP providers (Taiwan, UK)

- Software providers: Operating system (OS)/printed circuit design technique (MMI)/ graphic user interface (GUI) (India, Taiwan, USA)

- Foundries (China, Singapore, Taiwan)

- Chip packaging firms (China, South Korea, Taiwan)

- Tool vendors for design automation and testing (USA)

- Design support service providers (various Asian countries outside Japan)

Source: Based on Ernst (2009: 25, Table 2)

"captive" to or totally dependent on the U.S.-owned assembly plants but most were independent, local operations. Today, an increasing number of U.S. firms are locating engineering and manufacturing design, call centers, back-office operations, outsourcing centers, and software development in the three information technology (IT) zones in Manila (Eastwood City, Fort Bonifacio, and Filinvest). The U.S. and Philippine companies cluster together to minimize transactional costs and gain joint access to Manila International Airport. They are also at the center of a metropolitan labor market, which provides production workers who have considerable experience with the norms and rhythms of assembly work, and a pool of technicians and engineers with the necessary skills.

Attempts have been made in many LDCs to decentralize manufacturing activities to regional growth poles (see Chapter 7 for a discussion of the models of regional change on which these ideas are based). For example, beginning in 1967 the government of Indonesia established a set of tax incentives for "priority sectors" that were to be located in 11 industrial zones. The idea was to decentralize suitable industries away from the island of Java in general and the political capital of Jakarta in particular. The policy has failed. At a time when Indonesia's MVA growth rate was the highest in the world (1973–1981)—14.6 percent per annum—more than half of both domestic and foreign investment was in Java (56.8 percent of foreign and 64.6 percent of domestic investment). Only resource-based industries have pulled some investment away from the center even in the presence of significant tax incentives. Similar failures have been reported from Latin America, Africa, and other parts of Asia.

The possibility of industrialization throughout the periphery under present global conditions is limited. The process of industrial growth is not a linear diffusion process spreading throughout the globe. Indeed, there are a number of cases not only of a lack of any growth at all (largely, but not entirely, in Sub-Saharan Africa) but also of stagnation and even (as in Argentina) of deindustrialization. The causes of stagnation are numerous: For example, the migration of labor from Yemen to the oil-rich states of the Persian Gulf; the civil war during the late 1970s in Nicaragua; the pursuit of extreme free market monetarist policies by governments in Argentina. Consistent industrialization in the face of global downturns and increasingly competitive world markets seems to require at a minimum, in Petras' (1984: 199) words: "a cohesive industrializing class linked to a coherent policy, promoting an internal market and selective insertion in the international market." This together with a productive household farming sector and government provided financial protections for developing industries are what sets apart those countries that have successfully climbed up the "ladder" of global manufacturing industrialization from those that have not.

10.4 RISE AND FALL OF THE SOVIET MODEL OF INDUSTRIALIZATION

The experience of the former Soviet Union and its Eastern European satellites provides a very different model of industrial development (although in terms of spatial organization, the differences between them and capitalist industrial countries were fewer than might be expected) that was influential throughout the world, particularly in poorer countries, from the 1940s until the 1980s. It should be emphasized at the outset that the economies of the former Soviet Union and its satellites were *not* based on a true socialist or communist form of economic organization in which the working class had democratic control over the processes of production, distribution, and development. Rather, they seem to have evolved as something of a hybrid, in which a bureaucratic class used state power to exploit workers and to compete for power and economic advantage in the world economy. Social objectives aside, their experience can be interpreted as the pragmatic response of latecomers whose leadership felt they could only industrialize by disengaging from their semi-peripheral role in the world economy in order to pursue modernization and economic development through highly centralized and rigidly enforced government direction.

Eventually, the constraints imposed by excessive state control, the inherent disadvantages that resulted from (1) the absence of entrepreneurship and competition and (2) the dissent that resulted from the lack of democracy, combined to bring the experiment to a sudden halt (in 1990). By then, parts of the Soviet world empire had been able to achieve a relatively advanced stage of industrial development; but none of the survivors of the experiment, including Russia, has achieved better than semi-peripheral status in the world economy.

REVOLUTIONARY ECONOMIC REORGANIZATION

Since the time of Peter the Great (around 1700), Tsarist Russia had been attempting to modernize. By 1861, when Alexander II decreed the abolition of serfdom, Russia had built up an internal core with a large bureaucracy, a substantial intelligentsia, and a sizeable group of skilled workers. The abolition of feudal serfdom was designed to accelerate the industrialization of the economy by compelling the peasantry to raise crops on a commercial basis, the idea being that the profits from exporting grain would be used to import foreign technology and machinery. In many ways, the strategy seems to have been successful: Grain exports increased fivefold between 1860 and 1900, while manufacturing activity expanded rapidly. Further reforms, in 1906, helped to establish large, consolidated farms in place of some of the many small-scale peasant holdings. But the consequent flood of dispossessed peasants to the cities created acute problems as housing conditions deteriorated and labor markets became flooded.

These problems, to which the tsar remained indifferent despite the petitions of desperate municipal governments, nourished deep discontent, which eventually, aggravated by military defeats and the sufferings of the First World War, spawned the revolutionary changes of 1917.

It was not the peasantry or the oppressed provincial industrial proletariat, however, which emerged from the chaos to take control. It was the Bolsheviks, a radical element drawn largely from the professional and middle classes, whose orientation from the beginning favored a strategy of economic development in which the intelligentsia and skilled industrial workers would play the key roles.

In the first instance, however, the ravages of war and the upheavals of revolution precluded the possibility of planned economic reorganization of any kind. So the centralization of control over production and the nationalization of industry resulted as much from the need for national and political survival as from ideological beliefs. Similarly, it was rampant inflation that led to the virtual abolition of money, not revolutionary purism. By 1920 industrial production was still only 20 percent of the prewar level, crop yields were only 44 percent of the prewar level and national income per capita stood at less than 40 percent of the prewar level.

In 1921 the New Economic Policy was introduced in an attempt to catch up. Central control of key industries, foreign trade and banking was codified under *Gosplan*, the central economic planning commission. But in other spheres—and in agriculture in particular—a substantial degree of freedom was restored, with heavy reliance on market mechanisms operated by "bourgeois specialists" from the old intelligentsia. Improvement in national economic performance was immediate and sustained, with the result that recovery to prewar levels of production was reached in 1926 for agriculture and the following year in the case of industry.

Soon afterwards, however, there occurred a major shift in power within the Soviet Union. This power shift swept aside both the New Economic Policy and its "bourgeois specialists." They were replaced by a much more centralized allocation of resources: A command economy operated by a new breed of engineers/managers/*apparatchik*s drawn from the new intelligentsia that had developed among the membership of the Communist Party. With this shift there also came a more explicit strategy for industrial development. Like Japan, the Soviet Union chose to withdraw from the capitalist world-system as far as possible, relying instead on the capacity of its vast territories to produce the raw materials needed for rapid industrialization. As in Meiji Japan, the capital for creating manufacturing capacity and the required infrastructure and educational improvements was extracted from the agricultural sector. The foundation of Stalin's industrialization drive was the collectivization of agriculture. This involved the compulsory relocation of peasants into state or collective farms, on which their labor was expected to produce bigger yields. The state would then purchase the harvest at relatively low prices so that, in effect, the collectivized peasant was to pay for industrialization by "gifts" of labor.

In the event, the Soviet peasantry was somewhat reluctant to make these gifts, not least because the wages they could earn on collective farms could not be spent on consumer goods or services: Soviet industrialization was overwhelmingly geared towards manufacturing producer goods such as machinery and heavy equipment. It proved very difficult to regiment the peasants. Requisitioning parties and tax inspectors were met with violence, passive resistance, and the slaughter of animals. At this juncture, Stalin employed police terror to compel the peasantry to comply with the requirements of the Five-Year Plans that provided the framework for his industrialization drive. Severe exploitation required severe repression. Dissidents, along with enemies of the state uncovered by purges of the army, the bureaucracy and the Communist Party, provided convict *(zek)* labor for infrastructural projects. Together, some 10 million people were sentenced to serve in the *zek* workforce, to be imprisoned or to be shot. The barbarization of Soviet society was the price paid for the modernization of the Soviet economy.

ECONOMIC AND TERRITORIAL EXPANSION UNDER STATE SOCIALISM

The Soviet economy did modernize, however. Between 1928 and 1940 the rate of industrial growth increased steadily, reaching levels of over 10 percent per annum in the late 1930s: Rates that had never before been achieved and that have been equaled since only by Japan and China. The annual production of steel had increased from 4.3 million tons to 18.3 million tons; coal production had increased nearly five times; and the annual production of metal-cutting machine tools had increased from 2,000 to 58,400. In short: "An Industrial Revolution in the Western sense had been passed through in one decade" (Pollard, 1981: 299). When the Germans attacked the Soviet Union in 1941 they took on an economy that, in absolute terms (although not per capita), had output figures comparable with their own.

The Second World War cost the Soviet Union 25 million lives, the devastation of 1,700 towns and cities and 84,000 villages and the loss of more than 60 percent of all industrial installations. In the aftermath, the Soviet Union gave first priority to national security. The *cordon sanitaire* of independent east European countries that had been set up by the western allies after the First World War was appropriated as a buffer zone by the Soviet Union. Because this buffer zone happened to be relatively well developed and populous it also provided the basis of a Soviet world empire as an alternative to the capitalist world economy, which provided economic as well as military security.

However, the Soviet Union felt vulnerable to the growing influence and participation of the United States in world economic and political affairs, and Stalin felt compelled, in 1947, to intervene in Eastern Europe. In addition to the installation of the Iron Curtain, which severed most economic linkages with the west, this intervention resulted in the complete nationalization of the means of production, the collectivization of agriculture, and the imposition of rigid social and economic controls. The Communist Council for Mutual Economic Assistance (CMEA or COMECON) was also established to reorganize the eastern European economies in the Stalinist mold—even to the point of striving for autarky for individual members, each pursuing independent, centralized plans. This proved unsuccessful, however, and in 1958 COMECON was reorganized by Stalin's successor, Khrushchev. The goal of autarky was abandoned, mutual trade among the Soviet bloc was fostered, and some trade with Western Europe was permitted.

Meanwhile, the whole Soviet bloc gave high priority to industrialization. Between 1950 and 1955 output in the Soviet Union grew at nearly 10 percent per annum, although it subsequently fell away to more modest levels until the fall of communism. The experience of the east European countries varied considerably, but, in general, rates of industrial growth were high during the Soviet era. Equally important was the structural transformation of industry, for although producer goods remained dominant, the economic base of all Soviet bloc countries expanded to the point where per capita consumption of food, clothing, and other consumer goods came to be much closer to that of the capitalist core countries than to that of the peripheral and semi-peripheral countries of the capitalist world economy.

ECONOMIC GEOGRAPHY OF STATE SOCIALISM

As in western Europe, North America, and Japan, industrialization brought about radical changes in the economic landscapes of the Soviet Union and eastern Europe. But did the state socialism, or statism, that guided Soviet bloc industrialization result in qualitatively different economic landscapes? At face value, there were sound reasons for anticipating substantial differences. Central planning and control of economic development meant that ideological objectives could be translated into administrative fiat, while the absence of a competitive market eliminated risk and uncertainty, precluded the influence of powerful monopolies and facilitated the rapid dissemination of technological innovations.

In practice, however, spatial organization within the Soviet bloc did not exhibit any real distinctive dimension. As in the industrial core regions of the west and Japan, the industrial landscape came to be dominated by the localization of manufacturing activity, by regional specialization, by core–periphery contrasts in levels of economic development and by agglomeration and functional differentiation within the urban system. The reasons for this were several:

1. At the most fundamental level was the unevenness of natural resources and the consequent unevenness of population and economic development inherited from the pre-socialist era.
2. The principles of rationality and the primacy of *national* economic growth took precedence over ideological principles of spatial equality. As a result, Soviet planners applied the logic of agglomeration economies, developing *territorial production complexes*: Planned groupings of industries designed to exploit local energy resources and environmental conditions. As developed by the Tenth Five-Year Plan (1976–1980), these territorial production complexes were broadly defined, designed to foster broad-scale agglomeration economies among specialized sub-regional territorial production complexes.
3. The extensive bureaucracy required by command economies meant that a pronounced "control hierarchy" developed:

   > Central places in the spatial control hierarchy will have disproportionate numbers of high-level business and party posts. This, along with the tendency for the world of culture, the arts and education to concentrate spatially, will create local élites, as in Moscow, enjoying living standards substantially better than those of the mass of people.
   >
   > (D. Smith, 1979: 341–342)

   Conversely, places at the lower end of the control hierarchy offered limited occupational opportunities and limited access to upper level jobs.
4. Centralized economic planning was unable to redress unwanted spatial disparities because large parts of the system came to be characterized by inertia, insensitivity, conservatism, and compartmentalization. As a result, resource allocation was strongly conditioned by *incrementalism*, whereby those places already well endowed by past allocations got proportionally large shares of each successive round of budgeting.
5. Places and regions that were able to establish an initial advantage (by proximity to market or raw materials, for instance) were generally able to maintain a significant competitive advantage over other regions. Turnock, reviewing the outcomes of socialist location principles, observed that: "Despite oft-repeated assertions forecasting the impending elimination of backward regions, through appropriate allocations of investment under the system of central planning, growth rates continue to show wide spatial disparities" (1984: 316).

We can briefly illustrate both the extent of the resultant unevenness in economic development and the degree to which the economic landscapes of industrial socialism, like those of industrial capitalism, were dominated by core–periphery contrasts.

Within the former Soviet Union, the major contrast was always between, on the one hand, the richly endowed, relatively densely peopled, highly urbanized core of the Manufacturing Belt that stretches across Russia from St. Petersburg (formerly Leningrad) in the north and eastern Ukraine in the south, through the Moscow and Volga regions to the Urals and, on the other hand, the rest of the country. Within the latter, there are vast reaches in which physical isolation and harsh environmental considerations have prevented all but a veneer of modern economic development and where tribal folk still pursue local subsistence economies. Much of the rest of what was Soviet Central Asia and the southern portion of Kazakhstan also lagged well behind in terms of economic development, largely because of the fundamental problem of physical isolation. In addition, however, there were parts of the European portion of the former Soviet Union that remained some way behind the levels of development achieved in the center. These included Belorussia, eastern Latvia, Lithuania and western Ukraine—regions with large rural populations that were systematically excluded from Stalin's industrialization drive.

Such inequalities eventually contributed to the vulnerability of the Soviet system as an alternative model of economic development. The critical economic failure, however, was state socialism's inflexibility and its consequent inability to take advantage of the new technology system that was developing among capitalist core countries: "Soviet statism failed in its attempt ... to a large extent because of the incapacity of statism to assimilate and use the principles of informationalism embodied in new information technologies" (Castells, 2000: 13).

The dramatic failure of state socialism led to a period of radical change in the geography of the former Soviet Union with many of the constituent republics breaking away to form separate states. The former states of Yugoslavia and Czechoslovakia have been broken up into smaller entities; East Germany has been absorbed into Germany; and many eastern European countries (including Hungary, Poland, the Czech Republic, and the Baltic states) have been drawn rapidly into the European Union's sphere of influence.

Russia, meanwhile, has had to reconstitute its economic geography through a chaotic transition towards a market economy. In the process, all local and regional economies have been disrupted, leaving many Russian people to survive through a semi-formal "kiosk economy." Many accounts suggest that criminal business has flourished amid the chaos of Russia's transition, with one estimate putting the share of the country's GDP generated by organized crime at nearly 15 percent.

After well over a decade of transition and preoccupation with such pressing economic issues, Russia and the rest of the former Soviet Union have yet to address adequately the legacy of environmental degradation and health problems created by Soviet bloc industrialization (although the countries of eastern Europe, in particular, have benefited from European Union (EU) funding for environmental planning and remediation). In Russia, government funds for social and environmental programs have been limited as overall GDP fell dramatically: By 29 percent in 1992 and a further 13 and 13.5 percent each in 1993 and 1994, 4.2 percent in 1995 and 3.5 percent in 1996. After a slight recovery (0.9 percent) in 1997 GDP fell again (by 5 percent) in 1998 (at a time when Russia devalued the ruble and defaulted on its IMF loan), before recovering to 5.4 percent in 1999 and 8.3 percent in 2000, due partly to the rise in the price of oil and gas, Russia's two main exports. Material production fell, as industrial production declined 9.6 percent throughout the 1990s.

Foreign capital has flowed into Russia, but it has been targeted mainly at the fuel and energy sector, natural resources, and raw materials (which in 2000 accounted for more than two-thirds of Russia's total exports) rather than manufacturing industry. By January 2001 companies from the USA, the leading foreign direct investor into Russia, had invested $5.49 billion in the fuel, transportation, engineering, and communications sectors. As a result, Russia became increasingly dependent on oil and gas reserves in Siberia and further east (including off the coast of Sakhalin Island in the Pacific and along the Arctic Ocean coast), which together accounted for more than two-thirds of Russian exports by 2005. Industry is now almost entirely concentrated in European Russia and low levels of investment in manufacturing because of more attractive outlets in resources and foreign markets mean that the costly forced industrialization of the Soviet Union is now leading to a slow deindustrialization and increased interregional income inequalities with state employment as the only growth factor in many parts of contemporary Russia.

10.5 CHINA'S RISE IN THE WORLD ECONOMY

China is the economic development success story of the last few decades. With the largest population of any country in the world, 1.3 billion, China had a growth in GDP that averaged about 10 percent each year during the 1980s and 1990s respectively (against a 3.5 percent average for all LDCs during each of these decades). In the 1980s and 1990s China's rates of growth in GDP, agriculture, industry, manufacturing, and services were among the highest in the world. The average annual rate of growth in manufacturing was particularly impressive. Only South Korea, Indonesia, and Thailand had growth rates similar to China's 10.4 percent between 1980 and 1990. But these countries did not come close to matching China's 13.9 percent growth in manufacturing between 1990 and 1999. As a result, by 1994, China was already a little more advanced than South Korea had been in 1970 in terms of output per capita of some basic industrial products such as electricity, steel, cement, cotton fabrics, and cotton yarn. China, with one-fifth of the world's population, is now only a generation behind the older NIEs of East Asia in conventional indicators of economic development. As measured by total GDP, by 2013 China already had the world's second largest national economy, after the USA; almost 50 percent that of the USA. In per capita GDP terms, a more relevant comparison, China's economy is still relatively underdeveloped, ranking only 94th or so compared to all countries. This suggests the irony of the current situation: The world's most dynamic manufacturing hub is still a poor country. Unless the Chinese economy can successfully shift from an export-oriented to a national consumption model this is likely to continue but with serious social consequences in terms of keeping average wages down at the same time that social and geographical inequalities are growing and increasingly obvious.

Much of the growth of manufacturing has been concentrated in coastal China, especially in the zones around Shanghai and Hong Kong that have been opened to foreign investment and in the vicinity of the capital, Beijing (see Figure 10.9). Originally experimental, as the communist leadership in Beijing worked out a new model for maintaining political control over China while trying to absorb foreign capital, technology and management practices, the special economic zones (SEZs), such as Shenzhen in Guangdong Province on the border with Hong Kong, have become the nodal points for reforming the Chinese economy as a whole. Under the leadership of Deng Xiaoping from 1978, China embarked on a thoroughgoing reorientation of its economy away from the autarchic China of the 1960s when agriculture and national economic self-sufficiency were the priorities. The new model involved privatizing collective farms, closing or privatizing parts of state industry, dismantling central planning in favor of private entrepreneurship and market mechanisms, and fully integrating China into the world economy. Economic growth has been elevated above class struggle.

Some 54 percent of Chinese, 657 million people in 2011, still live on the land. The first economic reforms concentrated on privatizing peasant agriculture, taking a leaf out of the book of their East Asian neighbors Japan, Taiwan, and South Korea in privileging rural land reform. This increased farm production and rural incomes tremendously, providing capital for industrialization and improvements in infrastructure in some restricted areas. "World standards" were brought to bear on the economy through the establishment of SEZs to attract foreign technology and management. The dramatic success of Guangdong Province in southern

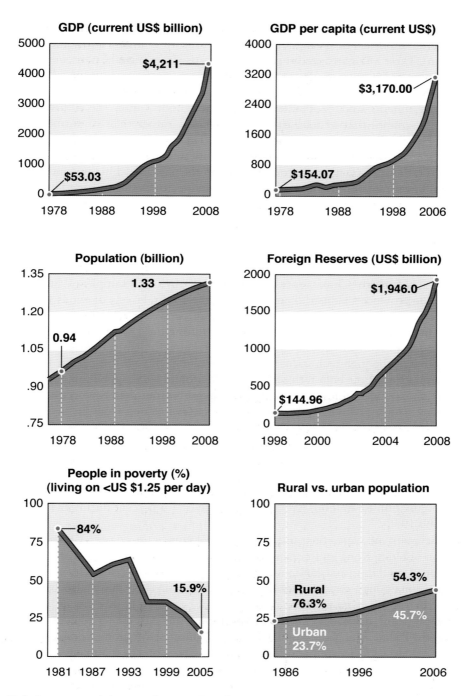

Figure 10.9 Economic and demographic growth in China

Source: Based on Wheatley (2009)

China transfixed managers and bureaucrats elsewhere. In 1979, three of the original four SEZs in China were established in Guangdong (with the fourth in Fujian Province across the strait from Taiwan). These SEZs attracted capital and expertise from outside, particularly from Hong Kong and Taiwan. Much of the investment was in joint ventures or inter-corporate alliances (such as those referred to in Chapter 3) between foreign firms and local enterprises operated by local governments and cooperatives. But throughout China the industrial structure increasingly resembles that of its East Asian neighbors rather than its former "ideological friends" in Eastern Europe and the former Soviet Union; Chinese manufacturing output is now dominated by a large number of small firms with mixed state and private (often foreign) participation rather than giant state companies as in the past. As China's economy transitions to an overall lower rate of growth in its manufacturing sector as this sector adapts increasingly to domestic rather than foreign demand for its products, the overall share of manufacturing in total GDP is slowing relative to that of services (see Figure 10.10). This signifies the maturing of the Chinese economy with a future perhaps somewhat less dependent on export-oriented manufacturing.

Nevertheless, manufacturing has been the driver of China's economic transformation. Guangdong in southern China represents the most obvious example of industrial transformation. The province's industrial output during the 1980s, largely of goods such as clothes, shoes, electrical appliances and toys, rose 15 percent per year (for an example, see Box 1.2 on Barbie). Industrial output continued to increase during the 1990s, by 16 percent annually. The promotion of high-tech sectors by the provincial government translated into an increase in output value of the high-tech industry of 50 percent each year during the 1990s. In 1999 the production value of the high-tech sector reached around US$25.6 billion, representing 13.9 percent of Guangdong's industrial output. Total exports from the province, funneled mostly through Hong Kong, accounted for 77.8 percent of Guangdong's output in 1999 and 36.9 percent of China's exports by 2000. During each year of the late 1990s, the province chalked up over US$100 billion annually in foreign trade, making it the top province in China as measured by foreign trade volume. With more than 86 million people, Guangdong experienced a growth in GDP of around 12.5 percent per year in the 1980s and more than 9 percent in the 1990s. This has continued if at somewhat lower rates in the 2000s. For comparison, Thailand (population 61 million) had a GDP growth of 7.6 percent during the 1980s and 4.7 percent in the 1990s.

From one point of view, this and similar development in coastal China represent the reintegration of Hong Kong and Taiwan with their continental hinterlands after a long period of separation. Low labor and land costs allied to high labor productivity have been major attractions. China's high export quotas to the EU and the USA have also been important. But more than "pure" economics has been at work. The surge of investment has been led by ethnic Chinese within networks that stretch from south and east China all over Southeast Asia and around the Pacific Rim. Common languages—Cantonese for Hong Kong and Guangdong, the Fujianese dialect for Taiwan and the closest provinces on the mainland—and a common culture eased the flow of money, managers, and trade. Taiwan and China are officially still on a war footing, Taiwan being the refuge for the nationalist government after the communist revolution in 1949 and until the past few years economic relations were carried on through Hong Kong. But in 1992 China displaced Malaysia to become the biggest single destination for Taiwanese foreign direct investment. With the spillover of investment and trade to the mainland, China is following economically where Taiwan led. And in 2001, under mounting pressure, the Taiwan government rescinded its 52-year-old ban on direct trade and investment with the mainland.

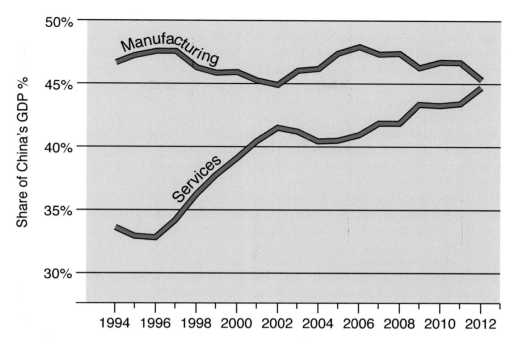

Figure 10.10 Percentage of China's GDP from industry and services
Source: Based on *Economist* (2013c)

The euphoria engendered by the recent economic transformation of China should be tempered somewhat by recognizing a number of serious barriers to the continuation of present trends. One is the poor condition of the transport infrastructure, particularly the railways in the interior: China has one of the world's smallest railway networks relative to population and arable land. The development along the south and east coasts, while supported by central government as a growth pole policy, has created significant and still growing disparities in growth and incomes between coastal and interior regions as the advantaged regions build up their external links. This spatial polarization has already generated internal political conflict. The millions of rural workers who migrated to coastal cities to find a job and a better standard of living have added to the growth pressures in these localities. Regional "resource wars" have erupted when some regions implemented questionable administrative and other measures in an effort to restrict interregional resource flows. Central and local government bargaining over development policies and resource allocations have become particularly heated when some disadvantaged areas felt they were receiving less favorable treatment by central government. As a result, reducing regional inequality, which is seen as a threat to China's future prosperity, stability and unity, has been propelled to the top of the central government's policy priorities. Be that as it may, poverty in China remains a major problem, with at least 15 percent of the population living below the national poverty line. This is estimated to be around 195 million people. Their condition contrasts markedly with that of those who have cashed in on the opening up of the Chinese economy (see Figure 10.9).

A second problem is the continuing drain on central government revenues of state-owned firms. In particular, coal and oil companies are important loss makers because of government insistence on subsidized energy prices. As a side effect this also contributes to the massive air pollution afflicting Chinese cities. Third, local governments have acquired considerable autonomy in pushing credit expansion and investment. This has led to an overheating of the national economy as expanding credit chases a shrinking money supply controlled from the center. Fourth, for all of the success in creating a private economy there is still a level of government regulation without a guaranteed rule of law, which makes for an arbitrary application of rules and encourages corruption among civil servants. Finally, China's manufacturing economy is increasingly dependent on importing resources from and exporting capital to elsewhere. This has led to charges that China is "buying the world." Although Chinese resource companies have been very active in land grabs and locked-in resource contracts in some places, particularly Sub-Saharan Africa, much Chinese investment is now following in the tracks of that of other economic "powers": To the businesses and stock markets of the world's financial centers such as London and New York. So, still a relatively, poor country, China has become a major source of capital to the "rich world" of the long established DCs.

At the root of many of these problems is the lack of political change paralleling the economic change. Unlike the countries of Eastern Europe and the former Soviet Union, which have undergone political change prior to economic reform, China has created an increasingly capitalist economy within a still formally socialist state. The consensus view among professional "China watchers" is that the Chinese Communist Party is now largely irrelevant to economic life and, as the revolt of the movement for greater political democracy that was violently suppressed in Tiananmen Square in Beijing in 1989 showed, its monopoly of state power is

Box 10.1 Dreaming the BRIC future

In 2007 economists at the U.S. investment bank, Goldman Sachs, predicted that over the next 40 years, Brazil, Russia, India, and China—what they termed the BRIC economies—could become a major force in the world economy, primarily as manufacturing centers (particularly in the case of China and India) but also as major resource economies (particularly in the case of Russia and Brazil). Using a variety of economic indicators, the Goldman Sachs economists projected GDP growth and incomes per capita for the BRICs to 2050. Although incomes are likely to be lower than the average for individuals in the G7 (the USA, Japan, Germany, France, UK, Canada, and Italy), GDP growth rates would be much higher so that by 2050 the BRIC economies together would be larger than the G7 (see Table 10.7). Relative growth in the size and skills of labor forces as well as the overall competitive advantage of the BRIC economies in crucial sectors are the motors driving this vision of the future. The major overall assumption is that the BRICs must at least maintain their present levels of openness to the world economy and follow development policies that privilege their respective competitive advantages, such as Russia in natural gas, Brazil in agribusiness and certain manufacturing sectors, India in manufacturing and producer services, and China in manufacturing. Of course, declining populations (particularly as dramatic as that in Russia), slower growth in the economies to which exports would be directed (such as Europe, the USA and Japan), and persistent problems such as massive underinvestment in road and other public infrastructure in India, rapidly increasing income inequalities between regions as in China and Brazil, massive corruption (in all countries), and the lack of symmetry between the shifts in the location of basic economic activities and the persistent dominance of the G7 countries (particularly the USA) in financial and monetary matters will limit the possibilities for a smooth transition to a truly new world economic order in which the core expands spectacularly to include the BRICs.

Table 10.7 The BRIC road to economic growth

	BRICs				G7								BRICs	G7
	Brazil	Russia	India	China	France	Germany	Italy	Japan	UK	Canada	US		BRICs	G7

Actual and projected GDP (constant US$ bill)

	Brazil	Russia	India	China	France	Germany	Italy	Japan	UK	Canada	US	BRICs	G7
2010	1,346	1,378	1,264	4,696	2,366	3,086	1,927	4,602	2,568	1,395	14,537	8,684	30,481
2030	3,720	4,269	6,748	25,652	3,306	3,764	2,407	5,814	3,627	2,071	22,821	40,389	43,810
2050	11,366	8564	38,277	70,605	4,592	5,028	2,969	6,675	5,178	3,164	38,520	128,812	66,126

Actual and projected GDP per capita (constant US$)

	Brazil	Russia	India	China	France	Germany	Italy	Japan	UK	Canada	US
2010	6,882	9,887	1,067	3,484	38,380	37,504	33,165	36,182	41,909	40,737	47,022
2030	16,694	34,402	4,403	17,521	52,327	47,301	43,479	49,959	56,398	52,918	62,727
2050	49,759	78,435	21,145	49,576	75,253	68,308	58,930	66,825	80,942	76,370	91,697

Source: Based on data in Goldman Sachs (2007: 149, Appendix)

unacceptable to many. As one commentator expressed it: "Communism in China will probably end not with a bang but a whimper" (MacFarquhar, 1992: 28). The capitalist genie unleashed in 1978 is probably too well established now to put back in the lamp of communist autarky.

Such internal political conflicts aside, China's rapid economic growth has produced a degree of political hysteria in some other countries, not least in the United States. Yet, access to China's presumably gargantuan market has long been the prize most celebrated in the U.S. idea of a Pacific Century. Now, however, the emergence of China as a major supplier of consumer goods to the USA, the huge buildup of U.S. dollar reserves by the Chinese Central Bank (see Figure 10.9) and the presumed base its high rate of economic growth will give China in the political–military competition of Great Power politics have conspired to produce an increasing sense of threat in some quarters. The bilateral trade deficits between the USA and Europe, on the one hand, and China, on the other, attract particular hostility when, in fact, this hostility reflects calculations that ignore the complexity of contemporary production chains and the imbalances between currencies, which, in turn, reflect high Chinese savings and considerable U.S. and European profligacy or consumer spending beyond available incomes. At the same time, China is seriously overrated as both an economic and a political challenge. At least during the Cold War, China was a "beacon for many in the LDCs. China now is a beacon for no one, and, indeed, an ally to no one. No other supposedly great power is as bereft of friends" (Segal, 1999: 25). China's reliance on U.S. and other foreign export markets, the close linkage of the Chinese currency, the yuan, to the U.S. dollar, and the heavy dependence of Chinese economic growth until recently on foreign investment and technology now all limit its external leverage. U.S. critics of China's rise should rest more easily.

SUMMARY

According to the World Bank (1983), the aggregate rate of growth of both industrial (mineral resources plus manufacturing) and manufacturing output were over 3 percent per annum for 34 low-income countries (for example, Bangladesh, Cambodia, and Sierra Leone) and over 6 percent per annum for 59 middle-income countries (for example, Mexico, Oman, and South Korea) over the period 1960–1981. These rates were higher than those for the DCs and, consequently, the share of the LDCs in the world's manufacturing output rose somewhat: From 17.6 percent in 1960 to 18.9 percent in 1981. Their combined share of world exports of manufactures rose from 3.9 percent to 8.2 percent in the same period. Or, from a slightly different perspective, the LDCs' share of the manufactured imports of all industrial countries rose from 5.3 percent in 1962 to 13.1 percent in 1978. At the same time, the share of the GDP of the low-income countries coming from the industrial sector rose from 25 percent in 1960 to 34 percent in 1981. For manufacturing the rise was only from 11 to 16 percent. In the middle-income group the changes were from 30 to 38 percent for industry, and from 20 to 22 percent for manufacturing alone. Much of this growth was concentrated in a relatively small group of NIEs such as South Korea and Taiwan.

With some notable exceptions to the general trends, including China, the East Asian NIEs, and some Southeast Asian countries, the growth of manufacturing industry in the periphery and its penetration of DC markets did not deepen much in the 1980s and 1990s, even though it spread somewhat beyond the older NIEs. Annual rates of growth in manufacturing averaged 4.9 percent in the 1980s and 5.8 percent in the 1990s for all LDCs (7.7 percent average in the 1980s and 2.7 percent in the 1990s for the low-income countries, and 4.6 percent in the 1980s and 6.3 percent in the 1990s for the middle-income ones). The OECD (major industrialized) countries averaged 3.3 percent per annum through the decade of the 1980s and 3.6 percent annually between 1993 and 2002. At the same time the LDCs' share of the manufactured imports into all industrial countries sank to 12.4 percent in 1999, having fallen from 12.9 percent in 1990, and 13.1 percent in 1978. In the 2000s the story has been one of significant manufacturing growth largely in East Asia and in the BRICs. So, even though the global manufacturing picture today is radically different from that in the 1960s, it is one in which only certain specific parts of the world have been added to the older map.

These figures call into question the generally optimistic conclusion of Warren (1980) and others (see Chapter 8) that a massive industrialization of the periphery was under way. Indeed, this chapter has suggested that there are significant constraints on the industrialization of the periphery and the 1970s to 2000 provided a time period uniquely favorable to the type of development that did take place. Indeed, even in that time period a number of economies stagnated or deindustrialized (for example, Argentina). The economic problems of the industrial core have not created inevitable advantages for industrialization in the global periphery. Any advantages have been created there rather than simply given from the outside. Indeed, as technology substitutes for labor in many manufacturing sectors, manufacturing can be reshored back to the DCs.

Some of the major conclusions are as follows:

- Industrialization plays a major role in national ideologies of modernization and is seen as a solution to various major practical economic and social problems.
- The spread and intensification of industrialization since the late 1960s coincided with the declining rate of profit for industrial companies in the industrial core.
- Much of the new industrialization is export oriented rather than directed (as in import substitution) to domestic markets.
- There are limits to the development of this industrialization: Such as improving profit rates in the industrial core, protectionist measures in export markets, technological changes that reduce the attractiveness of low-wage locations, incredible debt loads, and relatively limited employment effects. Most LDCs are either not industrializing or are industrializing only very slowly.
- Export processing zones (EPZs) represent one geographical form taken by the new international division of labor (NIDL), but offshore production of components or assembly is not the only feature of peripheral industrialization. There are now important industries engaged in the production of locally created final products.
- The new industrialization has favored existing metropolitan areas and coastal regions. Attempts at decentralization to growth poles have not met with much success.
- The profile of China's industrialization reflects a conscious decision by the Chinese government to re-engage with the world economy after a long period on its margins. As a geographical extension of the existing NIEs of East Asia, China represents the main wave of new industrialization in the world economy during the last few decades. This was anything but a global trend. But some commentators suggest that China's rapid growth will soon be followed by that of some other very large semi-peripheral economies (the three other BRIC economies) that benefiting from economies of scale also have resources and capacities that could lead to a major shift in the balance of power within the world economy.

KEY SOURCES AND SUGGESTED READING

Agnew, J.A., 2005. A New 'Pacific Century'? North America and the Pacific Rim, in *The USA and Canada 2005*, 7th edn. London: Europa.

Chen, X., 2005. *As Borders Bend: Transnational spaces on the Pacific Rim*. Lanham, MD: Rowman & Littlefield.

Economist, 2006. The Physical Internet: A survey of logistics, June 17.

Economist, 2011. Let a Million Flowers Bloom: Entrepreneurship in China, March 12.

Economist, 2013. China's overseas investment, January 19.

Goldman Sachs, 2007. *The BRICs and Beyond*. New York: Goldman Sachs.

Naughton, B. 2007. *The Chinese Economy: Transitions and growth*. Cambridge, MA: MIT Press.

Studwell, J., 2013. *How Asia Works: Success and failure in the world's most dynamic region*. London: Profile Books.

UNDP, 2007. *Human Development Report: Fighting climate change, human solidarity in a divided world*. New York: United Nations.

UNIDO, 2005. *Industrial Development: Global report*. New York: United Nations.

Wolf, M., 2006. The Answer to Asia's Rise is not to Retreat from the World, *Financial Times* March 15, 15.

Wood, T., 2007. Contours of the Putin Era, *New Left Review* 44, 53–68.

Chapter 11

Services: Going global?

Picture credit: Wikipedia

Concerns about outsourcing have received significant attention in the media and in political circles in developed countries such as the United States and the United Kingdom. Outsourcing in general, and the outsourcing of services in particular, has provoked a vocal reaction on the part of some politicians, unions, workers, and others. While the earlier trend associated with deindustrialization and the outsourcing of traditional manufacturing activities from the developed countries was treated initially with the same negative response, the first reports of service outsourcing were greeted not only with similar concern but also with profound shock. Unfounded as it was, the conventional wisdom had been that the developed countries would exploit their competitive advantage in service activities—impervious as they were to outsourcing—to more than fill any employment void left by outsourced industrial activity.

This misconception can partly be traced back to the 1940s and the Fisher-Clark thesis which suggested a three-sector division of economic activities into agriculture, industry and services (see Figure 11.1). This thesis maintained that increasing wealth over time in an economy will be associated with a shift from agriculture to manufacturing and then to service employment because wealthier societies consume more services such as entertainment, education and healthcare. Figure 11.2 shows this positive relationship between levels of per capita income and percentages of workers employed in services. Rich core countries such as Australia, Japan, Norway, the United Kingdom, and the United States tend to be concentrated at the high end of the graph while poor peripheral countries such as Bhutan, Cambodia, the Kyrgyz Republic, Liberia, and Uganda are at the low end of the graph.

Viewed as non-tradable—needing to be both produced and consumed in the same location—services were seen, therefore, as immune to outsourcing. Certainly, a look back at Figures 7.5 or 7.6 shows the strong and steady growth in service employment in absolute and percentage terms in developed countries such as the United States and the United Kingdom; the percentage of the workforce in service jobs has risen to about 80 percent in these countries, with most

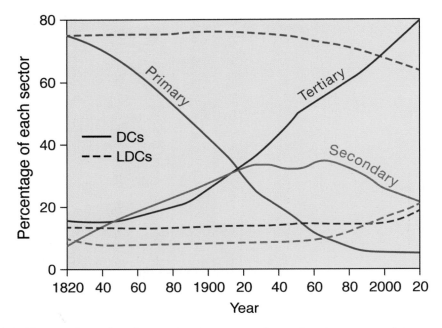

Figure 11.1 Changing share of employment in primary, secondary and tertiary sectors of the economy
Source: Adapted from Rubenstein (2005: 301, Figure 9–3)

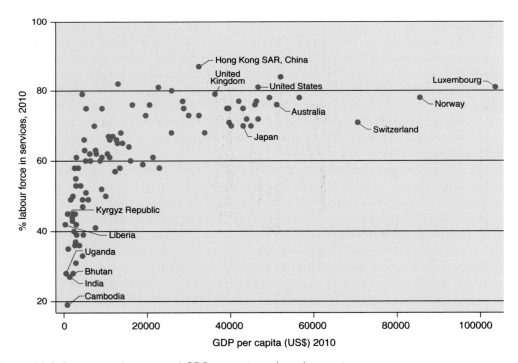

Figure 11.2 Service employment and GDP per capita, selected countries
Source: Based on World Bank, online World Development Indicators 2010; IMF, World Economic Outlook online Database 2010

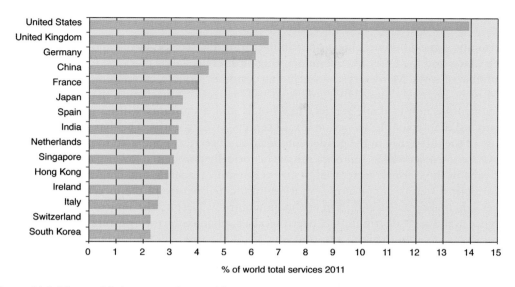

Figure 11.3 The world's largest service providers

Source: Based on World Trade Organization, WTO Statistics Database online 2011

new jobs being added in the service sector. Developed countries such as France, Germany, Japan, the United Kingdom, and the United States also rank highly in terms of the percentage of the world's total services produced (see Figure 11.3).

During the last few decades, however, advances in information and communications technologies, in association with the profit-seeking strategies of both large and small corporations, have resulted in an increasing amount of services becoming tradable—capable of being outsourced and produced in one location for consumption in another. In addition, the rather simplistic notion of a straightforward sequential shift from manufacturing to services is no longer valid given such insights as the interdependence of manufacturing and service activities, and the experience of LDCs such as China where manufacturing and service employment growth has been concurrent rather than sequential.

In this chapter, the shift to services and the contemporary geography of services in both the LDCs *and* DCs, and the interactions between them, are examined in six complementary ways. First, we begin with an attempt to define and theorize services, including the issue of whether the distinction between services and manufacturing is in fact redundant. Second, the national and global stimuli to service growth are described. Third, the benefits and drawbacks of service outsourcing for both the DCs and the LDCs are discussed. Fourth, important national and global constraints on the growth of services faced by certain LDCs are highlighted. Fifth, the geographical pattern of services is surveyed at global, regional, and urban scales. Sixth, a variety of services are profiled, including international retailing, tourism, financial services, and business services.

11.1 DEFINING AND THEORIZING SERVICES

The Fisher-Clark thesis encouraged a tendency to focus on the differences between manufacturing and services. Consequently, services have conventionally been defined negatively, comprising what remains after agriculture, mining, and manufacturing are excluded (CRIC, 2006).

Certainly, some differences between manufacturing and services have important implications for the LDCs and the DCs. The manufacturing and service labor markets, for example, differ in a number of ways. Many services—including customer services, entertainment, and education—tend to be relatively more labor intensive and less easy to mechanize than manufacturing. Labor-intensive non-tradable services like hotels and restaurants will be more secure from outsourcing than tradable services like call centers or data entry. This has implications also because of the difference in the gender composition of services compared to manufacturing. In the DCs, women, as well as minorities, have historically dominated lower paid so-called pink-collar jobs in the clerical, secretarial, retail, restaurant, teaching, and childcare fields. This is also an issue in terms of concerns about the bifurcated nature of the income distribution in services when contrasted with that for manufacturing. Despite the limited evidence for any difference in the wage distribution between services and manufacturing (Stutz and Warf, 2007), the loss of factory jobs in manufacturing which had allowed a middle-income lifestyle, and the bifurcation between high-skilled high-paying service jobs in producer services such as finance or research and development (R&D) versus low-skilled low-paying service jobs in restaurants and hotels, have been seen as promoting a polarization of income groups in the DCs.

The distinction between services and manufacturing also has implications for measuring and studying services. For example, should services be measured as a set of industries or a series of occupations? A secretary who works in a factory and one who works in a bank may have similar duties. Categorized by industry, the secretary in the bank would be in services and the one in the factory would be in manufacturing; categorized by occupation, however, both secretaries would be in the service sector.

Although there is no generally agreed on definition, there is general agreement on the major components of the service sector (Bryson, Daniels, and Warf, 2004: 7). The following seven components of the service sector can also be further categorized into private (marketed) and public (non-marketed) sectors. As a result of the private sector's concern for the profit motive of business operations, the internationalization of services has mostly entailed marketed services (Bagchi-Sen and Sen, 1997):

1. **Finance, insurance, and real estate (FIRE)** including commercial and investment banking, insurance of all kinds (property, medical, casualty), and the residential and commercial real estate business.
2. Business services including legal services, advertising and marketing, public relations, accounting, research and development, personnel training, recruitment, architecture and engineering, and consulting.
3. Transportation and communications, including electronic media, trucking, shipping, railroads, airlines, and local transportation (buses, taxis, etc.).
4. Wholesale and retail trade, including major wholesalers that supply major retailers. Eating and drinking establishments, personal services, and repair and maintenance businesses are closely affiliated.
5. Entertainment, hotels, and motels—part of tourism.
6. Public services at all government levels—only recently viewed as part of the service economy—including public servants, the armed forces, public school teachers, public healthcare professionals, and police and fire departments.
7. Nonprofit services, including churches, charities, museums, and nonprofit healthcare agencies.

The absence of a generally agreed on definition—given that many consumers in the DCs especially use a variety of services in their everyday lives—does not prevent most people from having some idea of what is meant by services. Conventionally, again following the Fisher-Clark thesis, services have been defined—in contrast to tangible manufactured goods—as involving the production and consumption of intangible inputs and outputs. What, for example, is more intangible than the voice of a teacher sharing knowledge or the touch of a doctor carrying out a medical examination on a patient? But the distinction between tangible manufactured goods and intangible services is not clear-cut. Many services come with tangible elements—what about the textbook written by the teacher or the injection given by the doctor?

Many manufactured goods provide a service; washing machines wash clothes, microwaves cook food, an automobile provides transportation. There are few manufactured goods that do not involve services in their production; indeed, an increasingly integral part of the manufacturing process depends on services such as research, design, and marketing. Similarly, most services depend on manufactured goods; for example, an airline flight requires a reservation and security check at the airport certainly, but making a reservation requires a computer and, since 9/11 especially, the security check requires an increasing amount of machinery to scan travelers and their luggage. The distinction between services and manufacturing has come to be seen increasingly as redundant:

> Neither manufacturing nor services is inherently better than the other; they are interdependent. Computers are worthless without software writers; a television has no value without programs ... Before long no one will care whether firms are classified under manufacturing or services. Future prosperity will depend not on how economic activity is labeled, but on economies' ability to innovate and their capacity to adjust.
>
> (*Economist*, 2005: 82)

Howells' (2003) notion of service encapsulation of goods and materials is useful for understanding how services are increasingly incorporated into manufactured products. Over time, many manufactured products have come to be offered not in their own right to consumers, but in terms of their wider service attributes. This has occurred in two ways. First, the manufactured product can be offered along with closely aligned service products in a single package. Figure 11.4 shows a variety of services that may be sold with a manufactured product over its lifetime, including those involved in purchasing and arranging delivery of the product, maintenance and repair, related support activities, and repurchase, disposal, or recycling.

Second, instead of buying a manufactured product in a single one-time purchase, a consumer can buy the service which the manufactured product provides as part of a continuing process involving long-term customer contact through service delivery. In the case of an automobile, for example, a customer can lease a car and use the vehicle without buying it. Another example, from the computer industry, is where, instead of purchasing computers to carry out certain tasks, a company pays for cloud-computing.

Encapsulation is a particularly helpful concept because it illustrates the interdependence between manufacturing and services; each can produce innovations not only in its own sector but also in the other. A manufactured product can generate service innovations. The cranes made by Liebherr, for example, now come with special software programming to better control and run these machines. Similarly, services can promote innovations in manufacturing, for example, through product improvements based on feedback from market research surveys.

Consequently, rather than defining services in terms of what they are not, or even in terms of what they are, a potentially better way to define services would be to ask what is changed

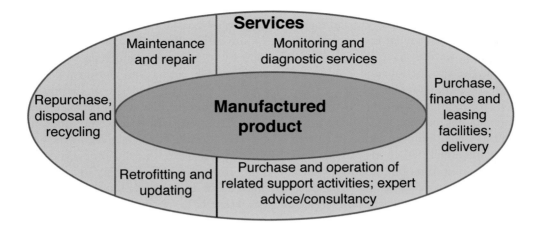

Figure 11.4 Service encapsulation

Source: Adapted from Howells (2003: 10, Figure 1)

by the service and how. Whereas manufacturers change raw materials and energy into products, services physically, spatially or temporally change a person, manufactured product or information. Some services transform people, as when a customer receives a haircut from a hairstylist or an operation in a hospital to remove an appendix. Other services can change manufactured products by repairing them or by transporting them from one location to another. Still other services, such as financial services, transform information, such as the balance in a person's bank account (CRIC, 2006).

11.2 NATIONAL AND GLOBAL STIMULI TO THE GROWTH OF SERVICES

Bryson *et al.* (2004: 11–14) suggest six main forces that drive the growth of services. These forces operate at a number of spatial scales, including national and global, involve a range of contemporary and historical factors, and play out differently depending on national and local contexts.

RISING PER CAPITA INCOMES

Rising incomes in most core and many semi-peripheral countries have contributed to an increase in service employment. Services have a high elasticity of demand—increases in personal income generate a significant increase in domestic demand for services. As per capita income rises, people tend to try to minimize the time they devote to everyday tasks. In core countries, for example, many people who can afford to will pay for services such as lawncare or home-cleaning; they may also eat in restaurants more frequently instead of cooking at home. The growth of tourism has been fueled by the consumption of services with a particularly high elasticity of demand, such as transportation, hotels, and entertainment.

GROWING DEMAND FOR HEALTHCARE AND EDUCATIONAL SERVICES

The growth in domestic demand for healthcare and educational services, particularly in core economies, has also contributed to the growth in services. Demographically, the aging of the population in many core economies has meant that demand for healthcare has risen on the part of middle-aged and elderly people who require relatively high levels of medical care. At the same time, factors such as the changing labor market, the need for advanced skills in the workplace, and the possibility of taking classes online, to name a few, have resulted in greater demand for educational services.

AN INCREASINGLY COMPLEX DIVISION OF LABOR

An increasingly complex division of labor has helped fuel the growth in services in general, and producer services in particular. An ever more complicated commercial environment has forced companies to depend on a variety of services, including accountancy, research and development, marketing and advertising, and public relations. High-tech equipment in modern office buildings requires skilled maintenance, repair, and security services.

THE SIZE AND ROLE OF THE PUBLIC SECTOR

The size and role of the public sector has been a factor in the growth of the service sector. First, despite efforts to curtail government employment through strategies such as privatization, public-sector employment remains significant; in some countries, public sector employees, from national to local levels, comprise the largest employee group. Second, government laws and regulations have created the need for legal, financial, and other experts to assist companies to negotiate the increasingly complex legal environment.

INCREASING INTERNATIONAL TRADE IN SERVICES

The increase of trade in services among countries has also contributed to the growth in services. Globally, trade in services, including tourism and business services, has grown to about 20 percent of all international trade. The opportunity for further major growth is considerable as more services become tradable (no longer needing to be produced and consumed in the same location).

THE GROWTH IN OUTSOURCING SERVICE FUNCTIONS

The growth in outsourcing service functions from core to semi-peripheral countries (and more recently from semi-peripheral to semi-peripheral countries) (see Box 11.5) by both large and small companies that have exploited advances in information and communications technologies has also led to the growth in services.

11.3 SERVICES OUTSOURCING: BENEFITS AND DRAWBACKS FOR ALL?

Before discussing the outsourcing of tradable services, it is necessary to decipher the often confusing terminology that is used in the academic and other literature. While it is helpful to differentiate between the terms in order to understand the outsourcing process itself, a look

Box 11.1 Bucking the Fisher-Clark thesis in China? Concurrent growth of manufacturing and services

Instead of a straightforward sequential shift from agriculture to manufacturing to services as suggested by the Fisher-Clark thesis, LDCs such as China have seen concurrent rather than sequential manufacturing and service growth. Between 1995 and 2012 industry and services in China each saw an average annual rate of growth of more than 10 percent. By 2012, industry as a percentage of GDP was at 45 percent (up from 43 percent in 1985); the percentage of GDP in services had also risen to 45 (up from only 29 in 1985). The percentage of the labor force in industry rose to 30 percent (up from 17 percent in 1985), while that for services rose from 12 to 36 percent.

A number of factors are responsible for the dramatic expansion of the service sector in China (Lu et al., 2002; Yang, 2004). First, like the DCs, as per capita incomes have risen in China, domestic demand for services has increased faster than the demand for food and daily necessities. Second, although most research has focused on foreign direct investment (FDI) in manufacturing, a significant amount of foreign capital is being invested in the service sector. Third, Chinese state policies associated with the economic reforms since 1978 have been an important factor in the growth of services. In particular, the large-scale privatization of state-owned enterprises has been a major reason underlying the strong growth in the service sector. Privatization is more straightforward for trade and service companies than for manufacturing ones because there are fewer government restrictions and lower startup costs. Fourth, urban planning in the central parts of larger Chinese cities that was designed to improve social and environmental conditions has restricted or removed industry while encouraging services.

The strong growth of China's service sector has led some to speculate that its IT services outsourcing industry may soon rival India's. Between 2005 and 2011, for example, Indian companies lost about 10 percent of the global business process outsourcing (BPO) market to countries such as China and the Philippines. Most analysts predict, however, that it will be several years at least before China poses a real threat to India's dominance. For example, even the largest Chinese IT services providers are small compared to their Indian counterparts. China's largest company, Pactera, employs fewer than 25,000 workers compared to the nearly 300,000 employed by India's largest company, Tata Consultancy Services. As such, many Indian companies have a much larger global clientele, including Tata, Infosys, and Wipro. Chinese IT service providers also face challenges including language skills; although the government is investing heavily in improving English proficiency. The Chinese government also still needs to make regulatory changes to protect the intellectual property of clients.

at Table 11.1 reveals the flashpoint for those concerned about the loss of employment in services (or manufacturing) in the DCs. This table captures how the different kinds of outsourcing can involve work undertaken *abroad* by foreign workers.

Outsourcing can be done domestically or abroad, and always involves work done externally, by an unaffiliated company (Table 11.1). In contrast, *offshoring* is always done abroad, but the work can be done either internally, by an affiliated company, or externally, by an

Table 11.1 Deciphering the outsourcing terminology

	Outsourcing/offshoring			
	Where		How	
	Domestically	Abroad	Affiliated company (internal)	Unaffiliated company (external)
Outsourcing	✓	✓		✓
Offshoring		✓	✓	✓
Captive outsourcing		✓	✓	
Offshore outsourcing		✓		✓

unaffiliated company. *External* outsourcing—where the work is done by unaffiliated companies (including independent foreign subcontractors, as in *offshore outsourcing*)—is common for tradable services that can be standardized easily, such as back-office work. *Internal* outsourcing—where the work is done by foreign affiliates as in *captive outsourcing*—is reserved for situations in which strong control of a core competency activity is vital (for example, in research and development), sensitive information is involved, internal interaction is crucial, or a company is attempting to capture savings and other advantages.

While the outsourcing of services is at an earlier stage than the outsourcing of manufacturing, it is seen as representing the leading edge of changes in global production. UNCTAD (2004) predicts that a tipping point is approaching that will reflect a shift to a new international division of labor in the production of services. OECD estimates place the total number of jobs that could potentially be affected by domestic or international outsourcing at close to 20 percent of total employment in DCs such as the United States, Canada, the United Kingdom, Germany, and Australia. Despite certain similarities in the outsourcing of both services and manufacturing activities, important differences are expected to fuel an acceleration in service outsourcing.

First, although the service sector is much larger than the manufacturing sector, just under 20 percent of service output currently enters international trade (compared with more than 50 percent for manufacturing). This means that there is substantial room for growth. Second, the rate of increase in the amount of services that has become tradable, and so capable of being outsourced, has been more rapid for services than for manufacturing. Third, while manufacturing companies have been the ones that have primarily carried out the outsourcing of goods production, companies in all sectors of the economy are outsourcing service functions. Fourth, skill levels are typically higher for outsourced services than for outsourced manufacturing, and as educational and skill levels continue to improve in many LDCs, there will be more opportunities for outsourcing white-collar jobs from the DCs. Fifth, services that are outsourced may be more mobile than outsourced manufacturing activities (because service activities may require lower capital investment, for example in buildings and machinery, compared to manufacturing).

As already mentioned, the outsourcing of services has received significant—mostly negative—attention in the media and in political circles in many DCs. It is important to consider, however, not only the potential drawbacks but also the potential benefits of service outsourcing for both the DCs and the LDCs. The likely drawbacks for the DCs, including the loss of service sector jobs in particular, have received much more attention than the possible benefits. Yet a number of scholars have pointed out that service outsourcing may allow companies in the DCs to enhance their competitiveness by reducing expenditures and improving quality and delivery—with positive benefits for the companies and their national economies.

In a 2004 article about service outsourcing from DCs such as the United States, Uday Karmarker asked whether companies in the service sector in the DCs can survive the outsourcing of service jobs. Karmarker acknowledged that there would be painful job losses for service workers in the DCs, but that the focus should not be on the loss of outsourced service jobs but on the benefits to the global competitiveness of companies in the service sector in the DCs. A 2004 study by the Information Technology Association of America (ITAA) has argued that outsourcing service jobs may ultimately create jobs, boost productivity, and lower inflation in the United States. The labor cost savings from outsourcing may allow companies to sell goods more cheaply or at a greater profit, allowing more capital for purchasing equipment, building facilities, and undertaking research and development. In addition, service outsourcing may allow the DCs to restructure toward more productive and higher value activities that create higher wage jobs. The underlying argument is that the outsourcing of certain service activities and jobs should not be met with cries for protectionist policies in the DCs, but embraced, because outsourcing may contribute to the competitiveness of service sector companies in the DCs by creating a new international division of labor in service production.

Although anxiety about the outsourcing from the DCs is high, the majority of service outsourcing is still currently taking place domestically (UNCTAD, 2004). Most outsourcing is done domestically, with much of the remainder going from the DCs to other DCs. Less than 10 percent of all **business process outsourcing (BPO)**—such as insurance claims processing, billing services, credit card services, telemarketing, and research and development—is done internationally. The top ten destinations for international services outsourcing are Bangalore, Mumbai, Delhi, Chennai, Hyderabad, and Pune in India, Manila and Cebu City in the Philippines, Dublin in Ireland, and Kraków in Poland.

The benefits of service outsourcing for LDCs such as India include the creation of higher skill jobs involving better pay, training, and transferable skills, and associated infrastructure investment that can contribute to further local job growth. The potential drawbacks for some LDCs include the possible relocation of outsourced service activities to other more competitive LDC locations unless worker skills and local infrastructure are continuously upgraded. There are also the perceived negative impacts on culture and tradition of such a rapid increase in employment opportunities for more educated young people, deepening any already entrenched social divisions. Generous salaries by LDC standards are creating a class of western-style consumers. In countries such as India, young, urban women with well-paid back-office jobs are now considering the possibility of a career, rather than the traditional route to financial security of early marriage.

11.4 LIMITS TO SERVICE EXPORT GROWTH IN THE SEMI-PERIPHERY AND PERIPHERY?

Not all services can be outsourced. Services with a low potential for being outsourced typically include the following features: a strong face-to-face servicing requirement; low information content; a work process not depending on telecommunications and the Internet; low wage differentials relative to similar occupations in the DCs; high setup barriers; and significant social networking requirements, including proximity to customers in order to gain a thorough knowledge of markets or a local presence in order to gain an understanding of technical requirements such as legal codes or healthcare regulations.

Consequently, despite the strong forces driving the growth of services, there are significant constraints—related to technology and infrastructure; education and training; government regulations and policies; and corporate strategies—that can limit the growth of services and prevent certain LDCs from capturing some of the service outsourcing market, especially for IT-enabled services.

TECHNOLOGY AND INFRASTRUCTURE

Technological limitations to the growth of service export growth in some LDCs include the fact that not all data can be converted to electronic form for use by computer and made amenable to outsourcing. In addition, as we saw in Chapter 7 when discussing the digital divide, in Africa, for example, fewer than 16 percent of people are Internet users (compared to a high of nearly 80 percent in North America) (see Table 7.3). Limited infrastructure, such as telecommunications, reliable power sources, and financial services and distribution logistics, can limit the growth of export services. While the type of infrastructure needed varies, most IT-enabled services require dependable telecommunications and Internet access.

The city of Mumbai and the southern states in India—which have enjoyed the greatest success in attracting outsourced services—have benefited from their proximity to the landing points of two submarine fiber-optic cables (see Figure 11.5). This has provided a competitive advantage because fiber optics are usually cheaper and more efficient than satellite links.

While the DCs are well-connected by submarine cables, many LDCs are still not linked into this telecommunications network, with the result that they are limited in their ability to develop competitive bases for service exports. Figure 11.5 shows how the United States, Europe, and East and Southeast Asia have good cable capacity, but that only one major cable connects

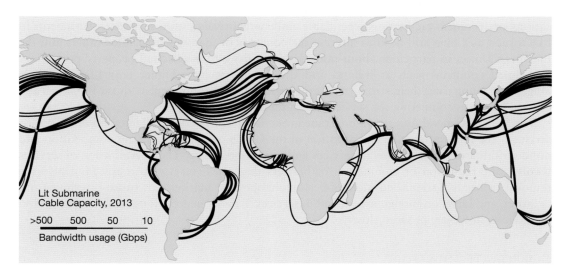

Figure 11.5 Submarine fiber-optic cable network

Source: Adapted from maps at http://submarine-cable-map-2013.telegeography.com/ and http://www.dailywireless.org/2012/07/10/new-submarine-cable-map/

parts of Africa to the rest of the world—the SAT-3 cable. In fact, in Sub-Saharan Africa, only Angola, Benin, Cameroon, Côte d'Ivoire, Gabon, Ghana, Nigeria, Senegal, and South Africa are linked directly to this cable.

EDUCATION AND TRAINING

Lack of education and training is a limiting factor in knowledge-intense services. While the kinds of skill needed differ depending on the kinds of service, most outsourced IT-enabled services involve information processing of various kinds. The strong software export performance of India partly reflects government education and training policies that have produced a large pool of technically trained English-speaking workers. Most software companies are in Mumbai and Bangalore, where the software industry initially developed. With other growing urban centers, specifically New Delhi and its surroundings, Andhra Pradesh and Tamil Nadu, these five areas contain about 50 percent of India's diploma-granting technical institutions.

Special skills are also needed for more routine services. Call centers need workers not only with good language abilities but also solid customer support skills, telesales abilities, data entry, and processing skills.

But sustained strong growth in service outsourcing to India, the Philippines, and South Africa depends on the continued availability and low cost of the necessary worker skills. Despite success so far, there is concern that LDCs such as India may not be able to keep pace with the demand for qualified workers; shortages of trained workers can force wages up and make a country less attractive as an outsourcing destination.

GOVERNMENT REGULATION AND POLICIES

The regulatory and legal framework in some less developed countries can place limits on the growth of export services. There is a need for a competitive regulatory environment that encourages competition among service providers, which includes a deregulated telecommunications environment facilitating dependable and competitively priced service. Governments in LDCs such as China still need to address the concerns of many in the United States and the European Union in particular about poor data security and intellectual property protection. Developed countries such as Ireland and Canada, as well as higher cost LDCs such as Singapore, emphasize their strong regulatory frameworks compared to those in China and even India when competing for service outsourcing work.

The WTO's General Agreement on Trade in Services (GATS) covers all internationally traded services. The goals of service liberalization in the GATS context are greater competition and nondiscrimination against foreign services and service providers. The liberalization of services involves the reduction or elimination of barriers that affect services certainly, but also the removal of legally established monopolies or oligopolistic market structures, discriminatory taxation, and limits on foreign investment in services. Further negotiations on the liberalization of services, however, proceed slowly.

CORPORATE STRATEGIES

Corporate decision making can result in limited opportunities for service export growth in some LDCs. Companies differ on their perception of risk and assessment of the benefits of internationally outsourcing services. In some situations, for example, the information that is

Box 11.2 A day in the life of a call center worker in India

Priya Chatterjee works at a large Indian call center in Bangalore that handles calls from customers of Fortune 500 companies in the United States and United Kingdom. She earns around US$3,000 a year, more than three times the national per capita income. When she started the job six months before, she had to successfully complete a four-week training course. Whereas a few years before, workers were trained to speak with either an American or British accent, she underwent accent neutralization so that she could be shifted around to answer calls from either of these markets with little additional training. Priya works nights to coincide with office hours in the United States and United Kingdom. She shares an apartment with three other female call center workers who all wake up in the afternoon. For breakfast, they eat *paratha* (a type of bread stuffed with vegetables), *daal* (lentils), *subzi* (a cooked vegetable dish), and *idli* (a fluffy rice cracker dipped in savory sauce). Then she usually meets some friends and goes shopping or to the cinema. She must be ready to get the company cab at 11.30 pm but it is always late and the journey to work takes 60 instead of 20 minutes because the streets are choked with traffic. The ten-hour shift begins at 1 am with a short team meeting on targets before she puts on her headset. Priya deals with credit card queries involving payment problems. She must aim for an average call-handling time of two and a half minutes—the system also monitors how often customers are put on hold (which implies she needed to ask for help). More than 2,000 people are employed in the call center and about 700 work on any given night. At 6 am she gets a 30-minute break for a meal in the crowded canteen serving similar dishes to what she ate for breakfast. She is allowed another two 15-minute breaks during her shift which ends at 11 am. She finds the shifts difficult to adjust to because they change every few weeks but always involve working late into the morning. She works a five-day week but her days off vary. She goes to bed as soon as she gets home.

Source: Based on *New Statesman*, January 30, 2006; Hartley and Walker (2012)

to be processed can be confidential; this can increase transaction costs and limit the desirability of outsourcing. Consequently, any assessment of the potential for service outsourcing needs to include an analysis of corporate strategies and organizational limitations.

11.5 GEOGRAPHY OF SERVICES

The major forces driving the growth in services and service outsourcing, combined with the constraints on service growth discussed earlier, mediated by local political, economic, and other contexts, have produced an uneven geographical pattern at global, regional, and urban spatial scales.

PATTERNS AND TRAJECTORIES

A milestone for services was reached back in 2006. For the first time, worldwide employment in services as a percentage of total employment reached 40 percent, surpassing the percentage in agriculture (which had decreased to 38.7 percent). Despite this overall increase in service

Table 11.2 Changing employment in services as a percentage of total employment

	Service employment as % of total employment	
	1996	2012
World	35.5	44.0
Developed countries (DCs) inc. European Union	66.4	73.9
Latin America and the Caribbean	56.5	62.6
Commonwealth of Independent States and Central and Eastern Europe (non-EU)	45.8	54.1
Middle East and North Africa	48.6	52.5
Southeast Asia and the Pacific	32.7	39.6
East Asia	20.7	37.1
Sub-Saharan Africa	22.9	29.3
South Asia	25.3	28.1

Source: Based on ILO (2007: 12, Table 5; 2012: 141, Table A11)

employment, the percentage of workers employed in services is uneven among different parts of the world, with implications for both the DCs and the LDCs (see Table 11.2).

The developed countries now have almost three-fourths of their workforce in service employment, followed by Latin America and the Caribbean with nearly two-thirds, and the central and eastern European countries, Russia, and the Middle East, and North Africa with more than half their workers in services. The remaining regions—Southeast Asia and the Pacific, East Asia, Sub-Saharan Africa, and South Asia—all have only between just over one-fourth and one-third of their workers in services (see Table 11.2).

Looking at a more detailed level, there is significant variation in the percentage of workers employed in services among the LDCs (see Figure 11.6). Countries in Latin America, such as Peru and Argentina, as well as many in the Middle East, including Jordan, Saudi Arabia, and Israel, have 75 percent or more of their workers in services. While Hong Kong and Singapore also have more than 75 percent, India and China are only at about 30 percent. Many countries in Sub-Saharan Africa have less than 50 percent of their workers employed in services.

While more and more services are becoming tradable, a significant amount remains non-tradable, especially in the LDCs. In this connection, although the official statistics from the World Bank and International Labour Organization (ILO) indicate that there is relatively low employment in services in the LDCs compared to that in the DCs, it is important to keep in mind that these data do not include jobs in services in the informal economy. In some of the poorest LDCs, the informal economy represents a large proportion of GNI and is associated with incredibly low per capita incomes. Low-income countries in Sub-Saharan Africa such as Zimbabwe, Tanzania, and Nigeria, which have 50 percent or more of their GNI generated within the informal economy, contrast sharply with the richest DCs including the United States, United Kingdom, and Australia (see Figure 11.7).

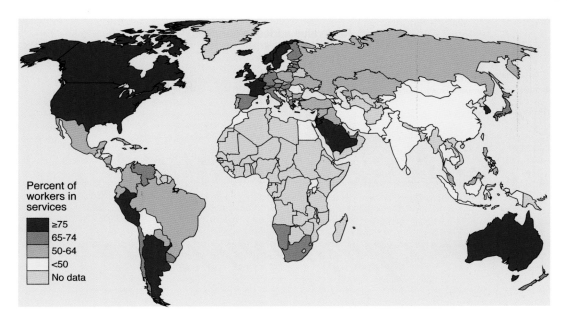

Figure 11.6 Percentage of workers in services

Source: Based on World Bank, online World Development Indicators 2010

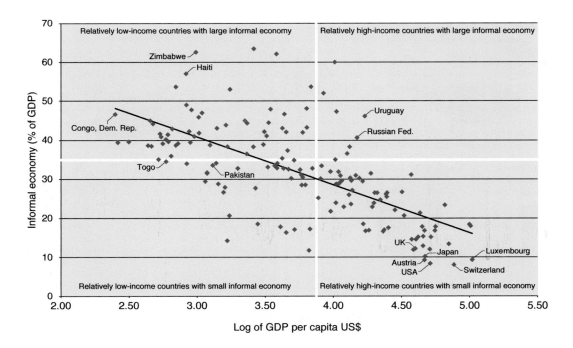

Figure 11.7 Informal economy and level of development

Source: Based on World Bank, online World Development Indicators 2013 and Schneider *et al.* (2010: 45–47, Appendix 5)

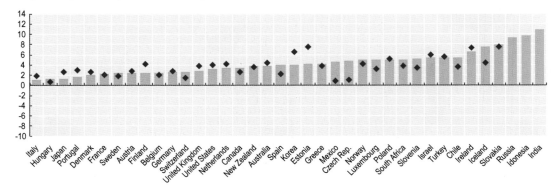

Figure 11.8 Average annual percentage growth in services

Source: Adapted from OECD (2010)

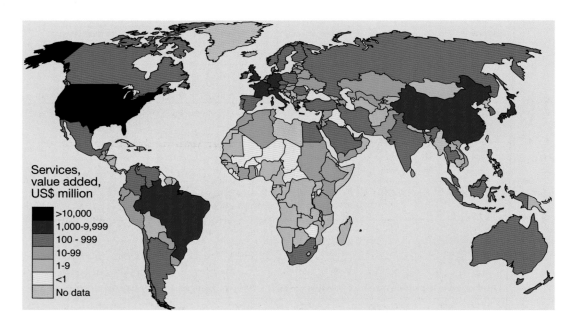

Figure 11.9 Service production

Source: Based on World Bank, online World Development Indicators 2010

In many cities in the LDCs, more than one-third of the population works in the informal sector; in some cities this figure is more than two-thirds. Across Africa, the ILO has estimated that informal-sector work is growing ten times faster than formal-sector employment. Although service jobs in the informal sector—including driving pedicabs, dress or shoe repair, and prostitution—may seem marginal from the point of view of the world economy, they support more than a billion people around the world. In many LDCs, the informal sector includes the world's most vulnerable workers—women and children.

Economic geographers recognize that the formal and informal sectors are interconnected —the informal sector represents an important resource for the formal sector. The informal sector provides a huge range of cheap services and goods that reduce the cost of living for employees in the formal sector, allowing employers to keep wages low. Although this arrangement does not contribute to economic growth or help alleviate poverty, it does keep many companies competitive within the global economic system. For export-oriented businesses, in particular, the informal sector provides a considerable indirect subsidy. And while this subsidy is often passed on to consumers in the DCs in the form of lower prices, the poorest households in the LDCs are forced to resort to increasingly drastic strategies for coping with worsening poverty.

The average annual percentage growth in services between 2007 and 2010 captures the relatively high rates of growth in services in LDCs such as India (10.8 percent) and Indonesia (9.7 percent) which started out with relatively low levels of services (see Figure 11.8). The relatively slower growth rates of the DCs reflect the fact that these countries already have large service economies to begin with.

Figure 11.9 shows the value of service production in 2010. The United States dominated with US$10.6 trillion, followed by Japan ($3.9 trillion) and China ($2.6 trillion). Germany, France, the United Kingdom, Italy, and Brazil were next (with between $1 and $2 trillion each), ahead of Canada, Spain, India, Australia, the Russian Federation, Mexico, South Korea, and the Netherlands.

INTERNATIONAL TRADE IN SERVICES

World service exports have risen to about 20 percent of total merchandise and service exports (up from 15 percent in 1980). Growing at an annual rate of about 10 percent, the value of service exports had risen to more than US$4,000 billion. In terms of world exports and imports of commercial services, and concerns about the outsourcing of services by some in the DCs, however, North America and Europe are the only world regions that export more services than they import; all other world regions import more services than they export (see Figure 11.10).

While countries differ in their service trade performance, fewer than two dozen countries— comprising a small number of DCs with an even smaller number of LDCs (China (with Hong Kong), India, Singapore, and South Korea)—account for three-fourths of total world exports, with the top four countries (United States, United Kingdom, Germany, and France) accounting for almost one-third of all world service exports.

A fast growing part of the world service sector comprises knowledge- and information-related services. Fast growing service exports are computer and information services (15 percent) and financial services (10 percent).

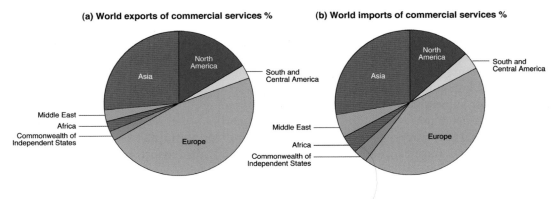

Figure 11.10 Exports and imports of services, 2010

Source: Based on World Trade Organization, WTO Statistics Database online 2010

TRANSNATIONAL INVESTMENT PATTERNS

The structure of foreign direct investment has shifted toward services. While severely impacted by the global financial crisis that began in 2007, the value of FDI projects in the services sector rebounded to $570 billion by 2011, representing 40 percent of world FDI projects.

The increase in the value and share of FDI projects in the services sector is due to a number of factors:

- delivery of many non-tradable services requires a physical presence in foreign markets
- companies are adopting internationalization strategies on the strength of their home market success in building and strengthening their competitive advantage for investment overseas
- many countries have relaxed their regulation of service industries and foreign service providers with the result that these countries are more open to FDI
- information and communications technologies have allowed more service companies to locate their facilities in lower cost locations worldwide (Bryson *et al.*, 2004).

Capturing some of the escalating global FDI is a priority for many less developed countries. Nevertheless, the growth of FDI in services initially occurred among the developed countries, as the historically dominant home countries for FDI in services. The LDCs joined the process during the second half of the 1980s after they began to open their service sector to FDI (particularly through privatization). Since the early 1990s, countries in central and eastern Europe as well as Russia have also joined the process.

FDI in greenfield services projects was about US$367 billion in 2011; the developed countries as destinations accounted for the largest share (nearly 40 percent); Southeast and East Asia as destinations together accounted for about 25 percent. Likewise, investment in greenfield services projects continues to be dominated by corporations from the developed countries, particularly the United States, the European Union, and Japan, which accounted for about 75 percent of the total compared to about 13 percent by East, Southeast and South Asian companies.

EXPORT PROCESSING ZONES (EPZS)

Many LDCs offer financial and other incentives to attract foreign investment, not only in manufacturing, but also in services. Subsidies are used to attract a variety of service industries, but are most common in tourism, transportation, and financial services.

While traditionally used to attract investment in manufacturing, export processing zones (EPZs) are increasingly being used to attract investment in export-oriented services. About three-fourths of EPZs for service industries are located in the LDCs. The kinds of service

Table 11.3 EPZs targeting services

	Business processing zones	Financial services zones	Commercial free zones
Physical characteristics	Part of city or zone	Entire city or "zone within zone"	Warehouse area often near airport or port
Economic objectives	Development of business processing center	Development of offshore banking, insurance, securities hub	Facilitation of trade and imports
Duty free products allowed	Capital equipment	Varies	All goods for storage and re-export of imports
Typical activities	Data processing, software development, etc.	Financial services	Warehousing, packaging, distribution, transshipment
Incentives • taxation • customs duties • labor laws • other	De-monpolization and deregulation of telecommunications; access to market-priced INTELSAT services; specific authority manages labor relations; trade union freedom restricted	Tax relief; strict confidentiality; deregulation of currency exchange and capital movements; free repatriation of profits	Exemption from import quotas; reinvested profits wholly tax free
Domestic sales	None	Limited to small proportion of activities	Unlimited, on payment of full duty
Typical examples	Bangalore, Caribbean	Bahrain, Dubai, Caribbean, Turkey, Cayman Islands	Jebel Ali, Colon, Mauritius, Iran

Source: Adapted from Engman *et al.* (2007: 15, Table 2)

attracted to these EPZs have grown rapidly, from commercial services and simple data entry to call centers, medical diagnoses, architectural, business, engineering, and financial services (see Table 11.3).

These EPZs offer a strong technology support infrastructure including modern communication technologies, reliable power supplies, and a highly skilled workforce. Incentives include 100 percent exemption from import duties and general sales taxes, full repatriation of earnings, and preferential customs clearance. The emphasis on the skills and language abilities of the workers for IT-enabled service jobs contrasts with the low- or semi-skilled workers advertised by the traditional manufacturing EPZs.

In India, many of the outsourced services have been attracted to dedicated technology parks for IT services that were established by individual states. India's first software technology parks were set up in 1990 in Bangalore, Bhubaneshwar, and Pune. There are now dozens of these parks, accounting for up to 75 percent of India's software exports.

11.6 VARIETY IN THE INTERNATIONALIZATION OF SERVICES

As we have seen, there is a wide variety of marketed services ranging from producer services (FIRE and business services) to transportation and communications, wholesale and retail trade, to entertainment, hotels and motels (part of tourism). In concluding this chapter, we briefly examine the internationalization of some leading services, namely, retailing, tourism, and financial and business services.

THE INTERNATIONALIZATION OF RETAILING

On the supply side, since the 1970s, the sales of the world's largest international retailers have grown considerably. In 1976 the total sales of the then largest retail company in the world (Sears Roebuck) were less than US$15 billion (equivalent to more than $60 billion in 2013 dollars). By 2013, Wal-Mart, the world's largest retailer today, had sales of almost $370 billion.

The largest retail companies have been more conservative than many other tradable services when entering foreign markets (Bryson *et al.*, 2004). The stores and receipts of many of the largest retailers are largely domestic (including, in the United States, Lowe's, Kroger, Walgreen, and Target). Based on the location of their stores, only a few of the world's largest retailers—Wal-Mart (USA), Carrefour, PPR and Auchun (France), Metro, Aldi, Tengelmann, Rewe, Lidl & Schwarz (Germany), Ahold (Netherlands), Delhaize (Belgium), IKEA (Sweden), and Tesco (UK)—can be considered truly global operators. Of these, only IKEA, Arhold, and Delhaize, have more than 75 percent of their sales from foreign markets.

With the exception of Wal-Mart, most of the largest international retailers (with annual international sales of more than US$1 billion and operations in ten or more countries) are from Europe. In fact, Wal-Mart remained a domestic U.S. company until as late as 1991 when it opened a Sam's Club near Mexico City (see Box 11.3). It currently derives almost one-third of its sales from foreign countries.

Williams (1992) identified five major reasons why retailers typically decide to internationalize their operations. First, growth-oriented goals in a highly competitive domestic market can lead companies to consider new foreign markets in order to maintain sales and profits. Second, there may be limited domestic market growth opportunities because of market maturity, saturation, exhausted, or unsuitable diversification possibilities and excessive government regulations. Third, there may be opportunities to implement internationally appealing and innovative retail concepts in some LDC markets. Fourth, more passive, reactive, or subjective opportunities for enhanced sales and profits may arise from offers of joint ventures from foreign partners or through acquisition of foreign retail competitors. Fifth, senior management may be driven to apply their retail know-how and techniques to foreign markets. In addition, technological breakthroughs, trade liberalization, and the opening-up of markets to FDI are vitally important policy changes that allow companies to expand internationally.

The international expansion of the largest retailers has been uneven. With the important exception of Africa, where smaller southern African retailers such as Shoprite and Pick'n'Pay dominate, the largest global retailers are extending their presence primarily into Latin America, East Asia, and South Asia. Within these regions, these large retailers target the most attractive markets with the largest consumer bases. In Latin America, for example, much of this foreign investment is going to Brazil, Chile, Uruguay, and Peru; in East and Southeast Asia, the investment is going to Malaysia, Sri Lanka, Indonesia, and, increasingly, China; and in South Asia, the investment is going to India.

At the same time that these large retailers have been expanding into foreign markets, they have been internationalizing their supply networks. Gereffi's (2001) notion of a buyer-driven commodity chain is useful in conceptualizing this arrangement (see Figure 1.4). A commodity chain refers to the entire range of activities involved in the design, production and marketing of a product. *Buyer-driven* commodity chains cover those industries in which large retailers, marketers, and branded manufacturers play pivotal roles in establishing decentralized production networks in a number of usually LDC exporters. This pattern of trade-led industrialization has become common in labor-intensive, consumer goods industries such as garments, footwear, toys, housewares, consumer electronics, and a variety of handicrafts. Production is usually undertaken by tiered networks of contractors in the LDCs who make finished goods for foreign buyers. The product specifications are supplied by the large retailers or marketers who order the goods:

> One of the main characteristics of the firms that fit the buyer-driven model, including retailers like Wal-Mart, Sears Roebuck, and J.C. Penney, athletic footwear companies like Nike and Reebok, and fashion-oriented apparel companies like Liz Claiborne and The Limited, is that these companies design and/or market—but do not make—the branded products they order. They are part of a new breed of "manufacturers without factories" that separate the physical production of goods from the design and marketing stages of the production process. Profits in buyer-driven chains derive not from the scale, volume, and technological advances as in producer-driven chains, but rather from unique combinations of high-value research, design, sales, marketing, and financial services that allow the retailers, designers, and marketers to act as strategic brokers in linking overseas factories and traders with evolving product niches in their main consumer markets.
>
> (Gereffi, 1999: 1)

In general, buyer-driven commodity chains are designed to keep costs down and involve forcing consolidation at all stages of the commodity chain in the less developed countries. Some of the negative outcomes for suppliers and workers in the LDCs are captured by this Oxfam (2004: 6) quote:

> [I]nternational mergers and acquisitions and aggressive pricing strategies have concentrated market power in the hands of a few major retailers, now building international empires. These companies have tremendous power in their negotiations with producers and they use that power to push the costs and risks of business down the supply chain. Their business model, focused on maximizing returns for shareholders, demands increasing flexibility through "just-in-time" delivery, but tighter control over inputs and standards, and ever-lower prices.

Box 11.3 Wal-Mart®

The U.S. giant, Wal-Mart, the largest retailer in the world and the 15th largest company globally, had net sales of almost US$370 billion in 2013. Based on the value of its annual sales, if it were a country, Wal-Mart would rank 28th in the world, ahead of Austria, South Africa, and Venezuela. The company employs 2.2 million workers worldwide, about the same number as the entire population of Slovenia or Lesotho and more than the combined populations of Luxembourg, Malta, and Cyprus.

Wal-Mart became an international operator only in 1991 when it opened a Sam's Club near Mexico City. Today, in addition to Wal-Mart's more than 3,800 stores in the United States, the company operates over 4,800 more in countries such as Brazil, Canada, China, Chile, Japan, Mexico, and the United Kingdom (see Table 11.4). The company usually enters

Table 11.4 Wal-Mart retail stores

Location	2002	2012	Change 2002–2012	
			No.	%
United States	2,744	3,823	1,079	39
Canada	196	333	137	69
Germany	95	0	−95	−100
United Kingdom	250	541	291	116
Argentina	11	88	77	700
Brazil	14	429	415	2,964
Chile	0	314	314	100
Costa Rica	0	200	200	100
Guatemala	0	198	198	100
Honduras	0	70	70	100
Mexico	505	1,600	1,095	217
Nicaragua	0	73	73	100
Puerto Rico	10	45	35	350
China	16	364	348	2,175
Japan	0	367	367	100
South Korea	9	0	−9	−100
South Africa	0	219	219	100
Sub-Saharan Africa	0	30	30	100
Total	**3,850**	**8,694**	**4,844**	**126**

Source: Wal-Mart 2002 and 2012 annual reports

new markets by acquiring foreign operators, such as ASDA in the United Kingdom and Seiyu in Japan.

The company is expected to concentrate increasingly on joint ventures since it was forced to pull out of Germany in 2006 where it faced stiff competition from much larger local rivals such as Aldi and Lidl & Schwarz. Wal-Mart sold its 85 German stores to Metro after it failed to repeat its extraordinary U.S. success in Europe's largest economy. Analysts concluded that Wal-Mart's U.S. approach to business did not translate well into German. In addition to having to conform to strict German labor laws, Wal-Mart's efforts to superimpose its own culture did not go smoothly. For example, Wal-Mart's initial requirement that its sales associates smile at customers was interpreted by male German shoppers as flirting; or requiring employees to bag groceries for customers was not appreciated by German customers who traditionally prefer to handle their own food and bag their own groceries. The consensus is that Wal-Mart learned important lessons in Germany.

Wal-Mart's sales strategy—based on offering products at "everyday low prices" —allows the company to minimize extensive advertising and promotional campaigns. The company can maintain low prices because it simplifies its purchasing activities and keeps costs down with suppliers by offering a relatively narrow choice of the most popular products combined with some particularly high volume items.

In 2002, Wal-Mart established its Global Procurement Center to manage the company's direct import business and factory direct purchasing. This unit of the company is responsible for identifying new suppliers, sourcing new products, building partnerships with existing suppliers and managing the global supply chain of Wal-Mart's direct imports. Wal-Mart sources its products from a large number of low-wage countries including Bangladesh, China, and Vietnam. Tight inventory management is maintained using computerization.

Wal-Mart has been criticized for a variety of practices including its extensive product sourcing in low-wage LDCs, and its low wages and poor worker benefits in its stores in the DCs, combined with its traditional resistance to union representation for its workers. A low degree of unionization, however, is not unique to Wal-Mart; it is something that distinguishes today's service employers from the traditional manufacturing enterprises in the developed countries in the past.

Reardon *et al.* (2003) offer a typology for considering the supply network practices for supermarkets that can usefully be applied to the buyer-driven commodity chains of the largest international retailers:

1. Centralized procurement using a distribution center serving multiple stores has replaced individual store procurement. This reduces administrative costs and increases the efficiency of the procurement network for retailers. It favors suppliers in the LDCs who can meet the retailers' delivery, volume and quality requirements.
2. Logistics improvements have accompanied procurement consolidation. Retailers are applying modern technologies to the supply chain in order to track inventory and delivery. Suppliers in the LDCs receive this technology transfer and training from the retailers themselves or from local consultants.

3. Retailers are increasingly using local specialized wholesalers—sidestepping or transforming the traditional wholesale system. Dealing directly with specialized wholesalers who are dedicated to and capable of meeting the specific needs of the retailers reduces the retailers' costs and ensures greater control over quality and delivery.
4. The increasing use of quasi-formal and formal contracts with price controls has formalized procedures for local suppliers. The large retailers typically draw up short-term contracts that give them the flexibility to adjust their supplier network and force LDC suppliers to bid for each new contract.
5. With little or no enforced public standards, the large retailers have imposed private certification and standards on their entire supplier network in order to harmonize product quality across LDC suppliers. Not surprisingly, many small LDC suppliers have found it impossible to meet these requirements and have been dropped from the procurement lists of the major retailers.

On the demand side, the national and global stimuli to the growth in services, including rising per capita incomes, have given rise to new shopping habits not only in the DCs but also in LDCs such as China and India. In addition, breakthroughs in information technology have allowed consumers to access services using the Internet. E-shopping, from the likes of Amazon and eBay, has added an additional dimension to retailing that does not involve traditional shopping venues, such as shopping malls and individual stores.

INTERNATIONAL TOURISM

The globalization of the world economy has been paralleled by a globalization of the tourist industry. Even in those parts of the world that do not have much of a base in primary commodities, are not an important part of manufacturing commodity chains, and are not closely tied into the global financial network, international tourism can offer the otherwise unlikely prospect of economic development. The growth of international tourism has been fuelled by rising incomes and the consumption of services with a particularly high elasticity of demand— including transportation, hotels and entertainment.

International tourism has reached an all-time high of about 1 billion international trips each year (up from just 147 million in 1970) (see Figure 11.11). Visitors from the more affluent countries of the world—including the United States, United Kingdom, Japan, Canada, France, Italy—make many of these trips, although the number of Chinese international tourists is rising sharply. Since 1990, despite economic downturns, international tourist arrivals have increased at an average rate of about 4 percent each year. About half of these trips are for leisure, recreation and vacations; one-sixth for business; and the remainder for other purposes including visiting friends and relatives, religious reasons such as pilgrimages, and health treatment. Top destinations for medical tourism by the more than 2 million patients from DCs seeking treatment, usually for elective procedures like cosmetic surgery, at lower cost in LDCs include India, Mexico, Singapore, and Thailand.

International tourism earnings have reached more than US$1,000 billion annually (see Figure 11.11). It is one of the world's largest industries, with up to one in every ten workers worldwide involved (directly and indirectly) in transporting, feeding, housing, guiding, or amusing tourists. Tourism exports represent about 30 percent of the world's exports of commercial services; for some LDCs, tourism is the main export and source of foreign exchange income. Tourism is estimated to contribute about 5 percent to world GDP.

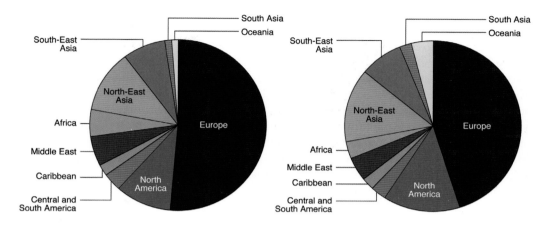

(a) International tourist arrivals, 2010 %   **(b) International tourism receipts 2011 US$ million %**

Figure 11.11 International tourist arrivals and receipts, 2006

Source: Based on tables in UNWTO (2012: 4–5)

What is most striking, however, is not so much the growth in the number of international tourists as the increased range of international tourism. Thanks largely to cheap long-distance flights, a significant proportion of tourism is now transcontinental and transoceanic. Europe (51 percent) and the Americas (16 percent) remain the most popular tourist destinations (see Figure 11.11). The top destinations in Europe are countries bordering the Mediterranean— France, Spain, Italy and Turkey. The United States alone offers tourist magnets such as Florida, Hawaii, Las Vegas, and New York. But countries in Asia, Africa, and the Pacific have grown to nearly one-third of the industry (see Figure 11.11). This, of course, has made tourism a central component of economic development in destinations with exotic wildlife (for example, Kenya), scenery (Nepal), beaches (the Seychelles), shopping (China and Hong Kong), culture (China, India, and Indonesia), or sex (Thailand).

But, although tourism can provide a basis for economic development in the less developed countries, it is often a mixed blessing. Tourism certainly creates jobs, but they are often seasonal. Dependence on tourism also makes for a high degree of economic vulnerability. Tourism, like other high-end aspects of consumption, depends very much on matters of style and fashion. As a result, once thriving tourist destinations can suddenly find themselves struggling for customers. Some places are sought out by tourists because of their remoteness and their "natural," undeveloped qualities, and it is these that are most vulnerable to shifts of style and fashion. Nepal is a recent example of this phenomenon: It is now a bit too "obvious" as a destination and consequently the tourist industry in Nepal is having to work hard to continue to attract sufficient numbers of tourists. Bhutan, Bolivia, Estonia, and Vietnam have been "discovered," and are coping with the growth in tourism. Tourism in more exotic destinations is also vulnerable in other ways: To political disturbances, natural disasters, outbreaks of disease or food poisoning, and atypical weather. International tourism certainly fell in the wake of the 9/11 terrorist attacks and the global economic downturn that began in 2007, with destinations that depended significantly on U.S. as well as European tourists suffering disproportionately.

Although tourism is a multibillion-dollar industry, the financial returns for tourist areas are often not as high as might be expected. The greater part of the price of a package vacation, for example, stays with the organizing company and the airline. Typically, the tourist region itself captures only 40 percent. If the package involves a foreign-owned hotel, this may fall below 25 percent. The costs and benefits of tourism, however, are not only economic. On the positive side, tourism can help sustain indigenous lifestyles, regional cultures, arts and crafts, and provide incentives for wildlife preservation, environmental protection, and the conservation of historic buildings and sites. On the negative side, tourism can adulterate and debase indigenous cultures, and bring unsightly development, pollution, and environmental degradation. Tourism can also involve exploitative relations that degrade traditional lifestyles and regional cultural heritages as they become packaged for outsider consumption. Traditional ceremonies, which formerly had cultural significance for the performers, are now enacted solely to be watched and photographed. In the process, indigenous cultures are edited, beautified, and altered to suit outsiders' tastes and expectations.

Such issues, coupled with the economic vulnerability of tourism, gave rise to ecotourism as a more sustainable strategy for economic development in the LDCs. Ecotourism emphasizes self-determination, authenticity, social harmony, preserving the environment, small-scale development, and greater use of local techniques, materials and architectural styles. Despite being a relatively poor country, Costa Rica, for example, has won high praise from environmentalists for protecting more than one-fourth of its territory in biosphere and wildlife

Box 11.4 Abu Dhabi, a tourist Mecca?

Many semi-peripheral countries have made a significant transition toward becoming service economies. Of these, some oil-rich states in the Persian Gulf have deliberately attempted to diversify their oil-dominated economies toward services such as banking and hosting conventions. They have also tried to become tourist Meccas by investing in expensive tourist amenities, including luxury hotels, shopping malls, and golf courses. Saudi Arabia, for example, with about 20 million international visitors annually, ranks among the top 20 international tourist destinations in the world.

The United Arab Emirates (UAE) is catching up following an aggressive policy to position itself as a major Persian Gulf tourist destination. In 2007 the head of Abu Dhabi's tourism agency and the French Culture Minister signed a 30-year agreement to open a Louvre Museum outlet in this Persian Gulf boomtown. France will receive US$525 million alone for the use of the Louvre brand name. The deal also involves a gift of $33 million to renovate a wing of the Paris Louvre, which will house Islamic art and be named for the longtime UAE ruler, Sheik Zayed bin Sultan Al Nahyan. In addition to the hundreds of millions of dollars in construction costs, a further $750 million will be spent to bring French staff and 300 loaned pieces of art to the Louvre Abu Dhabi for its opening in 2015. The waterfront museum is designed to be the centerpiece of a cultural district with the goal of attracting millions of well-heeled tourists each year.

Not everyone was happy about what some negatively saw as the globalization of French culture. In France, opponents argued that the French government was exploiting art for trade. The Louvre Abu Dhabi also has to overcome significant cultural barriers in the Islamic world where representations of the human body, even fully clothed, can be a religious taboo.

preserves. The payoff for Costa Rica is the escalating number of tourists who come to visit its active volcanoes, palm-lined beaches, cloud forests and tropical parks. Costa Rica receives more than 2 million tourists every year. Tourism is the country's largest source of foreign exchange, followed by bananas, pineapples, coffee, sugar, and, more recently, electronic components and medical equipment.

THE INTERNATIONALIZATION OF FINANCE

Having addressed the patterns of international finance in earlier chapters (see, for example, Chapters 2 and 3) we focus here on the internationalization of finance from the perspective of the LDCs in terms of transnational banks and FDI in financial services, foreign exchange transactions and offshore banking centers. Several factors—with implications for the LDCs—have reinforced the globalization of financial services. A particularly important factor has been the advances in information and telecommunications technologies, allowing the electronic flow of capital, which have significantly reduced the transaction and transmission costs associated with moving money. Another has been the institutionalization of savings in the developed countries (through pension funds and the like), which has established a large pool of capital managed by professional investors with few geographical allegiances or ties. Another has been the trend toward disintermediation—which involves borrowers (especially large corporations) raising capital and making investments without going through the traditional, intermediary channels of financial institutions. Yet another, and probably more important, factor was the deregulation of financial markets that began in many developed countries in the 1980s.

Financial services have traditionally accounted for a large share of FDI in greenfield services projects in many parts of the world. Despite the fallout from the global financial crisis, the value of FDI in greenfield financial services projects has reached more than $40 billion annually. In East and Southeast Asia, for example, the share of FDI greenfield services projects represented by finance has risen to about 25 percent each year. Financial services cross-border mergers and acquisitions (M&As), the primary means of foreign entry into LDC markets, have risen to more than $36 billion annually. Not only are two-thirds of all financial M&As within the DCs, Europe and North America dominate the list of the 20 largest financial TNCs. The European Union (specifically France, Germany, the United Kingdom, Italy, the Netherlands, and Spain) has 65 percent of the top 20 financial conglomerates, followed by Switzerland with 20 percent, the United States with 10 percent, and Japan with 5 percent (see Table 11.5).There has been a large increase in the presence of banks from the developed countries in the less developed countries, especially in East Asia and Latin America. Whereas foreign banks capture market shares (including deposits and profits) of about 20 percent in the DCs, the share is about 50 percent in the LDCs. Market shares are more than 75 percent in LDCs such as Cameroon, Senegal and Uganda in Africa and El Salvador, Nicaragua and Uruguay in Latin America.

At the same time, it is estimated that as many as one-third of the world's financial TNCs are from the less developed countries. Among these banks are some of the world's largest, including China's ICBC, China Construction Bank, Agricultural Bank of China and Bank of China, Brazil's Itaú Unibanco, Banco do Brasil, and Banco Bradesco, and Russia's Sberbank. Many LDC banks, however, tend to be relatively small by international standards and less internationally active (with a physical presence in relatively few foreign banking markets). A significant percentage of the foreign subsidiaries of these financial TNCs are concentrated within the region of the home LDC.

Nevertheless, just as some NIEs developed their own transnational corporations to compete with those of developed countries, so some have developed relatively large and aggressive banks with an increasing interest in international opportunities. Singapore's four largest banks, for example, have more than 50 overseas branches, 20 overseas subsidiaries and affiliates, and more than 30 representative offices in overseas locations including Beijing, Hong Kong, Jakarta, London, Los Angeles, Manila, New York, Seoul, Sydney, Taipei, Tokyo, and Vancouver.

In addition, the effects of the rapid spread of e-finance in recent years are not limited to the DCs. E-finance involves electronic financial services that are delivered online, for example through smartphones or smart cards (embedded with a microprocessor and/or a memory chip allowing information on the card to be added, deleted, or otherwise manipulated). Some LDCs with underdeveloped financial systems use e-finance to leapfrog ahead in some areas of finance, including banking. In some African countries, for example, electronic cash and smart cards are being offered as savings and payment services for low-income customers who do not have access to traditional banks and conventional bank accounts. Evidence from Brazil, for example, indicates that e-finance can be introduced quickly even when basic financial infrastructure is weak or non-existent.

At the same time, however, there is enormous variation in the extent of electronic banking and shopping across both the DCs and the LDCs (see Table 11.6). This variation is not closely related to level of development however. In some countries, DC as well as LDC, electronic delivery of financial services remains in its infancy. Meanwhile, other countries have experienced the rapid penetration of e-finance. In Sweden, e-finance accounts for more than one-third of financial transactions. In some LDCs, such as Mexico, e-finance penetration is quite high for some financial services (Table 11.6). At the same time, however, significant challenges to the spread of e-finance in some LDCs relate not only to the availability of technology and infrastructure (including access to smartphones and the Internet) but also to the enabling regulatory environment, involving security and related infrastructure for e-transactions, information and privacy, and contract enforcement.

In addition, the increasing use of electronic money has changed the nature of international financial investments in ways that may not benefit the LDCs. For example, foreign investments are already beginning to shift away from more tangible FDI to intangible portfolio investments such as stocks and bonds. The issue here is that whereas FDI can generate tangible levels of employment and facilitate technology transfer, portfolio investments typically create few jobs (Bryson *et al.*, 2004).

What is more, despite the dominance of electronic money flows, the majority of the foreign exchange transactions—more than 70 percent—are still made in only three currencies: the U.S. dollar, the euro, and the Japanese yen. Certainly the market opens each day in East Asia while it is still evening in the United States; financial transactions then look west, travelling along new or upgraded fiber-optic cables, typically including Tokyo to Shanghai or Hong Kong to Mumbai to Frankfurt or Zürich to London and then on to the United States (see Figure 11.12). But London has maintained its preeminence as the premier world center with about 35 percent of the volume of the more than US$4 trillion in foreign exchange transactions every day, followed by New York with 18 percent, and Tokyo with 6 percent.

Not all the needs of the global financial system, however, can be met by conventional financial services. The need for secrecy and the desire for shelter from taxation and regulation have resulted in the emergence of offshore banking centers. The incredible increase in the electronic flow of global capital has allowed many of these offshore banking centers to be located in the LDCs—in microstates and on islands (see Figure 11.13). There are five main specialized offshore

Table 11.5 World's largest financial TNCs, 2011

Rank 2011	Corporation	Home economy	Total assets (US$m)	Total employees	Number of affiliates Total	Number of affiliates Foreign
1	Allianz SE	Germany	832,726	141,938	720	595
2	Citigroup Inc.	United States	1,873,878	266,000	926	637
3	BNP Paribas	France	2,551,230	198,423	1,190	892
4	UBS AG	Switzerland	1,517,657	64,820	569	534
5	HSBC Holdings PLC	United Kingdom	2,555,579	288,316	1,178	816
6	Assicurazioni Générali Spa	Italy	549,191	81,997	461	408
7	Société Générale	France	1,533,597	159,616	497	345
8	Deutsche Bank AG	Germany	2,809,328	100,996	1,337	1,027
9	UniCredit spa	Italy	1,203,084	160,360	951	865
10	AXA S.A.	France	947,757	96,999	681	589
11	Zurich Insurance Group Ltd	Switzerland	385,870	52,648	338	328
12	Credit Suisse Group Ltd	Switzerland	1,121,981	49,700	299	262
13	Standard Chartered PLC	United Kingdom	599,070	86,865	200	137
14	Munich Reinsurance Company	Germany	321,396	47,206	522	274
15	Swiss Reinsurance Company	Switzerland	225,899	10,788	119	115
16	Credit Agricole SA	France	2,237,500	87,451	471	274
17	ING Groep NV	Netherlands	1,660,629	95,025	755	454
18	Morgan Stanley	United States	749,898	61,899	243	194
19	Banco Santander SA	Spain	1,624,667	193,349	469	345
20	Mitsubishi UFJ Financial Group	Japan	2,659,476	80,400	136	90

Source: Based on online UNCTAD data (World Investment Report 2012: Annex Tables at http://unctad.org/en/Pages/DIAE/World%20Investment%20Report/Annex-Tables.aspx)

Table 11.6 E-finance, 2009

	ATMs per 1 million pop.	Point-of-sale terminals per 1 million pop.	Cashless retail transactions, per capita
DCs	**780**	**20,880**	**102.0**
Australia	1,237	31,900	263.3
France	873	22,225	259.9
Germany	1,030	7,242	200.0
Italy	898	21,079	59.6
Japan	1,086	13,510	111.9
Netherlands	515	20,577	290.8
Norway	467	27,155	380.3
Sweden	356	19,739	306.0
United States	1,384	21,565	339.9
United Kingdom	1,006	19,069	232.3
LDCs	**230**	**1,700**	**4.0**
Brazil	855	17,589	94.5
China	161	1,810	38.9
El Salvador	197	2,693	2.2
India	52	420	4.7
Iraq	6	15	0.5
Mexico	316	4,159	15.3
Saudi Arabia	92	3,254	6.5
Sierra Leone	6	4	0.1
Singapore	431	16,336	73.9
Turkey	318	23,240	25.8
Difference magnitude: DCs/LDCs	**x 3.4**	**x 12.3**	**x 25.5**

Source: Based on World Bank (2011: 41: Table III.1; 54–55: Table III.9); Stein (2010: 12: Figure 7)

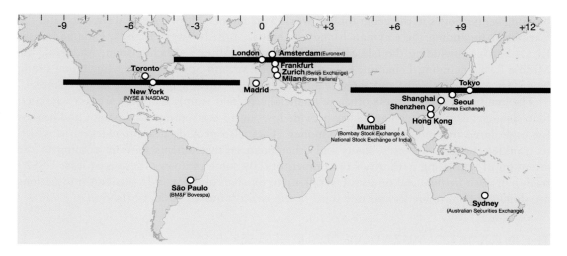

Figure 11.12 The world's major stock markets

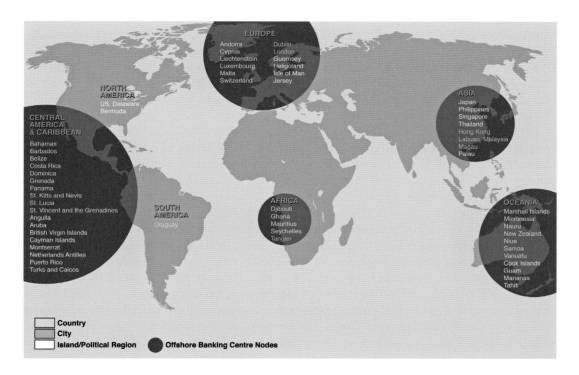

Figure 11.13 Major world areas of offshore banking

Source: Based on Grant Thornton 2010 infographic: *Locations of offshore tax jurisdictions* http://thinking.grant-thornton.co.uk/bespoke/index.php/article/ locations_of_offshore_tax_jurisdictions_infographic/

banking center nodes in the geography of worldwide financial flows: Central America and the Caribbean (including the Cayman Islands, Bahamas, British Virgin Islands, Panama, Costa Rica), Europe (including the Isle of Man, Jersey, Luxembourg, Liechtenstein, Andorra), the Middle East (including Lebanon, Bahrain, Israel), East and Southeast Asia (including Hong Kong, Singapore, Labuan) and the Oceania (including Nauru, Vanuatu, the Cook Islands).

The main attraction of these offshore financial centers is simply that they are less regulated than financial centers elsewhere. They provide low- or no-tax settings for savings, and havens for undeclared income and for "hot" money. They also provide discreet markets in which to transact currencies, bonds, loans and other financial instruments without coming to the attention of regulating authorities or competitors. By some estimates there is at least US$20 trillion in offshore financial centers. Overall, up to one-half of the world's business transactions are done in offshore banks and one-third of the world's money resides offshore.

THE INTERNATIONALIZATION OF BUSINESS SERVICES

We conclude this chapter by briefly examining the internationalization of business services in terms of information technology and business process outsourcing to the LDCs in general, and to India in particular. Instead of building in-house expertise, while still maintaining the secure and reliable provision of non-core IT services, large corporations in DCs such as the United States and United Kingdom initially outsourced to other large companies domestically. The largest IT and BPO service providers and their intermediaries, include companies in the DCs, particularly the United States, such as Computer Sciences Corporation (CSC), Hewlett-Packard/Electronic Data Systems (EDS), IBM and iGate. The largest IT and BPO service providers also include companies in the LDCs, particularly India, such as Tata Consultancy Services, Infosys, and Wipro (see Table 11.7).

The internationalization of IT-enabled services, including BPO, to LDCs such as India, has been driven by the corporate strategies, beginning in the early 1990s, of companies in the United States, Europe, and Japan particularly. U.S. companies began outsourcing to India the conversion of custom-made software programs from one operating system to another. This time-consuming and tedious operation could be outsourced easily to an LDC such as India. The Indian programmers were much less costly than their U.S. counterparts and had the necessary skills, speed, and attention to detail to perform the work (UNCTAD, 2003).

As the birthplace of BPO, the United States still dominates. For example, almost 60 percent of India's exports of software services go to the United States. European companies have shown less inclination, with just over 20 percent of India's software services going to the United Kingdom and only about 1 percent going to Germany. Some companies have decided not to outsource (yet) and, as in the United States, a few companies have moved some operations back in response to customer complaints. Of course, service outsourcing varies across Europe, with the United Kingdom, accounting for the largest share of BPO, most closely mirroring the United States.

The kinds of service activities involved now include, not only call centers, computer network support, legal services, accounting, and procurement, but also software development, research and development, and engineering services:

> BPO is a varied and flexible process. Service providers may provide rudimentary data entry services, or they may take over management functions or operations and become responsible for the entire process. Clients may outsource to several outsourcing providers. They may outsource data center management functions to one provider, network management functions to another, and business

Table 11.7 Major international business process outsourcing (BPO) corporations

Company and headquarters	Specialties	Low-cost locations	Outsourcing net revenues
Accenture (Ireland)	Software development; network support; finance and accounting (F&A); human resources (HR); procurement; insurance operations; general banking	India, Philippines, Spain, China, Czech Republic, Slovakia, Brazil, Mexico, Indonesia	Over US$10 billion
Cognizant Technology Solutions (USA)	Software development; network support	India, China, Hungary, Philippines, Argentina, Mexico	Over $5 billion
Computer Sciences Corp. (CSC) (USA)	Software development; insurance operations; demand management	Bulgaria, Czech Republic, Spain, India, Malaysia, China, South Africa, Brazil, Chile	Over $10 billion
HCL Technologies (India)	Software development; network support; R&D/ engineering; financial services	India, China, Malaysia, Mexico	$1–$5 billion
Hewlett-Packard/ Electronic Data Systems (EDS) (USA)	Software development; network support; F&A; HR; payroll; demand management; procurement; insurance; general banking; telecom; transportation; healthcare operations	Mexico, Brazil, Argentina, India, China, South Africa, Spain, Hungary	Over $10 billion
IBM (USA)	Software development; network support; F&A; HR; payroll; procurement; insurance operations	India, Philippines, China, Poland, Russia, Mexico, South Africa, Malaysia, Colombia, Hungary	Over $10 billion
iGate (USA)	Software development; network support; R&D/engineering	India, Brazil, Mexico	Over $10 billion
Infosys (India)	Software development; network support; banking; mortgage processing	India, Czech Republic, Poland, China, Mexico, Brazil, Philippines	$1–$5 billion
Tata Consultancy Services (TCS) (India)	Software development; R&D/engineering; F&A; telecom; transportation; hospitality operations	India, Hungary, Brazil, Uruguay, Chile, China, Malaysia, Morocco, South Africa	$1–$5 billion
Wipro (India)	Software development; R&D/engineering; demand management; mortgage processing; transportation operations; healthcare operations; banking; mortgage processing	India, Romania, Poland, China, Philippines, Mexico, Brazil	$1–$5 billion

Source: Based on company annual reports and Gartner, Inc. website, http://www.gartner.com

processes and help desk functions to still others. The BPO vendor may be a small local business or a large company, perhaps larger than the client.

(UNCTAD, 2003: 137)

Significant services outsourcing so far has tended to be concentrated in a relatively small number of LDCs. The necessary technology and infrastructure requirements, including fiber-optic submarine cables (see Figure 11.5), are less potentially ubiquitous than even a skilled workforce (which a national government's education and training policies can promote). AT Kearney ranked services outsourcing destinations based on financial attractiveness, people skills, and availability, and business environment; its top ranked countries for outsourcing IT and BPO are India, China, Malaysia, Egypt, Indonesia, Mexico, Thailand, Vietnam, the Philippines, and Chile. Other important services outsourcing destinations include Argentina, Brazil, Colombia, the Czech Republic, Israel, Pakistan, Poland, Russia, South Africa, and Ukraine.

Box 11.5 India's competitive advantage in BPO

The main outsourcing destination by far for business process outsourcing (BPO) is India. This country accounts for almost two-thirds of the global market in outsourced IT-enabled services and BPO. Large cities in India, particularly Bangalore, Delhi, and Mumbai, are attractive to BPO because of their large pool of skilled English-speaking workers, good information and communication technologies, convenient time zone difference for companies in the DCs, and relative political stability. In addition, low labor costs are a factor: In the United States, the average salary for a computer programmer is about US$70,000, versus $8,000 in India. India's established track record in the field of IT-enabled services including software development, has allowed this country to continue to benefit from its initial advantage in this area.

In addition, large established Indian corporations with expansive vertical coverage including mortgage processing, healthcare operations, and software development (see Figure 11.14)—such as Tata Consultancy Services, Infosys, and Wipro—now engage in BPO themselves (so-called double-sourcing), especially when the BPO does not depend on good language abilities or when the Indian companies need a presence in a foreign market close to its customers (see Table 11.7). In India, services as a percentage of GDP have risen to about 65 percent (up from 40 percent in 1985). Average annual growth in services between 1995 and 2011 was high by international standards: 8.8 percent. While still having a large percentage of its workers in agriculture (about 50 percent), India now has about 30 percent of its workers in services. Services outsourcing plays an important role in creating jobs, particularly for women, who are employed primarily, however, at the low end of BPO. Women tend to work in call centers (see Box 11.2) or perform data entry and other business operations that require lower level skills than their male counterparts.

India has recorded high growth rates in the export of services which have risen to more than US$12 billion annually. Growth has been particularly rapid in software, business, financial, and communication services. India's share of world commercial services' exports has risen to more than 3 percent, putting it among the top ten exporters globally (compared to less than 2 percent and 18th respectively for world merchandise exports). Concerns about India's continued service outsourcing dominance relate to whether the country can keep up with the growing demand for skilled workers, and address serious

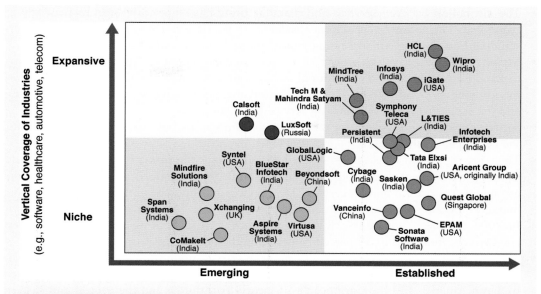

Figure 11.14 Global R&D service providers, 2012

Source: Adapted from Zinnov graph at http://indiatechonline.com/it-happened-in-india.php?id=798

congestion, power outages, rising costs, and competition from other LDC competitors such as the Philippines and South Africa.

Nevertheless, India's position appears to be relatively secure. This is because companies in the DCs are increasing their outsourcing to LDCs such as India of more complex, core competency service functions like R&D and software design that are associated with greater innovation and higher value. Especially in situations where companies have their own India-based development teams, the decision has increasingly been made to allow these teams to take sole responsibility for the development of certain service products, including conceptualizing and building new features and modules. In addition, whereas in the past, a constraint on greater innovation and value in BPO in India was distance from customers in the DCs, the Internet and cloud-computing have changed everything. Now that the Internet provides the customers for companies such as Google, access to the information and needs of customers has been equalized for the development teams in India. As a result, Google designed its development center in Bangalore to generate completely new service products, rather than to act merely as back offices for its U.S. operations.

SUMMARY

In this chapter, we examined the shift to services and the contemporary geography of services in both the less developed countries and the developed countries, as well as the interactions between them. We identified the following points as being of critical importance:

- Services can be categorized into a number of major components, including finance, insurance and real estate; business services; transportation and communications; wholesale and retail trade; entertainment, hotels, and motels; public services; and nonprofit services.

- The distinction between services and manufacturing has come to be seen increasingly as redundant. The notion of service encapsulation of goods and materials is useful for understanding how services are increasingly incorporated into manufactured products.
- Major forces that drive the growth of services include rising per capita incomes; growing demand for healthcare and educational services; an increasingly complex division of labor; the growing size and role of the public sector; increasing international trade in services; and the rapid growth in outsourcing service functions.
- Service outsourcing involves not only potential drawbacks but also potential benefits for the developed countries and the less developed countries.
- There are significant constraints—related to technology and infrastructure; education and training; government regulations and policies; and corporate strategies—that can limit the growth of services and prevent certain LDCs from capturing some of the service outsourcing market, especially for IT-enabled services.
- In 2006 for the first time, worldwide employment in services as a percentage of total employment increased to 40 percent, surpassing the percentage in agriculture.
- The percentage of workers employed in services is uneven among different parts of the world. The developed countries have almost three-fourths of their workforce in services; followed by Latin America and the Caribbean with nearly two-thirds; and the central and eastern European countries, Russia, and the Middle East and North Africa with more than half their workers in services. The remaining regions of the world have only between just over one-fourth and one-third of their workers in services.
- There is significant variation in the percentage of workers employed in services among the LDCs. LDCs including Peru and Argentina, as well as many in the Middle East, including Jordan, Saudi Arabia, and Israel, have 75 percent or more of their workers in services. While Hong Kong and Singapore also have more than 75 percent, India and China are only at about 30 percent. Many countries in Sub-Saharan Africa are below 50 percent.
- LDCs including India and Indonesia have seen relatively high rates of growth in services. Nevertheless, the United States has the highest production of services; followed by Japan, China, Germany, France, the United Kingdom, Italy, and Brazil, ahead of Canada, Spain, India, Australia, the Russian Federation, Mexico, South Korea, and the Netherlands.
- World service exports have risen to more than US$4,000 billion or 20 percent of total merchandise and service exports each year. The United States, United Kingdom, Germany, and France together account for almost one-third of world service exports.
- The structure of foreign direct investment has shifted toward services. The developed countries account for the largest shares of FDI in greenfield services projects.
- Export processing zones (EPZs) are increasingly being used to attract investment in export-oriented services by the LDCs.
- High value-added services, using skilled labor and tacit forms of knowledge, are highly agglomerated in world cities. In contrast, relatively low value-added service functions, such as back offices, call centers, and offshore banks, are increasingly dispersed to low wage LDCs.
- As the world's largest retailers such as Wal-Mart have been expanding into foreign markets, they have been internationalizing their supply networks. Their buyer-driven commodity chains create opportunities and drawbacks for suppliers in the LDCs. E-shopping has added an additional dimension to retailing that does not involve traditional shopping venues.
- International tourism reached an all-time high of more than 1 billion international trips annually. Although tourism can provide a basis for economic development in many LDCs,

it is often a mixed blessing. In contrast, ecotourism can offer a more sustainable strategy for economic development in some LDCs.

- The internationalization of finance has created opportunities and challenges for the LDCs in terms of the operation of transnational banks and FDI in financial services, the continued dominance of London, New York, and Tokyo over foreign exchange transactions despite the increasing use of electronic money, and the concentration of offshore banking centers in the LDCs.

- The internationalization of business services in terms of IT and business process outsourcing was begun by U.S. TNCs. The kinds of service activities involved now include not only call centers, computer network support, legal services, accounting, and procurement, but also software development, research and development, and engineering services. Major outsourcing destinations for BPO in the LDCs include India, the Philippines, and China.

KEY SOURCES AND SUGGESTED READING

Amiti, M. and Wei, S.-J., 2005. Fear of Service Outsourcing: Is it justified?, *Economic Policy* 42, 307–347. (http://www.oecd.org/dataoecd/44/24/35333668.pdf)

Bryson, J.R., Daniels, P.W., and Warf, B., 2004. *Service Worlds: People, organizations, technologies.* London and New York: Routledge.

CRIC, 2006. *Innovation in Services.* CRIC Briefing No. 2. Manchester: Centre for Research on Innovation and Competition, University of Manchester.

Cuadrado-Roura, J.R., Rubalcaba-Bermejo, L., and Bryson, J.R. (eds.), 2002. *Trading Services in the Global Economy.* Cheltenham: Edward Elgar.

Daniels, P.W. and Bryson, J.R., 2002. Manufacturing Services and Servicing Manufacturing: Knowledge-based cities and changing forms of production, *Urban Studies* 39, 977–991.

Economist, 2013. The Post-Industrial Future is Nigh, *Economist* February 19.

Gereffi, G. and Christian, M., 2009. The Impacts of Wal-Mart: The rise and consequences of the world's dominant retailer, *Annual Review of Sociology* 35, 573–591.

Karmarkar, U., 2004. Will you Survive the Services Revolution?, *Harvard Business Review* June, 82(6), 101–107.

Stutz, F.P. and Warf, B., 2012. *The World Economy: Geography, business and development,* 6th edn. Upper Saddle River, NJ: Pearson Education.

Tickell, A., 1999, 2001, 2002. *Progress in Human Geography* [progress reports on the geography of services]. Vol. 23: 633–639, Vol. 25: 283–292, Vol. 26: 791–801.

UNCTAD, 2004. *World Investment Report 2004: The shift towards services.* New York and Geneva: United Nations.

Part 4

Adjusting to the world economy

In this concluding part of the book, we examine some of the reactions to the emergence of ever larger and more powerful economic forces and the time–space compression that have come to characterize the world economy. In Chapter 12, we explore the changing role of national states within the world economy, emphasizing the relationships between economic change and geopolitics and, in particular, the spatial consequences of international and supranational political and economic integration that have occurred in response to the increased scale, sophistication and interdependence of the world economy. In Chapter 13, we examine the other side of the coin: Decentralist reactions to the changing world economy. Here, the focus is on regionalism and regional policy, nationalism and separatism and grassroots movements towards economic democracy. Finally, in Chapter 14, we review the key arguments that have shaped the book, emphasizing the dynamic interdependence of global and local change.

Picture credit:
Linda McCarthy

Chapter 12

International and supranational institutionalized integration

Picture credit: © European Community, 2013

In this chapter, we return to the theme of the relationship between economic development and the role of the state. As Parts 2 and 3 have shown, nation-states have been crucial, both in the struggle for domination over the world economy within the core and as peripheral and semi-peripheral economies have struggled to reduce their dependency on core economies. As the world economy has become more and more globalized, however, nation-states throughout the world economy have had to explore cooperative strategies involving international and supranational (trading bloc) political and economic integration of various kinds. This chapter outlines the rationale for these strategies, describes the scope of the major international and supranational organizations and illustrates some of the more important spatial implications of international and supranational institutionalized integration. Arguably the most successful of the attempts at supranational integration, the European Union, receives the most detailed examination in this chapter simply because its impacts on the world economy have been the greatest. Although it should be borne in mind that international institutions regulating trade, such as the World Trade Organization (WTO), have also become increasingly important if less visible arbiters of the global economic integration process. This is why we also examine the WTO in some detail.

12.1 ECONOMIC CHANGE AND GEOPOLITICS

In order to understand the emergence of international and supranational organizations, we must first remind ourselves of the shifting economic and geopolitical foundations of the world economy since the Second World War. In the aftermath of the war, the capitalist world economy was reordered as a more open system. It was a system without the economic barriers of the trading empires that had been set up in the years previously. Instead, it was based on free market capitalism with stable monetary relations and rapidly diminishing barriers to trade. This required, first of all, an *orderly* world, internally peaceful and secure from outside threats. Second, it required leadership in providing and furthering mechanisms for establishing a stable reserve standard for international currency exchange rates and for ensuring access to world trade markets. The one state that could provide military order—the United States—was also the only state economically strong enough to impose order on the economic system. The Soviet world empire had turned inward in its attempt to restructure its economy and society along different ideological lines, but its existence was extremely important because (until its dissolution in 1991) it served to mobilize an ideological reaction—anti-communism—that provided both an economic stimulus and political solidarity within the core economies.

In short, the world economy was characterized by the hegemony of the United States. Under U.S. hegemony, as we saw in Chapter 5, the world economy came to be characterized by Fordism, the socioeconomic system that links mass production with mass consumption. A tense but durable relationship among big business, big labor, and big government enabled Fordism to provide the basis for the long postwar boom and unprecedented rise in living standards throughout much of the capitalist world. This boom was also crucially dependent on the massive expansion of world trade and international investment flows made possible under the umbrella of U.S. financial and military power. Following the Bretton Woods Agreement of 1944 that made the U.S. dollar the world's reserve currency (see p. 54), Fordism was implanted in Europe and Japan, either directly, during the occupation phase, or indirectly, through the Marshall Plan and foreign direct investment (FDI) by U.S. companies. The consequent opening up of foreign trade, observes Harvey (1988: 4):

> [P]ermitted surplus productive capacity (and potentially surplus labor reserves) to be absorbed in the United States, while the progress of Fordism internationally meant the formation of global mass markets and the absorption of the mass of the world's population, outside the communist world, into the global dynamics of a new kind of capitalism . . . At the input end, the opening up of foreign trade meant the globalization of supply and often ever cheaper raw material. This new internationalism also brought a host of other activities in its wake—banking, insurance, hotels, airports, and, ultimately, tourism. It also meant a new international culture and a new global system of gathering and evaluating information.

The immediate postwar period (1947–1960) saw, therefore, the rise of a series of industries—automobiles, steel, petrochemicals, rubber, etc.—that acted as the propulsive engines of economic growth, coordinated through the collective powers of big labor, big business, and big government. And out of this there arose a series of grand production regions in the world economy—the Midwest of the United States, the West Midlands of Britain, the Ruhr in West Germany and the Tokyo-Yokohama production region in Japan—managed from worldwide financial, and governmental centers such as New York and London and reaching out to dominate an increasingly homogeneous world market.

The logic of Fordist production also fostered, as we have seen (Chapter 6), transnational corporations (TNCs) with the capacity to move capital and technology rapidly from place to place, drawing opportunistically on resources, labor markets, and consumer markets in different parts of the world. TNCs have now gone far beyond the point where they can be seen simply as extensions of a specific national economy; and even some small firms have now acquired both the capability and the propensity to operate globally. The significance of this is that, *although private companies are by no means absolute masters of their own fate, they do have the ability (as compared with governmental units) to redefine their commitments and objectives in response to the changing opportunities presented by the globalization of the world economy.*

Within this context of political and economic interdependence, regional and international shifts in economic and political power began to occur, as we saw in Chapters 5 and 6: Shifts that the policies of particular governments seemed powerless to prevent by normal means. The ascent of the NIEs has brought a new dimension to the world economy and effectively created a new hierarchical geopolitical system. Meanwhile, as the regional influence of NIEs has grown, so they have come to exert an independent effect on the landscapes of the world's core economies:

> None has had a greater political/psychological effect on the major powers than the omnipresence of persons, symbols and signs in Europe's great cities, and in such American cities as New York, Miami, New Orleans and Los Angeles. In Europe, billboards advertising Asian, African and Latin American Airlines, store signs in Arabic script, national airline offices and ethnic food restaurants from three continents, a plethora of Arabic-language newspapers displayed prominently in kiosks and, above all, the businessman, tourist, shopper, student and adolescent youth from these newly-powerful countries demonstrate that the world has changed. They have joined the overseas symbols of American power—the Hilton, the Holiday Inn, Hertz, Avis, ESSO, Mobil, IBM, the English-language newspaper, the American bar and restaurant, and the tourist, student and businessman—to share the landscape of sight, sound and taste with Americans.
>
> (Cohen, 1982: 227)

In the core economies, meanwhile, the prosperity associated with Fordism had been replaced by uncertainty, destabilization and crisis resulting from the conjunction of an extended episode of stagflation associated with the declining performance of many businesses at the same time that labor was increasingly militant (1965–1979) and the 1973 OPEC oil embargo. This created national economic management problems that could not be solved without accelerating the inflation that had undermined the role of the U.S. dollar as the international reserve currency.

As a consequence, the role and relative power of nation-states began to change significantly. Economic circumstances reduced the ability of governments to deliver full employment as well as a full range of welfare services; and the growth of the global financial system blunted the power of individual countries to pursue independent fiscal and monetary policies with any degree of success. In particular, the United States had to struggle hard to maintain its hegemony, running a mounting trade deficit and an enormous public debt and having persistently to devalue the dollar in order to maintain competitiveness with Japan and Germany.

In the decentralized, restructured and consolidated world economy that emerged in the 1980s, new communications technologies, new forms of corporate organization, and new business services were intensifying time–space compression, decreasing the time horizons of both public and private decision making and making it easier to spread those decisions over an ever wider space. As we saw in Chapter 7, one result has been the acceleration of shifts in the patterning of uneven development as more flexible corporate organization and flexible production systems have been able to quickly exploit particular local mixes of skills and resources. Another outcome is that local governments are being forced to be much more competitive with one another as they attempt not only to protect their economic base during a time of upheaval and transition, but also to identify and exploit some competitive edge with which to lure the newly flexible flows of finance and production. This intergovernmental competition has bred so-called entrepreneurial cities, whose governments have been drawn beyond questions of tax policies, infrastructure provision, and service delivery to explore public/private partnerships, foster favorable business climates and initiate controls on labor through contract negotiations with municipal workers.

12.2 INTERNATIONAL AND SUPRANATIONAL INSTITUTIONAL INTEGRATION

It is within this changing economic and geopolitical context that we have to see the various attempts to adjust to the world economy through strategies of international and supranational economic and political integration. We should nevertheless remember that, as Parts 2 and 3 of this book have shown, the dominant processes in both intra-core rivalry and in the struggle by the semi-periphery and periphery to escape from dependency have been dominated by conflict and competition. Economic nationalism, whether drawing on practical examples (for example, eighteenth-century Britain and nineteenth- and twentieth-century Japan; see Chapters 5 and 10 respectively), political ideology (for example, Juan Péron in Argentina and Getúlio Vargas in Brazil in the 1940s and 1950s) or development theory (for example, the import substitution industrialization espoused by Raoul Prebisch), continues to dominate global economics and geopolitics.

Having acknowledged this, however, we must also recognize the long-term trend among the world's national economies towards the progressive integration and interdependence of local, regional, and national economic systems. *What has happened is that the logic of the world economy has in many ways transcended the scale of nation-states.* The logic and apparatus of statehood are not conducive to international and supranational integration, economic or political; but the outcomes of flexible production have forced many states to explore cooperative strategies of various kinds. As a result, the world's economic landscapes now bear the imprint, in a variety of ways, of international and supranational economic and political integration.

THE LOGIC OF INTEGRATION

The increased scale, sophistication, and interdependence of the world economy would not have been feasible were it not for the fact that new technologies and new forms of corporate organization gradually made it possible to conquer several of the frictions that tend to operate against hierarchical flows of production and consumption. In addition to the obvious—for geographers—friction of distance itself, these include the frictions associated with spatial variations in social organization and culture. As railroads, the telegraph, automobiles, aircraft, computer networks, satellite communications systems, and fiber optics have successively "shrunk" the globe, Fordist principles of mass production have brought about a convergence of patterns of social organization and radio and television have undermined local and regional cultures and replaced them with an international culture characterized by the language and artifacts of consumerism: American Express, Benetton, Burger King, Coca-Cola, Gucci, Laura Ashley, Marlboro, Mercedes Benz, MTV, Rolex, Sony, Visa, and so on.

The framework of nation-states, however, is a source of friction that has persisted. The main reason for this, of course, is that the functional logic of statehood hinges on reinforcing *differences between* nations while reinforcing *similarities within* nations. In order to establish the required feelings of common identity, even the oldest states have had to engage in the process of creating and diffusing a distinctive identity. Much of the ideology and symbolism of nation-states in Europe, for example, has centered on the systematic mythologizing of history, reinforced by the stereotyping of outsiders. One very important outcome of this was the jingoism and xenophobia that set the context for the First World War, nurtured the ambitions of the German Third Reich, and hampered postwar attempts to establish common economic and legal frameworks.

Among the more explicit functions of nation-states that have contributed to the frictions affecting the world economy are those relating to national security and the promotion of homogeneous internal standards and conditions. The latter include controlling fiscal and monetary policy, upholding labor contracts, establishing standards for everything from education to ball bearings, and overseeing key industries such as telecommunications.

Once a significant amount of economic activity had spilled beyond national boundaries, however, countries had to confront the need to rethink these activities in order not to become isolated or to become even more vulnerable to underdevelopment. In short, *it was the international trade system that provided the major impetus for countries to be drawn into various forms of institutionalized integration.* For core countries, the objective was primarily to protect and consolidate existing advantages through increased international security, access to wider markets, investment opportunities, and labor markets. For peripheral and semi-peripheral countries, the objective was primarily to minimize or reduce dependency through harnessing more resources and more investment potential. In addition, most countries were able to subscribe, in public at least, to the more lofty ideals of good international relations and a more equitable international economic order.

The particular *advantages* of formalized international and supranational integration include:

1 potential for economies of scale, particularly for the smallest countries and the weakest national economies
2 potential for creating multiplier effects from the existence of enlarged markets
3 potential for strengthening regional interaction by easing the movement of labor, goods and capital.

The particular *disadvantages* of formalized international and supranational integration include:

1 potential loss of national sovereignty over a broad spectrum of issues
2 potential for the intensification of internal inequalities as a wider geographical context makes for more pronounced processes of uneven development.

TYPES AND LEVELS OF INTEGRATION

Figure 12.1 summarizes the "where" of international and supranational economic integration since 1945. In practice, integration can be pursued in a variety of ways and at different levels. It can be *formal*, involving an institutionalized set of rules and procedures (for example, United Nations, European Union (EU, formerly the European Community or EC), General Agreement on Tariffs and Trade (GATT), reorganized in 1995 as the World Trade Organization (WTO)); or *informal*, involving coalitions of interests (UN voting blocs). It can be *international*, involving attempts to foster integration between countries (North Atlantic Treaty Organization (NATO); the African Union (AU), formerly the Organization of African Unity (OAU); the World Trade Organization (WTO)); or *supranational*, involving a commitment to an institutionalized body with certain powers over member states (EU). It can be *economically* oriented (WTO, the North American Free Trade Agreement (NAFTA)), *strategic* (NATO), *political* (UN voting blocs), *sociocultural* (United Nations Educational, Scientific and Cultural Organization (UNESCO)), or *mixed* (EU, AU) in orientation. We distinguish all these institutions from, for example, the IMF and World Bank, which are not oriented so much towards encouraging integration among states as to regulating state macroeconomic policies

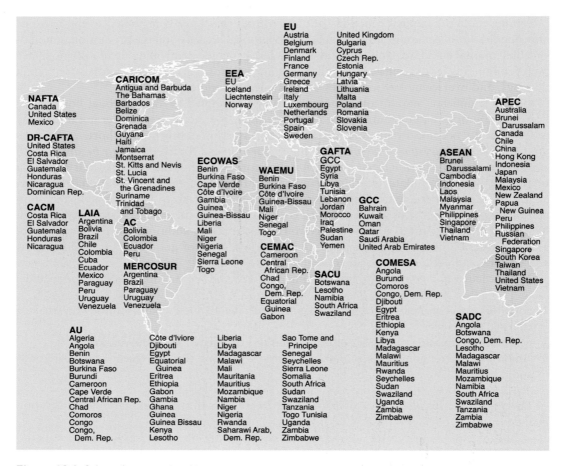

Figure 12.1 Selected supranational integration agreements

Source: Updated from World Bank (2002: 155, Figure 6.1)

and intervening within states to encourage certain kinds of development. In other words, they, like the UN General Assembly and Security Council, represent the global status quo insofar as they are based on a state logic of development even when, as the IMF and World Bank did in the 1970s and 1980s, they give advice and use coercive loans to open up national economies to global competition. They can be thought of as searching for a role for themselves in a world increasingly following a logic of globalization for which they themselves were never designed, dating as they do back to the era immediately after the Second World War.

THE GATT FRAMEWORK AND THE WTO

Our immediate concern here is with economically oriented integration schemes. Within the capitalist world, these have had to conform to the rules of the General Agreement on Tariffs and Trade (GATT), an international association of most of the world's trading countries formed to promote worldwide free trade and to untangle the complex trade restrictions in the aftermath of the Second World War. The original GATT agreement (in 1947) reduced the average tariff on goods from over 40 percent to less than 30 percent. Subsequent

Table 12.1 Average tariff levels (percent) for selected countries, 1988 and 2009–2011

	1988	2009–2011
Australia	15.6	1.8
Brazil	42.2	7.9
Canada	7.7	0.9
Chile	19.9	4.0
China	39.5	4.1
European Union	5.7	1.1
Hong Kong	0.0	0.0
India	79.1	8.2
Indonesia	18.1	2.6
Japan	5.9	1.3
Malaysia	13.6	4.0
Mexico	10.5	2.2
New Zealand	14.9	1.6
Philippines	27.9	4.8
Singapore	0.3	0.0
South Korea	19.2	8.7
Taiwan	12.6	1.5
Thailand	31.2	4.9
United States	5.7	1.6

Source: Based on World Bank (online World Development Indicators, 2012, http://data.worldbank.org/indicator)

rounds of renegotiation have brought the average tariff level down to below 5 percent (see Table 12.1).

Yet the GATT became the victim of its own success. As more countries have joined the agreement and the world economy became increasingly globalized and interdependent, so trade issues have become increasingly complex. The original agreement was written to deal primarily with trade in manufactured goods among developed countries, yet by 1990 and also in 2000 only about 60 percent of world export earnings came from manufactures. Services accounted for an increasing share of world trade; many of the NIEs were not fully subject to GATT rules; and foreign direct investment by TNCs was beyond the scope of the GATT. And although tariffs on manufactured goods were successfully reduced through the GATT, substantial non-tariff barriers (for example, import quotas, import licenses, exchange rate manipulation, government subsidies to domestic industries, special labeling, and packaging regulations, etc.) remained a problem. So whereas the early rounds of GATT renegotiation took several months, more recent rounds have taken several years. The Uruguay Round began in 1986 and was not concluded until December 1993. The chief obstacle was disagreement between the United States and the European Union over nontariff barriers in the form of various subsidies that both were paying to their farmers. The most recent Doha Round has so far achieved very little indeed, with major divisions once more between the USA and the EU over trade in services as well as agricultural goods and increasing hostility towards the GATT/WTO process from many of the world's poorer countries whose primary products continue to be excluded from most core country markets by exclusionary tariffs and other restraints on trade. Increased membership and the movement of debate into areas of trade such as services and agriculture long given the status of "national priorities" by many rich countries suggest that the WTO may be coming up against its political limits. For example, unable to persuade the U.S. delegation to cut its incredibly high subsidies to American farmers, the Indian commerce minister spent much of his time at the WTO's Doha Round of talks in July 2006 watching the soccer World Cup.

The crowning achievement of the Uruguay Round in 1993 was the creation of the World Trade Organization (WTO) in 1995 as a replacement for the GATT (which had become labeled by wags as the "General Agreement to Talk and Talk"). Whereas the GATT had little ability to enforce its decisions, the WTO is a global body with both judicial and regulatory power. Its framework is a series of lengthy agreements that extend beyond trade in manufactured goods to cover investment, services and intellectual property rights. In the words of the organization's former director-general, Renato Ruggiero, the WTO "is writing the constitution of a single global economy." A significant step towards this was taken in February 1997 when the 68 original members of the WTO signed an agreement to free up their markets to international competition in telecommunications. By January 2013 the WTO had a membership of 159 countries, including China and Russia, with a waiting list of 24 others. Whether the "constitution" Ruggiero refers to is worth the paper it is written on remains to be seen.

Many of the WTO's agreements are derived from GATT rulings, including the provision that each member state shall extend **most-favored-nation (MFN) treatment** to all other member countries. (So if, say, the USA were to lower its import duty on textile products from Canada, it would immediately have to extend that same reduced rate to every other WTO member.) There is, however, an exception to this principle for free trade associations and customs unions, members of which may reduce their tariffs against one another without extending such concessions to remaining WTO members. It is this exception that has provided the basis for regional economic integration within the world economy. To proponents of global free trade it is precisely this exclusion that encourages the substitution of bilateral and regional trading agreements for the multilateral ones that they see as the key to continued globalization. They worry that trading blocs are forming that will merely scale up protectionism from the national to the supranational level.

INSTITUTIONAL FORMS OF SUPRANATIONAL INTEGRATION

Supranational integration can take a number of different forms. In practice, some of these are also more or less successful in meeting their objectives. As yet many types of integration are relatively limited in both membership and efficacy. In a free trade association, member countries eliminate tariff and quota barriers to trade from other member states, but each individual member continues to charge its regular duties on materials and products coming from outside the association. Membership of the European Free Trade Association (EFTA) dwindled to Iceland, Liechtenstein, Norway, and Switzerland when six of the original members left to join the EU (Austria, Denmark, Finland, Portugal, Sweden, and the United Kingdom) (see Figure 12.1). The Southern African Development Community (SADC) countries signed a free trade agreement in 2000.

Meanwhile, Canada, Mexico, and the United States, with over 443 million consumers today, established a trading zone in 1994 with the completion of the North American Free Trade Agreement (NAFTA). This was not only an unprecedented economic integration of core countries and a semi-peripheral country but also the first instrument of economic integration to liberalize trade in services. Since the establishment of NAFTA, trade and investment between Canada, Mexico, and the USA has steadily increased. NAFTA's phasing out of tariffs and other trade and investment barriers between these countries over 15 years is leading to a reorganization of the economic geography of North America. Many labor-intensive manufacturing jobs have already been switched from Canada and the USA to Mexico; while the expanding Mexican market is now open to the kinds of product and service in which the USA and Canada have a competitive advantage, such as high-technology products, telecommunications, and financial services. Eventually, there will be total access in agricultural markets, which will rewrite the agricultural geography of Mexico and significantly modify agricultural patterns in the southwestern United States. The expected eventual expansion southwards of NAFTA to include countries such as Argentina, Brazil, Chile, Colombia, Paraguay, Peru, Uruguay, and Venezuela will lead to a further reorganization of the economic geography of these countries. Currently, however, this possibility meets with considerable opposition in Central and South America not least because the terms of trade are seen to favor U.S. businesses over local business and labor.

A customs union also involves the elimination of tariffs between member states, but has a common protective wall against non-members, as in the case of the Southern African Customs Union (SACU). Where, in addition, internal restrictions on the movement of capital, goods, labor and enterprise are removed, the result is a common market. Most customs unions have gone at least some way towards common market status. Examples include the Central American Common Market (CACM), the Southern Cone Common Market (or Mercado Común Sudamericano (MERCOSUR)), the Andean Common Market (ANCOM) of the Andean Community (AC) (formerly the Andean Pact), the Caribbean Community and Common Market (CARICOM), the Gulf Cooperation Council (GCC), the Economic Community of West African States (ECOWAS), and the Common Market for Eastern and Southern Africa (COMESA). The Arab Free Trade Area (AFTA) is considered an important step towards the ultimate goal of an Arab Common Market (Figure 12.1).

A still higher form of integration is the economic union, which, in addition to the characteristics of a common market, provides for integrated economic policies among member states. The members of WAEMU (the West African Economic and Monetary Union), for example, have moved some way towards economic union with their shared single currency and monetary policy. The Economic and Monetary Community of Central Africa (Communauté Économique et Monétaire de l'Afrique Centrale (CEMAC)) also share a single currency.

The highest form of integration possible involves some form of supranational political union, with a single monetary system and a central bank, a unified fiscal system, a common foreign economic policy and a supranational authority with executive, judicial and legislative branches. Except for the supranational political union of the EU, however, free trade associations and common markets have found it difficult to overcome the obstacles imposed by memberships that include countries at very different levels of development and that involve enormous distances and poorly developed transportation networks. It was in response to such problems that the GATT authorized, in 1971, the waiver of the Article I most-favored-nation (MFN) provision for LDCs offering concessions to other LDCs. As a result, Mexico, for example, could offer to reduce its duty on a product from Bolivia without having to extend the same lower rate to the USA. The GATT decision meant that LDCs were free to experiment with a variety of integration models without incorporating internal free trade as a legally binding obligation. The result has been the emergence of a series of trade preference associations such as the Association of Southeast Asian Nations (ASEAN) and the Latin American Integration Association (LAIA).

The increasing globalization of the world economy has broadened and deepened the trend towards regional economic integration. In 1989, for example, the ASEAN countries joined with Australia, Canada, China, Hong Kong, Japan, New Zealand, South Korea, Taiwan, and the United States to form the Asia Pacific Economic Cooperation group (APEC), with the objective of promoting the liberalization of trade and promoting cooperation in trade and investment around the Pacific Rim. APEC members have reduced or eliminated some trade and investment barriers as part of voluntary efforts in response to the 1994 Bogor Declaration (written by the Second Informal APEC Economic Leaders Meeting in Indonesia) and the Information and Technology Agreement that came out of the 1996 WTO Ministerial Conference in Singapore. But, so far, that is about it. In 1992 EFTA and the European Union agreed to establish a unified free trade zone, the European Economic Area (EEA), which has a combined market size today of over 495 million people. (The EEA took effect on January 1, 1993, without the Swiss, whose electorate voted against the agreement.) (See Figure 12.1.)

12.3 SPATIAL OUTCOMES OF ECONOMIC INTEGRATION

It follows from the basic principles of economic geography that the enlargement of markets and the removal of artificial barriers to trade will result in a realignment of patterns of economic activity. Two main sets of effects can, in fact, be anticipated. The first relates to patterns of trade. With international integration, the *removal of trade barriers should lead to a more pronounced spatial division of labor*, with each region in the larger association tending to specialize in those activities in which it has the greatest comparative advantage. In effect, production is thereby reallocated from high- to low-cost settings and a great deal of trade is generated within the association. At the same time, lower costs can, theoretically, be passed on to consumers, which contributes to improved levels of living. These effects of integration are generally referred to as trade creation effects.

Countries that do not belong to the association, however, tend to lose trade: The external tariff wall prevents them from competing effectively with higher cost internal producers whose output is able to circulate duty free within the association. To the extent that the old sources of supply were more efficient producers than the new ones, trade diversion will have taken place, with the result that consumption is shifted away from lower cost external sources to higher cost internal sources, consumers have to pay more for certain goods and levels of living may be depressed.

The extent to which trade creation might outweigh trade diversion depends on several factors, including the degree to which the range of goods produced in member states overlap and the degree of pre-integration reliance on trade with countries outside the association. If integration is successful in the long run in creating trade and accelerating economic growth, it is possible that consequent increases in demand for goods and raw materials will generate **spread effects**, which create a positive spillover effect for other economies.

The second set of effects relates to patterns of regional development. Because of the need to exploit new patterns of competitive advantage, a certain amount of relocation of production must take place, with related activities tending to cluster together in the most efficient settings. The corollary is the disinvestments that take place as production is withdrawn from less efficient locations. Given the logic of cumulative causation, the net effect in terms of regional development within the association will clearly be a tendency for *spatial polarization* as a result of **backwash effects**. Because of the political dimension inherent to integration, this in turn provides a powerful case for a strong *regional policy*.

Meanwhile, integration can also be expected to precipitate other changes in patterns of regional development. A reorganization of patterns of production may occur where changes in patterns of comparative advantage are not sufficient to write off past investment or to prompt relocation, but are sufficient to justify intra-industry specialization. Steel-producing regions, for example, may come to specialize in certain kinds of steel product rather than producing a broad spectrum of steel products for a domestic national market; or agricultural regions may move from mixed farming to a more specialized set of outputs.

Another important consequence of integration is the stimulus that is provided for foreign direct investment. Excluded by high external tariff barriers, foreign suppliers are likely to seek to open branch plants inside the association in order to get access to its market. If successful, this not only makes for a drain of capital when profits are repatriated, it also makes for a degree of external control of some local labor markets. Finally, we must consider the implications of integration for patterns of regional development *outside* the association. The most striking effects in this context will be those related to the dislocations experienced by specialized regions whose exports are no longer competitive within the protected market of the association.

These same principles and tendencies mean that we should expect integration to reinforce the dominant core–periphery patterns in the world's economic landscapes at the macro scale. Patterns of trade between core economies, for instance, are already so strong that integration is able to draw on a good deal of momentum. At the same time, *it is relatively easy for core states to meet the political, social and cultural prerequisites for successful economic integration.* These include:

- similarity in the power of units joining the association
- complementarity of élite value systems
- existence of pluralistic power structures in member countries
- positive perceptions concerning (a) the expected equity of the distribution of benefits from integration and (b) the magnitude of the costs of integration
- compatibility of states' decision-making styles
- adaptability, administrative capacity, and flexibility of member states' governments and bureaucracies.

The success of the European Union has dramatized how effective integration between core states can be. Between 1959 and 1971 trade between the six original members—Belgium, France, Italy, Luxembourg, the Netherlands, and the former West Germany—increased nearly sixfold; by 2000 the expanded Union of 15 countries accounted for one-fifth of all world trade, excluding intra-Union trade. In 2000 the EU was the world's largest exporter of goods and the second largest importer, after the USA. The EU was the largest importer of commercial services and the second largest exporter, again, after the USA. The enlargement of the EU in 2004 to include a set of semi-peripheral economies in eastern and southern Europe, however, has set the organization on a different course. These ten countries (plus the addition of Bulgaria and Romania in 2007) all have much lower standards of living than the previous 15 members, many of them were until recently Soviet-style economies and all have serious economic handicaps of one sort or another (political corruption, lack of legal transparency, outdated heavy industries, etc.). But just as the EU helped the economic growth of new members in prior rounds of enlargement, perhaps most notably Spain, Portugal, and Greece in their day, so too it will this time with a much larger group of new members. From one viewpoint, enlargement has been the most successful policy of the EU. Politically, its promise has stabilized states while they have been in transition from authoritarian rule and it has offered a broad range of potential economic benefits to countries whose economic growth has long lagged behind their potential, measured in terms of educational levels, adjacency to centers of economic growth, and technological sophistication. From another perspective, however, enlargement has undermined the process of "deepening" the central institutions of the EU by making them more transparent and accountable to the people already under the umbrella of the EU. The failure of Dutch and French referendums on a new European constitution in 2005 can at least, in part, be put down to fears of further expansion to include Turkey and countries in the western Balkans from which new immigrants might come, countries that are even more economically underdeveloped than most countries included in the recent enlargements.

In the case of peripheral economies, by the same token, patterns of trade offer little realistic scope for the reallocation of output following the removal of trade barriers in trade preference organizations, common markets, or free trade associations. As we have seen (Chapters 2, 9, and 10), most peripheral countries produce primary commodities that are exported to the core economies rather than to each other and most are so short of capital that even pooled resources are likely to be insufficient to trigger economies of scale of sufficient magnitude to be able to break free from their functional dependency on trade with core economies. Experience has shown, meanwhile, that it is difficult for peripheral and even semi-peripheral states to meet the political, social, and cultural preconditions for successful economic integration. Of course, the EU experiment in eastward enlargement and NAFTA suggest that collaboration across the core–semi-periphery divide can meet with some success. How great this is overall and who wins and who loses is something else again. The political strength of the EU and the redistributive policies this currently allows indicates that this may be the difference between these two cases. NAFTA has undoubtedly failed to solve the most pressing of Mexico's economic problems—providing sufficient new jobs for its growing population—and so one of its main goals, to reduce illegal immigration into the United States, remains unfulfilled. U.S. investment has never compensated for the loss of Mexican investment because of the removal of protective tariffs and increased competition from China in many of its previously most successful manufacturing sectors. Indeed, total manufacturing employment in Mexico has declined since 2000, suggesting that rather than helping Mexican development NAFTA has actually undermined it.

Some efforts at regional integration solely among semi-peripheral states suggest that simply signing up to the organization is never going to be enough to make them a success. ASEAN, for example, despite having generated a growing sense of regional identity, has been unable to progress beyond a preliminary stage of economic regionalism during the last three decades. Regional projects such as the Asian Highway and the Mekong Basin Project have been discussed and only tentative national responsibilities and commitments planned. In the early 1990s ASEAN began working towards the introduction of the ASEAN Free Trade Area (AFTA) and the share of intra-ASEAN trade rose from 20 to almost 25 percent of total trade as trade in electrical appliances and machinery, mineral products, and chemicals increased. A large proportion of intra-ASEAN trade, however, is accounted for by exports that are transshipped through Singapore with only marginal value added by processing or packaging. A significant obstacle to significant intra-ASEAN trade growth is that these economies are more complementary to those of Japan, the United States and Europe than they are to one another: ASEAN itself cannot absorb all the primary commodities it produces and it is still dependent on the core economies for capital, technology, and many consumer goods.

Consequently, as the ASEAN countries have struggled to recover from the Asian financial crisis in the late 1990s, a hot topic of discussion has been increased East Asian trade, not only with Japan, but also with countries such as China and South Korea. But ASEAN's continued commitment to a policy of non-interference in the affairs of other members and to decision-by-consensus continues to hinder progress. Until the political relationship between Japan and the rest of Asia, reflecting the failure of Japanese governments to come to terms with their country's imperial past in Asia, and the increasing military power of China are explicitly addressed, little progress in "deepening" ASEAN can be expected. This indicates the degree to which successful economic integration has to have at least minimal political foundations.

Even less successful than ASEAN as an example of supranational integration has been the Andean Pact, whose efforts at moving beyond the initial agreement between member states have been truly half-hearted. Members have been unwilling to build integration into their own economic planning and policymaking and have been unable to reach agreement about the harmonization of policies with regard to foreign trade, industrial development, or fiscal affairs. Although some progress has been made at diplomatic levels (on pronouncements in favor of human rights in Nicaragua, for instance) and some increase had been achieved in absolute levels of intra-market trade, the negative effects of spatial and socioeconomic polarization, particularly in Bolivia and Ecuador, have led to tensions. Meanwhile, there have been virtually no positive effects in terms of the promotion of new sectors of production or the strengthening of existing regions of production. As a result, the import substitution model was abandoned and the Andean Community (AC) replaced the Andean Pact in 1997. Beginning with the establishment of an executive body, council of presidents and council of foreign ministers, the AC is moving towards a common market and, in theory, eventually to a supranational political union modeled on the European Union with a parliament directly elected by the more than 96 million citizens of its four member countries, despite the exit of Venezuela (Figure 12.1).

One important response to the problems of development, trade and regional integration within the LDCs has been the so-called north–south dialogue. The most important platform for this dialogue has been the United Nations Conference on Trade and Development (UNCTAD), launched in Geneva in 1964. By the end of the Geneva meetings, a degree of political solidarity had emerged among LDCs. Under the banner of the Group of 77 they issued a declaration:

> The unity [of the developing countries in UNCTAD] has sprung out of the fact that facing the basic problems of development they have a common interest in a new policy for international trade and development. The developing countries have a strong conviction that there is a vital need to maintain, and further strengthen, this unity in the years ahead. It is an indispensable instrument for securing the adoption of new attitudes and new approaches in the international economic field.

The Group of 77, which now has more than 130 members, has succeeded in articulating demands for a "new international economic order" (not to be confused with the new international division of labor). Central to the new order envisioned by the Group of 77 are demands for fundamental changes in the marketing conditions of world trade in primary commodities. These changes would require a variety of measures, including price and production agreements among producer countries, the creation of international buffer stocks of commodities financed by a common fund, multilateral long-term supply contracts and the indexing of prices of primary commodities against the price of manufactured goods. Such changes have been at the center of discussions in a series of UNCTAD conferences, special sessions of the United Nations General Assembly, meetings of a specially convened Conference on International Economic Cooperation and successive meetings of the heads of state of the British Commonwealth. Throughout these discussions, however, the core countries in general, and the United States in particular, have been reluctant to do more than agree to general statements about the desirability of a new international economic order. As a result, Williams' observation (G. Williams, 1981: 99) remains true, that "the New International Economic Order is still a dream."

In practice, therefore, there have been two dominant sets of spatial outcomes of international and supranational economic integration. One has simply been the reinforcement of the dominant core–periphery structure of the world economy because of the relative success of economic integration between core states. The second has been the imprint of this success on particular regions. This imprint can be discerned: (1) in terms of the effects of trade creation, trade diversion, spatial polarization, regional policy and sociospatial tensions within core-based associations such as the EU, and (2) in terms of the dislocations experienced subsequently by non-member states. In the remainder of this chapter we illustrate the importance and complexity of the second of these sets of spatial outcomes—the consequences of the success of economic integration between core states—using the example of the European Union.

THE IMPRINT OF THE EUROPEAN UNION

The European Union had its origins in pragmatic responses to the changed economic climate of postwar Europe. The main objective was to recapture the core status within the world economy that Europe had forfeited as a result of the war. But there was also a more idealistic impulse to bind countries together so that the wars that had so bitterly divided Europe in the twentieth century would never reoccur. This political motivation is important because it has long made the EU more than simply a supranational *economic* organization. Although there was a good deal of popular concern over the dominance of U.S.-based TNCs in Europe's postwar economic recovery, the crux of the problem was that the center of technological advance had moved to the United States. As a result, "The real challenge was to ensure that Europe did not remain dependent on imported capital goods and that it began to generate its own research so as to pre-empt the United States in any future technological cycle of production" (George, 1991: 59).

The European Community (now European Union) was formed in 1957 by an amalgamation of three institutions that had been set up in the 1950s in order to promote progressive economic integration along particular lines for six countries (Belgium, France, West Germany, Italy, Luxembourg, and the Netherlands): Euratom, the European Coal and Steel Community (ECSC) and the European Economic Community (EEC). This amalgamation fostered the recovery of core status:

> [B]y providing favorable conditions for multinational investment, so bringing jobs and prosperity back to Europe; by encouraging the emergence of European multinationals that had cultural reasons

for situating their headquarters and research facilities in Europe; by providing the conditions in which capital accumulation could proceed to the point where research and development funds were available for profits; but also by an injection of public funding into the process, through Euratom, and through EEC industrial research programs.

(George, 1991: 59)

Having expanded from its six original members to include Denmark, the Republic of Ireland, and the United Kingdom in 1972, Greece in 1981, Portugal, and Spain in 1984 and Austria, Finland, and Sweden in 1995, the European Union (as by then it had been renamed) then boasted a population of nearly 380 million, with a combined GDP in 2000 of over US$7.8 trillion (which was nearly 80 percent of the United States' almost US$9.9 trillion). In 2004 Cyprus, the Czech Republic, Estonia, Hungary, Latvia, Lithuania, Malta, Poland, Slovakia, and Slovenia joined, with Bulgaria and Romania coming into the fold in 2007. In 2013 Croatia joined. As of 2012 the EU had a population of 501 million people and a 2012 GDP larger than that of the USA, US$16.5 trillion compared to the United States at $14.5 trillion. The EU has developed into a large, sophisticated, and powerful institution with a pervasive influence on patterns of economic and social well-being within its member states and a significant impact on certain aspects of economic development within many non-member countries (see Figure 12.2).

The initial cornerstone of the European Community was a compromise worked out between the strongest two of the original six members. The former West Germany wanted a larger but protected market for its industrial goods; France wanted to continue to protect its highly inefficient (but large and politically important) agricultural sector from overseas competition. The result was the creation of an internal tariff-free market, a common external tariff, and a Common Agricultural Policy (CAP) to bolster the agricultural sector. Given the nature of this compromise, it should be no surprise that the European Community performed very unevenly.

Meanwhile, the rest of the world economy had changed significantly, intensifying the challenge to Europe. By the early 1980s the U.S. and Japanese economies, having accomplished a large measure of restructuring, were becoming increasingly interdependent and prosperous on the basis of globalized producer services and high-tech industries. London's once pre-eminent financial services were losing ground to those of New York and Tokyo; and even the former West Germany, with the Community's healthiest economy, faced the prospect of being left behind as a producer of obsolescent capital goods and consumer goods. In response, the European Community re-launched itself, beginning in the mid-1980s with the ratification of the Single European Act 1985, which affirmed the ultimate aim of economic and political harmonization within a single supranational government. The measures to introduce the Single European Market in 1992 with the Treaty of European Union (the Maastricht Treaty), conferred on the Union many of the major functions of a sovereign nation-state, including:

- creation of a single currency, the euro
- coordination, supervision, and enforcement of economic policies
- maintenance of a completely free internal market
- preservation of law and order
- protection of fundamental rights of individual citizens
- maintenance of equity and, where necessary, redistribution of wealth between regions
- management of a common external policy covering all areas of foreign policy and a common defense policy.

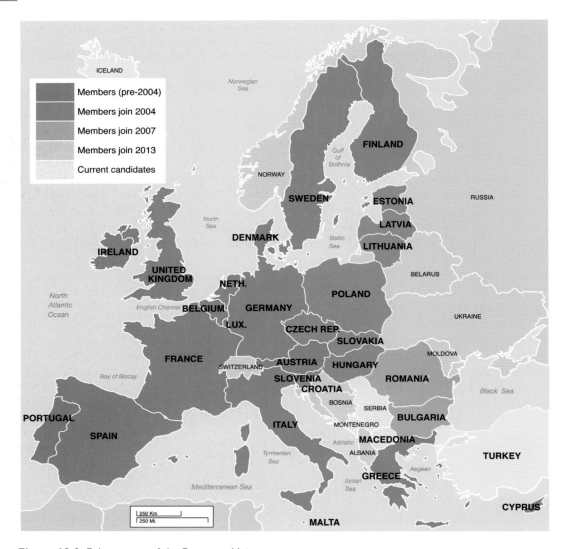

Figure 12.2 Enlargement of the European Union

This re-launching represented an impressive achievement, particularly since it had to be undertaken at a time when there were major distractions: Having to manage a changing relationship with the United States through GATT renegotiations, having to cope with the reunification of Germany and the breakup of the former Soviet sphere of influence in eastern Europe and, not least, having to cope with a widespread resurgence of nationalism (see Chapter 13). Ratification of the Maastricht Treaty was in fact achieved in 1993 only after last-minute maneuverings prompted by the concerns of Danish and UK voters over aspects of nation-state sovereignty. Despite such misgivings, however, the economic benefits of EU membership are widely recognized. Turkey had opened discussions about future membership, although its application has long faced considerable hostility because of its economic and political distinctiveness to say nothing of its religious difference (its population is largely Muslim) and geographical "distance" from the core of Europe. Other countries, such as Iceland, Macedonia,

Montenegro, and Serbia are also in the queue for membership at some time in the future. Among existing member states there is something of an "enlargement fatigue" particularly associated with incorporating 12 new members during 2004–2007, the failure to make much progress on making the European Commission in Brussels (the main bureaucracy of the EU) more responsive to public opinion or representative of the varying population size of the different member states as the organization expands in membership and the increasingly liberal and less redistributive policies pursued in recent years that seem to augur, at least for some people, the specter of increased uneven development within the EU as businesses from rich areas search out opportunities such as lower wages and standards of living in the poorer areas. Western contracts and investment have undoubtedly flooded into Eastern Europe. The biggest fuel for this boom has been wage costs that are typically still half that of western levels. At the same time there is also concern that membership is also increasingly à la carte with some members committed to all EU policies whereas others continue to opt out on such key issues as a common currency (for example, Britain, Sweden) and common immigration and travel rules (for example, Britain). The common currency (now covering 17 member states) has made it impossible for member governments to use monetary policies such as devaluations and interest rate shifts to manipulate their economies but it has boosted trade among them, if less than originally envisaged. The fact that the countries that opted out of the euro—Britain, Denmark and Sweden—have seen equivalent gains to within-EU trade suggests that in the future many one-size-fits-all policies may face increasing pressure for opt-outs from member states.

For its part, the euro entered into a long crisis in the immediate aftermath of the global financial crisis that began in 2007. The European Central Bank was faced with a run on the sovereign bonds of some Eurozone countries (Ireland, Greece, Spain, Portugal, and Cyprus) whose banks had taken on massive debts and whose general economic conditions had deteriorated as the banking crisis took a toll on national finances. This persisting crisis, allied to a perception that the EU could only survive the crisis through further institutional integration such that fiscal and monetary policies could be managed together, had led by 2013 to increasing hostility towards the EU from both previous enthusiasts for the project and its nationalist enemies across member states.

TRADE CREATION AND DIVERSION

The economic benefits of membership were soon felt in the years after the EC's creation. Even by 1970, trade between member countries was 40–50 percent higher, overall, than it would have been had the Community not been formed; by 1980 the figure had risen to a gain of between 100 and 125 percent. While many trade liberalization measures to implement the Single European Market in 1992 have still to take full effect, the most recent analyses by the EU indicate an increase in intra-EU trade of 4 to 5 percent in 1994, for example, due to the Single Market. The net benefits of these increases are far from clear, however, since it is generally acknowledged that the overall increase in intra-Union trade has been the product of a high degree both of *trade creation* and of *trade diversion*. While it is very difficult to isolate the effects of Union membership from other effects, such as TNC activity, overall increases in intra-Union competition and trade have generated economies of scale for EU producers; these have, in turn, stimulated further competition and trade, accelerated changes in industrial structure and corporate organizations and brought about efficiency gains in both importing and exporting countries. Owen (1983) estimated that these economic benefits could be more than half as great as the value of trade itself.

SPATIAL POLARIZATION

It is also clear that these benefits have been associated with a significant amount of regional change within the Union, although once again it is difficult to isolate the effects of supranational political union from others. In overall terms, the removal of internal barriers to labor, capital and trade has worked to the clear disadvantage of peripheral regions within member states and in particular to the disadvantage of those furthest from the European core (see Figure 5.3) that is increasingly the "center of gravity" in terms of both production and consumption. At the same time, integration has accelerated and extended the processes of concentration and centralization, creating structural as well as spatial inequalities. As Holland (1980: 8) put it:

> [T]he market of the Community is essentially a capitalist market, uncommon and unequal in the record of who gains what, where, why and when. Its mechanisms have already disintegrated major industries and regions in the Community and threaten to realize an inner and outer Europe of rich and poor countries.

Slower growing member states with economies dominated by inefficient primary or manufacturing industries are, in short, in danger of remaining problem regions within a prosperous EU. Evidence on trends in personal incomes supports this prognosis. Clusters of richer and poorer member states are identifiable in terms of per capita income relative to the EU average. At the top end of the range, with above average incomes, are Austria, Belgium, Denmark, France, Germany, Italy, Luxembourg, the Netherlands, and Sweden. Just below the average are Finland, Ireland, and the United Kingdom, although the UK's position has fallen during the last 15 years while Ireland's has risen. Per capita incomes in Spain, Portugal, and Greece, not to mention most of the most recent new members, have remained consistently well below the EU average. Of course, variations in income persist among the regions *within each member* country, such as between the richer north and poorer south of Italy (see Figure 7.8). Indeed, the enlargement of the EU to include a large number of countries in eastern Europe whose GDPs are much lower than the EU average is rewriting the pattern of spatial polarization within the EU.

EFFECTS OF THE COMMON AGRICULTURAL POLICY (CAP)

The most striking changes in the regional geography of the EU, however, have been those related to the operation of the CAP. It is the CAP that dominates the EU budget. For a long time, it accounted for more than 70 percent of the EU's total expenditures, and it still accounts for more than 40 percent. Its operation has had a significant impact on rural economies, rural landscapes, and rural levels of living and has even influenced urban living through its effects on food prices.

The basis of the CAP was a system of support for farmers' incomes that was operated through the artificial support of wholesale prices for agricultural produce. While motivated mainly by political considerations, the CAP provided a relatively risk-free environment in which investment for farm modernization could be encouraged. At the same time, stable, guaranteed prices provided security and continuity of food supplies for consumers. Assured markets also allowed trends in product specialization and concentration by farm, region, and country to proceed at a faster rate than might otherwise have occurred, as Bowler (1985) showed in his survey of the geography of agriculture under the CAP. Not all products have been subject to CAP support, however. While regions specializing in crops and livestock subject to price guarantees, intervention, and market regulation have been able to intensify their specialization, other regions have been subject to Union-wide competition.

The overall result has been a *realignment of production patterns*, with a general withdrawal from mixed farming. Ireland, the United Kingdom, and Denmark, for example, have increased their specialization in the production of wheat, barley, poultry, and milk; while France and Germany have increased their specialization in the production of barley, maize, and sugar beet. It is at regional and sub-regional scales that these changes have been most striking. CAP support for oilseeds, for example, made rapeseed a profitable break crop in cereal-producing regions of the United Kingdom, with the bright yellow flowers of the crop bringing a remarkable change to the summer landscapes of the countryside.

However, the reorganization of Europe's agricultural geography under the CAP also brought some *unwanted side-effects*:

1. Environmental problems occurred because of the speed and scale of modernization, combined with farmers' desire to capitalize on generous levels of guaranteed prices for arable crops. In particular, moorlands, woodlands, wetlands, and hedgerows have come under threat and some "vernacular" landscapes of small farms have been replaced by the prairie-style settings of specialized agribusiness.

2. Another serious problem with geographical implications concerns the large surpluses fostered by the price support system. These "mountains" of beef, butter, wheat, sugar, and milk powder and "lakes" of olive oil and wine had to be sold off at a loss to neighboring countries, dumped on world markets, donated as famine relief, or "denatured" (rendered unfit for human consumption) at a considerable cost.

3. A third set of problems arose from the income transfers caused by CAP policies. Price support mechanisms involve a transfer of income from taxpayers to producers and from consumers to producers. There is plenty of evidence to show that these transfers are regressive within member countries and inequitable between them. Expenditure on food generally accounts for a larger proportion of disposable income in poorer households than in better off households. Producers, contrariwise, benefit from price support policies in proportion to their total production, so that the larger and more prosperous farmers receive a dispro-portionate share of the benefits. Spatial inequity arises because countries or regions that are major producers of price-supported products receive the major share of the benefits while the costs of price support are shared among member states according to the overall size of their agricultural sector. Furthermore, the CAP pricing system made no concessions for a long time to the variety of agricultural systems practiced on farms of different sizes and in different regions. As a result, areas with particularly large and/or intensive or special-ized farm units (such as northern France and the Netherlands) benefited most, together with regions specializing in the most strongly supported crops (cereals, sugar beet and dairy products). Effectively, this has meant that the most prosperous agricultural regions have benefited most from the CAP, so that farm income differentials within member countries have been maintained, if not reinforced.

4. In addition to all this, the budgetary cost of the CAP escalated. By 1983 budgetary problems had become acute; but reform of the CAP was hampered by domestic political considerations in member countries that were the biggest beneficiaries of the CAP. The CAP became a source of serious disharmony, particularly in the United Kingdom, where, before EU membership, food policies had been progressive, subsidizing lower income households. Embracing the CAP meant a higher and regressive system of food prices without any compensatory benefits: Peasant farming and inefficient agricultural practices had been purged from the UK economy long before.

Meanwhile, EC agricultural subsidies had become a serious issue in GATT and WTO negotiations; and the re-launch of the EC/EU in the mid-1980s required a more open and competitive approach to internal markets in every sector, including agriculture. Together with increasing awareness of the unwanted side-effects of the CAP, these considerations have led to ongoing reforms of the CAP. Since 1992 the guaranteed prices that farmers receive for arable crops, beef, dairy, and wine products have been cut gradually. To offset the lower guaranteed prices, direct payments to farmers have been increased, but the member states are now allowed to target these payments to achieve specific national or regional production priorities.

REGIONAL POLICY

The United Kingdom's accession to the Community in 1972 highlighted the lop-sidedness of Community policy in favor of rural interests compared with those of industrial areas. Although the Community had effectively operated regional policies through the ECSC and the European Investment Bank (EIB) for some time, there had been no comprehensive, coordinated framework within which to operate. The ECSC was limited to the "re-adaptation" of workers and the "conversion" of local economies in depressed coalmining and steel-producing regions. The EIB was a Community banking system designed to reduce intra-Community disparities in economic development by disbursing loans to selected projects in priority regions; but although it was particularly influential in sponsoring projects in marginal, cross-border regions, it was simply not equipped to deal with the casualties of regional economic restructuring within an expanding common market.

The entry of the United Kingdom to the Community not only made for a significant increase in the scope and intensity of regional restructuring processes but also brought a legacy of chronic regional problems and, with them, a certain political resolve. Following an examination of the issues (Commission of the European Communities, *Thompson Report*, 1973), the Community launched the structural funds in 1975 with a relatively modest budget.

The addition of Portugal and Spain to the Community in 1984 changed both the nature and intensity of regional problems. The proportion of the EC population living in "least favored" regions (those where gross domestic product was under half the EC average) doubled, with most of the increase being accounted for by depressed rural regions. At the same time, the re-launch of the EC, with more open internal markets, brought the probability of intensified spatial polarization. This was recognized by the Single European Act (SEA), which raised "economic and social cohesion" to the status of a new policy objective within the Community. The SEA doubled the funding—in real terms—for regional development assistance from the structural funds (from 7 billion European currency units (ECUs) to 14 billion ECUs at 1988 prices). This was further reinforced by the budget for 1993–1997 (the so-called Delors II Package), which contained a real increase of 30 percent in the European Union's budget, including 10 billion ECUs over five years for a cohesion fund to help Greece, Ireland, Portugal, and Spain achieve comparable levels of economic development to the rest of the EU by financing transportation and environmental infrastructure projects. By 2000 the structural funds accounted for 35 percent of the overall EU budget.

Meanwhile, the re-launch of the Union, the reform of the CAP and the accessions of 15 new members between 1995 and 2007 (Figure 12.2) provided the impetus for ongoing reforms of EU regional policy objectives. There are now three priority objectives to guide the disbursement of regional development assistance grants from the structural and cohesion funds between 2007 and 2013, as follows:

- *Convergence*: To promote the development of the most disadvantaged regions (with a GDP per capita less than 75 percent of the EU average) whose development is lagging behind (equivalent to the previous Objective 1 regions).
- *Regional competitiveness and employment*: To strengthen the competitiveness, employment and attractiveness of selected regions (other than those which are the most disadvantaged).
- *European territorial cooperation*: To strengthen cross-border and interregional cooperation (based on the previous INTERREG initiative).

A number of funds address these priority objectives. The structural funds comprise the European Regional Development Fund (ERDF), which supports productive investment including transportation and communications infrastructure, and the European Social Fund (ESF), which supports education and training. The cohesion fund is intended to reduce social and economic disparities. The *convergence and regional competitiveness and employment* objectives have an explicit regional dimension (see Figure 12.3).

These policies clearly represent a serious response to the spatial implications of economic integration. It is difficult to assess how effective they can be in redressing the regional restructuring and spatial polarization that have accompanied the creation and enlargement of the European Union. Yet it is debatable whether regional policies can, in fact, do so, particularly since the new reach and flexibility of TNCs can exploit cost advantages elsewhere in the world that the EU structural funds could never hope to match. What is clear, however, is that the funds are not always put to use. For the period 2000–2006, percentages actually used varied from 48 percent in Portugal to only 16 percent in the Netherlands. Within countries, usage also varies, seemingly in line with administrative capacity.

EXTERNAL EFFECTS OF THE EU

Meanwhile, the scale of the EU and its maintenance of a strongly protectionist agricultural policy, as well as a protected EU market, have inevitably had a significant impact on non-member countries: Diverting trade and creating complex new layers of interdependence. Much of this complexity relates to the "pyramid of privilege" that has arisen from the EU's trade agreements with different groups of non-member countries. At the base of the pyramid is a generalized system of preferences negotiated through UNCTAD. This allows access to the EU market for a broad range of products from LDCs. Bilateral trade agreements also exist with some countries as a result of attempts by the EU to extend and diversify its trading patterns. The most favorable trading privileges are extended to a large group of countries in Africa, the Caribbean, and the Pacific (the ACP states), most of them former colonial territories of member states. Originally established as the Youndé Convention in 1963 and later extended at Lomé Conventions in 1975 and 1979, these privileges allow access to the EU market for tropical agricultural products without having to provide reciprocal privileges to EU members or abandon trading agreements with other LDCs. They also involve an export revenue stabilization scheme—STABEX—that covers nearly 50 key primary products and raw materials.

In detail, the mechanics of these privileges are complex and it is very difficult to assess their impact on patterns of trade and development. It is clear, however, that the "privileges" extended to non-members are essentially designed to enhance the position of the EU rather than to contribute to a "new international economic order." "Sensitive" products (specifically those that compete directly with EU agricultural and industrial products), for example, are excluded from preferential treatment or are subject to seasonal restrictions. Moreover, the net effect of

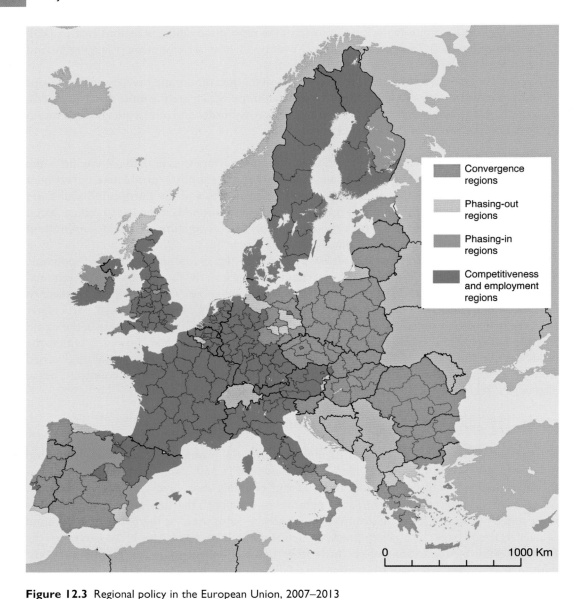

Figure 12.3 Regional policy in the European Union, 2007–2013

Source: Adapted from European Union, available at http://ec.europa.eu/regional_policy/atlas2007/fiche_index_en.htm

the Lomé "privileges" has been to increase the dependency of many countries on exports of a narrow range of primary produce to the EU market. Particular examples include:

- Burundi (coffee)
- Chad (cotton)
- Côte d'Ivoire (groundnuts)
- Ghana (cocoa)
- Senegal (wood, groundnuts)
- Sudan (coffee)

- Tonga (copra)
- Uganda (coffee).

As a result, EU relations with ACP countries have been interpreted as neo-imperialist, effectively extending the core–periphery structure of the world economy.

It is not only peripheral countries that have been affected by the EU, however. The EU represents an outcome of the struggle for economic power within the core and to preserve power in relation to the semi-periphery as much as it is an attempt to consolidate power in relation to the periphery. EU trade relations with the United States have been fractious, while the EU has been forced to mount a "diplomatic assault" on Japan in an attempt to stem the impact of Japanese direct investment in sophisticated manufacturing industries (automobiles, electronics, etc.) within the EU.

Core and semi-periphery countries have also been directly affected by the trade-diverting effects of the EU's protection of temperate climate agricultural products. Trade diversion effects are particularly evident where EU subsidies have produced large surpluses for export. For example, EU exports of beef rose from 5 percent of world trade in 1977 to over 20 percent in 1980 and in so doing displaced Australian and Argentinian exports to Egypt and Uruguayan exports to Ghana. Of course, the outbreaks of so-called mad cow disease and foot and mouth disease, which caused EU beef exports in 2001 to reach their lowest level for 20 years, created serious disruptions in trade for the EU's primary beef export markets of Egypt and Russia, which, like other countries, put a temporary ban on EU beef.

Other striking cases of trade diversion have occurred as specialist producers of temperate climate products with strong traditional ties to European markets found themselves largely excluded by the EU's external tariff wall. New Zealand is a good example. The United Kingdom used to take nearly all New Zealand's butter, cheese, and lamb, so that after the United Kingdom joined the EU, New Zealand agriculture had to be restructured, new products had to be developed (a notable success here being the kiwi fruit) and new markets had to be penetrated in Latin America, India, and Japan—in the face of competition from the subsidized surpluses of dairy produce from the EU.

Box 12.1 Genetically modified foods and U.S.–EU trade

"Not since the Green Revolution of the 1960s, when high-yielding wheat and rice varieties were developed that increased harvests in Asia two-, five- and even tenfold, have technological advances had the potential to so affect world agricultural trade." These optimistic words touting the "promise of technology" came from James M. Murphy, Assistant U.S. Trade Representative in 1999. The proponents of bioengineered seeds and food—genetically modified through biotechnology—argue that, compared to traditional crops, genetically modified ones require less water and fewer herbicides, produce higher yields and not only can taste better but may also be more nutritious and easily digested. Yet the opponents of genetically modified food for human consumption contend that the technology has not been adequately tested scientifically. They point to the potential for new allergens and toxins to be produced, which could cause mild to potentially deadly allergic reactions or for antibiotic-resistant genes to be transferred to humans, leading to the growth of antibiotic-resistant disease strains.

Since the mid-1990s, genetically modified foods have become a hot-button issue, not only in Europe, but also, increasingly, in the USA. In 1999 Cornell University researchers reported in *Nature* that laboratory tests showed that the use of genetically modified Bt-corn killed not only intended pests such as the corn borer, but also Monarch butterfly larvae. In September 2000 bioengineered food hit the newspaper headlines when Kraft Foods recalled tainted taco shells from supermarket shelves because they contained Star Link, a kind of genetically modified corn that was not approved for human consumption by the U.S. Food and Drug Administration (FDA).

In the USA, however, consumer awareness, while growing, remains low. The FDA does not require testing of genetically modified foods and has not indicated that it intends to change its existing voluntary labeling guidelines. U.S. agribusinesses (including Monsanto, the major supplier of genetically modified seeds in the USA) remain staunchly against mandatory labeling of genetically modified foods. Yet polls report that only between about 20 and 30 percent of those surveyed are aware that genetically modified foods are being sold in U.S. supermarkets. At the same time, over 80 percent of those surveyed support the mandatory labeling of genetically modified foods and would avoid buying them if they were clearly labeled. It is estimated that between 60 and 70 percent of the foods eaten by U.S. consumers contain some genetically modified organisms (GMOs).

Europe since the mid-1980s has seen increasingly widespread public support for stricter health safety standards in conjunction with the strengthening of EU environmental and consumer protection standards. More recently, there has been an undermining of public confidence following the highly visible failure of EU and national regulations to respond quickly enough to particular high-profile health and safety crises, most notably the outbreak of so-called mad cow disease. In response to protests and shoppers' objections, some supermarket chains in Britain and other European countries have refused to stock genetically modified foods. A number of companies, including Frito Lay, Gerber, and McDonald's, announced that they would stop using foods that came from genetically altered seeds in their products that are sold in Europe and the USA.

In 2001 the European Commission approved new rules requiring labeling of food or animal feed that contains more than 1 percent genetically modified ingredients. The rules also establish mechanisms to trace GMOs through production to distribution—"from farm to table." While these rules, considered to be the toughest in the world, remain to be approved by EU member countries and the European Parliament, EU decisions such as this are expected to have a major impact on U.S. farmers and producers who export food and animal feed to Europe. The USA produces a significant proportion of the genetically modified food that is marketed throughout the world and is the world's largest exporter of wheat for animal as well as human consumption. More than three-fourths of the U.S. soybean crop and nearly 40 percent of the corn crop is grown from genetically modified seeds. Agribusiness in the USA has already lost millions of dollars in exports because individual countries (such as Austria, France, and Luxembourg) have refused to accept shipments of genetically modified crops and food despite the fact that they were approved for sale by the EU and their actions were in violation of WTO rules against vetoing products without a clear scientific basis. In 2007, however, the EU trade commissioner officially urged member countries to comply with EU rules or face serious countermeasures in the WTO.

Showing greater concern for the food and environmental plight of less developed countries where poor soils and hunger are serious problems, the UN has called for a more balanced approach (UNDP, 2001: 3–4):

> The current debate in Europe and the United States over genetically modified crops mostly ignores the concerns and needs of the developing world. Western consumers who do not face food shortages or nutritional deficiencies or work in fields are more likely to focus on food safety and the potential loss of biodiversity, while farming communities in developing countries are more likely to focus on potentially higher yields and greater nutritional value, and on the reduced need to spray pesticides that can damage soil and sicken families.

SUMMARY

In this chapter, we have shown how the imprint of international and supranational economic and political integration has begun to affect the world's economic landscapes as nation-states have responded to the changing economic and geopolitical context of the world economy. This response has resulted in a variety of forms and levels of integration, but, in practice, the basic principles of economic geography have resulted in three main outcomes:

1. reinforcement of the dominant core–periphery structure of the world economy
2. spatial reorganization of production as trade creation and trade diversion affect both member and non-member states
3. creation and intensification of regional polarization as the economies of scale and multiplier effects in regions most favored by integration create backwash effects elsewhere.

Yet, while it is important to acknowledge international and supranational integration as a response to the globalization of the world economy, it would be unwise to overstate its effects. International and supranational organizations are not about to replace countries and we must recognize their limits as contributors to the constant rewriting of the world's economic landscapes. The further integration of the EU, for example, is seriously hampered by a number of issues that transcend its territory and jurisdiction, including its inability to reduce unemployment and significantly reduce income disparities between the richest and poorest regions. Moreover, as Nairn pointed out, spatial polarization within the EU has:

> [S]ought out and found the buried fault lines of the area . . . Nationalism in the real sense is never a historical accident, or a mere invention. It reflects the latent fracture lines of human society under strain.
>
> (Nairn, 1977: 69)

Indeed, nationalism and localism can be seen to be intensifying, not only in response to the backwash effects of international and supranational integration but also in response to the overall globalization of the economy, the internationalization of culture and society, and the insecurity and instability generated by the transition to advanced capitalism.

In the next chapter of the book, therefore, we turn to an examination of decentralist reactions to the changing world economy.

KEY SOURCES AND SUGGESTED READING

Agnew, J., 2001. How many Europes? The European Union, eastward enlargement and uneven development, *European Urban and Regional Studies* 8, 29–38.

Artis, M. and Lee, N. (eds.), 1994. *The Economics of the European Union*. Oxford: Oxford University Press.

Foster, J.W., 2005. NAFTA after its First Decade, in *The USA and Canada 2005*. London: Europa Publications.

Gibb, R. and Michalak, W. (eds.), 1994. *Continental Trading Blocs: The growth of regionalism in the world economy*. Chichester: John Wiley & Sons.

Grant, C., 2006. *Europe's Blurred Boundaries: Rethinking enlargement and neighbourhood policy*. London: Centre for European Reform.

Hardy, S., Hart, M., Albrechts, L., and Katos, A. (eds.), 1995. *An Enlarged Europe: Regions in competition?* London: Regional Studies Association.

Hoekman, B.M. and Kostecki, M.M., 1996. *The Political Economy of the World Trading System: From GATT to WTO*. New York: Oxford University Press.

Jacoby, W., 2004. *The Enlargement of the European Union and NATO: Ordering from the menu in central Europe*. Cambridge: Cambridge University Press.

Milio, S., 2007. Can Administrative Capacity explain Differences in Regional Performances? Evidence from structural funds implementation in southern Italy, *Regional Studies* 41, 429–442.

Peet, R., 2003. *Unholy Trinity: The IMF, World Bank and WTO*. London: Zed Books.

Pinder, D., 1998. *The New Europe*. Chichester: John Wiley & Sons.

Vogel, D., 2001. *The Regulation of GMOs in Europe and the United States: A case-study of contemporary European regulatory politics*. Paper prepared for a workshop on trans-Atlantic differences in GMO regulation sponsored by the Council on Foreign Relations.

Chapter 13

Reassertion of the local in the age of the global: Regions and localities within the world economy

Picture credit: Linda McCarthy

A persisting theme of this book is the existence of trends towards ever more powerful states and ever larger corporate structures. In the broader sweep of change within the world economy, these trends can appear to be inexorable and irreversible. Similarly, the increasing prominence of international and supranational institutions and initiatives (such as the EU, NAFTA, and the WTO) and transnational corporations (TNCs) can suggest a pervasive bureaucratization of life under the control of fewer and fewer organizations and individuals. Yet, while trends toward centralization, homogenization and standardization are real enough, there is also evidence for persisting and even increasing differentiation and decentralization: The peripheral industrialization that has come with the new international division of labor (NIDL) and the growth of the NIEs, the apparent reversal of previously depressed or underdeveloped local economies (for example, the Sunbelt phenomenon in the USA), and the revival or creation of regional–national identities (for example, Ukraine, Quebec, Scotland, Catalonia, Lombardy, Punjab), for example.

The two sets of phenomena are often related. So, for example, it is the centralization of economic power in TNCs that has often led to a decentralization of their productive activities (as noted in Chapters 3, 6, and 10); and it is attempts at political and cultural homogenization through international and supranational political unification that have generated resistance at the local or regional level (as noted in Chapters 3 and 8). The end of the Cold War and the collapse of the former Soviet Union gave an added stimulus to economic decentralization and political fragmentation. It is still too early to say whether this signals for eastern Europe and the former Soviet Union a permanent trend or a temporary hiatus prior to renewed political–economic centralization as evidenced by the accession of a number of east European countries to the EU and Russia's inclusion on the waiting list to become a member of the WTO.

However, a general trend all over the world in the wake of the increased integration of the world economy has been an enhanced differentiation between places. So, even as the world has shrunk in real terms with respect to flows of goods, services and investments, small differences in economic characteristics and cultural practices have taken on greater significance. As a result pressures towards a localizing of political decision-making power and political identities have increased. Three kinds of decentralist reaction have been increasingly common since the 1970s.

First, national governments have had to satisfy local and regional constituencies that they represent their best interests. When faced by geographically differentiated patterns of economic growth and decline, regional policies and regional devolution have been important responses. Since 1998, for example, certain powers formally vested in the UK's parliament have been devolved to legislative bodies in Scotland, Wales, and Northern Ireland. Under regional devolution, Scotland, Wales, and Northern Ireland have gained some measure of self-government, while remaining, with England, constituent parts of the UK and its national institutional framework. In Scotland, the responsibilities of the Scotland Office now include health, education, crime, housing, and economic development; the UK government has retained responsibility for a range of other issues for Scotland including employment, fiscal and economic policy, taxation, and social security.

Second, many countries are internally divided along cultural lines with regional/geographical bases. This has sometimes led to nationalist–separatist movements directed towards achieving autonomy or independence for disaffected regions. Basque nationalist separatists consider seven provinces that straddle the Pyrenees, four in Spain and three in France, to be the Basque country. The almost 3 million people in this region are believed to be the oldest indigenous ethnic group in Europe. During Franco's dictatorship in Spain (1939–1975), the Basques lost any political autonomy they had previously enjoyed, their culture was suppressed, and the use of their language was forbidden. In an effort to accommodate the region's separate national identity, the Spanish government has recognized since the early 1980s three Basque Provinces as an autonomous region, with its own parliament, police force, and separate language. While the majority of Basques in this region oppose the use of violence, up to 40 percent continue to support independence from Spain.

Third, and most generally, the growing globalization of the world economy has encouraged decentralization rather than centralization in the location of economic activities; whether in the form of the branch plants of big corporations or the localization and clustering of specialized small firms inherent in industrial districts. In particular, small-scale production has become of increasing importance (batch production, etc., as noted in Chapters 6, 7, and 10). Ideas such as "basic needs," "appropriate technology," and "local control" have become increasingly attractive in this context in framing the basic demands of new political movements calling for economic as well as political democracy. Although usually dismissed as Utopian, such ideas have become especially attractive as global resource and pollution problems arising as by-products of constant increases in global production and consumption have attracted more attention.

13.1 REGIONALISM AND REGIONAL POLICY

As we have shown in previous chapters, processes of economic growth and decline are not geographically neutral in their impact. In particular, the locational requirements of high-tech manufacturing production under the market-access regime and the growing service industries (such as finance) are likely to differ from those of established production (see Chapters 3, 6, and 10). It is in this context that appeals for governmental action arise to "help" a particular region or set of regions either "adjust" to a changing economic situation or encourage compensating investment by means of fiscal measures such as tax breaks or relocation allowances.

In the 1950s and 1960s there was widespread acceptance in many core countries of the need to encourage regional "balance" in economic growth at a time when established regional economies were beginning to experience challenges to their competitive advantage, and "poorer" regions (such as the Italian south or the U.S. South) were seen as lagging behind other regions. It is no coincidence that this acceptance flourished at a time of relative prosperity: Quite simply, affluent societies could afford to indulge in redistributive policies. In some countries this took the form of revitalizing or establishing lower "regional" tiers of government. Regional governments were viewed as agents for maintaining or attracting private investment. In some countries, such as Italy, Norway, and, to a lesser degree, the United Kingdom, regional authorities were introduced to encourage regional economic planning and foster local industrial regeneration. In many countries, especially those with federal political systems (such as the United States, Canada, Australia, Switzerland, and Germany), lower tier governments have traditionally played an important role in stimulating economic growth within their territories.

Two questions are especially pertinent with respect to the history of regional policy. One concerns the extent to which regionalism, or explicit commitment to spatial or regional planning, has inspired regional policy. The second involves the impact, if any, of regional and local development policies organized by lower tiers of government.

REGIONAL PLANNING

With regard to the first, some countries, such as France, Germany, and Italy, have long-established traditions of regionalism. In particular, French programs of "territorial management" and the Italian Cassa del Mezzogiorno (Southern Development Agency, replaced in 1987 by several smaller agencies) provide well-known examples. The United Kingdom and the United States acquired formal regional policies in the 1930s but in neither case has there been the same political consensus in favor of such policies (or anything that smacks of formal planning) as in other countries in Europe and elsewhere (for example, Japan, Brazil, India). In both cases earlier initiatives have largely been abandoned since the late 1960s in favor of either very localized programs, such as enterprise zones, or lower level government rather than national-level policies.

This reflects, in part, the coming to national power of governments ill disposed on ideological grounds to government intervention in the direction of economic activities to particular places. But it also reflects a negative appraisal of the effects of previous planning activities. If regional unemployment rates can be used as an indicator of regional "economic well-being," the fact that such rates are highly correlated over time and across countries irrespective of the commitment to regional policies suggests that such policies do not make much difference (Chisholm, 1990: 167–169). Incidentally, however, this also suggests a lack of evidence for the long-run spatial equilibrium in the distribution of economic activities assumed in most static models of regional development (discussed in Chapters 7 and 10), and so for the ideology that has often inspired the abandonment of regional policies.

REGIONAL AND LOCAL DEVELOPMENT POLICIES

The trend throughout the core countries over the past 30 years has been away from formal regionalism sponsored by national governments and towards the adoption of competitive spatial policies by regional and local governments. This has older roots, particularly in the United States where it dates back to the years immediately after the Second World War when southern states such as Tennessee and Mississippi began "attracting" firms from the northeast and Midwest with a mix of low production costs and fiscal advantages (especially low taxes). This approach spread widely in the USA with the onset of the massive restructuring of industry in the early 1970s. At the same time the narrow focus on attracting industry shifted to a broader concern with general local economic competitiveness, primarily through improving the overall "climate" for business and creating a mix of incentives for stimulating "new" industries with potential multiplier effects.

Individual U.S. states and municipalities have created economic development agencies to attract industry and foster endogenous economic development. Particularly conspicuous have been public–private partnerships and government offices established abroad, in London, Tokyo, or Brussels, to entice foreign business to particular locations in the USA. This latter strategy paid off handsomely for some states, such as Ohio and Kentucky, which succeeded in beating off other U.S. states in attracting major Japanese auto-assembly plants to their jurisdictions. Local government intervention of a similar type began to appear in the UK and other European countries in the 1980s, in part taking a leaf out of the United States book but also reflecting the availability of EU structural funds for providing grants and cheap loans to prospective employers.

The overall effectiveness of these local development efforts remains in doubt. While "success stories," such as that of Kentucky in the USA or the local government/small business linkages present in many parts of central and northeast Italy, are well known, evidence for the positive impact of these efforts in general is mixed. Frequently, local programs of tax abatements and subsidized plant and equipment merely relocate industry from another state or municipality rather than building fresh capacity and employment nationally. Ironically, given the usual association of "good" business climate with low taxes and limited public services, some evidence suggests that in the USA, state and local education, training, and infrastructure expenditures are more beneficial in generating fresh investment and new businesses than are subsidies to individual firms. Systematic fiscal reform, in the sense of aiming for low tax rates and broad tax bases, combined with efficient service delivery seems to offer the best formula for successful local development efforts in the USA. Specific tax subsidies to firms (including so-called enterprise zone experiments) seem a much less successful route to job growth and overall economic development in local economies. The theoretical framework outlined in Chapter 3 would suggest that the best local policies would be those that assist clustering by firms so as to increase transactional, learning, work training and other linkages. Otherwise competition between states and localities might encourage a "race to the bottom" with jurisdictions such as Mississippi becoming the norm against which states with long traditions of active government intervention to regulate workplaces and encourage economic development based on high-quality public services would find themselves at a disadvantage in the scramble to attract inward investment.

THE BALANCE SHEET ON REGIONAL PLANNING AND COMPETITION POLICIES

A basic dilemma remains unresolved, however. When local policies successfully promote economic development, both the capital and the labor (when educated or skilled) that benefited from the policies will probably act to undermine what the policies initially achieved. Capital will do this through takeovers and moving investment elsewhere; labor by migration to higher wage areas. This is the paradox of planning regional and local initiatives in the context of a world economy that is dynamic and "placeless" in its orientation to securing improvements in rates of return on investment. Yet, at the same time, rather than "rooting capital" in weak as well as strong local economies, local policies will, as the model of regional cumulative causation described in Chapter 7 might suggest, produce deepening spatial inequalities. Richer localities will have advantages in revenues and infrastructure that poor ones lack. The end result of this will be greater geographical concentration of productive economic activities.

Whatever the strength of this logic, however, after the downturn of the world economy beginning in the mid-1970s local governments in many countries became relatively focused on economic development efforts. What they were able to achieve, however, was constrained by their relative autonomy and by the need for national governments to curb public spending and reduce taxation. In the UK until 1994, the absence of a formal regional tier of government could be seen as a particular drawback to local initiative. Localities, such as city governments, are often too small to create effective economic development policies. Their control over physical (land use) planning is likewise too parochial when the environmental and labor market impacts of "new" industries extend beyond jurisdictional boundaries. Indeed, from one point of view the UK had the worst of all worlds because of the absence of a regional tier of government: the national government monopolized most controls over economic development and local government exercised control over physical planning. There was no intersection of authority or coordination of powers. This disadvantaged the UK as a whole in a context of increased international competition in which administrative regions elsewhere were able to offer "packages" of advantages not available in the UK. Partly in response to this and to the difficulties of coordinating the EU's regional programs and funding from London, the main state departments (employment and education, environment, trade and industry, transportation) were integrated in 1994 to create a regional government office for each region. These regional government offices (for the northeast, northwest, West Midlands, etc.) now attempt to coordinate the economic development activities of the local governments within their regions in their dealings with potential inward investors. But even in Italy, usually presented as a "showcase" of successful economic regionalism, the regions are relatively weak institutions with limited powers and a low popular profile. The Italian administrative regions have served mainly as spending agencies for national government policies. They also vary substantially in size, competence, and legitimacy; with those in the south particularly disadvantaged. Strong "regional motors" (such as those in central and northern Italy), therefore, do not necessarily generate a parallel strong model of regional or local governance.

One trend that goes against the tendency for enhanced competition between regions and localities is the emergence of "compacts" or agreements between regions in different countries with respect to technology licensing and plant establishment. So, such regions as Baden-Württemberg in Germany and Wales in the United Kingdom have arranged contacts to encourage the flow of technology and investment from the former to the latter. Of course, such cooperation can be regarded as simply a way of encouraging competitive advantage for the regions concerned rather than something totally different from more typical competitive strategies.

The fiscal crises experienced by many national governments beginning in the 1970s— increasing expenditures on social welfare programs (social security, healthcare) and defense were not matched by corresponding increases in revenues—led to increased pressure on local governments to provide compensatory spending on education and other public services. This has at one and the same time increased the fiscal problems of local governments, especially apparent in the USA, and reduced their ability to negotiate special "deals" with increasingly mobile businesses. Indeed, a strong case can be made that there has been a rolling back of government in many countries irrespective of the level at which it operates. Regions and localities are as caught up in the pressures of the market-access regime as are national governments. The primacy of market forces leads to an emphasis on entrepreneurialism, the imperative of flexibility and a "business localism" in which there is little or no popular oversight or control.

Direct central government intervention has still remained an important dimension of regional policy, if in some countries more than others. Substantial regional variations in economic development (especially when they are reflected in high regional unemployment rates) have been widely viewed as creating a national political problem, threatening the social and economic cohesion of the state itself. As a result, and beginning in the 1930s, national governments have felt it necessary to intervene in the economic geography of their territories by manipulating the costs of production.

The policies adopted have varied over time and by country. Some have sought to reduce the costs of fixed capital in declining regions by undertaking government investment in infrastructure, providing subsidies for private investment, etc. (German and Italian policies, in particular, have emphasized such strategies.) Others have sought to reduce the costs of capital in declining regions through subsidies for labor costs in those regions or through artificially increasing costs in "overheated" regions (in the 1960s French and British policies tended in this direction). The increased globalization of the world economy, however, has challenged the relevance of such conventional regional policies. In the first place, the largest firms no longer choose sites from among a single national set. There is no longer much correspondence between the scale of economic–locational decision making and the spatial scale over which governments can exert their fiscal powers. Indeed, in this context "the difficulties of guaranteeing full employment nationally mean that the state has increasingly to focus its attention on the national rather than the regional crisis; regional policy is in large part irrelevant" (Johnston, 1986: 274).

In the second place, the cost of subsidies raises government spending and produces a macroeconomic environment that is unattractive to global capital. Regional policy is then viewed as both an expensive luxury and an increasing liability. This was very much the view adopted by the Thatcher/Major governments in the United Kingdom and the Reagan/Bush administrations in the United States. Reducing state spending on regional policy was seen as a necessity for improving national competitiveness in a world economy. In the UK between 1979 and 1985, for example, regional aid was cut from £842 million to £560 million, a cut in real terms of exactly one-third. Moreover, the areas eligible for aid were "rolled back" considerably (see Figure 13.1).

Third, and finally, as Doreen Massey (1984: 298) points out: "No longer is there really a 'regional' problem in the old sense. No longer is there a fairly straightforward twofold division between central prosperous areas and a decaying periphery." Rather, because "it is not regions which interrelate, but the social relations of production which take place over space" (1984: 122); a new spatial division of labor based on spatial division of firms' activities has given rise to a changing and more localized pattern of spatial inequality. Her summary argument is as follows:

> The old spatial division of labor based on sector, on contrasts between industries, has gone into accelerated decline and in its place has arisen to dominance a spatial division of labor in which a more important component is the inter-regional spatial structuring of production within individual industries. Relations between economic activity in different parts of the country are now a function less of market relations between firms and rather more of planned relations within them.
>
> (Massey, 1984: 295)

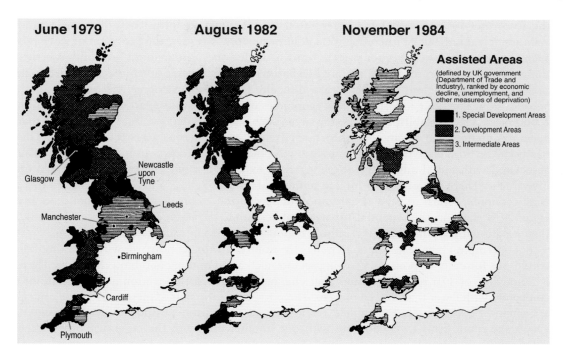

Figure 13.1 The Thatcher government's rolling back of regional aid, 1979–84

Source: Based on Martin (1986: 273, Figure 8.4)

From this perspective regional policy no longer engages with economic reality. Rather than discrete regional economies in competition with one another: "[A] new spatial division of labor made possible by new information and communication technologies and new non-spatial scale economies has created a localized pattern of restructuring" (Agnew, 1988: 131). In a world in which exogenous links have become central to local development, area-based policies for endogenous development often can appear less and less fruitful.

As decentralist reactions, therefore, regional policy and, especially, regionalism appear increasingly problematic. As Martin and Hodge (1983: 319) have argued with respect to the United Kingdom:

> [H]owever much regional policies of the conventional type are strengthened, if they are pursued against a background of continued unfavorable macro-economic conditions, their "social" role will be largely limited to simply spreading the misery of mass unemployment more "fairly" around the country while providing little boost to total economic activity or employment.

It may be too early to write off regional policy entirely, even if regionalism, the commitment to planning in a nationally coordinated fashion the economic development of fixed regional units, is largely in retreat. For one thing, to counterbalance its commitment to improving the overall efficiency of European industry in global competition through enhanced European competition, the EU has placed renewed emphasis on regional incentives to compensate for losses of traditional industries and employment (see Chapter 12). After declining steadily from the late 1950s, regional disparities in income increased throughout the EU from the mid-1970s, leveled off and fell slightly during the late 1980s, before rising again since 1990. In response, the European Commission has concentrated its spending on infrastructural projects and vocational training, while its competition policy attempts to restrict the provision of state aids such as grants and subsidies for private firms to within the "less favored" regions. Whether these are sufficient to rebalance regional disparities is questionable given the continued high rates of job loss in regions facing structural difficulties or where development is lagging behind. Second, some emerging industries enjoy economies of agglomeration (particularly localization economies). Once established in specific areas such industries have a local dependence that generates competitive advantage. Unfortunately, not everywhere can benefit from this. This has been the problem with proliferating growth poles, trying to mass produce the dynamism associated with initial advantage at great public expense but without much of a return on investment. There can only be a limited number of Silicon Valleys and Silicon Fens. So there is an economic return to early organization of specialized complexes. If a national economy is going to share in the potential for national economic development represented by these, then a national government will require a regional policy to encourage early response.

For example, Japanese regional policy is now centrally concerned with creating the conditions for early response in growing sectors such as computing and biotechnology. There is a network of 170 regional centers that channel support for innovation and research into companies with fewer than 300 employees. Three-fourths of the staff of these *kohsetsushi* centers are engineers who carry out applied research, and offer training and advice. The centers encourage small firms to collaborate with one another and with the large firms they often supply with components and services. The regional system provides an organizational framework for Japanese economic innovation. Countries such as the United Kingdom and the United States have suffered from the absence of such government "priming" of the pump of innovation while waiting for firms to do it themselves. Firms, however, have had no incentive to look beyond their own short-run interests to the interest of the country as a whole. From one point of view, that is what governments are for.

13.2 NATIONALIST SEPARATISM

The growth of industrial capitalism in the nineteenth century was accompanied by promotion of the nation-state and the growth of nationalism. The conviction grew among élites and populations at large that each state should be clearly bounded geographically, it should be organized as an economy, and it should be as linguistically and culturally homogeneous as possible. To many in Europe and North America, for example, the blessings of material abundance and personal freedom became associated with the interrelated development of capitalism and the nation-state.

In the twentieth century, however, nationalism was regularly perverted into fascism. Dreadful wars were fought. National independence has been no guarantee of national prosperity. Even in the original founding states of Europe, such as Spain, France, and the UK, political movements rejecting established national claims and asserting political and economic rights for regional and ethnic populations have become widespread.

STATES VERSUS NATIONS

This last trend reflects the fact that state making and nationalism have never redounded equally to the benefit of all the nominal citizens of a country. In particular, since their earliest formation, nation-states have contained a diversity of cultural groups within their boundaries. In the social science literature, the term ethnicity has come to signify the organization of cultural diversity within states. So an ethnic group can be defined as "a collectivity of people who share some pattern of normative behavior, or culture, and who form a part of a larger population, interacting within the framework of a common social system" (Cohen, 1974: 92).

Ethnic groups, however, are not simply primordial groupings, even though they usually draw on myths of common ancestry and cultural distinctiveness. They are differentiated from one another and integrated internally through such mechanisms as a cultural division of labor, political favoritism, and historically created economic roles. Indeed, ethnicity can be viewed as a mechanism for allocating wealth and power within states that have not inherited or successfully imposed a unifying set of cultural practices and symbols on their populations. A clear example of the use of ethnicity in this way would be Northern Ireland between 1920 and 1972 where a dominant élite of British Protestant landowners and businessmen maintained their hegemony over the region through a web of mutual obligations, customs, duties and economic favors that bound Protestant workers and small farmers to them while excluding the Irish Roman Catholic population. This process of ethnic competition for control over the fruits of economic growth and government policy is extremely widespread the world over; the more so the greater the number of ethnic groups and the weaker the alternative means of political mobilization (for example, social class).

Students of ethnicity have noted that in recent years the level and intensity of conflicts between ethnic groups have been on the increase. Writing of Indonesia, Clifford Geertz (1973: 244–245) has provided a particularly vivid description:

> Up until the third decade of this century, the several ingredient traditions—Indic, Sinitic, Islamic, Christian, Polynesian—were suspended in a kind of half-solution in which contrasting, even opposed styles of life and world outlook managed to coexist, if not wholly without tension, or even without violence, at least in some sort of workable, to-each-his-own sort of arrangement. This modus vivendi began to show signs of strain as early as the mid nineteenth century, but its dissolution got genuinely under way only with the rise, from 1912 on, of nationalism; its collapse, which is still not complete, only in the revolutionary and post-revolutionary periods [1945 on]. For then what had been parallel traditionalisms became competing definitions of the essence of the new Indonesia. What was once, to employ a term I have used elsewhere, a kind of "cultural balance of power" became an ideological war of a peculiarly implacable sort.

Some, for example Kedourie (1960), have seen this ethnic schismogenesis as a worldwide process associated with the diffusion of the idea of nationalism from Europe along with colonialism. Others have emphasized modernization or industrialization as universal processes laying the material foundations for the politics of nationalism (for example, Gellner, 1964). In fact, ethnic nationalism seems to have developed in different ways and with different causes in different parts of the world. Rokkan and Urwin (1983), for example, argue that processes of economic, military–administrative and cultural "system building" have combined in different ways to produce different effects in different European localities. In turn, ethnic nationalism has been accommodated in some settings (such as the Celtic fringe of the UK (Scotland, Wales, Northern Ireland) and the Basque provinces of Spain), and discouraged elsewhere (for example, in Alsace, France, and the South Tyrol, Italy).

ETHNIC CONFLICT AND NATIONALIST SEPARATISM

Whatever its precise origins in particular cases, however, ethnic conflict and nationalist separatism (when ethnic groups are geographically concentrated) have become increasingly marked features of the contemporary world. From Ireland to the former Yugoslavia to

Lebanon to India to Sri Lanka to Canada, to name just a few of the best known cases, ethnic groups and ethnic conflict have become major elements in national political life. Three factors seem to be especially important in this trend. One of these is the increased economic–geographical differentiation within states and its relationship to ethnic divisions. It is not that ethnic conflict always involves increasingly poor regions rebelling against more affluent ones. It is difference *per se* generating a sense of deprivation or exploitation. In Spain, for example, it is the Basque and Catalan regions—the most prosperous in the country—that are the most rebellious. Likewise in the former Yugoslavia, where the relatively well-off Slovenians and Croatians demonstrated their impatience with "subsidizing" the ethnic groups (Serbs, ethnic Albanians, etc.) that occupy other regions, by agitating for and achieving political independence. Yet within Serbia today, it is the poorer ethnic Albanians in the province of Kosovo who have engaged in the most active nationalist separatism efforts.

The breakup of the former Yugoslavia illustrates a second factor of singular importance in the explosion of nationalist separatism in 1989–1993: The collapse of the Soviet Union, the exhaustion of state socialism and the end of the Cold War. The demise of strong central governments and the exhaustion of state socialism as an ideology have opened the way for a re-emergence of political identities based on ethnic divisions. Formerly communist states such as Yugoslavia and the Soviet Union were organized administratively around geographical units that reproduced ethnic cleavages. Even though some groups, such as Russians in the Soviet case, are to be found scattered in considerable numbers outside their own republics, the dominant identity of particular administrative units remained that of the historically dominant ethnic group.

Within Russia itself the absence of the distinct groups that underpin the political divisions in other countries, such as organized labor or religious traditions, has led to political organization by entrenched vested interests from the bureaucracies and by ethnic group. One of the early results of the breakup of the Soviet Union within Russia was the shift in power from the center to the regions. Local governments were formerly the instruments of central rule but by 1992 had become major protagonists in political–economic development. Some of the 27 million non-Russians in Russia even declared sovereignty within their local government units. These units have different economies, some are raw materials producers and others are industrial, and so have different interests in terms of pricing and macroeconomic policies. So there is the possibility within Russia of ethnic and economic differences becoming mutually reinforcing as they did within the former Soviet Union as a whole (see Figure 13.2). Yet, this has proved to be a passing phenomenon. Since the first election of Vladimir Putin as President in 1999, the regions have been brought to heel under a revived central government in Moscow, even if there is still continuing hostility to the center in such parts of Russia as the North Caucasus.

Well beyond Russia's confines, however, the end of the Cold War with the United States has had a loosening effect on what had become "historic" international borders. The "freezing" of political boundaries that both sides in the Cold War had quietly accepted no longer can be tied to an overriding global conflict. Each and every territorial dispute is no longer a potential spark for a Third World War. This opens the way for a possible proliferation of nationalist separatisms (and also expansionist claims by existing states) in all world regions as established political boundaries lose their previous inviolability.

The third factor is the growing globalization of political and economic activity. The shift of power and control over local economies to ever more distant locations provides an incentive for regional counter-mobilization. The development of the EU in Europe may have been one stimulus; the growing importance of TNCs may be another. At the same time the increase in

Figure 13.2 Governmental decentralization of Russia, 1993

Source: Based on *Economist* (1992c: 25)

Box 13.1 International terrorism

One strategy used by some nationalists, following in the path of guerrilla movements in former colonies, has been to use terrorist methods to try to make the "occupation" of their lands by others intolerable. These tactics have ranged from the bombs planted by IRA (Irish) and ETA (Basque) separatists to fully fledged warfare, as in Chechnya and other territories in the Northern Caucasus region of Russia.

Since the late 1990s, however, this largely nationalist terrorism has been supplemented by the growth of a brand of international terrorism interpreted by many, not least by the U.S. administration of G.W. Bush, as a war against the United States and the world economy. Associated with such shadowy groups as the Islamic jihadist *al Qaeda* (which has been linked to the attacks on the World Trade Center in New York on September 11, 2001), this type of terrorism has been seen by some as a violent response to the spread of globalization, particularly by extreme elements in the Arab and Islamic worlds, even as it makes use of the very technologies, such as cellphones and planes, that globalization entails.

Not only has terrorism taken on a global reach, however, it is also apparent that cities have become a preferred location for large-scale terrorist attacks. Even before the terrorist attacks on the World Trade Center in 2001, cities had become the central venues of terrorist attacks. Between 1993 and 2000, for example, there had been over 500 terrorist incidents in more than 250 cities around the world.

There are several reasons for this. First, cities—especially world cities—have considerable symbolic value. They are not only dense agglomerations of people and buildings but also symbols of national prestige and military, political, and financial power. A bomb in London's underground or a poison gas release in a Tokyo metro arouses international alarm. This kind of event will be communicated instantly to a world audience. Second, the assets of cities—densely packed and with a large mix of industrial and commercial infrastructure—make them rich targets for terrorists. Third, cities have become nodes in vast international networks of communications. This is a reflection not only of their power, but also of their vulnerability. A well-placed explosion can produce enormous reverberations, paralyze a city, and spread fear and economic dislocation. Finally, word gets around more quickly and socialization proceeds more rapidly in high-density localities. These kinds of environment can be an abundant source of recruits for terrorist organizations.

Terrorism takes a toll on cities in a variety of ways. The impacts of the 2001 terrorist attacks on New York City have been significant—and not limited to the death and destruction that was targeted on lower Manhattan. The time and cost of doing business in New York has gone up, even for workers and companies quite distant from the attack site.

In this connection, central London has sought to reduce the threat of terrorist attacks. Physical and increasingly technological approaches to security have been adopted at increasingly expanded scales. In 1993 a security cordon was put into place, securing all entrances to the central financial zone of the City of London (the Square Mile). The 30 entrances to the City were reduced to seven, with roadblocks manned by armed police. Over time the spatial scale of the security cordon was increased to cover 75 percent of the Square Mile. The security cordon, as a territorial approach to security, was augmented by retrofitting a closed circuit TV (CCTV) system. The police, through their Camera Watch partnership effort, encouraged private companies to install CCTV. At the seven entrances to the security cordon, 24-hour automated number plate recording (ANPR) cameras, linked to police databases, were installed. Within a decade the City of London had been transformed into the most surveiled space in the United Kingdom, and perhaps in the world, with more than 1,500 surveillance cameras in operation, many of which are linked to the ANPR system.

Ultimately, the response to international terrorism mirrors the response to the nationalist use of terrorist methods that have been attempted in the past. Although bloody in terms of loss of civilian lives, terrorism of all kinds is rather like the strategic bombing campaigns carried out by air forces during the Second World War: It does not work to intimidate populations into surrender. Although often the seemingly only available "weapon of the weak," it tends to create increased animosity and violence rather than achieve the political outcomes desired.

the flow of international migrants, especially into Europe, introduced ethnic groups, such as Indians in the United Kingdom, Turks in Germany, and Algerians in France, which both stimulate the demands of indigenous ethnic groups and provide additional "out-groups" for new rounds of ethnic conflict and nationalist politics.

With the exception of the former Soviet Union and some parts of eastern Europe (former Yugoslavia and Czechoslovakia) outright separatist movements have not met with success, at least as measured in terms of political independence. During the Cold War the world superpowers generally refused to back separatist movements, perhaps for fear of stimulating them at home or within their own spheres of influence. Often, political changes short of outright independence have proved satisfactory responses to regional–ethnic revolt. These include federalism, regional devolution, and *consociationalism* (power sharing among ethnic groups as in Switzerland and the Netherlands). What is clear, however, irrespective of the prospects for nationalist separatism as such, is that "there is little likelihood of an abatement of ethnic nationalism in the near future" (Williams, 1982: 36).

13.3 GRASSROOTS REACTIONS

SMALL IS BEAUTIFUL?

A peculiar paradox of the growing globalization of the world economy has been the stimulus it has provided to the destruction of some specialized local/regional economies and the decentralization/localization of single plants or use of subcontractors at disparate locations. Breakthroughs in information and transportation technologies have made it possible to decentralize production operations to lower cost locations or ones with special advantages in terms of access to technology, labor skills, or markets at the same time that central corporate control is maintained or enhanced. This process has been brought about by increased competitive pressures on large firms from the appearance of foreign competitors. Many large firms, especially in Europe and the United States, now face a global marketplace far more competitive than the more geographically restricted ones they had known previously. Of course, as argued in general in Chapters 1 and 3, and then repeatedly in subsequent chapters, branch–plant industrialization is only one among a number of strategies for re-establishing firm competitiveness. And, it does not signify that there are not continuing and new pressures for the clustering of production facilities. The point to be made here is more that, today, there are greater incentives and technological possibilities for firms—even small firms—maintaining control over production at a distance than was the case in the past.

Coincidentally, union/management conflicts in established production facilities and changes in market conditions, especially the increased demand in many developed countries for customized rather than mass-produced goods, have also encouraged decentralization of production. In Italy, for example, one can see evidence of both causes. Large firms in Piemonte (around Turin) and Lombardia (around Milan) have increasingly contracted out to small firms for parts and services that used to be provided onsite at large factories. Moving production in this way both undermines the power of workers in large factories, where solidarity is more easily achieved than in scattered small factories, and protects the large firm from the need to shed labor cyclically. In Emilia-Romagna (around Bologna) and elsewhere in central Italy many small firms provide customized products (both producer and consumer goods) to domestic and export markets that are better served by the flexible response to shifts in fashion that these small firms can provide. Whether they can continue to do so in the face of intense competition from Chinese and other low cost producers in the same sectors remains to be seen. Some Italian firms have already joined the rush to invest in China and Eastern Europe to reduce their wage bills once more, this time by foreign outsourcing rather than by relocating or decentralizing production within Italy.

To some commentators, however, the previous development of a highly competitive Third Italy (see Box 6.4) signified a wholehearted shift away from mass to customized production, with local rather than long-distance connections central to flexible production. In fact, although local clusters of small firms have been very important to this process, they are strongly tied into the longstanding industrial system and its main urban centers. So in Italy during the 1990s the provinces that had the highest levels of foreign exports all cluster in the vicinity of Milan in the "traditional" industrial northwest rather than in the Third Italy of central and northeastern Italy (see Figure 13.3).

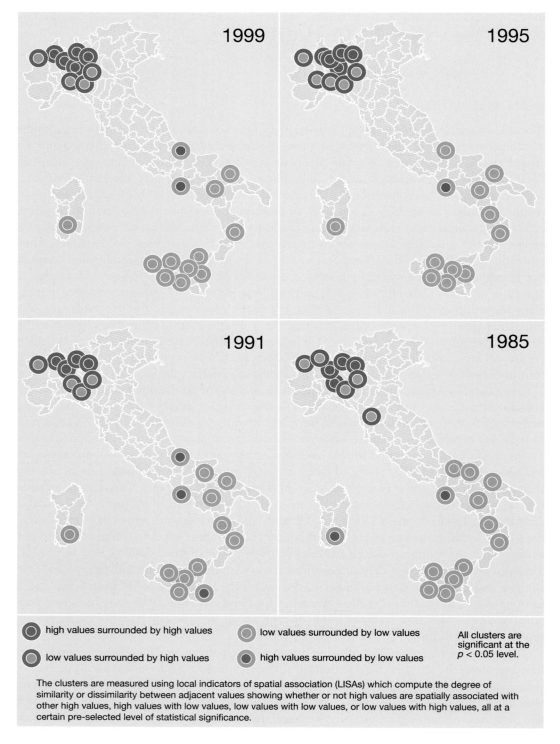

Figure 13.3 Geographical clustering of export shares in Italy, by province, 1985–1999

Source: Adapted from Agnew *et al.* (2005: 96, Figure 4)

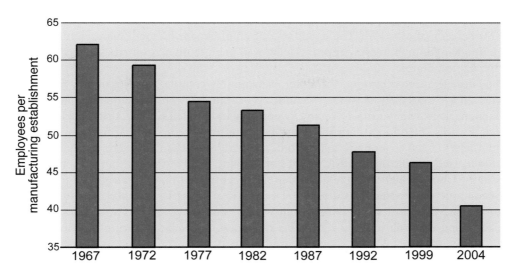

Figure 13.4 How plant size in the United States has shrunk

Source: Based on U.S. Census Bureau, Statistics of U.S. Businesses (2004) Manufacturing data http://www.census.gov/epcd/susb/2004/us/US31.HTM#table0

In the United States, both types of decentralization to small firm clusters—away from large, unionized plants and in the interest of customized production—are also increasingly common, although the small firm as supplier to the plants of a large firm has been most characteristic since the late 1960s. Companies with fewer than 500 workers added 1.2 million jobs in the United States between 1976 and 1984, while larger companies lost 300,000 jobs. By 1999, Census Bureau statistics for U.S. businesses indicate that companies with fewer than 500 workers had risen to 91 percent of all manufacturing firms and accounted for 41 percent of manufacturing jobs. Paralleling this, production facilities have also shrunk in size. Plants in the USA during the 1960s had, on average, 30 percent more workers compared to plants now (see Figure 13.4). These data cast doubt on the claim that not much has changed in either the scale of production or the average ownership size in U.S. manufacturing industries. What perhaps has not changed in the USA is the *directing* role of large firms, only now in relation to a variety of production arrangements with other firms rather than solely in terms of vertical integration of all stages of production within one firm (see Chapter 3).

A similar fragmentation of production has been noted in the United Kingdom. Since the late 1970s average employment in plants of firms employing more than 999 people has shrunk and the average number of plants operated by these firms has increased significantly. In 1973 average employment per plant was 459 and there were 12 plants per firm. The comparable figures for 1982 were 338 and 15. Yet, at the same time, there was a decrease in the number of firms in the category of 999+ employees, indicating an increasing concentration of ownership. By 1996, for example, only 0.1 percent of the manufacturing businesses registered for value-added tax (VAT) in the UK had 999 employees, while 69.4 percent of manufacturing firms had between one and nine employees. In the British case, reduction in the size of plants has not involved a reduction in the concentration of ownership of manufacturing industry. However, some of the decrease in the average size of all workplaces (as opposed to just those of the larger firms) is due to the expansion of employment in small firms. Between 1980 and 1991, for example, UK manufacturing firms with fewer than 100 employees were responsible for at least 50 percent of job creation, while firms with over 100 employees accounted for just over 60 percent of jobs lost. The expansion of employment in small manufacturing firms in the United Kingdom, however, appears to lag well behind other countries such as Italy, Japan, and the United States.

The "decentralized economy," however, is incredibly volatile. Patterns of new firm creation/ destruction are extremely sensitive to minor cyclical fluctuations in demand from larger firms and final markets. Many firms have tiny inventories and limited equity. Although advantageous financially, this introduces major limitations in terms of employee security and long-term commitment to local economies. In this context, local rather than aggregate national conditions are of increased significance for the welfare of populations. Political parties such as the Labour Party in the United Kingdom have supported the decentralization of some governmental powers over economic development as a reaction (in part, at least) to the increased localization of economic activities. In Italy, the Communist Party (and since 1992 its successor party on the left), excluded from national governments on a permanent basis during the Cold War, has been a major sponsor of both economic and political decentralization in the central regions (Emilia-Romagna, Toscana and Umbria) where it dominates local government. Without supportive national government policies, however, local economic initiatives organized on a geographical basis are problematic. Above all, they face the dilemma of protecting current employment while attempting to create local economies that can capture the benefits of "emerging" industries.

Perhaps local initiative and control or "small is beautiful" can only work when decision-making power is no longer vested in giant transnational corporations, as it still is even when small firms proliferate to serve them (as is the case even in many industrial districts). For example, the strategy for procuring components used by Ford in Europe involves the increased use of widely scattered and expendable suppliers. This approach is hardly amenable to either government controls or local economic development strategies geared towards long-term stability in employment. The lack of a "natural evolution" of economies towards greater security of employment and increased participation in the benefits of economic growth has fuelled the revival of thinking about the possibility of wider popular participation in economic decision making. The theme of economic democracy is especially strong in some recent proposals for stimulating the U.S. economy. Economic democracy refers to an egalitarian form of political–economic structure in which a serious attempt is made to democratize the economic sphere in general and workplaces in particular. The major point is to challenge the political and economic position of global capital and the commitment to it and its international role by the major states (the UK, USA, etc.). It builds on the view characteristic of movements for participatory democracy that, to be more than a sham, democracy should be extended beyond episodic political activities such as voting into the economic sphere.

ECONOMIC DEMOCRACY

Economic democracy differs from democratic capitalism (in both laissez-faire and welfare state manifestations), in which democracy is limited to periodic involvement in electoral politics and the means of production are largely privately owned. It also differs from conventional state socialism, especially of the now defunct Soviet variety, in which markets are prohibited (in public), there is little meaningful electoral politics and the means of production are owned by the state.

There are a number of reasons why interest in both the theory and practice of economic democracy has increased in recent years. First, as the heavy "smokestack" industries in the

United States, Europe, and other DCs have become less profitable there have been numerous attempts by their employees to save their jobs by buying failing factories. Changes in tax laws have also made employee stock option plans (ESOPs) more attractive to businesses. Between 1976 and 2002 the number of ESOPs grew in the United States from fewer than 300 to about 11,000. While owning stock is hardly the same as control, ESOPs do raise the question of where employees' participation should stop. But it is probably the competitive environment for businesses and the prospect for bankruptcy that do most to encourage talk about and proposals for producer cooperatives and worker self-management.

Second, interest in economic democracy reflects consideration of actual practices in a number of countries that have had high levels of economic growth and high standards of living. In Germany, for example, "codetermination" allocates positions on corporate boards to employees. In addition, the success of several large-scale prototypes has given advocates of economic democracy a ready reply to those who assert that self-management by employees is inherently Utopian. The extensive network of Basque financial, industrial, and distribution cooperatives in Spain, Mondragon, employs over 60,000 workers and has for many years enjoyed better profitability due in part to higher worker productivity compared to some of its more conventional competitors.

Third, and finally, especially in the LDCs, economic democracy can be seen as an alternative both to U.S.-style corporate capitalism and socialist-style central planning. Under nationalistic pressure to avoid becoming satellites of the major world powers, the rhetoric and sometimes the substance of economic democracy arise; but the pressure is also immediately practical. Considerable evidence suggests that rapid economic growth in LDCs does not necessarily improve the welfare of large numbers of their people. Markets, based on effective demand, which means the given distribution of income, have generally failed to allocate resources to the basic human needs of the poor in many LDCs—mass poverty, unemployment and malnutrition are the consequences. Directing attention to local production for "basic needs" has been one response, based on the use of indigenous knowledge and traditions of production as opposed to imported ones. Discussions of "appropriate technology" and "alternative" development strategies, however, are both inspired by similar concerns. They also reflect increasing concern over the limits to the growth of the world economy. Can the world's natural resources and increasingly fragile physical environment support the levels of production that it would take for the entire population of the world to enjoy U.S. or Swiss levels of consumption?

There is, of course, a range of possible grassroots reactions within the general confines of economic democracy. Among advocates of economic democracy are some who are committed to a vision of large firms with powerful, central councils, while others look to smaller, decentralized firms embedded in non-state networks of social association. Comisso (1979) compared this difference to that between federalists (such as Hamilton) and Jeffersonians at the time of the founding of the United States. To take one example, Hirst (1994) has argued for what he calls "associative democracy" resting on two major principles: A decentralized economy based on cooperation and mutuality and a system of governance based on self-governing associations of mutual interests. State power is to be weakened through a pluralizing and federalizing of political authority and economic power is to be redistributed to local economies. Hirst draws on the examples of manufacturing success in Italian industrial districts, German regions and Japanese firms to propose an associative regional economy linked into others by federal channels of authority. Hirst claims that such a model would both stimulate economic growth and redistribute its fruits in a more egalitarian fashion than is the case with the present world economy.

One important objection to economic democracy of all types, but especially the more decentralized ones, is that they are Utopian dreams since those in power will not hear of them. The world is now bureaucratized; bureaucracies will not seriously consider participatory democracy. In particular, most states are run by oligarchies determined to fix the best deals for themselves from transnational corporations and the most powerful states. Williams (1981), however, has defended the need for utopias. He argues that:

> [T]he purpose of a radical utopia is to create a tension in our souls . . . We must imagine something better. That defines us as people who offer our fellow citizens a meaningful choice about how we can define and live our lives . . . Radicals must confront centralized nationalism and internationalism and begin to shake it apart, break it down, and imagine a humane and socially responsible alternative. It simply will not do to define radicalism as changing the guard of the existing system.
>
> (W.A. Williams, 1981: 95, 98)

In essence, this is what grassroots reactions inspired by a vision of economic democracy are really all about.

A second objection is more by way of a critique of possibilities of successful long-term local development in the absence of a relatively strong state presence in economic regulation. Amin (1996: 309–310), for example, is critical of proponents of local "associative democracy" (particularly Hirst) for "failing to distinguish between different forms of state economic intervention and state practice," and for undervaluing "the strategic and developmental role played by the state in some of the most successful economies in the world." In a somewhat different vein, Donahue (1997) has claimed that a state defines something of a commons in which externalities across local and regional boundaries are so intense as to vitiate against the possibility of ever successfully separating out groups of people into discrete geographical communities that can be run as if the others did not exist. Federalism is about achieving a balance between the common and the particular. For Donahue (1997: 42), the "devil in devolution" is that, in the U.S. case, the dominant consensus:

> [I]n favor of letting Washington [the U.S. federal government] fade while the states take the lead is badly timed. The public sector's current trajectory—the devolution of welfare and other programs, legislative and judicial action circumscribing Washington's authority, and the federal government's retreat to domestic role largely defined by writing checks to entitlement claimants, creditors, and state and local governments—would make sense if economic and cultural ties reaching across state lines were *weakening* over time. But state borders are becoming more, not less permeable.

SUMMARY

In this chapter, three types of decentralist reaction to the impact of the world economy have been described in the context of the trend towards globalization of economic activities.

Regionalism and regional policy under state sponsorship were common up until the 1960s but their relevance has increasingly been questioned in the face of the changed relationship between national and world economies. In many countries, regional and local tiers of government have tended to displace the regionalism carried out under the sponsorship of national governments.

Nationalist separatism challenges existing states, international and supranational organizations, and the existing distribution of economic activities. But most separatist movements will usually settle for something less than complete independence.

Finally, recent trends in the world economy have generated renewed interest in the possibility of economic democracy. Disillusionment with both U.S.-style corporate capitalism and Soviet-style socialism in the face of an increasingly volatile world economy has directed attention to the possibility of people taking control of their economic activities and putting them to work for them.

Whether or not decentralist reactions increase in importance depends in part on whether the world economy recovers from its present problems, especially the debt crisis, the slowing of growth in world trade and the lack of congruence between global production and global consumption, and whether or not the free market ideologies antithetical to many of these reactions and associated for many years with the Reagan/Bush administrations in the USA, and the Thatcher/Major governments in the United Kingdom continue to find support around the world and in international organizations such as the World Bank. Without some dramatic rebalancing of global patterns of production and consumption, and the geographical spread of the benefits of globalization, the trend towards decentralist reactions may prove inexorable as people begin to take their fate into their own hands.

KEY SOURCES AND SUGGESTED READING

Amin, A. and Thrift, N., 1992. Neo-Marshallian Nodes in Global Networks, *International Journal of Urban and Regional Research* 16, 571–587.

Brusco, S. and Sabel, C., 1981. Artisan Production and Economic Growth, in F. Wilkinson (ed.), *The Dynamics of Labour Market Segmentation*. London: Academic Press.

Eisenschitz, A. and Gough, J., 1992. *The Politics of Local Economic Policy: The problems and possibilities of local initiative*. London: Macmillan.

Gause, F.G., 2005. Can Democracy stop Terrorism?, *Foreign Affairs* 84(5), 62–76.

Harvie, C., 1994. *The Rise of Regional Europe*. London: Routledge.

Johnston, R.J., 1986. The State, the Region, and the Division of Labor, in A.J. Scott and M. Storper (eds.), *Production, Work, Territory: The geographical anatomy of industrial capitalism*. Boston, MA: Allen & Unwin.

Rokkan, S. and Urwin, D., 1983. *Economy, Territory, Identity: Politics of west European peripheries*. London: Sage.

Scott, A.J., 1998. *Regions in the World Economy*. Oxford: Oxford University Press.

Shin, M.E., Agnew, J., Breau, S., and Richardson, P., 2006. Place and the Geography of Italian Export Performance, *European Urban and Regional Studies* 13, 195–208.

Chapter 14

Conclusion

Picture credit: © European Union, 2013

In this book, the task of making sense of the geographical pattern of economic activities around the world has been pursued using the geographical metaphor of core and periphery; this provides a basic model or way of seeing, against which actual patterns of economic activities at a range of scales can be compared. Even at scales finer than the global, or world region like Europe, the core–periphery conceptualization can be useful in thinking about how the core and periphery intersect locationally. The social, economic, and locational polarization of migrant workers in world cities such as Los Angeles, for example, reflects a kind of periphery or semi-periphery existing within the core. Significant pockets of wealth in some NIEs and oil rich semi-peripheral countries, in contrast, could be viewed as part of the core within the semi-periphery.

In this connection, geographers Stephen Graham and Simon Marvin identified an important tendency in cities—splintering urbanism. Splintering urbanism is characterized by an intense geographical differentiation, with individual cities and parts of cities engaged in different— and rapidly changing—ways in ever broadening and increasingly complex circuits of economic and technological exchange. The uneven evolution of networks of information and communications technologies is forging dynamic landscapes of innovation, economic development, and cultural transformation, while at the same time intensifying social and economic inequalities within and between cities.

In this book, a number of claims about the processes creating and recreating economic landscapes have been investigated using a historical–geographical framework deriving from an evolutionary perspective on the development of the world economy. In this short concluding chapter, the key elements in the argument are drawn together as a way of summarizing this perspective and pointing the reader back to the themes of the introductory chapters. A number of controversial issues at the heart of contemporary academic and policy debates are also introduced to highlight the open-ended and exciting character of globalization.

First, the world economy is an open system that evolves over time. Although there is an obvious geographical path dependence to its motion, as illustrated by the return to initial advantage produced through increasing returns to scale, there are also locational reversals and shifts in the way it works in different contexts. So although the core of the world economy was the first to industrialize on a massive scale, some other parts of the world have recently experienced a substantial expansion in industrial and, more recently, service activities. This reflects the shifts in the operation of the world economy detailed in Chapters 4–11, especially the disintegration of Fordism that was important in the core for much of the twentieth century. The workings and outcomes of the world economy, therefore, are not set in stone but evolve and change over time. In particular, the evolution of the world economy is not best thought of as a cyclical repetition of what has happened previously only with different technologies and countries. The mechanisms driving the world economy have also changed. Today, for example, it is misleading to suppose, as some scholars still do, that the globalization of production and finance is essentially indistinguishable from the territorialization of production and trade within empires that characterized the world economy in the late nineteenth and early twentieth centuries. These are different processes and should be seen as such.

Second, this dynamic understanding of the world economy allows us to combine a focus on general economic forces with a concern for the local variability that characterizes the world economy. The expansion of the world economy has incorporated regions with distinctive economic histories that are themselves changed in different ways as they engage with the interests and influences emanating from organizations that span ever larger geographical areas. The growth of the world economy has produced difference rather than homogenization. The uniqueness of different places is the result of interaction over space rather than of a singularity produced by isolation. This approach challenges the assertion that globalization portends the demise of uneven development or a progressive reduction of economic differences between places. The "end of geography" remains nowhere in sight. Indeed, recent trends indicate a deepening of differences between regions and localities within countries as well as between countries at a global scale. These reflect not only the impact of decisions by transnational corporations and major economic differences between countries but also the relative success of localities and regions in inserting themselves into the world economy.

Third, the evolution of the world economy has followed a number of long-term cyclical fluctuations that correlate strongly with the emergence of distinctive technological systems. A vital dimension of change bringing about these shifts has been the transformation of the nature of capitalism. The earliest phase, that of competitive capitalism, lasted from the eighteenth century until the end of the nineteenth. The second phase, that of organized capitalism based on close coordination between business, government and labor, dominated throughout much of the twentieth century. However, we have now shifted to a third phase, that of advanced capitalism, in which the geographical coincidence between production and consumption that characterized the previous phase has unraveled. Rather than a sudden shift, however, the transition has been gradual with some "old" and many more "emerging" sectors representing most clearly the new forms of organization and production.

The focus on cycles can lead to the overemphasis of sharp breaks and the exaggeration of the suddenness of change. The idea of total change is open to doubt, as argued in the latter part of Chapter 3 where the new "regional motors" of the world economy are placed in the context of a range of locational outcomes depending on the mix of externalities and spatial transaction costs associated with different economic sectors. Mass production is still important, particularly in relation to LDC industrialization such as that in manufacturing export processing zones (EPZs). But it is a mistake to identify the continuance of some mass production with the persistence of Fordism or with the absence of *any* break with the past in the essential structure of economic organization. What is also clear, however, is that the flexible production systems of TNCs today (see Chapters 3 and 7) do not necessarily portend a stable and irreversible pattern that will extend indefinitely into the future.

Fourth, as the world economy has evolved so have the economic and locational principles that govern its operation. The perfect competition, transportation costs, factor endowments, and comparative advantage that were important in competitive capitalism and, to a lesser degree, with organized capitalism, have faded in relative significance. Today, monopolistic competition (sectors dominated by small numbers of large companies), macroeconomic regulation, the market access form of international relations, position in the global urban hierarchy, global financial networks, regional motors, industrial districts, economies of scope and coordination, and competitive advantage are all more important (Chapters 3–11). This means that the traditional models of economic activities that identified, for example, transportation costs as *the* key factor in determining location are misleading in contemporary circumstances. Such static models—based on the presumed eternal significance of this or that factor, and so deaf to history and geography—need to be adjusted or supplemented by more dynamic conceptualizations of change (in this connection, see Ron Martin's (1999) critique of the "new economic geography" and the work of economists including Paul Krugman and Michael Porter). And during the last few years, there have been a number of promising theoretical reorientations in economic geography (including the "relational turn," as well as attention for the **cultural economy** (see Chapter 7)).

Fifth, although neglected by economic analysts in the past, the daunting environmental problems associated with such trends as the destruction of tropical rainforests, climate change, and increasing air, soil and water pollution are vitally important in understanding the world economy. Despite the disproportionate per capita use of non-renewable resources by the core countries, most of these threats are greatest in the world's LDCs, and will intensify the contrasts between rich and poor regions. Environmental problems are inseparable from processes of economic development and human welfare. These issues are alarming not only for the people in the affected regions but also for the people in the DCs; it is clear that environmental problems are increasingly enmeshed in matters of national security and regional conflict. The main concern is that the continued prosperity of the DCs and the growing prosperity of the LDCs may depend on processes of globalization that are being disrupted by large-scale environmental disasters, unmanageable mass migrations, or breakdowns of economic and political stability in the world as a whole.

Sixth, political regulation of economic activities is fundamentally important in understanding the world economy, yet was typically neglected or ignored in conventional economic explanations. The achievement of competitive advantage by companies rests importantly on political organization and coordination. The real world economy is not one of laissez-faire economics but of political economy. Macroeconomic policies regulating agriculture, industry, and services have significant and often determining impacts on economic landscapes. The rise of different national economies within the world economy both historically and currently by the NIEs reflects in part the organizational and mobilization capabilities of governments. Financial systems are particularly important in mediating between states and the investment decisions of firms that are so important in influencing patterns of economic location (Chapter 3). Today governments face the challenge of coping with trends towards globalization and localization that make economic management much more problematic than in the past. One reaction—creating trading blocs such as the European Union (see Chapter 12)—threatens both to reinforce economic inequities within the world trading system and, through its centralization of decision making in distant seats of power, further stimulate the decentralist politics of various types already under way around the world (see Chapter 13). Success in many sectors of the world economy also seems related to certain social and cultural attributes, such as high levels of social trust and well-developed social networks supporting, for example, the links between dominant firms and subcontractors. This strongly suggests that global analysts can no longer neglect consideration of the institutional, sociological and cultural bases of economic activity (see Chapter 7).

As we have tried to show throughout this book, the world economy is a complex and changing intermeshing of institutions and markets producing different effects in different places, rather than a closed global isotropic plain governed by the same invariant determinants everywhere and always. This is the central message we hope you take away from reading this book.

KEY SOURCES AND SUGGESTED READING

Flint, C.R. and Taylor, P.J., 2007. *Political Geography: World-economy, nation-state and locality*. Upper Saddle River, NJ: Pearson Education.

Jacobs, J., 2004. *Dark Age Ahead*. New York: Random House.

Stutz, F.P. and Warf, B., 2012. *The World Economy: Geography, business and development*, 6th edn. Upper Saddle River, NJ: Pearson Education.

Glossary

advanced capitalism (disorganized capitalism, globalized capitalism) Label given to the most recent or advanced phase of capitalism that uses **flexible production systems**. Under advanced capitalism, relationships between capital, labor, and government are more flexible, largely because a great deal of corporate activity has escaped the framework of nation-states and their institutions that still constrain organized labor and most government functions.

African Union (AU) The 53 members of the AU, formerly the Organization of African Unity (OAU), represent nearly all African countries. A treaty that entered into force in 1994 put these countries on the road to a continent-wide African Economic Community (AEC).

agglomeration Clustering together of functionally related activities. This clustering allows **agglomeration economies**—cost advantages that accrue to individual firms because of their location among functionally related activities.

agglomeration economies Cost advantages that accrue to individual firms because of their location clustered together among functionally related activities.

agribusiness An integrated, corporate system, involving all aspects of agriculture, from food production, to processing and distribution. The direct corporate involvement of **TNCs** in agriculture is an aspect of the **new international division of labor (NIDL)**.

Asian Tigers The four Asian Tigers—Hong Kong, Singapore, South Korea, Taiwan—are so called because they adopted an export-driven model of economic development that allowed them to maintain high rates of economic growth and industrialization during the later decades of the twentieth century.

autarky National economic self-sufficiency and independence.

back-office functions Record keeping and analytical functions that do not require frequent personal contact with clients or business associates and so can be located in back offices in lower rent areas instead of the high-rent locations necessary for the front offices of companies, such as banks.

backwash effects The negative spillover effects on a region (or regions) of the economic growth of some other region.

batch production Manufacturing production involving small batches (rather than continuous mass production) of similar items. Small-batch, **just-in-time (JIT) production** and distribution systems can allow producers to reduce the costs of raw materials stockpiles, parts inventories, and warehousing.

BPO *See* **business process outsourcing**

branch plant A factory, often located in a low-cost country or region, that belongs to a company that is headquartered elsewhere.

branch–plant industrialization The growth of manufacturing employment in declining industrial regions to which companies have moved some or all of their operations in order to take advantage of reserves of skilled manual labor, cheap factory space, and an established infrastructure. Branch–plant industrialization proper typically involves activities that require significant inputs of technology and of skilled (or at least experienced) labor and that also require a certain degree of centrality in order to assemble and distribute raw materials and finished products, such as light industries (car batteries, cash registers, and cameras to tires, watches, and light engineering), white-collar information processing and wholesaling functions.

Bretton Woods Agreement An effort to stimulate international trade by stabilizing currency fluctuations between countries. The allies met for a conference in Bretton Woods, New Hampshire, in 1944 in order to make financial arrangements for the postwar period following the anticipated defeat of Germany and Japan. They established a system of fixed exchange rates that lasted until 1971. The U.S. dollar served as the convertible medium of currency with a fixed relationship to the price of gold. The representatives of 44 countries, including the Soviet Union, also agreed to establish the **International Bank for Reconstruction and Development** (of the **World Bank**) and the **International Monetary Fund** (IMF).

BRICs The term comes from 2003 when economists at the U.S. investment bank, Goldman Sachs, predicted that during the following 50 years, Brazil, Russia, India and China—termed the BRIC economies—could become a major force in the world economy, primarily as export-oriented manufacturing centers (particularly in the cases of China and India) but also as major export-oriented resource economies (particularly in the case of Russia and Brazil).

business process outsourcing (BPO) BPO is a cost-saving measure that involves the contracting out of business tasks—such as insurance claims processing, billing services, credit card services, telemarketing—to another company either domestically or internationally.

business services A subset of producer services. Business services include legal services, advertising and marketing, public relations, accounting, research and development, personnel training, recruitment, architecture and engineering, and consulting.

capital flight The withdrawal of liquid assets from a national economy by domestic and overseas investors or transnational corporations.

capital goods Producer goods that are "fixed," e.g., machinery and heavy equipment, and used to produce other goods.

capitalization The process whereby capital-intensive inputs such as technology are deployed by large firms and replace labor-intensive methods associated with smaller scale production.

captive outsourcing Captive outsourcing involves "internal" outsourcing where a company from a country where labor and other costs are high has certain manufacturing or services activities undertaken by an affiliated company in another country where the work can be done more cheaply. *See also* **offshore outsourcing, offshoring, outsourcing**

carrying capacity Used in the context of food and agriculture, this term refers to the maximum population that can be supported within a given territory on a minimum daily diet, given the quality of local soils and local climatic conditions and assuming the availability of appropriate forms of mechanization.

cartel An organization of independent producers or traders who agree to restrict output, fix prices, or divide markets in order to improve the profitability of its members. A well-known international cartel is the **Organization of Petroleum Exporting Countries (OPEC)**.

cash crops Crops grown by farmers for trade or sale (as opposed to subsistence crops where most of what is produced is consumed by the farmers and their families).

Celtic Tiger Comparing Ireland's incredible economic boom to that of the **Asian Tigers** (Hong Kong, Singapore, South Korea, Taiwan), economist Kevin Gardiner, working for the U.S. investment bank Morgan Stanley, coined this term to describe Ireland's astonishing growth rates of between 5 and 10 percent of gross domestic product during the 1990s. By the end of the decade unemployment was down to 4.5 percent, the national debt was down and the country's GDP per capita had outstripped that of the UK and even Germany.

central place systems and central places Central places are urban centers (hamlets, villages, towns, cities) that provide goods and services to their surrounding hinterlands (market areas). A central place system comprises a hierarchy of central places—ranging from a small number of very large central places (cities) that offer higher order goods (expensive and infrequently purchased items, such as designer furniture and jewelry) at the top of the hierarchy, down to a large number of small central places (hamlets) that offer low-order goods (inexpensive, frequently purchased, everyday necessities, such as newspapers and milk).

centralization A trend towards corporate economic integration that involves the ownership of private enterprise by progressively fewer corporations. It is the result of corporate mergers and takeovers and is characterized by diversified, conglomerate, **transnational corporations** (TNCs).

client states Countries that are dependent, economically or militarily, on larger and more powerful countries.

collectivization of agriculture The creation of communal (collective ownership) farms.

COMECON (also known as the Council for Mutual Economic Assistance (CMEA)) A communist international organization established in 1949 to reorganize the eastern European economies in the Stalinist mold—even to the point of striving for autarky for individual members, each pursuing independent, centralized plans. This proved unsuccessful, however, and in 1958 COMECON was reorganized by Stalin's successor, Khrushchev. The goal of autarky was abandoned, mutual trade among the Soviet bloc was fostered and some trade with western Europe was permitted. COMECON was disbanded in 1991 because the fall of communism and the shift to democracy and capitalism in eastern Europe made it redundant.

command economy An economy commanded from a central administration, in which the means of production are publicly owned. The central administration plans what and how many goods will be produced and sets prices. Communist countries such as the former Soviet Union, North Korea, Vietnam, and Cuba established command economies; the command economy model was imposed on the countries of eastern Europe that had been liberated from German occupation by Soviet forces.

commodity chain A network of labor and production processes beginning with the extraction or production of raw materials and ending with the delivery of a finished commodity.

common market A form of international economic integration that involves the elimination of **tariffs** and other trade barriers between members states, the removal of internal restrictions on the movement of **factors of production** and the creation of a common set of trade agreements with non-member countries.

comparative advantage The principle used to explain patterns of trade and specialization: If each region or country specializes in those economic activities they perform relatively better than others and imports goods for which their own production costs are relatively higher, each is likely to gain (transport costs and **terms of trade** notwithstanding).

competitive advantage The advantage acquired in economic competition by some locations because of the benefits that accrue from an early start in production of a particular good and the continuing defense of that historic base through superior organization and adaptability. Other locations can suffer from a **competitive disadvantage** due to the absence of an **initial advantage** that would have allowed them to develop and maintain an ongoing competitive advantage.

competitive capitalism The early phase of industrial capitalism, lasting until about 1890, characterized by comparatively high levels of free market competition, with many small-scale producers, consumers, and workers who acted almost completely independently, and with relatively little government intervention.

competitive disadvantage The disadvantage suffered in economic competition by some locations because of the absence of an initial advantage that would have allowed them to develop and maintain an ongoing competitive advantage.

concentration A trend towards the reduced numbers of firms in any given industry or economic sector. It is partly the result of the elimination of smaller weaker firms through competition and partly the result of corporate mergers and takeovers.

consumer durables Goods intended to be used for a number of years, e.g., household appliances.

consumer goods Commodities purchased by individuals or households for final consumption. Consumer goods comprise **consumer durables** (goods intended to be used for a number of years, e.g., household appliances, automobiles) and consumer non-durables (goods intended for relatively immediate consumption, e.g., fresh fruit, ice cream).

consumer services Personal services, including retailing, medical care, personal grooming, leisure and recreation, culture and entertainment.

containerization Automated cranes efficiently load and unload massive standardized containers filled with large amounts of cargo between the ships and the flatbeds of nearby trains and trucks. Before container cargo handling, ships, trains, and trucks sat in port while major items of cargo were loaded and unloaded individually by a sizeable intensive workforce. Containerization has improved the operation of cargo shipping and handling by reducing the turnaround time at ports.

cordon sanitaire A chain of independent buffer countries set up to form a barrier (literally, a "sanitary line") around a country that is considered hostile militarily or dangerous ideologically.

creative destruction The withdrawal of investments from activities (and regions) that yield low rates of profit, in order to reinvest in new activities (and new places).

crossover system of trade A system of multilateral trade established in the late nineteenth century, which lingered into the 1930s. Europe and North America bought raw materials from less developed countries (LDCs). In return, Britain imported manufactures from and exported capital to Europe and North America. Britain's assets were boosted by the return on foreign investment and the export of manufactured goods to the LDCs (both colonies and independent states).

cultural economy The cultural economy can be defined as a group of sectors (cultural products industries) that produce goods and services—including jewelry, live theatre, music recording, film production—whose symbolic value to consumers is high relative to their practical purpose.

cumulative causation A spiral buildup of advantages that occurs in specific geographic settings as a result of the development of **agglomeration economies, external economies of scale,** and **localization economies.**

customs union A form of international economic integration that involves the elimination of some (but not necessarily all) trade barriers between member states and the creation of a common set of trade barriers to non-member states.

debt trap The cycle of borrowing that results when the productivity gains from investments undertaken with borrowed capital are insufficient to meet interest repayments. Further loans and debt rescheduling provide temporary relief, but in the long term make it even more difficult to achieve increases in productivity sufficient for self-sustaining growth.

deindustrialization Many authors use this term rather loosely. At the heart of the concept is a *relative* decline in industrial *employment* in a country or region in which industry has traditionally been a significant component of the economy. It may be the result of climacteric changes or of secular shifts in an economy that are related to technological change and/or the globalization of the economy. In some instances, such trends may involve not just a relative decline but an *absolute* one; and may involve declining industrial *output* as well as employment.

demographic transition The evolution of vital rates—birth rates and death rates—over time, from high to low levels. The demographic transition model posits improved diets, public health and scientific medicine as causing a steady decline in death rates with increasing levels of economic development over time. Birth rates decline later, and more slowly, as sociocultural practices take time to adjust to these new circumstances. The result is a sharp increase in population growth, until birth rates fall to relatively low levels. This model is based on the experience of DCs and so all LDCs should not be expected to follow this exact demographic path.

dependence or dependency A high level of dependence by a country on foreign enterprises, investment or technology. External dependence for a country can mean that it is highly dependent on levels of demand and the overall economic climate of other countries. Dependency for a LDC, for example, can result in a narrow economic base in which the balancing of national accounts and the generation of foreign exchange are dependent on the export of one or two agricultural or mineral resources.

deskilling A reduction in the range and level of skills within a local labor market that is the result of two trends: Increased mechanization and computerization of production processes (including management and management support functions) and the geographic consolidation and localization of higher skilled activities in **world cities**, major control centers and centers of innovation (which leaves other labor markets with a preponderance of routine jobs in local offices and **branch plants**).

diagonal integration A form of business organization in which a company tries to diversify its interests by using corporate mergers or acquisitions of firms that are engaged in separate and distinct enterprises, producing different goods or services for different markets. An automobile manufacturer, for example, may buy into energy, advertising, or entertainment companies.

differential of contemporaneousness In regional economic development, new technologies, ideas, and market conditions may reach different regions at the same time, but they impact the regions in very different ways because they were *differently equipped to respond to them.*

diffuse industrialization The growth of manufacturing employment in rural regions, in which companies decentralize some or all of their activities from established centers of operation, such as a major city, in response to the increasing shortage, cost, and militancy of labor there and the availability of reserves of relatively cheap and non-unionized unskilled and semi-skilled labor, less expensive/more available land, and lower taxes in rural areas. Typically, diffuse industrialization involves activities in which labor costs are an important part of overall production costs and in which there has been little scope for reducing labor costs through technological change.

digital divide The digital divide refers to the gap in opportunities between individuals, households, businesses, and areas at different socioeconomic levels to access advanced information and communication technologies (ICTs) for a variety of activities. The digital divide exists both within countries (for example, between urban and rural areas or between richer and poorer neighborhoods) and between countries or groups of countries (for example, between the LDCs and the DCs). The primary concern is that lack of access to and development of information, communication and e-commerce technology will prevent many people from benefiting from the new knowledge-based economy.

diminishing returns (law of) The tendency for productivity to decline, after a certain point, with the continued application of capital and/or labor to a given resource base. A simple example is provided by agricultural productivity: A large farm will yield progressively higher levels of output with the addition of more farm hands and more machinery, but there will be a point at which productivity decreases as some of the labor and machinery is under-employed, people get in one another's way, and coordination of activities becomes costly.

disorganized capitalism (advanced capitalism) The label given to the most recent or advanced phase of capitalism that uses flexible production systems. Under disorganized capitalism, the relationships between capital, labor, and government are more flexible, largely because a great deal of corporate activity has escaped the framework of nation-states and their institutions that still constrain organized labor and most government functions.

ecological footprint The World Wide Fund for Nature's (WWF) measure of the human pressures on the natural environment from the consumption of renewable resources and the production of pollution. The ecological footprint indicates how much space a population needs compared to what is available. It changes in proportion to population size, average consumption per person, and the resource intensity of the technology being used. It is measured in "area units," where one area unit is equivalent to one hectare of biologically productive land with world average productivity. As land varies in productivity, a hectare of highly productive cropland would represent more area units than the same amount of less productive grazing land.

economic union A form of international economic integration that involves the removal of all internal barriers to trade and the movement of **factors of production**, the creation of a common set of trade barriers, and trade agreements with non-member states and the coordination of integrated economic policies within the union.

economies of scale Cost advantages for firms from large-scale production. Economies of scale in production are equivalent to **increasing returns to scale**. *See also* **economies of scope, internal economies of scale**

economies of scope Cost advantages from large-scale flexible organization. Economies created by the capacity to provide entirely new products and/or services through the flexible use of the same production or service network.

elasticity of demand The degree to which levels of demand for a product or service change in response to changes in price. Where a relatively small change in price induces a significant change in demand, elasticity is high; where levels of demand remain fairly stable in spite of price changes, demand is said to be inelastic.

encapsulation Howells' (2003) notion of service encapsulation of goods and materials illustrates how services are increasingly incorporated into manufactured products. Manufactured products are offered to consumers not in their own right but in terms of their wider service attributes. Either the manufactured product is offered along with closely aligned service products in a single package (e.g., finance, insurance, maintenance warranties with an automobile purchase) or the consumer may not be offered the manufactured product itself in a single one-time purchase, but may be offered the service that the manufactured product offers in a continuing process involving long-term customer contact through service delivery (e.g., automobile lease instead of purchase).

enterprise zone An officially designated economically distressed area where public fiscal and other incentives are available as an inducement to private companies to locate and create jobs there.

entrepôt A port that specializes in the trade of goods for re-export. Entrepôts operate primarily as intermediary trading centers—they receive goods from foreign countries for re-export to other countries. Hong Kong, Singapore, and Rotterdam are the world's top three entrepôts.

eurodollars U.S. currency that is held in banks located outside the United States, traditionally mostly in Europe, hence the name. Currently, China holds the largest amount—in the hundreds of billions. It is a pool of currency for which there is a distinctive and independent market, because it conventionally represented a fairly stable, hard currency that is beyond the control of the U.S. government and its financial institutions. At the same time, the strength of the U.S. economy and the value of the dollar affect investor confidence in eurodollars.

export processing zones (EPZs) Small, closely definable areas within which especially favorable investment and trading conditions are created by governments in order to attract export-oriented industries, usually foreign owned. These conditions include the absence of foreign exchange controls, the availability of factory space and warehousing at subsidized rents, low tax rates, and exemption from tariffs and export duties.

external control The term refers to situations in which employment opportunities and decisions about investment and production in a given plant or locality are controlled by corporate managers based in other cities, regions or countries.

external economies of scale The specific benefits that accrue to producers from associating with similar producers in places that offer services that they need.

externalities Side-effects or consequences (of an industrial or commercial activity) that affect other parties without being reflected in the current cost of the goods or services involved.

factors of production The fundamental components of any economic system: Land, labor, and capital. Land includes not only space or territory but also the associated soils and natural resources. Labor includes not only the size of the available workforce but also its skills, experience, and discipline. Capital includes not only money capital but also everything deliberately created for the purpose of production, such as factories and machinery (i.e., "fixed" capital). A fourth factor, enterprise, is often recognized, although it may be missing from some forms of economic organization (e.g., subsistence economies), while it may legitimately be regarded as one aspect of labor.

"fallow" agriculture or shifting cultivation Involves sowing or planting on scorched land using slash-and-burn methods (cutting down the natural vegetation (e.g., forest) and burning it to release its nutrients into the soil). No special tools are required for such a system, neither is weeding or fertilization necessary, provided that cultivation is shifted in a couple of years to another burned plot after a few crops have been taken from the old one, which is then abandoned (left fallow) for a period of time.

feudalism and feudal systems Forms of economic organization based on a mode of production in which the surplus product (i.e., the outputs of productivity in excess of subsistence levels) is appropriated through a hierarchy of sociopolitical ranks by institutionalized coercion. In classical feudal systems, lords allocate land to vassals in return for military service and/or labor on the lord's estate.

finance, insurance, and real estate (FIRE) A subset of **producer services**. Finance, insurance, and real estate (FIRE) includes commercial and investment banking, insurance of all kinds (property, medical, casualty) and the residential and commercial real estate business.

FIRE *See* **finance, insurance, and real estate services**

flexible accumulation A phase of capitalist development (a **regime of accumulation**, in the terminology of regulation theory) characterized by a set of production technologies, labor practices, inter-firm relations, and consumption patterns that have evolved in order to allow greater economic and geographic flexibility in economic affairs. Because it succeeded **Fordism**, flexible accumulation is also known as post- or **neo-Fordism.**

flexible production systems and **flexible production** Various practices whereby manufacturing operations achieve flexibility in what they produce, when they produce it, how they produce it and where they produce it. These practices include the exploitation of various kinds of enabling technology, greater use of subcontracting, the exploitation of different labor markets, the exploitation of different market niches for products, and the development of new labor processes using flexible working hours, part-time workers, etc.

Fordism (Fordist regime of accumulation) A **regime of accumulation** that centers on the mutual reinforcement of mass production and mass consumption. Named after Henry Ford because of his innovations and philosophy concerning automobile manufacture, it features a highly specialized and differentiated division of labor with assembly-line production geared to the provision of standardized, affordable goods for mass markets.

foreign direct investment (FDI) Direct investment in a company or companies in one or a number of foreign countries (e.g., takeovers, new subsidiaries, etc., rather than portfolio investment) in order to achieve managerial and production control.

fracking A process of producing fractures in the rock formation that stimulate the flow of natural gas or oil, increasing the volumes that can be recovered. Fractures are created by pumping large quantities of fluids at high pressure down a wellbore and into the target rock formation. Once the injection process is completed, the internal pressure of the rock formation causes fluid to return to the surface through the wellbore.

franchise A business that has a license to manufacture or sell a product or service under the name of the original company, e.g., a retail outlet that is not owned by Benetton but that has a franchise agreement to use the company's name for the store and to sell the company's products.

free trade association A form of international economic integration that involves the elimination of some (but not necessarily all) trade barriers between member states, but where each member state continues to set its own **tariffs** and quotas as trade barriers to non-member states.

geoengineering The amelioration of global warming by deliberately altering the climate of the earth.

geographical path dependence The historical relationship between the present economic activities associated with a place and its past experience.

global currencies Currencies used in international transactions. The U.S. dollar is the most important one today, although its use is challenged to a certain extent by the euro and the Japanese yen.

global sourcing Use of multiple sources in different countries for the components of a particular product that is assembled elsewhere.

globalized capitalism (advanced capitalism or disorganized capitalism) The label given to the most recent or advanced phase of capitalism that uses **flexible production systems**. Under globalized capitalism, the relationships between capital, labor, and government are more flexible, largely because a great deal of corporate activity has escaped the framework of nation-states and their institutions that still constrain organized labor and most government functions.

Green Revolution An agricultural program in the 1960s and 1970s to tackle hunger in the LDCs by transferring methods of modern agricultural technology to traditional farming regions. A package of measures to improve yields of food crops such as wheat, maize, and rice involved the application of greater mechanization, chemical fertilizers and pesticides, and irrigation water to genetically improved high-yield seeds. While many Asian and some Latin American countries achieved significantly increased crop yields, the strategy was not embraced to the same extent in Sub-Saharan Africa. The Green Revolution has been criticized for benefiting wealthier farmers at the expense of smaller landowners who cannot afford the costs of initial inputs, especially fertilizers and irrigation water. Production has also leveled out in recent years.

gross domestic product (GDP) When **gross national income (GNI)** is adjusted to remove the value of profits from overseas investments and the "leakage" of profits accruing to foreign investors, the result is a measure of gross domestic product (GDP).

gross national income (GNI) Also known as gross national product (GNP), this is a measure of the market value of the production of a given economy in a given period (usually a year). It is based on the market price of finished products and includes the value of subsidies; it includes the value of profits from overseas investments and profits accruing to foreign investors but does not take into account the costs of replacing fixed capital.

gross national product (GNP) *See* **gross national income (GNI)**

gross value added (GVA) Gross value added is equal to GDP minus taxes on products plus subsidies on products. In other words, GVA plus taxes on products minus subsidies on products is equal to GDP.

growth pole A growth pole, which can be unplanned or planned, benefits from **agglomeration economies** and can spread prosperity to nearby regions through **spread effects**. Examples of efforts to plan growth poles to stimulate regional development by generating spread effects include the industrial complexes located in Taranto and Bari in the Mezzogiorno (south) of Italy or the eight *metropoles d'équilibre* (balancing metropolises) in France, such as Lyon, Marseille, and Bordeaux, which were intended also to redirect some economic activity away from Paris and reduce its **primacy**.

hearth area An area of origin of people, ideas, or technologies that then spread to other areas.

hegemony A difficult and controversial concept, hegemony is often applied to the dominance of one country or region or group over others. Greek for "leadership," this term was originally applied to the dominance of one Greek city-state over others. In addition to military dominance, the hegemonic power must also have economic and cultural dominance to set and enforce the rules of conduct that it prefers.

horizontal integration A form of business organization in which a company tries to capture the market for a single stage of production, a single good or service, or an entire industry and achieve **economies of scale**, by using corporate mergers or acquisitions of firms that formerly competed in the same market(s) with similar goods or services. A successful automobile manufacturer, for example, might buy out other automobile manufacturers.

hydraulic fracturing. *See* **fracking**

imperialism The extension of the power of a country through direct or indirect control of the economic and political life of other territories.

import penetration This describes the result of a significant share of domestic markets for a particular product or service being lost by domestic firms in the face of competition from foreign sources.

import substitution Development of domestic firms capable of producing goods or services formerly provided by foreign firms.

increasing returns to scale Cost advantages from large-scale production. Increasing returns to scale in production are equivalent to economies of scale. An increase in inputs (raw materials, labor, etc.) by x percent results in an increase in output by more than x percent.

industrial production In addition to manufacturing, industrial production includes mining and power generation.

infant industries Industries at an early stage of development that are protected by tariffs and other trade barriers until they can survive foreign competition without protection.

inflation A decline in the value of money because prices keep rising. A number of economic theories have been formulated to explain why inflation occurs. In the eighteenth century Hume theorized that prices rise as the supply of money rises. And a reduction in the purchasing power of money can occur when a national government increases the supply of money. Keynes assumed that inflation takes place when demand outstrips the supply of goods and services (resulting in the need for a national government to intervene to control inflation by adjusting spending, tax and interest rate levels). In the cost–push theory, the price–wage spiral causes inflation—worker demands for higher wages necessitate an increase in prices, which result in demands for higher wages, and so on. Structural theory suggests that structural factors can cause a country's currency to lose value so that prices rise, such as when an LDC has poor terms of trade where the price of imports keep rising relative to the price of exports.

informal sector Economic activities that are undertaken without any formal systems of regulation or remuneration. In addition to domestic labor, these activities include strictly illegal activities such as drug peddling and prostitution as well as a wide variety of legal activities such as casual labor in construction crews, on docks or on farms; domestic piece work; street trading; scavenging; and providing personal services such as shoe shining or letter writing.

information economy A mode of economic production and management in which productivity and competitiveness rely heavily on generating new knowledge and accessing and processing appropriate information.

initial advantage Advantage acquired in economic competition by some locations because of the benefits that accrue from an early start in production of a particular good.

intermediate goods Producer goods for manufacturing, processing or resale (e.g., raw materials and semi-finished items for use in the production of other goods or services rather than for final consumption), such as shoelaces for shoe manufacturing, hinges for furniture manufacturing, and tires for automobile manufacturing.

intermodal transportation Transportation using more than one means of conveyance, e.g., truck and ship, air and rail, etc.

internal economies of scale The specific benefits that accrue to producers from large-scale production within a company. As a company increases production, the average cost of each product begins to fall for that company.

international division of labor The idea of the organization of spatial divisions of labor, organized principally at the national scale until the late twentieth century, in which each country specialized in certain sectors of the economy, such as industry in the UK or agriculture and raw materials in many African countries.

International Monetary Fund (IMF) The IMF is a United Nations affiliate established in 1945 to help encourage international monetary cooperation, to ensure international currency exchange stability, to promote economic and employment growth and, while not a development bank, to provide temporary economic assistance to countries experiencing balance of payment problems. In early 2007 the IMF had about U.S.$28 billion in credit and loans to 74 of its more than 180 member countries. The IMF and the World Bank are different organizations but share the same member countries.

involuntary labor The term for labor that is obtained under threat or penalty and that the worker does not offer voluntarily, such as bonded, indentured, or forced labor.

isotropic Homogenous in all respects.

just-in-time (JIT) production "Lean" production employing vertical disintegration within large formerly functionally integrated firms such as automobile manufacturers in which daily and even hourly deliveries of parts and other supplies from smaller (often non-union) subcontractors and suppliers now arrive "just in time" to maintain "last-minute" and "zero" inventories. A computer system is used to adjust deliveries at short notice to meet changing demand. The goal is to reduce costs by eliminating waste from overproduction and minimizing warehousing.

keiretsu Keiretsu is a Japanese form of corporate organization. A *keiretsu* is a grouping of affiliated companies that form a business network to work towards each other's mutual benefit. The *keiretsu* system also involves central government setting up favorable trade policies, technology policies, and fiscal policies to help Japanese industry compete successfully in the world economy.

Keynesianism Specifically, this refers to a doctrine of macroeconomic management that is closely associated with British economist, John Maynard Keynes, who advocated the use of fiscal policy (e.g., budget deficits) and the exploitation of economic multiplier effects in order to achieve and maintain full employment. The term is also used to denote the mode of regulation associated with the Fordist regime of accumulation.

kin-ordered system (or societies) A system of social organization based on relationships between people who are related either biologically or by other arrangements such as through marriage or adoption.

Knowledge economy. *See* information economy

Kondratiev cycles Cyclical waves of 50–55 years in duration that have characterized the rate of change in price inflation within the capitalist world economy for the past 250 years. Their origins and significance remain controversial, but in recent years they have been widely recognized to be closely tied in to distinctive phases of political–economic development.

Kuznets cycles Business cycles of approximately 25 years in duration that have characterized the pattern of acceleration and deceleration in economic growth. Named after Ukrainian-born economist Simon Kuznets, who established their existence in the 1920s, they are cycles of activity in investment and building.

land tenure A system of land use rights and transfer mechanisms. The major types of land tenure include owner occupation, cash tenancy, share cropping (a form of tenancy in which rent is paid in kind), use rights (where there is no codified legal owner and a person or group establishes a right to the land by using it), and collectivism (in which individual farmers work in cooperatives that own the land).

latifundia Large farms or estates farmed by laborers—found predominantly in Latin America, where they originated as imperial land grants to new settlers.

less developed countries (LDCs) Peripheral and semi-peripheral countries within the world-system, with low economic output and per capita incomes, which tend to have politically weak states and low-wage, labor-intensive production. Since the end of the Cold War and the economic growth of the **newly industrializing economies (NIEs)** and other formerly peripheral countries, the designation of "Third World" is no longer as useful in distinguishing the LDCs from either the capitalist, economically developed countries such as the U.S. and the UK (the "First World") or the former communist countries in what was the Soviet bloc (the "Second World").

localization economies Cost savings that accrue to firms as the output of their particular industries increase as a result of clustering together at a specific location.

machinofacture A form of organization of industrial production that is capital intensive, with labor tending machines rather than operating them directly. It was the basis for the **regime of accumulation** that preceded **Fordism**.

maquiladoras Literally, "mills" in Spanish. These are **export processing zones (EPZs)** in Mexico, located mainly near the U.S. border.

Marshall Plan A U.S.-financed program that provided almost US$13 billion in economic aid to its war-torn European allies after the Second World War. Conceived as a self-help plan that would foster a healthy world economy, the funds were used to promote economic recovery and political stability in Europe.

Mercantilism The basis of this economic ideology adhered to from the sixteenth to the early 18th century in most European countries was that national wealth was to be measured in terms of gold or silver and that the fundamental source of economic growth was a persistently favorable balance of trade. This was the economic "logic" that justified not only overseas colonization but also the coercion of plantation labor and the prohibition of manufacturing in the colonies. It was also the logic that, on the domestic front, promoted thrift and saving as a means of accumulating capital for overseas investment. It required a high degree of economic regulation, sponsorship, and protection by the government.

merchant capitalism The label given to the initial phase of capitalism. As the feudal system disintegrated, it was replaced by an economy that was dominated by market exchange, in which communities came to specialize in the production of the goods and commodities that they could produce most efficiently in comparison with other communities. The key actors in this system were the merchants who supplied the capital required to initiate the flow of trade—hence the label merchant capitalism.

minisystems Local societies with a simple division of labor within a single cultural framework, such as hunter–gatherer and some agricultural societies.

mode of production A fundamental form of economic organization, such as **feudalism** or capitalism, which has distinctive relationships between the main **factors of production** (land, labor, and capital). The concept is derived from Marxian economics (a body of theory originally derived from the work of Karl Marx) but now has much wider use.

mode of regulation The terminology of regulationist theories for a collection of structural forms (political, economic, social, cultural) and institutional arrangements that define the "rules of the game" for individual and collective behavior within a specific **regime of accumulation** or phase of economic development. The mode of regulation gives expression to, and serves to reproduce, fundamental social relations.

monetarism A doctrine of macroeconomic management that disavows demand management and regards the money supply as the most important determinant of economic stability. Important in the U.S. and UK in the late 1970s and early 1980s, it reasserted the relevance of price theory and the importance of free markets.

monoculture The agricultural practice in which one crop is grown intensively over a large area of land.

most-favored-nation (MFN) treatment This non-discrimination principle means that a country treats all trading partners in a manner equal to that accorded to the most favored nation. This is a method for promoting free trade by ensuring equal trading opportunities among countries, especially as they relate to import duties and freedom of investment. Members of the World Trade Organization (WTO) negotiate this kind of tariff and trade arrangement.

multiplier effects The extra industries, firms, incomes and employment generated by a new activity can be said to result from that activity's multiplier effects. Such effects can be localized and give rise to a **growth pole** or occur in a more diffuse manner.

nation A group of people with a common identity based on such shared characteristics as origins, history, customs, and, frequently, language—a nationality. The terms country, nation, **nation-state**, and **state** are often used interchangeably.

nation-state A state that corresponds for the most part with the people of only one nation.

nationalism Devotion and loyalty to one's **nation**. Nationalist feelings and movements, for example, often arose in opposition to colonialism in African and Asian countries in the twentieth century based on a desire for national independence.

nationalist separatism A nation or group's autonomy or independence from a larger group or political unit.

neoclassical economics Forms the basis of a particular conceptualization of how economic activity operates in capitalist society. The economy comprises many small producers and consumers. All act rationally, although none is large enough to affect significantly the operation of the market. Firms are seen as atomistic agents with full information in a world of pure markets (with no entry barriers) and all have exactly the same resources, technological capability and market power with deviations regarded as market "imperfections." Firms utilize **factors of production** (land, labor, capital) in order to maximize their profits. Consumers sell their factors of production (especially labor) in order to purchase goods and services that maximize their individual preferences. The market is an abstract space in which firms and consumers set the prices. The forces of supply and demand cause economic resources to be used in the most efficient way possible. Neoclassical economics involves normative model building—constructing simplified versions of how the real world ought to operate.

Neo-Fordism Sometimes referred to as "post-Fordism" or **flexible accumulation**, this term identifies the regime of accumulation that has succeeded **Fordism** within parts of the world's economies. Rather than being predicated on the mutual reinforcement of mass production and mass consumption, it depends on **flexible production systems** to exploit specific market segments and/or niches.

neoliberalism Neoliberalism involved a shift in the 1970s away from the egalitarian liberalism (and the Keynesian welfare state) that had dominated public policy in DCs like the U.S. and the UK since the 1930s and a selective return to the ideas of classical liberalism. The process involves "rollback" neoliberalization (e.g., deregulation of finance and industry, cutbacks in welfare programs) and "rollout" neoliberalization (e.g., **public–private partnerships**, workfare requirements, privatization of government services). The net effect has been to "hollow out" the capacity of the central governments while forcing local governments to become entrepreneurial in pursuit of jobs and revenues and pro-business in their expenditures. The "free markets" associated with neoliberalism have intensified uneven relationships among places, the inevitable result being an intensification of economic inequality at every scale, from the neighborhood to the **nation-state**.

Neolithic period Literally, the "New Stone" age, between about 7000 and 5500 BCE, characterized by the use of stone tools produced by grinding or polishing, and an agricultural revolution involving the switch from hunting and gathering to food production based on the domestication of plants and animals.

New Deal The domestic program in response to the Great Depression introduced by President Franklin D. Roosevelt in the 1930s to promote economic recovery. Roosevelt promised "a new deal for the American people" in his acceptance speech for the 1932 presidential nomination. New Deal legislation included the establishment of the Civil Works Administration to address unemployment, the National Recovery Administration to restore industrial production and the Agricultural Adjustment Administration to bolster farm production.

New Deal era The 1933–1938 period of the New Deal in the United States during which President Franklin D. Roosevelt introduced a domestic program to promote economic recovery in response to the Great Depression. *See also* **New Deal**

new international division of labor (NIDL) The idea of the reorganization of spatial divisions of labor, formerly organized principally at the national scale, to a global scale based on international production and marketing systems.

newly industrializing economies (NIEs) Countries, formerly peripheral within the world-system, that have acquired a significant industrial sector, usually through **foreign direct investment (FDI)**.

non-tariff barriers Policy instruments (other than import taxes) designed to protect domestic industry from foreign competition (e.g., import quotas, import licensing requirements, special standards and regulations, exchange rate manipulation, government subsidies to domestic industries, and special labeling and packaging regulations).

offshore financial centers Islands or micro-states that have become specialized nodes in the geography of worldwide financial flows. They are attractive because they provide no- or low-tax settings for savings and are less regulated than financial centers elsewhere.

offshore outsourcing External outsourcing, where a company has certain manufacturing or services activities undertaken by an unaffiliated company in another country where the work can be done more cheaply. *See also* **captive outsourcing, offshoring, outsourcing**

offshoring Offshoring involves a company having certain manufacturing or services activities undertaken by an affiliated or unaffiliated company in another country where the work can be done more cheaply. *See also* **captive outsourcing, offshore outsourcing, outsourcing**

Organization for Economic Cooperation and Development (OECD) The OECD is an organization of 30 industrialized countries that include the U.S., Canada, Japan, Australia, New Zealand, South Korea, and European countries such as the UK, France, Switzerland, and the Czech Republic. It was founded in 1961 to stimulate economic growth and world trade (taking over from the Organization for European Economic Cooperation (OEEC), which had been established to administer U.S. and Canadian postwar reconstruction aid in Europe under the **Marshall Plan**). The OECD defines itself as an international organization of countries that share a commitment to democratic government and the market economy. Its mission is to help governments tackle the economic, social and governance challenges of the global economy. Based in Paris, the OECD is probably best known as a source of economic statistics and publications.

Organization of Petroleum Exporting Countries (OPEC) OPEC is an international cartel of 12 oil-exporting countries. It was created in 1960 to coordinate oil output and fix prices in an effort to improve profitability and reduce market volatility. Its members are all LDCs: Algeria, Angola, Indonesia, Iran, Iraq, Kuwait, Libya, Nigeria, Qatar, Saudi Arabia,

UAE, and Venezuela. In 2005 OPEC produced nearly 31 million barrels of crude oil each day or almost 43 percent of the world's output of nearly 72 million daily barrels. OPEC holds more than 78 percent of the world's estimated proven crude oil reserves of more than 1 trillion barrels.

organized capitalism The later phase of industrial capitalism that was characterized by comparatively highly structured relationships between labor, government, and corporate enterprise. These relationships were mediated through legal and legislative instruments, formal agreements, and public institutions.

outsourcing In the global economy, outsourcing typically involves a company, often in a DC where labor and other costs are high, having certain manufacturing or services activities undertaken by an unaffiliated company (e.g., an independent subcontractor) either domestically or in another country. *See also* captive outsourcing, offshore outsourcing, offshoring

outsourcing, captive *See* captive outsourcing

overaccumulation A distinctive phase in the long-term dynamics of capitalist economies, characterized by unused or underutilized capital and labor. It is an inevitable outcome of the difficulty of matching supply to demand under changing conditions and it represents a critical moment for the political economy of capitalism. It can be recognized by the appearance of idle productive capacity, excess inventories, gluts of commodities, surplus money capital, and high levels of unemployment.

positional goods Consumer goods that are acquired (in part, at least) in order to denote affluence, social status, and/or style.

primary commodities *See* primary production

primary production Primary production and its commodities are derived from natural resources, as in agriculture, mining, forestry, and fishing.

primate cities/primacy A primate city is a country's leading city as evidenced by measures of this primacy including its significantly larger population compared to other cities (being more than twice as large as the country's second largest city) and by other characteristics reflecting the primate city's national importance and influence, such as economic activity, political activity, and power.

privatization The sale of government assets (such as key industries) to private owners and the contracting out of services formerly provided by government to private companies in an effort to increase government efficiency and save public money.

producer goods Manufactured goods used in the production of other goods and services rather than for final consumption. Producer goods comprise intermediate goods (e.g., raw materials and semi-finished items) and fixed capital goods (e.g., machinery and heavy equipment).

producer services Services that enhance the productivity or efficiency of other firms' activities or that enable them to maintain their specialized roles. Usually subdivided into (a) business services, including legal services, advertising and marketing, public relations, accounting, research and development, personnel training, recruitment, architecture and engineering, and consulting and (b) finance, insurance, and real estate (FIRE), including commercial and investment banking, insurance of all kinds (property, medical, casualty), and the residential and commercial real estate business.

product lifecycle Locational requirements for production change as products move from being novel and expensive to being standardized and cheaper. In particular, labor costs can become more important than adjacency to markets.

Progressive Era The 1890–1920 period in the United States of social activism and political reform.

public–private partnerships Coalitions of private sector businesses and/or business leaders and public sector officials and agencies, with others such as unions and chambers of commerce, which seek to promote economic growth and the well-being of an area.

purchasing power parity (PPP) When making international comparisons of economic prosperity, it is vital to take into account differences in national price levels. PPP does this by measuring how much of a common "market basket" of goods and services each country's currency can purchase locally.

rank redistribution and **rank-redistributive societies** A form of economic organization that is dominated by the redistribution of surplus product from one social group to another (e.g., **feudalism**), usually through an institutional framework such as one established by the **state**.

Reaganomics The application of **supply-side** economics to the management of the U.S. economy in the 1980s. Supply-side economics is similar to monetarism in its disavowal of demand management; rather, the key to economic stability and well-being is seen to be the enhancement of aggregate supply. Reaganomics consisted of a set of objectives that included tax reduction, deregulation of business, increased government spending on defense, and decreased government spending on social welfare.

regime of accumulation The terminology of regulationist theories for a particular way of organizing economic production, income distribution, consumption, and public goods and services.

regional devolution The transfer of certain powers from national government to one or more regional units of government within a country. Under devolution, the region or regions remain part of the country while gaining some measure of self-government within the overall national institutional framework (in contrast to independence where the regions would no longer be constituent parts of the country).

regionalism The commitment to planning in a nationally coordinated fashion the economic development of fixed regional units that are designated as the basis for allocating economic activities by central government. Regionalism also refers to the ideology of political movements.

resource curse Rather than a blessing, overreliance on a primary commodity, such as oil, for which there is major world demand creates negative effects such as encouragement of corruption (particularly when state-owned companies have a monopoly), a rise in the exchange rate between the country's currency and others that can then raise inflation and squeeze out investment in agriculture and manufacturing, and pressure to share revenues by investing in prestige projects and providing subsidies that do not stimulate long-term economic growth.

resource grabbing When outside forces (either national or international, such as TNCS) take valuable resources (such as water) for their own profit, and, as a result deprive local communities of livelihoods that depended on these resources or of the benefits of using or controlling these resources themselves.

returns to scale, increasing *See* **increasing returns to scale**

sexual divisions of labor The division, specialization, and different rewards between occupations predominantly occupied by either men or women.

share cropping A type of farming in which the rent for the land that tenants pay to the landowner is in agricultural produce rather than in cash. Landlords often provide inputs such as seeds and fertilizer in return for a fixed percentage of what is produced.

shifting cultivation or **"fallow" agriculture** Involves sowing or planting on scorched land using slash-and-burn methods (cutting down the natural vegetation (e.g., forest) and burning it to release its nutrients into the soil). No special tools are required for such a system, neither is weeding or fertilization necessary, provided that cultivation is shifted in a couple of years to another burned plot after a few crops have been taken from the old one, which is then abandoned (left fallow) for a period of time.

socialism A system of social and economic organization in which private property and income distribution are subject to social control. In state socialism, or statism, the central government has responsibility for social control. In the former Soviet Union, for example, markets were prohibited (in public), there was little meaningful electoral politics and the means of production were owned by the state.

sovereignty Sovereignty is a notion that is interrelated with other concepts, such as state, government, independence, democracy, **nation-state**, and **nationalism**, to name a few. Sovereignty in government is the ultimate and independent authority—the absolute right to govern—as held or claimed by a **state** or **nation**. It involves the international independence of a state or nation—the right and authority to regulate its internal affairs without outside interference.

space–process relationship The idea that different types of firm activity are carried out at different locations within a hierarchy of places, from **world cities** to various peripheries.

spatial division of labor Regional economic specialization, based on the distribution of resources and markets and on the exploitation of **agglomeration economies, economies of scale,** and **localization economies.**

spread effects The positive spillover effects on a region (or regions) of the economic growth of some other region.

stagflation Episodes of economic recession accompanied by comparatively high rates of price inflation.

state Political organization of society (requiring a state bureaucracy, a state religion, a judicial apparatus, a military establishment, and a police force) with a defined territory over which it has complete **sovereignty** to use the rule of law to maintain order and security on behalf of the nation or nations inhabiting that territory.

strategic alliances Commercial agreements between transnational corporations, usually involving shared technologies, marketing networks, market research, or product development.

structural adjustment program Structural adjustment programs involve policy changes that are stipulations for getting new loans from the **International Monetary Fund (IMF)** or the **World Bank** or for obtaining lower interest rates on existing loans. Structural adjustment programs were created with the primary goal of reducing the borrowing country's macroeconomic imbalances. In general, loans from both the World Bank and the IMF are designed to promote economic growth, to generate income and to pay off the debt that the countries have accumulated. Structural adjustment programs are based on fiscal and monetary restraint, combined with deregulation and liberalization of national markets. They have been criticized for exacerbating the hardships experienced by ordinary citizens during the structural adjustments.

subsistence economies An economic system, usually of farming, in which the producers (farmers), and their families, consume most of what is produced, leaving little surplus for trade or sale.

supply side Related to the economic theory that increasing the availability (supply) of capital for investment in an economic system by reducing marginal tax rates will promote long-term growth by providing the incentive necessary to increase overall economic activity, productivity, and income.

supranational political union A form of international economic integration that extends beyond economic union to a unified fiscal and monetary system controlled by a supranational authority with executive, judicial, and legislative powers.

sustainable development A pattern of resource use and economic development that does not jeopardize nonrenewable resources, damage existing ecosystems or harm individual species.

tariffs The schedules of duties or taxes imposed by a government on exported or, more typically, imported goods and services.

Taylorism The name given (after analyst F.W. Taylor) to forms of organization in manufacturing industries wherein the planning and control of work are given over entirely to management, leaving production workers to be allocated specialized tasks that are subject to careful analysis—"scientific management"—using techniques such as time-and-motion studies.

technology systems Distinctive "packages" of technologies, energy sources, and political–economic structures that represent the most efficient means for the organization of production at any given phase of economic development. Based on key sets of interdependent technologies, they represent the underpinnings of successive *modes of regulation* and *regimes of accumulation*.

technopole A planned development, within a concentrated area, for technologically innovative, industrial-related production. Technopoles include science parks, science cities, and other high-tech industrial complexes.

terms of trade The ratio of the prices at which exports and imports are exchanged. When the price of exports rises relative to the price of imports, the terms of trade reflect an improvement for the exporting country.

time–space compression The reduction in barriers to decision making and action as a result of transportation and communications improvements that have allowed the pace of life to accelerate. Increasingly, key economic activities, such as the circulation of capital and goods, can occur more rapidly. Time–space compression involves more than the traditional notion of *time–space convergence*, which identified how new systems of transportation and communications, such as the railway replacing the stagecoach, while not changing absolute distance over space, made places closer when distance is measured in time.

TNCs *See* **transnational corporations**

trade creation effects The positive effects of international economic integration, resulting from the free movement of factors of production and free trade, which allows each country or region to specialize according to its comparative advantage, so leading to a greater overall productivity and internal trade.

trade diversion The displacement of pre-existing trade flows as a result of international and supranational economic integration.

trade preference associations Loose forms of international economic integration that involve reduced trade barriers between member states.

trading blocs Groups of countries with formalized systems of trading agreements.

transnational corporations (TNCs) Also known as multinational corporations (MNCs), these companies operate in a number of countries. Many of the headquarters of the largest TNCs are concentrated in the world cities of London, New York, and Tokyo. Production is carried out at a global scale in such a way as to maximize profits. For example, as part of a global assembly system, a low-skill labor-intensive stage of the production process may be located in a less developed country where wages and unionization levels are low.

undeveloped technically recoverable resource (UTRR) Oil and gas that may be produced as a consequence of natural pressure, artificial lift, pressure maintenance, or other secondary recovery methods, but without any consideration of economic viability. Primarily located outside known fields.

undiscovered resources Resources postulated, on the basis of geologic knowledge and theory, to exist outside known fields or accumulations.

unequal exchange Biases in the international trade system promote the unequal exchange of commodities between countries, which results in some countries gaining more and others less. Goods produced in peripheral countries (especially agricultural products), for example, command low prices on the international market compared to the amount of intensive labor that went into producing them, thereby transferring value from peripheral countries to core countries.

uneven development The spatial outcome, within and between countries, of the continuous see-sawing of capital from one set of opportunities to another on the basis of particular local mixes of skills and resources. Capital is invested unevenly over time and across space because, whenever possible, development will occur wherever businesses judge that their investment will yield the highest return. When businesses try to exploit differences between places, they create a continuously variable geometry of labor, capital, production, markets, and management.

unproven reserves Quantities of hydrocarbon resources that are assessed based on geologic and engineering information similar to that used in developing estimates of proven reserves, but technical, contractual, economic, or regulatory uncertainty precludes such reserves from being classified as proved.

UTRR *See* **undeveloped technically recoverable resource**

vertical disintegration A form of business organization in which specialized firms are created and operate as part of a network of subcontractors and suppliers within industries formerly dominated by large, functionally integrated firms. As part of a **flexible production system**, an automobile manufacturer, for example, may subcontract parts and other supplies to smaller specialized firms.

vertical integration A form of business organization in which a company tries to control all aspects of the same industry or enterprise and capture a greater proportion of the final selling price by using corporate mergers or acquisitions of firms that were formerly engaged in different stages of the same industry or enterprise (from production to sale). A car manufacturer, for example, may take over companies that make specialized components such as engines or car navigation systems or that distribute or sell automobiles.

wage labor When people work in exchange for monetary payment rather than bartered goods, military protection or as a result of enslavement.

World Bank The World Bank (and its main component, the International Bank for Reconstruction and Development) is a United Nations affiliate established in 1948 to finance productive projects that further the economic development of its more than 180 member countries from Afghanistan to Uganda. In fiscal year 2005 the World Bank loaned U.S.$22.3 billion to less developed countries for 278 projects.

world city One of the cities that dominate world finance and serve as headquarters to transnational corporations. Typically, London, New York, and Tokyo are identified as the leading tier of world cities, although other cities such as Chicago, Frankfurt, Paris, Los Angeles, and Zürich also have important global roles.

world-system Any spatially extensive economic system that has a single division of labor but multiple cultural systems.

xenophobia *Xenos* and *phobos* are the Greek words for "stranger" and "fear": Xenophobia is a rejection of strangers or foreigners or of anything that is foreign or unknown based simply on feelings of fear or hatred.

Bibliography

Aarebrot, F.H. 1982. On the structural basis of regional mobilization in Europe, in B. De Marchi and A.M. Boileau (eds.) *Boundaries and Minorities in Western Europe*. Milan: Franco Angeli.

Achebe, C. 1975. *Morning Yet on Creation Day*. London: Faber.

Adelman, I. 1984. Beyond export-led growth, *World Development* 12, 937–949.

Agnew, J.A. 1987. *The United States in the World Economy: A regional geography*. Cambridge: Cambridge University Press.

Agnew, J.A. 1988. Beyond core and periphery: the myth of regional political–economic restructuring and a new sectionalism in American politics, *Political Geography Quarterly* 7, 127–139.

Agnew, J.A. 1992. The United States and American hegemony, in P.J. Taylor (ed.) *The Political Geography of the Twentieth Century*. London: Belhaven Press.

Agnew, J.A. 1993. Trading blocs or a world that knows no boundaries? in C.H. Williams (ed.) *The Political Geography of the New World Order*. London: Belhaven Press.

Agnew, J.A. and Corbridge, S. 1995. *Mastering Space: Hegemony, territory and international political economy*. London: Routledge.

Agnew, J.A. and Grant, R.J. 1997. Falling out of the world economy? Theorizing "Africa" in world trade, in R. Lee and J. Wills (eds.) *Geographies of Economies*. London: Edward Arnold.

Agnew, J.A., Shin, M., and Richardson, P. 2005. The saga of the "Second Industrial Divide" and the history of the "Third Italy": evidence from export data, *Scottish Geographical Journal* 121, 83–101.

Albrechts, L. and Swyngedouw, E. 1989. The challenges for regional policy under a flexible regime of accumulation, in L. Albrechts, F. Moulaert, R. Roberts, and E. Swyngedouw (eds.) *Regional Policy at the Crossroads*. London: Jessica Kingsley.

Allen, J. 1988. Fragmented firms, disorganised labour? in J. Allen and D. Massey (eds.) *Restructuring Britain: The economy in question*. London: Sage.

Amin, A. (ed.) 1994. *Post-Fordism. A reader*. Cambridge, MA: Blackwell.

Amin, A. 1996. Beyond associative democracy, *New Political Economy* 1, 309–333.

Amin, A. and Robins, K. 1990. The re-emergence of regional economies? The mythical geography of flexible accumulation, *Society and Space* 8, 7–34.

Amin, A. and Thrift, N. 1992. Neo-Marshallian nodes in global networks, *International Journal of Urban and Regional Research* 16, 571–587.

Amin, A. and Thrift, N. (eds.) 1994. *Globalization, Institutions, and Regional Development in Europe*. Oxford: Oxford University Press.

Amiti, M. and Wei, S.-J. 2005. Fear of service outsourcing: is it justified? *Economic Policy* 42, 307–347; available at http://www.oecd.org/dataoecd/44/24/35333668.pdf

Amsden, A. 1989. *Asia's Next Giant: South Korea and late industrialization*. New York: Oxford University Press.

Anderson, A.B. 1990. Smokestacks in the rainforest: industrial development and deforestation in the Amazon Basin, *World Development* 18, 1191–1205.

Andrews, J. 1992. A change of face: a survey of Taiwan, *Economist*, October 10.

Aoyama, Y. 2000. Networks, keiretsu, and locations of the Japanese electronics industry in Asia, *Environment and Planning A*, 32, Pion Limited, London, 223–244.

Armstrong, W. and McGee, T. 1986. *Theatres of Accumulation: Studies in Asian and Latin American urbanization*. London: Methuen.

Arthur, W.B. 1989. Competing technologies, increasing returns, and lock-in by historical events, *Economic Journal* 99, 116–131.

Artis, M. and Lee, N. (eds.) 1994. *The Economics of the European Union*. Oxford: Oxford University Press.

Aryeetey-Attoh, S. (ed.) 1997. *Geography of Sub-Saharan Africa*. Upper Saddle River, NJ: Prentice Hall.

Auty, R.M. 1991. Third world response to global processes: the mineral economies, *Professional Geographer* 43, 68–76.

Baer, W., da Fonseca, M.A.R., and Guilhoto, J.J.M. 1987. Structural changes in Brazil's industrial economy, 1960–80, *World Development* 15, 275–286.

Bagchi-Sen, S. and Sen, J. 1997. The current state of knowledge in international business in producer services, *Environment and Planning A* 39, 1153–74.

Bain, J.S. 1959. *Industrial Organization*. New York: John Wiley & Sons.

Bairoch, P. 1982. International industrialization levels from 1750 to 1980, *European Journal of Economic History* 11, 269–333.

Balassa, B. 1979. *The Changing International Division of Labor in Manufactured Goods*. Washington, DC: World Bank, Working Paper 329.

Baldwin, R.E. 1994. *Towards an Integrated Europe*. London: Centre for Economic Policy Research.

Banco de España. 2012. *Nota Informativa: Balanza de pagos en marzo 2012*, May 31; available at http://www.bde.es/f/webbde/GAP/Secciones/SalaPrensa/NotasInformativas/12/Arc/Fic/presbe2012_17.pdf

Bank for International Settlements. 2010. *Triennial Central Bank Survey: Report on global foreign exchange market activity in 2010*. Basel: BIS, Monetary and Economic Department.

Barff, R. 1995. It's gotta be da shoes, *Environment and Planning A* 27, 55–79.

Barnes, M. and Haskel, J. 2001. *Job Creation, Job Destruction and Small Firms: Evidence from the UK*. Queen Mary, University of London Research Paper.

Barnet, R.J. and Cavanagh, J. 1994. *Global Dreams. Imperial corporations and the new world order*. New York: Simon & Schuster.

Barratt Brown, M. 1993. *Fair Trade: Reform and realities in the international trading system*. London: Zed Books.

Bates, R.H. 1983. *Essays on the Political Economy of Rural Africa*. Cambridge: Cambridge University Press.

Bayoumi, T. 1989. *Saving-Investment Correlations*. Washington, DC: IMF Working Paper 89/66.

Bayoumi, T. and Lipworth, G. 1997. Japanese foreign direct investment and regional trade, *Finance & Development* September, 11–13.

BEA (Bureau of Economic Analysis). 2012. *Summary Estimates for Multinational Companies*. New release. Washington, DC: US Department of Commerce, April 18; available at http://www.bea.gov/news releases/international/mnc/mncnewsrelease.htm

Beenstock, M. 1983. *The World Economy in Transition*. London: Allen & Unwin.

Belderbos, R. and Zou, J. 2006. Foreign investment, divestment and relocation by Japanese electronics firms in East Asia, *Asian Economic Journal* 20, 1–27.

Bellini, N. 1996. Regional economic policies and the non-linearity of history, *European Planning Studies* 4, 63–73.

Belussi, F. 1996. Local systems, industrial districts and institutional networks: towards a new evolutionary paradigm of industrial economics?, *European Planning Studies* 4, 5–26.

Beneria, L. 1981. Conceptualising the labour force: the underestimation of women's activities, *Journal of Development Studies* 17, 10–27.

Benko, G. and Dunford, M. (eds.) 1991. *Industrial Change and Regional Development: The transformation of new industrial spaces*. London: Belhaven Press.

Berg, E.J. 1965. The development of the labor force in sub-Saharan Africa, *Economic Development and Cultural Change* 13, 394–412.

Berry, B.J.L. 1991. *Long Wave Rhythms in Economic Development and Political Behavior*. Baltimore, MD: Johns Hopkins University Press.

Berry, B.J.L., Conkling, E.C., and Ray, D.M. 1976. *The Geography of Economic Systems*. Englewood Cliffs, NJ: Prentice Hall.

Beyon, J. and Dunkerley, D. (eds.) 2000. *Globalization: The reader*. New York: Routledge.

Biersteker, T.J. 1995. The "triumph" of liberal economic ideas in the developing world, in B. Stallings (ed.) *Global Change, Regional Response: The new international context of development*. Cambridge: Cambridge University Press.

Bingham, R.D. and Hill, E.W. (eds.) 1997. *Global Perspectives on Economic Development*. New Brunswick, NJ: Center for Urban Policy Research.

Birch, D.L. 1979. *The Job Generation Process*. Cambridge, MA: MIT Program on Neighborhood and Regional Change.

BISNIS (Business Information Service for the Newly Independent States). 2001. *Russia: Fact sheet, July 2001*; available at http://www.bisnis.doc.gov/bisnis/country/RussiaFactsheet-2001.htm

Blinder, A.S. 2006. Offshoring: the next industrial revolution?, *Foreign Affairs*, March/April.

Bluestone, B. and Harrison, B. 1982. *The Deindustrialization of America*. New York: Basic Books.

Boggs, J.S. and Rantisi, N.M. 2003. The "relational turn" in economic geography, *Journal of Economic Geography* 3, 109–116.

Bonacich, E., Chen, L., Chinchilla, N., Hamilton, N., and Ong, P. 1994. *Global Production: The apparel industry in the Pacific Rim*. Philadelphia, PA: Temple University Press.

Borchert, J.R. 1967. American metropolitan evolution, *Geographical Review* 57, 301–332.

Borchert, J.R. 1978. Major control points in American economic geography, *Annals of the Association of American Geographers* 68, 214–232.

Boserup, E. 1981. *Population and Technology*. Oxford: Blackwell.

Bourgin, F. 1989. *The Great Challenge: The Myth of Laissez-Faire in the Early Republic*. New York: George Braziller.

Bowler, I. 1985. *Agriculture under the Common Agricultural Policy*. Manchester: Manchester University Press.

Boyte, H. 1980. *The Backyard Revolution: Understanding the new citizen movement*. Philadelphia, PA: Temple University Press.

Bradford, C.I. 1987. Trade and structural change: NICs and next-tier NICs as transitional economies, *World Development* 15, 299–316.

Bradley, P.N. and Carter, S.E. 1989. Food production and distribution—and hunger, in R.J. Johnston and P.J. Taylor (eds.) *A World in Crisis?*, 2nd edn. Oxford: Blackwell.

Bradshaw, M. 1991. *The Soviet Union: A new regional geography?* London: Belhaven Press.

Braudel, F. 1972. *The Mediterranean and the Mediterranean World in the Age of Phillip II* (trans. S. Reynolds). New York: Harper & Row.

Brenner, R. 1977. The origins of capitalist development: a critique of neo-Smithian Marxism, *New Left Review* 104, 25–91.

Brockett, C.D. 1988. *Land, Power, and Poverty: Agrarian transformation and political conflict in Central America*. Boston, MA: Unwin Hyman.

Brundenius, C. 1984. *Revolutionary Cuba: The challenge of economic growth with equity*. Boulder, CO: Westview.

Brunn, S. (ed.) 2006. *Wal-Mart World: The world's biggest corporation in the global economy*. New York and London: Routledge.

Brusco, S. and Sabel, C. 1981. Artisan production and economic growth, in F. Wilkinson (ed.) *The dynamics of labour market segmentation*. London: Academic Press.

Bryson, J.R., Daniels, P.W., Henry, N., and Pollard, J. (eds.) 2000. *Knowledge, Space, Economy*. London: Routledge.

Bryson, J.R., Daniels, P.W., and Warf, B. 2004. *Service Worlds: People, organisations, technologies*. London and New York: Routledge.

Bryson, J.R., Henry, N., Keeble, D., and Martin, R. (eds.) 1999. *The Economic Geography Reader*. New York: John Wiley & Sons.

Buchanan, K. 1972. *The Geography of Empire*. London: Spokesman Books.

Burbach, R. and Flynn, P. 1980. *Agribusiness in the Americas*. New York: Monthly Review Press.

Busse, M. 2002. *Competition Intensity, Potential Competition and Transaction Cost Economics*. HWWA Discussion Paper 183. Hamburg: Hamburg Institute of International Economics.

Buzzetti, L. 1996. Efforts to reorganize the state within the new international framework, in A. Vallega, R.C. de Azevedo, L. Buzzetti, A. Celant, P. Landini, B. Cardinale *et al.* (eds.) *The Geography of Disequilibrium: Global issues and restructuring in Italy*. Rome: Societa Geografica Italiana.

Cameron, R. 1973. The logistics of European economic growth: a note on historical periodization, *Journal of European Economic History* 2, 145–158.

Carter, H. 1983. *An Introduction to Urban Historical Geography*. London: Edward Arnold.

Castells, M. 1988. High technology and urban dynamics in the United States, in M. Dogan and J. Kasarda (eds.) *The Metropolis Era. Vol. 1: A world of giant cities*. Newbury Park, CA: Sage.

Castells, M. 2000. *The Information Age: Economy, society and culture. Vol. I: The rise of the network society*, 2nd edn. Oxford: Blackwell.

Castells, M. and Hall, P. 1994. *Technopoles of the World. The making of 21st century industrial complexes*. London: Routledge.

Celant, A. 1996. Italy's foreign trade: economic and territorial characteristics, in A. Vallega, R.C. de Azevedo, L. Buzzetti, A. Celant, P. Landini, B. Cardinale *et al.* (eds.) *The Geography of Disequilibrium: Global issues and restructuring in Italy*. Rome: Societa Geografica Italiana.

Champion, A.G. and Townsend, A.R. 1990. *Contemporary Britain*. London: Edward Arnold.

Chandler, A.D. 1992. Organizational capabilities and the economic history of the industrial enterprise, *Journal of Economic Perspectives* 6, 79–100.

Chaunu, P. 1969. *L'expression européenne du XIIIe au XVe siècle*. Collection Nouvelle Clio 26. Paris: Presses Universitaires de France.

Chernesky, R.J. 2006. *Strategic Alliances*. Dayton, OH: Chernesky, Heyman & Kress PLL.

Cheshire, P.C., D'Arcy, E., and Giussani, B. 1992. Purpose built for failure? Local, regional and national government in Britain, *Environment and Planning C: Government and Policy* 10, 355–369.

Childe, V.G. 1950. The urban revolution, *Town Planning Review* 21, 3–17.

China Sourcing Blog. 2012. Textile and apparel sourcing: the rise of South East Asia. *China Sourcing Blog*. January 30; available at http://www.chinasourcingblog.org/2012/01/textile-and-apparel-sourcing-t-1.html

Chisholm, M. 1982. *Modern World Development*. Totowa, NJ: Barnes & Noble.

Chisholm, M. 1990. *Regions in Recession and Resurgence*. London: Unwin Hyman.

Christopher, A.J. 1984. *Colonial Africa*. Beckenham: Croom Helm.

Christopherson, S. 1989. Flexibility in the US service economy and the emerging spatial division of labour, *Transactions of the Institute of British Geographers* 14, 131–143.

Christopherson, S. 1995. Changing women's status in a global economy, in R.J. Johnston, P.J. Taylor, and M.J. Watts (eds.) *Geographies of Global Change*. Oxford: Blackwell.

CIA (US Central Intelligence Agency). 1999. *Handbook of International Economic Statistics*. Washington, DC: Directorate of Intelligence.

CIA. 2001. *World Factbook 2001*. Washington, DC: Directorate of Intelligence.

CIA. 2012. *World Factbook 2012*. Washington, DC: Directorate of Intelligence; available at https://www.cia.gov/library/publications/the-world-factbook/geos/xx.html

Cipolla, C. 1981. *Before the Industrial Revolution. European society and economy, 1000–1700*, 2nd edn. London: Methuen.

Claessens, S., Glaessner, T., and Klingebiel, D. 2002. *Electronic Finance: A new approach to financial sector development?* World Bank Discussion Paper No. 431. Washington, DC: World Bank.

Claessens, S. and Jansen, M. (eds.) 2000. *The Internationalization of Financial Services: Issues and lessons for developing countries*. London: Kluwer Law International.

Clark, C. 1977. *World Prehistory in New Perspective*. Cambridge: Cambridge University Press.

Clark, G.L. 1993. Global interdependence and regional development: business linkages and corporate governance in a world of financial risk, *Transactions of the Institute of British Geographers* 18, 309–325.

Clark, G., Feldman, M., and Gertler, M.S. (eds.) 2000. *The Oxford Handbook of Economic Geography*. New York: Oxford University Press.

Cline, W.R. 1982. Can the East Asian model of development be generalized? *World Development* 10, 81–90.

Coaffee, J. 2004. Rings of steel, rings of concrete and rings of confidence: designing out terrorism in Central London pre and post September 11th, *International Journal of Urban and Regional Research* 28, 201–211.

Coffey, W.J. 2000. The geographies of producer services [progress report], *Urban Geography* 21, 170–183.

Coghlan, A. 1993. Dying for innovation, *New Scientist* 137, 12–14.

Cohen, A. 1974. *Two-Dimensional Man: An essay on the anthropology of power and symbolism in complex society*. Berkeley, CA: University of California Press.

Cohen, R.B. 1981. The new international division of labor, multinational corporations and the urban hierarchy, in M. Dear and A.J. Scott (eds.) *Urbanization and Urban Planning in Capitalist Society*. London: Methuen.

Cohen, S.B. 1982. A new map of global geopolitical equilibrium: a developmental approach, *Political Geography Quarterly* 1, 233–241.

Cohen, S.S. and Zysman, J. 1987. *Manufacturing Matters: The myth of the post-industrial economy*. New York: Basic Books.

Comisso, E. 1979. *Workers' Control under Plan and Market*. New Haven, CT: Yale University Press.

Commission of the European Communities. 1973. *Report on the Regional Problem in the Enlarged Community (Thompson Report)*. Brussels: Commission of the European Communities.

Competitiveness Clusters Agency of the French Government. 2011. *Competitiveness Clusters in France*. Paris: Competitiveness Clusters Agency of the French Government; available at http://competitivite. gouv.fr/

Conzen, M.P. 1981. The American urban system in the nineteenth century, in D.T. Herbert and R.J. Johnston (eds.) *Geography and the Urban Environment: Progress in research and applications*. *Vol. IV*. Chichester: John Wiley & Sons.

Cook, P. and Kirkpatrick, C. 1997. Globalization, regionalization and third world development, *Regional Studies* 31, 55–66.

Cooke, P. 1988. Flexible integration, scope economies, and strategic alliances: social and spatial mediations, *Society and Space* 6, 281–300.

Cooke, P. 1992. Regional innovation systems: competitive regulation in the new Europe, *Geoforum* 23, 65–82.

Cooper, F. 1980. *From Slaves to Squatters: Plantation labour and agriculture in Zanzibar and coastal Kenya, 1890–1925*. New Haven, CT: Yale University Press.

Cooper, W.H. 2013. *EU-U.S. Economic Ties: Framework, scope, and magnitude*. Washington, DC: Congressional Research Service.

Corbridge, S. 1982. Urban bias, rural bias, and industrialisation: an appraisal of the work of Michael Lipton and Terry Byres, in J. Harriss (ed.) *Rural Development: Theories of peasant economic and agrarian change*. London: Hutchinson.

Corbridge, S. 1986. *Capitalist World Development: A critique of radical development geography*. London: Macmillan.

Corbridge, S. 1993. *Debt and Development*. Oxford: Blackwell.

Corbridge, S.E. and Agnew, J.A. 1991. The US trade and budget deficits in global perspective: an essay in geopolitical economy, *Society and Space* 9, 71–90.

Cornford, A. 2004. *The WTO Negotiations on Financial Services: Current issues and future directions*. Geneva: UNCTAD.

CorpTech. 2000. *CorpTech Directory of Technology Companies*, 15th US edn. Woburn, MA: Corporate Technology Information Services.

Council of Economic Advisors. 1991. *Economic Report of the President*. Washington, DC: US Government Printing Office.

Cowhey, P.F. and Aronson, J.D. 1993. *Managing the World Economy: The consequences of corporate alliances*. New York: Council on Foreign Relations Press.

Cox, K.R. (ed.) 1997. *Spaces of Globalization: Reasserting the power of the local*. New York: Guilford Press.

CRIC. 2006. *Innovation in Services*. CRIC Briefing No. 2. Manchester, Centre for Research on Innovation and Competition, University of Manchester.

Crook, C. 1991. Sisters in the wood: a survey of the IMF and the World Bank, *Economist*, October 12.

Crossley, J.C. 1983. The River Plate countries, in H. Blakemore and C.T. Smith (eds.) *Latin America: Geographical perspectives*, 2nd edn. London: Methuen.

Crow, B. and Thomas, A. 1985. *Third World Atlas*. Milton Keynes: Open University Press.

Cuadrado-Roura, J.R., Rubalcaba-Bermejo, L., and Bryson, J.R. (eds.) 2002. *Trading Services in the Global Economy*. Cheltenham: Edward Elgar.

Cumings, B. 1984. The origins and development of the Northeast Asian political economy: industrial sectors, product cycles and political consequences, *International Organization* 38, 1–40.

Cunningham, S. 1986. Multinationals and restructuring in Latin America, in C.J. Dixon, D.W. Drakakis-Smith, and H.D. Watts (eds.) *Multinational Corporations and the Third World*. Boulder, CO: Westview.

Currah, A. and Wrigley, N. 2004. Networks of organizational learning and adaptation in retail TNCs, *Global Networks* 4, 1–23.

Curtin, P., Feierman, S., Thompson, L., and Vansina, J. 1978. *African History*. Boston, MA: Little, Brown.

Dalton, D.H. and Serapio, M.G. 1999. *Globalizing Industrial Research and Development*. Washington, DC: US Department of Commerce, Technology Administration Office of Technology Policy.

Daniels, P.W. 1985. *Service Industries: Growth and location*. London: Methuen.

Daniels, P.W. 1991. A world of services?, *Geoforum* 22, 359–376.

Daniels, P.W. and Bryson, J.R. 2002. Manufacturing services and servicing manufacturing: knowledge-based cities and changing forms of production, *Urban Studies* 39, 977–991.

Daniels, P. and Lever, W.F. (eds.) 1996. *The Global Economy in Transition*. New York: Addison Wesley Longman.

Dawson, A.H. 1993. *A Geography of European Integration*. London: Belhaven Press.

Day, G. and Rees, G. (eds.) 1991. *Regions, Nations, and European Integration: Remaking the Celtic periphery*. Cardiff: University of Wales Press.

De Janvry, A. 1984. The role of land reform in economic development: policies and politics, in C.K. Eicher and J.M. Staatz (eds.) *Agricultural Development in the Third World*. Baltimore, MD: Johns Hopkins University Press.

De Vries, J. 1976. *Economy of Europe in an Age of Crisis, 1600–1750*. Cambridge: Cambridge University Press.

De Vroey, M. 1984. A regulation approach to interpretation of the present crisis, *Capital and Class* 23, 45–66.

DeGeer, S. 1927. The American manufacturing belt, *Geografiska Annaler* 9, 233–359.

Deininger, K. and Byerlee, D. 2011. *Rising Global Interest in Farmland: Can it yield sustainable and equitable benefits?* Washington, DC: World Bank; available at http://siteresources.worldbank.org/DEC/Resources/Rising-Global-Interest-in-Farmland.pdf

Destler, I.M. 1992. *American Trade Politics*, 2nd edn. Washington, DC: Institute for International Economics.

Deudney, D. and Ikenberry, G.J. 1991/92. The international sources of Soviet change, *International Security* 16, 74–118.

Diakosavvas, D. and Scandizzo, P. 1991. Trends in the terms of trade of primary commodities, 1900–1982: the controversy and its origins, *Economic Development and Cultural Change* 39, 231–264.

Diamandi, Z., Dubey, A., Pleasance, D., and Vora, A. 2011. *Winning in the SMB Cloud: Charting a path to success*. San Francisco: McKinsey & Company; available at http://www.google.com/url?sa=t&rct=j&q=&esrc=s&source=web&cd=1&ved=0CDkQFjAA&url=http%3A%2F%2Fwww.mckinsey.com%2F~%2Fmedia%2FMcKinsey%2Fdotcom%2Fclient_service%2FHigh%2520Tech%2FPDFs%2F

Winning_in_the_SMB_Cloud.ashx&ei=RJS3Uc3gG4nA9gSqpIH4AQ&usg=AFQjCNENYj9YqQIskL
2hLbjQlQ19gDDhuA&bvm=bv.47534661,d.eWU

Diamond, J. 1997. *Guns, Germs, and Steel*. New York: W.W. Norton.

Dicken, P. 1992. International production in a volatile regulatory environment: the influence of national regulatory policies on the spatial strategies of transnational corporations, *Geoforum* 23, 303–316.

Dicken, P. 1994. Global-local tensions: firms and states in the global space-economy, *Economic Geography* 70, 101–127.

Dicken, P. 1998. *Global Shift: Transforming the world economy*, 3rd edn. New York and London: Guilford Press.

Dicken, P. 2007. *Global Shift: Mapping the changing contours of the world economy*, 5th edn. New York and London: Guilford Press.

Dicken, P. and Thrift, N. 1992. The organization of production and the production of organization: why business enterprises matter in the study of geographical industrialization, *Transactions of the Institute of British Geographers* 17, 279–291.

Dobb, M. 1963. *Studies in the Development of Capitalism*. London: Routledge & Kegan Paul.

Donahue, J.D. 1997. The devil in devolution, *American Prospect* 32, 42–47.

Dosi, G. 1988. Sources, procedure and microeconomic effects of innovation, *Journal of Economic Literature* 26, 1–12.

Duignan, P. and Gann, L.H. 1985. *The United States and Africa: A history*. Cambridge: Cambridge University Press.

Duncan, K. and Rutledge, I. 1977. Introduction: patterns of agrarian capitalism in Latin America, in K. Duncan and I. Rutledge (eds.) *Land and Labor in Latin America*. Cambridge: Cambridge University Press.

Dunford, M. 1990. Theories of regulation, *Society and Space* 8, 297–321.

Dunford, M. 2002. *Cohesion and Enlargement*, paper presented at a Round Table on European Regional Disparities and the Cohesion and Structural EU Policy, Jean Monnet Graduate School in European Law and Economics, Belvedere di San Leucio, San Leucio, Caserta, Italy, May 8.

Dunford, M. and Kafkalas, G. 1992. The global-local interplay, corporate geographies and spatial development strategies in Europe, in M. Dunford and G. Kafkalas (eds.) *Cities and Regions in the New Europe*. London: Belhaven Press.

Dunford, M. and Perrons, D. 1983. *The Arena of Capital*. London: Macmillan.

Dunford, M. and Smith, A. 2000. Catching up or falling behind? Economic performance and regional trajectories in the "New Europe," *Economic Geography* 76, 169–195.

Dunning, J.H. 1979. Explaining changing patterns of international production: in defence of the eclectic theory, *Oxford Bulletin of Economics and Statistics* 41, 269–296.

Dunning, J.H. 1983. Changes in the level and structure of international production: the last one hundred years, in M. Casson (ed.) *The Growth of International Business*. London: Allen & Unwin.

Dunning, J.H. (ed.) 1997. *Governments, Globalization, and International Business*. New York: Oxford University Press.

Dunning, J.H. and Norman, G. 1987. The location choice of offices of international companies, *Environment & Planning A* 19, 613–631.

Durham, K.F. 1977. *Expansion of Agricultural Settlement in the Peruvian Rainforest: The role of the market and the role of the state*, paper presented at the Latin American Studies Association, Houston, TX, November 2–5.

Economist. 1983. Hard pounding this, gentlemen: a survey of the world economy, September 24.

Economist. 1987. The rag trade: on the road to Mandalay, June 27, 67–68.

Economist. 1992a. Net financial flows to developing countries, September 14, 120.

Economist. 1992b. Zero inflation: how low is low enough?, November 7, 23–26.

Economist. 1992c. Survey: Russian reborn, September 24, 25.

Economist. 1993a. The global firm: RIP, February 6, 69.

Economist. 1993b. The final frontier, February 20, 63.

Economist. 1996a. Getting together, June 29, 42–43.

Economist. 1996b. How poor is China?, October 12, 35–36.

Economist. 1996c. The buying and selling of Brazil Inc., November 9, 83–84.

Economist. 1997. The last emperor, February 22, 21–25.

Economist. 2001. The technology industry: big is beautiful again, July 21, 51–52.

Economist. 2005. Industrial metamorphosis, September 29, 81–82.

Economist. 2006. The euro and trade, June 24, 90.

Economist. 2010a. Small island for sale, March 25.

Economist. 2010b. Leaving home, November 18.

Economist. 2012a. Space invaders, January 7.

Economist. 2012b. The new maker rules, November 24.

Economist. 2013a. O for a beaker full of the warm south, January 19.

Economist. 2013b. Herd instinct, January 19.

Economist. 2013c. The post-industrial future is nigh, February 23.

Economist. 2013d. The next big thing, January 19.

Edquist, C. 1985. *Capitalism, Socialism, and Technology: A comparative study of Cuba and Jamaica.* London: Zed Books.

Edwards, C. 1985. *The Fragmented World: Competing perspectives on trade, money and crisis.* London: Methuen.

EIA (Environmental Information Agency). 2011a *Analysis and Projections: World shale gas resources: An initial assessment of 14 regions outside the United States.* Washington, DC: US Department of Energy, April; available at http://www.eia.gov/analysis/studies/ worldshalegas/ archive/2011/pdf/full report.pdf?zscb=73375579

EIA. 2011b. *International Energy Outlook 2011.* Report Number: DOE/EIA-0484(2011). Washington, DC: US Department of Energy, September 19; available at http://www.eia.gov/forecasts/ieo/ world.cfm

EIA. 2012. *Energy in Brief: What is shale gas and why is it important?* (website), December 5; available at http://www.eia.gov/energy_in_brief/article/about_shale_gas.cfm

Eichengreen, B. 1996. *Globalizing Capital: A history of the international monetary system.* Princeton, NJ: Princeton University Press.

Eicher, C.K. 1984. Facing up to Africa's food crisis, in C.K. Eicher and J.M. Staatz (eds.) *Agricultural Development in the Third World.* Baltimore, MD: Johns Hopkins University Press.

Eisenschitz, A. and Gough, J. 1992. *The Politics of Local Economic Policy: The problems and possibilities of local initiative.* London: Macmillan.

Eisenschitz, A. and Gough, J. 1996. The contradictions of neo-Keynesian local economic strategy, *Review of International Political Economy* 3, 434–458.

Eisold, E. 1984. *Young Women Workers in Export Industries: The case of the semiconductor industry in South East Asia.* Geneva: Working Paper, ILO World Employment Programme.

Elbaum, B. 1990. Cumulative or comparative advantage? British competitiveness in the early 20th century, *World Development* 18, 1255–1272.

Emmanuel, A. 1972. *Unequal Exchange: A study of the imperialism of trade.* London: New Left Books.

Engman, M., Onodera, O., and Pinali, E. 2007. *Export Processing Zones: Past and future role in trade and development.* OECD Trade Policy Working Paper No. 53. Paris: OECD.

Epstein, G. 1992. Political economy and comparative central banking, *Review of Radical Political Economics* 24, 1–30.

Ernst, Dieter. 2009. *A New Geography of Knowledge in the Electronics Industry? Asia's role in global innovation networks.* Policy Studies, No. 54. Honolulu, Hawaii: East-West Center.

Esser, S. 2001. Globalizing the board of directors, *Corporate Board* 22, 1–5.

Ettlinger, N. 1991. The roots of competitive advantage in California and Japan, *Annals of the Association of American Geographers* 81, 391–407.

Ettlinger, N. 1997. An assessment of the small-firm debate in the United States, *Environment and Planning A* 29, 419–442.

Eurostat. 2012a. *Eurostat Regional Yearbook 2012*. Luxembourg: Publications Office of the European Union; available at http://epp.eurostat.ec.europa.eu/portal/page/portal/ publications/regional_yearbook Eurostat. 2012b. Regional GDP per capita in 2009: seven capital regions in the first ten places. News Release 38/12. March 13; available at http://epp.eurostat.ec.europa.eu/cache/ITY_PUBLIC/ 1–13032012-AP/EN/1–13032012-AP-EN.PDF

Evans, D. and Alizadeh, P. 1984. Trade, industrialization and the visible hand, in R. Kaplinsky (ed.) *Third World Industrialisation in the 1980s: Open economies in a closing world*. London: Cass.

Evans, P. 1979. *Dependent Development: The alliance of multinational, state, and local capitalism in Brazil*. Princeton, NJ: Princeton University Press.

Evans, P. 1987. Dependency and the state in recent Korean development: some comparisons with Latin American NICs, in K.D. Kim (ed.) *Dependency Issues in Korean Development: Comparative perspectives*. Seoul: Seoul National University Press.

Evenson, R.E. 1984. Benefits and obstacles in developing appropriate agricultural technology, in C.K. Eicher and J.M. Staatz (eds.) *Agricultural Development in the Third World*. Baltimore, MD: Johns Hopkins University Press.

FAO (Food and Agricultural Organization of the United Nations) 1992. *World Food Conference: Report*. Rome: FAO.

FAO. 1995. *World Agriculture: Towards 2010*. Rome: FAO; available at http://www.fao.org/ docrep/v4200e/v4200e00.htm

FAO. 2000. *The State of Food Insecurity in the World, 2000*. Rome: FAO.

FAO. 2001. *The State of Food Insecurity in the World, 2001*. Rome: FAO.

FAO. 2004. *The State of Agricultural Commodity Markets*. Rome: FAO.

FAO. 2005. *Summary of World Food and Agricultural Statistics*. Rome: FAO.

FAO. 2010. *Global Forest Resources Assessment*. Rome: FAO.

FAO. 2011. *The State of Food Insecurity in the World*. Rome: FAO.

Farrell, D. (ed.) 2005. *The Emerging Global Labor Market*. San Francisco: McKinsey Global Institute.

Feder, E. 1978. *Strawberry Imperialism*. Mexico City: Editorial Campesina.

Fiala, R. and Kamens, D. 1986. Urban growth and the world polity in the nineteenth and twentieth centuries: a research agenda, *Studies in Comparative International Development* 21, 23.

Fine, B. 1994. Towards a political economy of food, *Review of International Political Economy* 3, 519–545.

Fisher, P.S. and Peters, A.S. 1997. Tax and spending incentives and enterprise zones, *New England Economic Review* March/April, 109–130.

Fisher, R.C. 1997. The effects of state and local services on economic development, *New England Economic Review* March/April, 53–67.

Flannery, K.V. 1969. Origin and ecological effects of early domestication, in P.J. Ucko and G.W. Dimbleby (eds.) *The Domestication and Exploitation of Plants and Animals*. London: Duckworth.

Florida, R. and Smith, D.F. Jr. 1993. Venture capital formation, investment and regional industrialization, *Annals of the Association of American Geographers* 83, 434–451.

Foek, A. 1997. Sweat-shop Barbie, *Humanist* 57, 9–13.

Forbes, D. and Thrift, N. 1987. International impacts on the urbanization process in the Asian region: a review, in R.J. Fuchs, G.W. Jones, and E.M. Pernia (eds.) *Urbanization and Urban Policies in Pacific Asia*. Boulder, CO: Westview.

Forbes Global 2000; available at http://www.forbes.com/global2000/list/

Fortune. 2002. *Fortune* magazine's Global 500 in 2000 (the world's largest corporations); available at http://www.fortune.com

Fothergill, S. and Guy, J. 1990. *Retreat from the Regions: Corporate change and the closure of factories*. London: Regional Studies Association.

Frank, A.G. 1967. *Capitalism and Underdevelopment in Latin America*. New York: Monthly Review Press.

Frankman, M.T. 1974. Sectoral policy preferences of the Peruvian government, 1946–1968, *Journal of Latin American Studies* 6, 289–300.

Freund, C. and Spatafora, N. 2005. *Remittance: Transaction costs, determinants, and informal flows.* World Bank Policy Research Working Paper 3704, September. Washington, DC: World Bank.

Fridell, G. 2006. Fair trade and neoliberalism: assessing emerging perspectives, *Latin American Perspectives* 33, 8–28.

Friedman, B. 1989. *Day of Reckoning: The consequences of American economic policy.* New York: Random House.

Friedmann, H. 1991. Changes in the international division of labor: agro-food complexes and export agriculture, in W.H. Friedland, L. Busch, F.H. Buttel, and A.P. Rudy (eds.) *Towards a New Political Economy of Agriculture.* Boulder, CO: Westview.

Friedmann, J. 1956. Locational aspects of economic development, *Land Economics* 32, 213–227.

Friedmann, J. 1986. The world city hypothesis, *Development and Change* 17, 69–84.

Friedmann, J. 1995. Where we stand: a decade of world city research, in P.L. Knox and P.J. Taylor (eds.) *World Cities in a World-system.* Cambridge: Cambridge University Press.

Friedmann, J. and Weaver, C. 1979. *Territory and Function.* Berkeley, CA: University of California Press.

Fröbel, F., Heinrichs, J., and Kreye, O. 1977. *Die neue internationale Arbeitsteilung.* Reinbeck: Rowoholt.

Fuchs, R.J. and Pernia, E.M. 1987. External economic forces and national spatial development: Japanese direct investment in Pacific Asia, in R.J. Fuchs, G.W. Jones, and E.M. Pernia (eds.) *Urbanization and Urban Policies in Pacific Asia.* Boulder, CO: Westview.

Galbraith, J.K. 1977. *The Affluent Society.* Boston, MA: Houghton-Mifflin.

Galbraith, R. 2001. Leaner, more innovative, Italian footwear springs back to fore, *International Herald Tribune*, June 27.

Galor, O. and Weil, D. 2000. Population, technology, and growth: from Malthusian stagnation to the demographic transition and beyond, *American Economic Review* 90, 806–828.

Gann, L.H. and Duignan, P. 1978. *The Rulers of British Africa, 1870–1914.* Stanford, CA: Stanford University Press.

Geertz, C. 1973. *The Interpretation of Cultures.* New York: Basic Books.

Gellner, E. 1964. *Thought and Change.* London: Weidenfeld & Nicolson.

George, S. 1991. European political cooperation: a world-systems perspective, in M. Holland (ed.) *The Future of European Political Cooperation. Essays on theory and practice.* New York: St. Martin's Press.

George, S. 1992. *The Debt Boomerang: How third world debt harms us all.* London: Pluto Press.

Gereffi, G. 1995. Global production systems and third world development, in B. Stallings (ed.) *Global Change, Regional Response: The new international context of development.* Cambridge: Cambridge University Press.

Gereffi, G. 1999. *A Commodity Chains Framework for Analyzing Global Industries.* Durham, NC: Duke University.

Gereffi, G. 2001. Shifting governance structures in global commodity chains, with special reference to the internet, *American Behavioural Scientist* 44, 1616–1637.

Gereffi, G. and Fernandez-Stark, K. 2010. *The Offshore Services Value Chain.* North Carolina: Duke University, Center on Globalization, Governance, and Competitiveness (CGGC); available at http://www.cggc.duke.edu/pdfs/CGGC_CORFO_The_Offshore_Services_Global_Value_Chain_March_1_2010.pdf

Gereffi, G. and Korzeniewicz, M. (eds.) 1993. *Commodity Chains and Global Capitalism.* Westport, CT: Greenwood Press.

Gertler, M.S. 1986. Discontinuities in regional development, *Society and Space* 4, 71–84.

Gertler, M.S. 1992. Flexibility revisited: districts, nation-states, and the forces of production, *Transactions of the Institute of British Geographers* 17, 259–278.

Gibb, R. and Michalak, W. (eds.) 1994. *Continental Trading Blocs: The growth of regionalism in the world economy.* Chichester: John Wiley & Sons.

Gibson, M.L. and Ward, M.D. 1992. Export orientation: pathway or artifact?, *International Studies Quarterly* 36, 331–344.

Gifford, P. and Louis, W.R. (eds.) 1971. *France and Britain in Africa: Imperial rivalry and colonial rule.* New Haven, CT: Yale University Press.

Gill, S. 1990. *American Hegemony and the Trilateral Commission.* Cambridge: Cambridge University Press.

Glasmeier, A. 1990. The role of merchant wholesalers in industrial agglomeration formation, *Annals of the Association of American Geographers* 80, 394–417.

Glyn, A., Hughes, A., Lipietz, A., and Singh, A. 1990. The rise and fall of the Golden Age, in S.A. Marglin and J.B. Schor (eds.) *The Golden Age of Capitalism: Reinterpreting the postwar experience.* Oxford: Clarendon Press.

Goldberg, M. 1987. *The Chinese Connection: Getting plugged in to the Pacific Rim real estate, trade and capital markets.* Vancouver: University of British Columbia Press.

Goldman Sachs. 2007. *The BRICs and Beyond.* New York: Goldman Sachs.

Goldsbrough, D. 1985. Foreign direct investment in developing countries: trends, policy issues, and prospects, *Finance and Development* 22, 31–34.

Gomes, C. and Perez, A. 1979. The process of modernization in Latin American agriculture, *CEPAL Review* 8, 55–74.

Goodman, M.K. 2004. Reading fair trade: political ecological imaginary and the moral economy of fair trade foods, *Political Geography* 23, 891–915.

Gordon, D.M. 1979. *The Working Poor: Toward a state agenda.* Washington, DC: Council of State Planning Agencies.

Gordon, D. 1988. The global economy: new edifice or crumbling foundations?, *New Left Review* 68, 24–64.

Gordon, L. 2001. *Brazil's Second Chance: En route toward the first world.* Washington, DC: Brookings Institution Press.

Gore, C. 1984. *Regions in Question: Space, development theory and regional policy.* London: Methuen.

Gough, J. 1996. Not flexible accumulation—contradictions of value in contemporary economic geography: 1. Workplace and interfirm relations, *Environment and Planning A* 28, 2063–2079.

Graham, S. and Marvin, S. 2001. *Splintering Urbanism.* New York: Routledge.

Grant, R.J. and Agnew, J.A. 1996. Representing Africa: the geography of Africa in world trade, 1960–1992, *Annals of the Association of American Geographers* 86, 729–744.

Grant, R.J., Papadakis, M.C., and Richardson, J.D. 1993. Global trade flows: old structures, new issues, empirical evidence, in C.F. Bergsten and M. Noland (eds.) *Pacific Dynamism and the International Economic System.* Washington, DC: Institute for International Economics.

Green, A.E. 1988. The north–south divide in Great Britain: an examination of the evidence, *Transactions of the Institute of British Geographers* 13, 178–198.

Greener, R. 2004. The impact of HIV/AIDS on poverty and inequality, in M. Haacker (ed.) *The Macroeconomics of HIV/AIDS.* Washington DC: International Monetary Fund.

Griffin, K. 1974. *The Political Economy of Agrarian Change.* Cambridge, MA: Harvard University Press.

Griffith-Jones, S. 1980. The growth of multinational banking, the Eurocurrency markets and their effects on the developing countries, *Journal of Development Studies* 16, 96–109.

Griffith-Jones, S. and Rodriguez, E. 1984. Private international finance and industrialization in LDCs, in R. Kaplinsky (ed.) *Third World Industrialization in the 1980s: Open economies in a closing world.* London: Cass.

Griffith-Jones, S. and Stallings, B. 1995. New global financial trends: implications for development, in B. Stallings (ed.) *Global Change, Regional Response: The new international context of development.* Cambridge: Cambridge University Press.

Grindle, M.S. 1986. *State and Countryside: Development policy and agrarian politics in Latin America.* Baltimore, MD: Johns Hopkins University Press.

Grote, M.H. and Täube, F.A. 2006. Offshoring the financial services industry: implications for the evolution of Indian IT clusters, *Environment and Planning A* 38, 1287–1305.

Gugler, J. 1980. A minimum of urbanism and a maximum of ruralism: the Cuban experience, *International Journal of Urban and Regional Research* 4, 516–536.

Gugler, J. and Flanagan, W.G. 1978. *Urbanization and Social Change in West Africa*. Cambridge: Cambridge University Press.

Haggett, P. 1983. *Geography: A modern synthesis*, 3rd edn. London: Harper & Row.

Hall, D. and Danta, D. (eds.) 2000. *Europe Goes East: EU enlargement, diversity and uncertainty*. London: Stationery Office.

Hall, P. 1981. The geography of the fifth Kondratieff cycle, *New Society* 55, 535–536.

Hall, T.D. 1986. Incorporation in the world-system: toward a critique, *American Sociological Review* 51, 390–402.

Hamilton, C. 1983. Capitalist industrialisation in the four little tigers of East Asia, in P. Limqueco and B. McFarlane (eds.) *Neo-Marxist Theories of Development*. Beckenham: Croom Helm.

Hamilton, F.E.I. 1978. Multinational enterprise and the European Economic Community, in F.E.I. Hamilton (ed.) *Industrial Change: International experience and public policy*. London: Longman.

Hamilton, F.E.I. 1984. Industrial restructuring: an international problem, *Geoforum* 15, 349–364.

Hammond, R. 1994. A geography of overseas aid, *Geography* 79, 210–221.

Hancock, M.D., Logue, J., and Schiller, B. 1991. *Managing Modern Capitalism: Industrial renewal and workplace democracy in the United States and Europe*. Westport, CT: Praeger.

Hansen, G. and Prescott, E. 2002. Malthus to Solow, *American Economic Review* 94(2), 1205–1217.

Hansen, N. 1981. Mexico's border industry and the international division of labor, *Annals of Regional Science* 15, 1–12.

Hansson, K. 1952. A general theory of the system of multilateral trade, *American Economic Review* 42, 59–88.

Hardy, S., Hart, M., Albrechts, L., and Katos, A. (eds.) 1995. *An Enlarged Europe: Regions in competition?* London: Regional Studies Association.

Hargittai, E. and Centeno, M.A. 2001. Introduction: defining a global geography, *American Behavioral Scientist* 44, 1545–1560.

Harrigan, J. and Martin, P. 2002. Terrorism and the resilience of cities, *Economic Policy Review* 8, 97–116.

Harris, C. 1982. The urban and industrial transformation of Japan, *Geographical Review* 72, 50–89.

Harris, N. 1978. *The Mandate of Heaven: Marx and Mao in modern China*. London: Quartet.

Harris, N. 1986. *The End of the Third World: Newly industrializing countries and decline of an ideology*. London: I.B. Tauris.

Harris, N. 1996. *The New Untouchables: Immigration and the new world worker*. London: Penguin.

Harrison, B. 1992. Industrial districts: old wine in new bottles?, *Regional Studies*, 26, 469–483.

Harrison, B. 1994. *Lean and Mean: The changing landscape of corporate power in the age of flexibility*. New York: Basic Books.

Harrison, B. 1997. Comments on "The effects of state and local public policies on economic development," *New England Economic Review*, March/April, 140–141.

Harrison, B. and Bluestone, B. 1990. Wage polarization in the US and the "flexibility" debate, *Cambridge Journal of Economics* 14, 351–373.

Hart, K. 1982. *The Political Economy of Agriculture in West Africa*. Cambridge: Cambridge University Press.

Hartley, M. and Walker, C. 2012. The culture shock of India's call centers, *Forbes Magazine*, December 16.

Harvey, D. 1990. *The Condition of Postmodernity*. Oxford: Blackwell.

Harvey, D.W. 1988. The geographical and geopolitical consequences of the transition from Fordist to flexible accumulation, in G. Sternlieb and J.W. Hughes (eds.) *America's New Market Geography: Nation, region, and metropolis*. New Brunswick, NJ: Center for Urban Policy Research.

Harvie, C. 1994. *The Rise of Regional Europe*. London: Routledge.

Hauchler, I. and Kennedy, P.M. 1994. *Global Trends*. New York: Continuum.

Hayami, Y. 1984. Assessment of the green revolution, in C.K. Eicher and J.M. Staatz (eds.) *Agricultural Development in the Third World*. Baltimore, MD: Johns Hopkins University Press.

Hayes, S.L. and Hubbard, P.M. 1990. *Investment Banking: A tale of three cities*. Cambridge, MA: Harvard Business School Press.

Haywood, J. 1998. *Historical Atlas of the 19th Century World 1783–1914*. New York: Barnes & Noble.

Heim, C.E. 1991. *Dimensions of Decline: Industrial regions in Europe, the US, and Japan in the 1970s and 1980s,* paper presented at the Social Science History Association, Annual Meeting, New Orleans, November.

Held, D. and McGrew, A. (eds.) 2000. *The Global Transformations Reader*. Malden, MA: Blackwell.

Hellerstein, R., Daly, D., and Marsh, C. 2006. Have U.S. import prices become less responsive to changes in the dollar?, *Current Issues in Economics and Finance* 12(6).

Henderson, J. 1989. *The Globalization of High Technology Production: Society, space and semiconductors in the restructuring of the modern world*. London: Routledge.

Henry, N. and Pinch, S. 2000. (The) industrial agglomeration (of Motor Sport Valley): a knowledge, space, economy approach, in J.R. Bryson, P.W. Daniels, N. Henry, and J. Pollard (eds.) *Knowledge, Space, Economy*. London: Routledge.

Hepworth, M. 1992. *Geography of the Information Economy*. London: Belhaven Press.

Herzog, H.W. and Schlottmann, A.M. (eds.) 1991. *Industry Location and Public Policy*. Knoxville, TN: University of Tennessee Press.

Hewett, E.A. and Gaddy, C.G. 1992. *Open for Business: Russia's return to the global economy*. Washington, DC: Brookings Institution.

Hicks, J.R. 1959. *Essays in World Economics*. Oxford: Oxford University Press.

Hill, P. 1986. *Development Economics on Trial: The anthropological case for the prosecution*. Cambridge: Cambridge University Press.

Hirschman, A.O. 1958. *The Strategy of Economic Development*. New Haven, CT: Yale University Press.

Hirschman, A. 1992. Industrialization and its manifold discontents: west, east and south, *World Development* 20, 1225–1232.

Hirst, P. 1994. *Associative Democracy*. Cambridge: Polity Press.

Hirst, P. and Zeitlin, J. (ed.) 1989. *Reversing Economic Decline: Industrial structure and policy in Britain and her competitors*. Oxford: Berg.

Hobsbawm, E.J. 1968. *Industry and Empire*. New York: Pantheon.

Hoekman, B.M. and Kostecki, M.M. 1996. *The Political Economy of the World Trading System: From GATT to WTO*. New York: Oxford University Press.

Holland, S. 1980. *UnCommon Market*. London: Macmillan.

Holloway, S.R. and Pandit, K. 1992. The disparity between the level of economic development and human welfare, *Professional Geographer* 44, 57–71.

Hopkins, A.G. 1973. *An Economic History of West Africa*. London: Longman.

Hopkins, M. 1983. Employment trends in developing countries, 1960–80 and beyond, *International Labour Review* 122, 461–478.

Hopkins, T.K. and Wallerstein, I. 1979. Cyclical rhythms and secular trends of the capitalist world-economy, *Review* 2, 483–500.

Hossain, M. 1982. Agrarian reform in Asia—a review of recent experience in selected countries, in S. Jones, M. Murmis, and C. Joshi (eds.) *Rural Poverty and Agrarian Reform*. New Delhi: Allied.

Howells, J. 2003. *Innovation, Consumption and Knowledge: Services and encapsulation*. CRIC Discussion Paper No. 62. Manchester: University of Manchester, Centre for Research on Innovation and Competition.

Howells, J. and Wood, M. 1993. *The Globalisation of Production and Technology*. London: Belhaven Press.

Hsu, R.C. 1982. Agricultural financial policies in China, 1949–80, *Asian Survey* 22, 638–658.

Hudson, R. 1983. Regional labour reserves and industrialization in the EEC, *Area* 15, 223–230.

Hugill, P. 1993. *World Trade Since 1431. Geography, technology, and capitalism*. Baltimore, MD: Johns Hopkins University Press.

Hugill, P. 1999. *Global Communications since 1844: Geography, technology, and capitalism.* Baltimore, MD: Johns Hopkins University Press.

Hutton, W. 1995. *The State We're In.* London: Jonathan Cape.

Hutton, W. and Giddens, A. (eds.) 2000. *Global Capitalism.* New York: New Press.

Hymer, S. 1972. The multinational corporation and the law of uneven development, in J.N. Bhagwati (ed.) *Economics and World Order.* London: Macmillan.

IFAD (International Fund for Agricultural Development). 2001. *Rural Poverty Report 2001: The challenge of ending rural poverty.* Rome: IFAD.

IFAD. 2002. *Regional Strategy Paper: IFAD strategy for rural poverty reduction in Asia and the Pacific.* Rome: IFAD.

ILO (International Labour Office). 2000. *HIV/AIDS: A threat to decent work, productivity and development.* Geneva: ILO.

ILO. 2000. *Yearbook of Labour Statistics.* Geneva: ILO.

ILO. 2004. *World Employment Report 2004–05.* Geneva: ILO.

ILO. 2007. *Global Employment Trends.* January brief. Geneva: ILO.

ILO. 2010. *International Labour Migration: A rights-based approach.* Geneva: ILO.

IMF (International Monetary Fund). 1991. *World Imports and Exports.* Washington, DC: International Monetary Fund.

IMF. 2001. *World Economic Outlook: The information technology revolution.* Washington, DC: IMF Publication Services.

IMF. 2011. *Changing Patterns of Global Trade.* Washington, DC: IMF; available at http://www.imf.org/external/np/pp/eng/2011/061511.pdf

Independent Commission on International Development Issues. 1980. *North–South: A programme for survival (the Brandt Report).* Oxford: Oxford University Press.

Independent Commission on International Development Issues. 1983. *Common Crisis: North–south cooperation for world recovery (Brandt II).* London: Pan Books.

Jackson, R.H. 1986. Conclusion, in P. Duignan and R.H. Jackson (eds.) *Politics and Government in African States, 1960–1985.* Beckenham: Croom Helm.

Jackson, R.H. 1990. *Quasi-States.* Cambridge: Cambridge University Press.

Jacobs, J. 1969. *The Economy of Cities.* New York: Random House.

Jacobs, J. 1984. Cities and the wealth of nations, *Atlantic Monthly* March, 41–66.

Jacquemin, A. 1987. *The New Industrial Organization.* Oxford: Clarendon Press.

JAMA (Japan Automobile Manufacturers Association, Inc). 2012. *Japan Automobile Manufacturers: Creating economic growth and jobs in America.* Tokyo, Japan: JAMA; available at http://www.jama-english.jp/publications/30025_JAMA_Spreads.pdf

Jenkins, R.O. 1988. *Transnational Corporations and Uneven Development: The internationalization of capital and the third world.* London: Methuen.

Jessop, B. 1992. Post-Fordism and flexible specialization. Incommensurable, contradictory, or just plain different perspectives?, in H. Ernste and V. Meier (eds.) *Regional Development and Contemporary Industrial Response.* London: Belhaven Press.

Joekes, S.P. 1987. *Women in the World Economy.* New York: Oxford University Press.

Johnson, A.W. and Earle, T. 1987. *The Evolution of Human Societies: From foraging group to agrarian state.* Stanford, CA: Stanford University Press.

Johnston, R.J. 1980. *City and Society.* Harmondsworth: Penguin.

Johnston, R.J. 1984. The world is our oyster, *Transactions of the Institute of British Geographers* 9, 443–459.

Johnston, R.J. 1986. The state, the region, and the division of labor, in A.J. Scott and M. Storper (eds.) *Production, Work, Territory: The geographical anatomy of industrial capitalism.* Boston, MA: Allen & Unwin.

Johnston, R.J. 1992. The rise and decline of the corporate-welfare state: a comparative analysis in global context, in P.J. Taylor (ed.) *The Political Geography of the Twentieth Century.* London: Belhaven Press.

Johnston, R.J., Taylor, P.J., and Watts, M. (eds.) 2002. *Geographies of Global Change: Remapping the world*, 2nd edn. Cambridge, MA: Blackwell.

Jones, R.S., King, R.E., and Klein, M. 1993. Economic integration between Hong Kong, Taiwan and the coastal provinces of China, *OECD Economic Studies* 20, 115–144.

Jones, S. 1982. Introduction, in S. Jones, M. Murmis, and C. Joshi (eds.) *Rural Poverty and Agrarian Reform*. New Delhi: Allied.

Jones, S., Murmis, M., and Joshi, C. (eds.) 1982. *Rural Poverty and Agrarian Reform*. New Delhi: Allied.

Kaimowitz, D. 1995. *Livestock and Deforestation in Central America in the 1980s and 1990s: A policy perspective*. EPTD Discussion Paper No. 9. Washington, DC: Environment and Production Technology Division, International Food Policy Research Institute.

Kai-Sun, K., Leung-Chuen, C., Lui, F.T., and Qiu, L.D. 2001. *Industrial Development in Singapore, Taiwan, and South Korea*. River Edge, NJ: World Scientific.

Kaplinsky, R. 1984. *Automation: The technology and society*. London: Longman.

Kapur, D. and Ramamurti, R. 2001. India's emerging competitive advantage in services, *Academy of Management Executive* 15, 20–32.

Karmarkar, U. 2004. Will you survive the services revolution?, *Harvard Business Review* 82, 101–107.

Katouzian, H. 1983. Shi'ism and Islamic economics: Sadr and Bani Sadr, in N.R. Keddie (ed.) *Religion and Politics in Iran: Shi'ism from quietism to revolution*. New Haven, CT: Yale University Press.

Kedourie, E. 1960. *Nationalism*. London: Hutchinson.

Kirchner, J. 1983. Supercomputing seen as key to economic success, *Computerworld*, October 3, 8.

Knox, P.L. 1984. *The Geography of Western Europe: A socio-economic survey*. Beckenham: Croom Helm.

Knox, P.L. 1994. *Urbanization: An introduction to urban geography*. Englewood Cliffs, NJ: Prentice Hall.

Knox, P.L. 1995. World cities in a world-system, in P.L. Knox and P.J. Taylor (eds.) *World Cities in a World-System*. Cambridge: Cambridge University Press.

Knox, P.L., Bartels, E.H., Bohland, J.R., Holcomb, B., and Johnston, R.J. 1988. *The US: A contemporary human geography*. London: Longman.

Knox, P.L. and Taylor, P.J. (eds.) 1995. *World Cities in a World-System*. Cambridge: Cambridge University Press.

Kopytoff, I. 1987. *The African Frontier: The reproduction of traditional African societies*. Bloomington, IN: Indiana University Press.

Kornhauser, D. 1989. *Japan: Geographical background to urban and industrial development*, 2nd edn. London: Longman.

Kosters, M.H. 1992. *Workers and Their Wages*. Washington, DC: American Enterprise Institute.

Krebs, G. 1982. Regional inequalities during the process of national economic development: a critical approach, *Geoforum* 13, 71–81.

Krueger, A. 1978. Alternative trade strategies and employment in developing countries, *American Economic Review* 68, 523–536.

Krugman, P. 1990. *The Age of Diminished Expectations: US economic policy in the 1990s*. Cambridge, MA: MIT Press.

Krugman, P. 1991. *Geography and Trade*. Cambridge, MA: MIT Press.

Krugman, P. 1994. The myth of Asia's miracle, *Foreign Affairs* 73, 62–78.

Krugman, P. 1995. The localization of the world economy, *New Perspectives Quarterly* Winter, 34–38.

Laird, S. and Yeats, A. 1990. Trends in non-tariff barriers of developed countries, 1966–1986, *Weltwirtschaftliches Archiv. Review of World Economies* 126, 299–325.

Land Matrix. 2012a. Public database of land deals; available at http://landportal.info/

Land Matrix. 2012b. *Transnational Land Deals for Agriculture in the Global South*; available at http://landportal.info/landmatrix/get-the-detail#analytical-report

Landes, D.S. 1999. *The Wealth and Poverty of Nations*. New York: W.W. Norton.

Landsberger, H.A. 1969. *Latin American Peasant Movements*. Ithaca, NY: Cornell University Press.

Langley, L.D. 1983. *The Banana Wars: An inner history of American empire, 1900–1934*. Lexington, KY: University Press of Kentucky.

Lanvin, B. 1991. Services and new industrial strategies: what is at stake for developing countries?, in P.W. Daniels and F. Moulaert (eds.) *The Changing Geography of Advanced Producer Services: Theoretical and empirical perspectives*. London: Belhaven Press.

Lardy, N.R. 1984. Prices, markets, and the Chinese peasant, in C.K. Eicher and J.M. Staatz (eds.) *Agricultural Development in the Third World*. Baltimore, MD: Johns Hopkins University Press.

Lash, S. and Urry, J. 1987. *The End of Organized Capitalism*. Cambridge: Polity Press.

Latham, A.J.H. 1978. *The International Economy and the Undeveloped World, 1865–1914*. Beckenham: Croom Helm.

Lavrov, S.B. and Sdasnyk, G.V. 1982. The growth pole concept and the regional planning experience of developing countries, *Development Dialogue* 3, 15–27.

Lechner, F.J. and Boli, J. (eds.) 2000. *The Globalization Reader*. Malden: Blackwell.

Lee, R. and Wills, J. (eds.) 1997. *Geographies of Economies*. London: Edward Arnold.

Lele, U. 1984. Rural Africa: modernization, equity, and long term development, in C.K. Eicher and J.M. Staatz (eds.) *Agricultural Development in the Third World*. Baltimore, MD: Johns Hopkins University Press.

Lessard, D.R. and Williamson, J. 1987. *Capital Flight and Third World Debt*. Washington, DC: Institute for International Economics.

Lesser, A. 1961. Social fields and the evolution of society, *Southwestern Journal of Anthropology* 17, 40–48.

Leung, C.K. and Chin, S.S.K. (eds.) 1983. *China in Readjustment*. Hong Kong: Centre for Asian Studies, University of Hong Kong.

Levy, F. and Murnane, R.J. 1992. US earnings levels and earnings inequality: a review of recent trends and proposed explanations, *Journal of Economic Literature* 30, 1333–1381.

Levy, F. and Murnane, R.J. 2004. *The New Division of Labor: How computers are creating the next job market*. New York: Russell Sage Foundation.

Lewis, P. 1988. Third world funds: wrong way flow, *New York Times*, 11 February, D1, D9.

Lewis, W.A. 1978. *Growth and Fluctuations 1870–1913*. London: Allen & Unwin.

Leyshon, A. and Thrift, N. 1997. *Money/Space: Geographies of monetary transformation*. London: Routledge.

Light, I. 1983. *Cities in World Perspective*. New York: Macmillan.

Linge, G.J.R. and Hamilton, F.E.I. 1983. Industrial systems, in F.E.I. Hamilton and G.J.R. Linge (eds.) *Spatial Analysis, Industry and the Industrial Environment*. Chichester: John Wiley & Sons.

Lipietz, A. 1986. Behind the crisis: the exhaustion of a regime of accumulation. A "regulation school" perspective on some French empirical works, *Review of Radical Political Economics* 18, 13–32.

Little, I. 1982. *Economic Development: Theory, policy, and international relations*. New York: Basic Books.

Lloyd, P. and Meegan, R. 1996. Contested governance: European exposure in the English regions, *European Planning Studies* 4, 75–97.

Locatelli, R.L. 1985. *Industrializacao, Crescimento e Empregno: Uma avaliacao da experiencia Brasileira*. Rio de Janeiro: IPEA/INPES.

Lonsdale, J. 1985. The European scramble and conquest in African history, in R. Oliver and G.N. Sanderson (eds.) *The Cambridge History of Africa. Vol. 6: From 1870 to 1905*. Cambridge: Cambridge University Press.

Loup, J. 1983. *Can the Third World Survive?* Baltimore, MD: Johns Hopkins University Press.

Lu, M., Fan, J., Liu, S. and Yan, Y. 2002. Employment restructuring during China's economic transition, *Monthly Labor Review* August, 25–31.

McCarthy, L. 2000. European integration, urban economic change, and public policy responses, *Professional Geographer* 52, 191–205.

McDowell, L. 1991. Life without father or Ford: the new gender order of post-Fordism, *Transactions of the Institute of British Geographers* 16, 400–419.

MacFarquhar, R. 1992. Deng's last campaign, *New York Review of Books*, December 17, 22–28.

McGowan, K. 2001. Lessons from around the world, *American Demographics* 23, 50–53.

McGranahan, D.V., Richard-Proust, C., Sovani, N.V., and Subramanian, M. 1970. *Content and Measurement of Socio-Economic Development*. Geneva: UNRISD.

MacLaughlin, J.G. and Agnew, J.A. 1986. Hegemony and the regional question: the political geography of regional industrial policy in Northern Ireland 1945–1972, *Annals of the Association of American Geographers* 76, 247–261.

McMichael, P. 1994. Agro-food restructuring in the Pacific Rim: a comparative-international perspective on Japan, South Korea, the United States, Australia, and Thailand, in R.A. Palat (ed.) *Pacific-Asia and the Future of the World-system*. Westport, CT: Greenwood Press.

McMichael, P. 1996. *Development and Social Change: A global perspective*. Thousand Oaks, CA: Pine Forge Press.

McMichael, P., Petras, J., and Rhodes, R. 1974. Imperialism and the contradictions of development, *New Left Review* 85, 83–104.

Maddison, A. 1983. A comparison of levels of GDP per capita in developed and developing countries, 1700–1980, *Journal of Economic History* 43, 27–41.

Maddison, A. 2001. *The World Economy: A millennial perspective*. Paris: OECD Development Centre Studies.

Malassis, L. 1975. *Agriculture and the Development Process*. Paris: UNESCO.

Malecki, E. 1979. Locational trends in R&D by large US corporations, 1965–1976, *Economic Geography* 55, 308–323.

Malecki, E. 1991. *Technology and Economic Development. The dynamics of local, regional and national change*. London: Longman.

Malmberg, A. and Maskell, P. 1997. Towards an explanation of regional specialization and industry agglomeration, *European Planning Studies* 5, 25–41.

Mankiw, N.G. and Swagel, P. 2006. *The Politics and Economics of Outsourcing*. Working Paper 12398. Cambridge, MA: National Bureau of Economic Research.

Marks, S. and Rathbone, R. (ed.) 1982. *Industrialization and Social Change in South Africa*. London: Longman.

Markusen, A. 1983. High tech jobs, markets, and economic development prospects, *Built Environment* 9, 18–27.

Marshall, A. 1920. *Principles of Economics*. London: Macmillan.

Marston, S.A., Knox, P.L., and Liverman, D.M. 2008. *World Regions in Global Context: Peoples, places, and environments*, 3rd edn. Upper Saddle River, NJ: Pearson Education.

Martin, R.L. 1986. Thatcherism and Britain's industrial landscape, in R.L. Martin and B. Rowthorne (eds.) *The Geography of Deindustrialization*. London: Methuen.

Martin, R.L. 1988. The political economy of Britain's north–south divide, *Transactions of the Institute of British Geographers* 13, 389–418.

Martin, R.L. 1999. The new "geographical turn" in economics: some critical reflections, *Cambridge Journal of Economics* 23, 65–91.

Martin, R.L. and Hodge, J.S.C. 1983. The reconstruction of Britain's regional policy, 2: towards a new agenda, *Environment and Planning C: Government and Policy* 1, 317–340.

Martin, R.L. and Rowthorn, B. (eds.) 1986. *The Geography of De-Industrialisation*. London: Macmillan.

Mason, M. 1992. *American Multinationals and Japan: The political economy of Japanese capital controls, 1899–1980*. Cambridge, MA: Harvard University Press.

Massey, D. 1984. *Spatial Divisions of Labour*. London: Methuen.

Massey, D. 2005. *For Space*. London: Sage.

Massey, D. and Meegan, R. 1989. Spatial divisions of labour in Britain, in D. Gregory and R. Walford (eds.) *Horizons in Human Geography*. Totowa, NJ: Barnes & Noble.

Mattera, P. 1985. *Off the Books: The rise of the underground economy*. New York: St. Martin's Press.

Meadows, D.H., Meadows, D.L., Randers, J., and Behrens, W.W. 1972. *The Limits to Growth*. London: Earth Island.

Meier, G.M. and Baldwin, R.E. 1957. *Economic Development*. New York: John Wiley & Sons.

Menzel, U. and Senghaas, D. 1987. NICs defined: a proposal for indicators evaluating threshold countries, in K.D. Kim (ed.) *Dependency Issues in Korean Development: Comparative perspectives*. Seoul: Seoul National University Press.

Metcalf, D. 1969. *The Economics of Agriculture*. Harmondsworth: Penguin.

METI. 2001. *White Paper on International Trade 2001: External economic policy challenges in the 21st century*. Tokyo: Ministry of Economy, Trade and Industry.

Meyer, D.R. 1983. Emergence of the manufacturing belt: an interpretation, *Journal of Historical Geography* 9, 145–174.

Mintz, S. 1985. *Sweetness and Power: The place of sugar in modern history*. New York: Viking.

Mishra, P. 2007. Impasse in India, *New York Review of Books*, June 28, 48–51.

Mittelman, J.H. 2000. *The Globalization Syndrome: Transformation and resistance*. Princeton, NJ: Princeton University Press.

Morello, T. 1983. Sweatshops in the sun?, *Far Eastern Economic Review*, September 15, 88–89.

Morris, M.D. 1980. *Measuring the Condition of the World's Poor*. New York: Pergamon.

Moudoud, E. 1986. *The Rise and Fall of the Growth Pole Approach*. Department of Geography Discussion Paper No. 89. Syracuse, NY: Syracuse University.

Münz, P., Straubhaar, T., Vadean, F., and Vadean, N. 2007. *What are the Migrants' Contributions to Employment and Growth? A European approach*. Hamburg: Hamburg Institute of International Economics, Migration Research Group.

Murata, K. 1980. *An Industrial Geography of Japan*. London: Bell & Hyman.

Murphey, R. 1980. *The Fading of the Maoist Vision*. London: Methuen.

Myrdal, G. 1957. *Economic Theory and Underdeveloped Regions*. London: Duckworth.

Nairn, A. 1981. Guatemala, *Multinational Monitor* 2, 12–14.

Nairn, T. 1977. Super-power or failure? in T. Nairn (ed.) *Atlantic Europe?* Amsterdam: Transnational Institute.

Nakane, C. 1970. *Japanese Society*. London: Weidenfeld & Nicolson.

NASA, Earth Observatory. 2008. *Aquatic Dead Zones*, January 1; available at http://earthobservatory.nasa.gov/IOTD/view.php?id=44677

National Science Board. 2012. *Science and Engineering Indicators 2012*. (NSB 12–01) Arlington, VA: National Science Foundation.

Nau, H.R. 1990. *The Myth of America's Decline: Leading the world economy into the 1990s*. New York: Oxford University Press.

Nayar, A. 2012. African land grab hinders sustainable development, *Nature.com*; available at http://www.nature.com/news/african-land-grabs-hinder-sustainable-development-1.9955

Nelson, R. 1995. Recent evolutionary theorizing about economic change, *Journal of Economic Literature* 33, 1089–1098.

Nelson, R. and Winter, S. 1982. *An Evolutionary Theory of Economic Change*. Cambridge, MA: Belknap Press of the Harvard University Press.

Neuman, S.G. 1984. International stratification and third world military industries, *International Organization* 38, 167–198.

New Economics Foundation. 2012. *The Happy Planet Index: 2012 report: A global index of sustainable well-being*. London: NEF.

Noor, M. 2008. Many families limiting themselves to a single car, *New York Times*, July 27; available at http://www.nytimes.com/2008/07/27/nyregion/nyregionspecial2/27Ronecar.html

Newby, H. 1980. Rural sociology, *Current Sociology* 28, 1–141.

Nolan, P. 1983. De-collectivization of agriculture in China, 1979–82: a long-term perspective, *Cambridge Journal of Economics* 7, 381–403.

Oakland Institute. 2011. Understanding land investment deals in Africa; available at http://www.oaklandinstitute.org/sites/oaklandinstitute.org/files/OI_brief_land_grabs_leave_africa_thirsty_1.pdf

O'Brien, P. 1982. European economic development: the contribution of the periphery, *Economic History Review* 35, 1–18.

O'Brien, R. 1992. *Global Financial Integration: The end of geography*. New York: Council on Foreign Relations Press.

O'Connor, A.M. 1978. *The Geography of Tropical African Development: A study of spatial patterns of economic change since independence*, 2nd edn. Oxford: Pergamon.

ODCCP (United Nations Office for Drug Control and Crime Prevention). 2000. *Global Illicit Drug Trends 2000*. Vienna: ODCCP.

OECD (Organization for Economic Cooperation and Development). 1979. *The Impact of the NICs on Production and Trade in Manufactures*. Paris: Organization for Economic Cooperation and Development.

OECD. 1996. *The Knowledge-Based Economy*. Paris: OECD.

OECD. 2001a. *OECD in Figures, 2001: Statistics on the member countries*. Paris: OECD Observer.

OECD. 2001b. *Understanding the Digital Divide*. Paris: OECD.

OECD. 2006. *OECD in Figures, 2006–07 Edition: Statistics on the member countries*. Paris: OECD Observer.

OECD. nd. Poverty reduction and social development: migration and the brain drain phenomenon; available at http://www.oecd.org/dev/poverty/migrationandthebraindrainphenomenon.htm

OECD iLibrary. 2010. *OECD Factbook 2010: Economic, environmental and social statistics*, Paris: OECD; available at http://www.oecd-ilibrary.org/sites/factbook-2010-en/02/04/02/02–04–02-g3.html?contentType=&itemId=/content/chapter/factbook-2010–20-en&containerItemId=/content/serial/18147364&accessItemIds=&mimeType=text/html

O'Farrell, P.N. and Wood, P.A. 1998. Internationalisation by business service firms: towards a new regionally based conceptual framework, *Environment and Planning A* 30, 109–128.

O hUallacháin, B. 1996. Vertical integration in American manufacturing: evidence for the 1980s, *Professional Geographer* 48, 343–356.

O'Loughlin, J. 1989. World power competition and local conflicts in the third world, in R.J. Johnston and P.J. Taylor (eds.) *A World in Crisis? Geographical perspectives*, 2nd edn. Oxford: Blackwell.

O'Meara, P., Mehlinger, H.D., and Krain, M. (eds.) 2000. *Globalization and the Challenges of a New Century*. Bloomington, IN: Indiana University Press.

Orwell, G. 1962 (1937). *The Road to Wigan Pier*. Harmondsworth: Penguin.

Owen, N. 1983. *Economies of Scale, Competitiveness, and Trade Patterns within the European Community*. Oxford: Clarendon Press.

Oxfam. 2002. *Rigged Rules and Double Standards: Trade, globalisation, and the fight against poverty*. Oxford: Oxfam International.

Oxfam. 2004. *Trading Away our Rights: Women working in global supply chains*. Oxford: Oxfam International.

Parboni, R. 1988. US economic strategies against western Europe: from Nixon to Reagan, *Geoforum* 19, 45–54.

Parker, J. 1992. Survey: Russia reborn, *Economist*, December 5, 58.

Pauly, L.W. and Reich, S. 1997. National structures and multinational corporate behavior: enduring differences in the age of globalisation, *International Organization* 51, 1–30.

Payer, C. 1974. *The Debt Trap*. Harmondsworth: Penguin.

Pearse, A. 1975. *The Latin American Peasant*. London: Cass.

Pearse, A. 1979. *Seeds of Plenty, Seeds of Want*. London: Oxford University Press.

Peck, J. and Tickell, A. 2002. Neoliberalizing space, in N. Brenner and N. Theodore (eds.) *Spaces of Neoliberalism. Urban restructuring in North America and western Europe*. Oxford: Blackwell.

Peet, R. 1987. Industrial devolution, underconsumption and the third world debt crisis, *World Development* 15, 777–788.

Peet, R. 1991. *Global Capitalism: Theories of societal development*. New York: Routledge.

Perez, C. 1983. Structural change and assimilation of new technologies in economic-social systems, *Futures* 10, 357–375.

Perkins, D. and Yusuf, S. 1984. *Rural Development in China*. Baltimore, MD: Johns Hopkins University Press.

Perroux, F. 1955. Note sur la notion de pole de croissance, in I. Livingstone (ed.) (pub. 1979) *Development Economics and Policy: Selected readings*. London: Allen & Unwin.

Perroux, F. 1961. La firme motrice dans la région et la région motrice, *Théorie et Politique de la Expansion Regionale*. Liège: Université de Liège.

Petras, J. 1984. Toward a theory of industrial development in the third world, *Journal of Contemporary Asia* 14, 182–203.

Pew Charitable Trusts. 2013. *Who's Winning the Clean Energy Race?* Pew: Philadelphia and Washington, DC: Pew.

Picciotto, S. 1991. The internationalization of the state, *Review of Radical Political Economics* 22, 28–44.

Pinder, D. 1998. *The New Europe*. Chichester: John Wiley & Sons.

Piore, M. and Sabel, C. 1984. *The Second Industrial Divide*. New York: Basic Books.

Plant, R. 1998. *Indigenous Peoples and Poverty Reduction: A case study of Guatemala*. Washington, DC: Inter-American Development Bank.

Polanyi, K. 1957. The place of economies in societies, in K. Polanyi, C. Maynadier Arensberg, and H.W. Pearson (eds.) *Trade and Markets in the Early Empires*. Glencoe, IL: Free Press.

Pollard, S. 1981. *Peaceful Conquest: The industrialization of Europe, 1760–1970*. Oxford: Oxford University Press.

Popke, J. 2006. Geography and ethics: everyday mediations through care and consumption, *Progress in Human Geography* 30, 504–512.

Porter, M.E. (ed.) 1986. *Competition in Global Industries*. Cambridge, MA: Harvard Business School Press.

Porter, M.E. 1990. *The Competitive Advantage of Nations*. New York: Free Press.

Powell, A. 1991. Commodity and developing country terms of trade: what does the long run show?, *Economic Journal* 101, 1485–1496.

Puyana de Palacios, A. 1982. *Economic Integration among Unequal Partners: The case of the Andean group*. New York: Pergamon.

Rada, J. 1984. *International Division of Labor and Technology*. Geneva: ILO.

Ranger, T. 1976. From humanism to the science of man: colonialism in Africa and the understanding of alien societies, *Transactions of the Royal Historical Society* 26, 115–141.

Reardon, T., Timmer, C.P., Barrett, C.B., and Berdegué, J. 2003. The rise of supermarkets in Africa, Asia and Latin America, *American Journal of Agricultural Economics* 85, 1140–1146.

Reich, R. 1991. *The Work of Nations: Preparing ourselves for 21st century capitalism*. New York: Vintage Books.

Reid, M. 1996. Mercosur: remapping South America, *Economist*, October 12.

Reitsma, H.A. and Kleinpenning, J.M.G. 1985. *The Third World in Perspective*. Totowa, NJ: Rowman & Allanheld.

Reynolds, R. 1961. *Europe Emerges. Transition toward an industrial world-wide society*. Madison, WI: University of Wisconsin Press.

Richardson, R. and Belt, V. 2001. Saved by the bell? Call centres and economic development in less favoured regions, *Economic and Industrial Democracy* 22, 67–98.

Roberts, S. 1994. Fictitious capital, fictitious spaces: the geography of offshore financial flows, in S. Corbridge, R. Martin, and N. Thrift (eds.) *Money, Power, and Space*. Oxford: Blackwell.

Rodriguez, E. and Griffith-Jones, S. (eds.) 1992. *Cross-Conditionality, Banking Regulation and Third World Debt*. London: Macmillan.

Rogoff, K. 1992. Dealing with developing country debt in the 1990s, *The World Economy* 15, 475–492.

Rokkan, S. 1980. Territories, centres and peripheries, in J. Gottmann (ed.) *Centre and Periphery*. London: Sage.

Rokkan, S. and Urwin, D. 1983. *Economy, Territory, Identity: Politics of west European peripheries*. London: Sage.

Rondinelli, D.A. 1983. Dynamics of growth of secondary cities in developing countries, *Geographical Review* 73, 42–57.

Roosevelt, A.C. 1992. Secrets of the forest: an archaeologist appraises the past—and the future—of Amazonia, *The Sciences*, November/December, 22–28.

Rose, R. 1984. *Understanding Big Government*. London: Sage.

Ross, R. 1982. Regional illusion, capitalist reality, *Democracy* 2, 93–99.

Rowher, J. 1992. When China wakes: a survey, *Economist*, November 28.

Rowher, J. 1993. A billion consumers: a survey of Asia, *Economist*, October 30.

Rubenstein, J.M. 2005. *The Cultural Landscape: An introduction to human geography*, 8th edn. Upper Saddle River, NJ: Pearson/Prentice Hall.

Rudolph, L.I. and Rudolph, S.H. 1987. *In Pursuit of Lakshmi: The political economy of the Indian state*. Chicago, IL: University of Chicago Press.

Ruggie, J.G. 1983. International regimes, transactions and change: embedded liberalism in the postwar economic order, in S.D. Krasner (ed.) *International Regimes*. Ithaca, NY: Cornell University Press.

Sabel, C.F. 1982. *Work and Politics: The division of labor in industry*. New York: Cambridge University Press.

Sachar, A. and Öberg, S. (eds.) 1990. *The World Economy and the Spatial Organization of Power*. Brookfield, VT: Gower.

Sadler, D. 1992a. Industrial policy of the European Community: strategic deficits and regional dilemmas, *Environment & Planning A* 24, 1711–1730.

Sadler, D. 1992b. *The Global Region*. Tarrytown, NY: Pergamon.

Samuelson, P. 1964. *Economics*. New York: McGraw-Hill.

Sanders, R. 1987. Towards a geography of informal activity, *Socio-Economic Planning Sciences* 21, 229–237.

Santiago, C.E. 1987. The impact of foreign direct investment on employment structure and employment generation, *World Development* 15, 317–328.

Sassen, S. 2001. *The Global City: New York, London, Tokyo*, rev. edn. Princeton, NJ: Princeton University Press.

Savitch, H.V. and Ardashev, G. 2001. Does terror have an urban future?, *Urban Studies* 38, 2515–2533.

Saxenian, A. 1994. *Regional Advantage: Culture and competition in Silicon Valley and Route 128*. Cambridge, MA: Harvard University Press.

Sayer, A. 1985. Industry and space: a sympathetic critique of radical research, *Society and Space* 3, 3–29.

Sayer, A. and Walker, R. 1992. *The New Social Economy: Reworking the division of labor*. Cambridge, MA: Blackwell.

Schejtman, A. 1982. Land reform and entrepreneurial structure in rural Mexico, in S. Jones, M. Murmis, and C. Joshi (eds.) *Rural Poverty and Agrarian Reform*. New Delhi: Allied.

Schiffer, J. 1981. The changing post-war pattern of development: the accumulated wisdom of Samir Amin, *World Development* 9, 515–537.

Schoenberger, E. 1990. US manufacturing investments in western Europe: markets, corporate strategy, and the competitive environment, *Annals of the Association of American Geographers* 80, 379–393.

Scott, A.J. 1986. Industrial organization and location: division of labor, the firm, and spatial process, *Economic Geography* 62, 215–231.

Scott, A.J. 1987. The semi-conductor industry in Southeast Asia: organization, location, and the international division of labour, *Regional Studies* 21, 143–160.

Scott, A.J. 1988a. Flexible production systems and regional development. The rise of new industrial spaces in Europe and North America, *International Journal of Urban and Regional Research* 12, 171–186.

Scott, A.J. 1988b. *New Industrial Spaces*. London: Pion.

Scott, A.J. 1992. The role of large producers in industrial districts: a case study of high technology systems houses in southern California, *Regional Studies* 26, 265–275.

Scott, A.J. 1993. *Technopolis: High-technology industry and regional development in southern California*. Berkeley and Los Angeles: University of California Press.

Scott, A.J. 1996. Regional motors of the global economy, *Futures* 28, 391–411.

Scott, A.J. 1998. *Regions in the World Economy*. Oxford: Oxford University Press.

Scott, A.J. 2000. *The Cultural Economy of Cities: Essays on the geography of image-producing industries*. London: Sage.

Scott, A.J. 2001. *Global City-Regions: Trends, theory, policy*. New York: Oxford University Press.

Scott, A.J. 2005. *On Hollywood: The place, the industry*. Princeton, NJ: Princeton University Press.

Scott, A.J. 2010. Creative cities: the role of culture, *Revue d'Économie Politique* 1, 181–204.

Scott, A.J. and Storper, M.J. (eds.) 1986. *Production, Work, Territory*. London: Allen & Unwin.

Seattle Times. 2006. Boeing's outsourcing for the 787 Dreamliner, *Seattle Times*, September 29.

Schneider, F., Buehn, A., and Montenegro, C.E. 2010. *Shadow Economies All over the World: New estimates for 162 countries from 1999 to 2007*. Washington, DC: World Bank.

Seers, D. 1979. The periphery of Europe, in D. Seers, B. Schaeffer, and M.-L. Kiljunen (eds.) *Underdeveloped Europe: Studies in core–periphery relations*. Hassocks: Harvester Press.

Segal, G. 1999. Does China matter?, *Foreign Affairs* 78, 24–36.

Seiber, M.J. 1982. *International Borrowing by Developing Countries*. New York: Pergamon.

Sen, A. 1981a. *Poverty and Famines: An essay on entitlement and deprivation*. Oxford: Clarendon Press.

Sen, A. 1981b. Public action and the quality of life in developing countries, *Oxford Bulletin of Economics and Statistics* 43, 111–123.

Sender, J. and Smith, S. 1986. *The Development of Capitalism in Africa*. London: Methuen.

Shaw, D. 1999. *Russia in the Modern World*, 2nd edn. Oxford: Blackwell.

Shiba, T. and Shimotami, M. (eds.) 1997. *Beyond the Firm: Business groups in international and historical perspective*. New York: Oxford University Press.

Shunzan, Y. 1987. Urban policies and urban housing programs in China, in R.J. Fuchs, G.W. Jones, and E.M. Pernia (eds.) *Urbanization and Urban Policies in Pacific Asia*. Boulder, CO: Westview.

Shutt, J. and Whittington, R. 1987. Fragmentation strategies and the rise of small units: cases from the north west, *Regional Studies* 21, 13–23.

SIA. 2007. Global chip sales hit record $247.7 billion in 2006, Semiconductor Industry Association February 2; available at http://www.sia-online.org/pre_release.cfm?ID=426

SIPRI (Stockholm International Peace Research Institute). 2013. SIPRI military expenditure database; available at http://www.sipri.org/research/armaments/milex/milex_database

Sirbu, M.A. Jr, Treitel, R., Yorsz, W., and Roberts, E.B. 1976. *The Formation of a Technology-Oriented Complex*. Cambridge, MA: MIT Center for Policy Alternatives.

Skocpol, T. 1976. France, Russia, China: a structural theory of social revolution, *Comparative Studies in Society and History* 18, 181–196.

Slater, D. 1992. On the borders of social theory: learning from other regions, *Society and Space* 10, 307–327.

Smeeding, T., O'Higgins, M., and Rainwater, L. (eds.) 1990. *Poverty, Inequality, and Income Distribution in Comparative Perspective: The Luxembourg income study*. London: Harvester Wheatsheaf.

Smith, D. 1979. *Where the Grass is Greener*. Harmondsworth: Penguin.

Smith, I. 1979. The effects of external takeover and manufacturing employment change in the northern region, 1963–1973, *Regional Studies* 13, 421–436.

Smith, R. and Walter, I. 1996. *Global Banking*. New York: Oxford University Press.

Smith, T. 1979. The underdevelopment of development literature: the case of dependency theory, *World Politics* 31, 247–288.

Smith, T. 1981. *The Pattern of Imperialism: The United States, Great Britain, and the late-industrializing world since 1815*. Cambridge: Cambridge University Press.

South, R.B. 1990. Transnational "maquiladora" location, *Annals of the Association of American Geographers* 80, 549–570.

Stallings, B. 1990. The role of foreign capital in economic development, in G. Gereffi and D.L. Wyman (eds.) *Manufacturing Miracles: Paths of industrialization in Latin America and East Asia*. Princeton, NJ: Princeton University Press.

Stein, A.A. 1984. The hegemon's dilemma: Great Britain, the United States, and the international economic order, *International Organization* 38, 355–386.

Stein, P. 2010. *Inclusive Finance*, paper presented at the Korea-World Bank High Level Conference on Post-Crisis Growth and Development, June, Busan, South Korea; available at http://siteresources.world bank.org/DEC/Resources/84797–1275071905763/Inclusive_Finance-Stein.pdf

Stokes, E. 1959. *The English Utilitarians and India*. London: Oxford University Press.

Storper, M. 1987. The new industrial geography, *Urban Geography* 8, 585–598.

Storper, M.J. and Harrison, B. 1991. Flexibility, hierarchy and regional development: the changing structure of industrial production systems and their forms of governance in the 1990s, *Research Policy* 20, 407–422.

Storper, M.J. and Scott, A.J. (eds.) 1992. *Pathways to Industrialization and Regional Development*. London: Routledge.

Streeten, P. 1968. A poor nation's guide to getting aid, *New Society* 18, 154–156.

Studwell, J. 2013. *How Asia Works: Success and failure in the world's most dynamic region*. London: Profile Books.

Stutz, F.P. and Warf, B. 2007. *The World Economy: Resources, location, trade and development*, 5th edn. Upper Saddle River, NJ: Pearson Education.

Suhartono, F.X. 1987. *Growth Centers in the Context of Indonesia's Urban and Regional Development Program*, MA thesis, Social Science Program, Syracuse University.

Sukhotin, I. 1994. Stabilization of the economy and social contrasts, *Problems of Economic Transition*, November, 44–61.

Sutcliffe, R.B. 1971. *Industry and Underdevelopment*. London: Addison-Wesley.

Sutcliffe, R.B. 1984. Industry and underdevelopment re-examined, in R. Kaplinsky (ed.) *Third World Industrialization in the 1980s: Open economies in a closing world*. London: Cass.

Svedberg, P. 1991. The export performance of sub-Saharan Africa, *Economic Development and Cultural Change* 39, 549–566.

Swindell, K. 1985. *Farm Labour*. Cambridge: Cambridge University Press.

Sylvester, D. 2001. Recession seen as painful cure for excesses of dot.com era, *Mercury News*, October 1.

Szczepanik, E. 1969. The size and efficiency of agricultural investment in selected developing countries, *FAO Monthly Bulletin of Agricultural Economics and Statistics*, December, 2.

Tata, R.J. and Schultz, R.R. 1988. World variation in human welfare: a new index of development status, *Annals of the Association of American Geographers* 78, 580–593.

Taussig, M. 1978. Peasant economies and the development of capitalist agriculture in the Cauca Valley, Colombia, *Latin American Perspectives* 18, 62–91.

Taylor, A. 2000. Bumpy roads for global automakers, *Fortune*, December 18.

Taylor, M.J. and Thrift, N.J. 1983. Business organisation, segmentation and location, *Regional Studies* 17, 445–465.

Taylor, P.J. (ed.) 1992. *Political Geography of the Twentieth Century: A global analysis*. Harlow: Pearson Education.

Taylor, P.J. 2000. *Political Geography: World-economy, nation-state, and locality*, 4th edn. London: Longman.

Taylor, P.J. 2004. *World City Network: A global urban analysis*. London: Routledge.

Taylor, P.J. 2007. Cities within spaces of flows: theses for a materialist understanding of the external relations of cities, in P.J. Taylor, B. Derudder, P. Saey, and F. Witlox (eds.) *Cities in Globalization: Practices, policies and theories*. London: Routledge.

Taylor, P.J., Derudder, B., Saey, P., and Witlox, F. (eds.) 2007. *Cities in Globalization: Practices, policies and theories*. London: Routledge.

Tempest, R. 1996. Barbie and the world economy, *Los Angeles Times*, September 22, A1, A12.

Terlouw, C.P. 1989. World-system theory and regional geography, *Tijdschrift voor Economische en Sociale Geografie* 80, 206–221.

Terlouw, C.P. 1992. *The Regional Geography of the World-System: External arena, periphery, semiperiphery, core*. Utrecht: Netherlands Geographical Studies, No. 144.

Tharpa, G. and Gaiha, R. 2011. *Smallholder Farming in Asia and the Pacific: Challenges and opportunities*. Rome, Italy: International Fund for Agricultural Development (IFAD).

Thiesenhusen, A. 1989. *Searching for Agrarian Reform in Latin America*. Boston, MA: Unwin Hyman.

Thomas, T. 1996. Africa for the Africans, *Economist*, September 7.

Thrift, N. 1989. The geography of international economic disorder, in R.J. Johnston and P.J. Taylor (eds.) *A World in Crisis?*, 2nd edn. Oxford: Blackwell.

Thrift, N.J. 2002. A hyperactive world, in R.J. Johnston, P.J. Taylor, and M. Watts (eds.) *Geographies of Global Change: Remapping the world*, 2nd edn. Oxford: Blackwell.

Thurow, L. 1993. *Head to Head: The coming economic battle among Japan, Europe, and America*. New York: Warner Books.

Tickell, A. 1999; 2001; 2002. *Progress in Human Geography* progress reports on the geography of services, 23, 633–639; 25, 283–292; 26, 791–801.

Tickell, A. and Peck, J.A. 1992. Accumulation, regulation and the geographies of post-Fordism, *Progress in Human Geography* 16, 190–218.

Tilly, C. (ed.) 1975. *The Formation of National States in Western Europe*. Princeton, NJ: Princeton University Press.

Tilly, C. 1992. *Coercion, Capital, and European States*. Cambridge, MA: Blackwell.

Timmer, C.P. and Falcon, W.P. 1975. The political economy of rice production and trade in Asia, in L.G. Reynolds (ed.) *Agriculture in Development Theory*. New Haven, CT: Yale University Press.

Tosh, J. 1980. The cash crop revolution in Africa: an agricultural reappraisal, *African Affairs* 79, 79–94.

Toyota. 2001. *Corporate Profile 2001*, Tokyo: Toyota; available at http://toyota.irweb.jp/IRweb/corp_info/datacenter/2001/2001databook.pdf

Tunstall, J. 1986. *Communications Deregulation*. Oxford: Blackwell.

Tuong, H.D. and Yeats, A.J. 1981. Market disruption, the new protectionism, and developing countries: a note on empirical evidence from the US, *The Developing Economies* 19, 107–118.

Turits, R. 1987. Trade, debt, and the Cuban economy, *World Development* 15, 163–180.

Turnock, D. 1984. Postwar studies on the human geography of eastern Europe, *Progress in Human Geography* 8, 315–345.

Turnock, D. (ed.) 2001. *East Central Europe and the Former Soviet Union: Environment and society*. London: Arnold.

Turton, A. 1982. Poverty, reform and class struggle in rural Thailand, in S. Jones, M. Murmis, and C. Joshi (eds.) *Rural Poverty and Agrarian Reform*. New Delhi: Allied.

Tylecote, A. 1992. *The Long Wave in the World Economy: The current crisis in historical perspective*. London: Routledge.

Tyler, W.G. 1976. Manufactured exports and employment creation in developing countries: some empirical evidence, *Economic Development and Cultural Change* 24, 355–373.

Tyler, W.G. 1981. Growth and export expansion in developing countries: some empirical evidence, *Journal of Development Economics* 9, 121–130.

Tyler, W.G. 1986. Stabilization, external adjustment, and recession in Brazil: perspectives on the mid-1980s, *Studies in Comparative International Development* 21, 5–33.

Tyson, L.D. 1992. *Who's Bashing Whom? Trade conflict in high technology industries*. Washington, DC: Institute for International Economics.

UK Office of the Deputy Prime Minister. 2004. *The English Indices of Deprivation 2004 (revised)*. London: HMSO; available at http://www.viral.info/iod/iodpdf/odpm_urbpol_029534.pdf

UN (United Nations). 2000. *Economic Vulnerability Index: Explanatory note*. New York: UN Committee for Development Policy; available at http://www.un.org/esa/analysis/devplan/cdp00p21.pdf

UN. 2001a. *2001 Report on the World Social Situation*. New York: Economic and Social Council.

UN. 2001b. *World Economic and Social Survey 2001: Trends and policies in the world economy*. New York: UN Department of Economic and Social Affairs.

UN. 2010. *World Population Prospects 2009*; available at http://esa.un.org/unpd/wpp/Excel-Data/population.htm

UN Centre for Human Settlements (Habitat). 2001. *Cities in a Globalizing World: Global report on human settlements 2001*. Nairobi: UN.

UN Centre for Human Settlements (Habitat). 2012. *State of the World's Cities 2010/2011: Cities for all: Bridging the urban divide*. New York: United Nations.

UN Economic Commission for Europe 1972. *Economic Survey of Europe in 1971*. New York: United Nations.

UN Economic Commission for Europe 1979. *The European Economy in 1978*. New York: United Nations.

UN Economic Commission for Europe. 1989. *Economic Survey of Europe in 1988–89*. New York: United Nations.

UNAIDS (Joint United Nations Programme on HIV/AIDS). 1999. *The UNAIDS Report*. Geneva: UNAIDS.

UNAIDS. 2003. *HIV/AIDS: It's your business*. Geneva: UNAIDS.

UNAIDS. 2006. *Report on the Global AIDS Epidemic*. Geneva: UNAIDS.

UNAIDS. 2010. *Global Report*. Geneva: UNAIDS.

UNAIDS. 2011. *World Aids Day Report*. Geneva: UNAIDS.

UNCTAD (United Nations Conference on Trade and Development). 1987. *Commodity Yearbook*. Geneva: UNCTAD.

UNCTAD. 1994. *Commodity Yearbook*. Geneva and New York: UNCTAD.

UNCTAD. 2000. *Handbook of Statistics*. New York and Geneva: UN.

UNCTAD. 2001. *Economic Development in Africa: Performance, prospects and policy issues*. New York and Geneva: UN (UNCTAD/GDS/AFRICA/1TB/B/48/12).

UNCTAD. 2001. *World Investment Report 2001*. New York and Geneva: United Nations.

UNCTAD. 2003. *E-Commerce and Development Report 2003*. New York and Geneva: United Nations.

UNCTAD. 2004a. *World Investment Report 2004: The shift towards services*. New York and Geneva: United Nations.

UNCTAD. 2004b. *Globalisation and Development: Facts and figures*. New York and Geneva: United Nations.

UNCTAD. 2006. *World Investment Report 2006: FDI from developing and transition economies: Implications for development*. New York and Geneva: United Nations.

UNCTAD. 2011. *World Investment Report: Non-equity modes of international production and development*. New York and Geneva: United Nations.

UNDP (United Nations Development Programme). 2001. *Human Development Report 2001: Making new technologies work for human development*. New York and Oxford: Oxford University Press.

UNDP. 2005. *Human Development Report 2005: International cooperation at a crossroads*. New York: UNDP.

UNDP. 2006. *Human Development Report 2006: Beyond scarcity: Power, poverty and the global water crisis*. New York and Oxford: Oxford University Press.

UNDP. 2007. *Human Development Report: Fighting climate change, human solidarity in a divided world*. New York: United Nations.

UNDP. 2009. *Human Development Report 2009: Overcoming barriers: Human mobility and development*. New York: United Nations Development Programme; available at http://hdr.undp.org/en/media/HDR_2009_EN_Chapter4.pdf

UNEP (United Nations Environment Programme). 2002. *Vital Water Graphics: An overview of the state of the world's fresh and marine waters*. New York: UNEP.

UNFPA (United Nations Population Fund). 2001. *The State of World Population 2001*. New York: UNFPA.

UNIDO (United Nations Industrial Development Organization). Various issues. *Industrial Development: Global report*. Oxford: Oxford University Press.

UNIDO. 1981. *A Statistical Review of the World Industrial Situation, 1980*. Vienna: United Nations Industrial Development Organization.

UNIDO. 2000. *UNIDO Round Table—Marginalization versus Prosperity*. Vienna: UNIDO.

UNWTO (UN World Tourism Organization). 2012. *UNWTO Tourism Highlights: 2012 edition.* Madrid: UNWTO.

Urry, J. 1985. Social relations, space and time, in D. Gregory and J. Urry (eds.) *Social Relations and Spatial Structures.* London: Macmillan.

US Bureau of Labor Statistics. 2010. *International Comparisons of Manufacturing Productivity and Unit Labor Cost Trends 2009, Supplementary Tables 1950–2009.* Washington, DC: US BLS.

US Census Bureau. 1999. *Statistics of US Businesses: 1999, manufacturing.* Washington, DC: US Department of Commerce; available at http://www.census.gov/epcd/susb/1999/us/US31.HTM

US Census Bureau. 2007. *International Data Base.* Washington, DC: US Department of Commerce; available at http://www.census.gov/

US Census Bureau. 2013. Characteristics of new housing: characteristics of new single family housing completed: median and average square feet by location; available at http://www.census.gov/construction/chars/pdf/medavgsqft.pdf

US Department of Agriculture (USDA). 2002. *Agricultural Outlook.* AGO-293, August. Washington, DC: USDA Economic Research Service.

US Department of Commerce, Bureau of Economic Analysis. 1999. *Preliminary Results from 1999 Foreign Direct Investment in the United States: Operations of US affiliates of foreign companies.* Washington, DC: BEA.

US Department of Commerce. 2000. *Digital Economy 2000.* Washington, DC: Economics and Statistics Administration.

US Department of Housing and Urban Development. 1982. *The President's National Urban Policy Report 1982.* Washington, DC: US Government Printing Office.

US Department of Labor. 2001a. *International Comparisons of Hourly Compensation Costs for Production Workers in Manufacturing, 2000.* Washington, DC: Bureau of Labor Statistics.

US Department of Labor. 2001b. *International Comparisons of Manufacturing Productivity and Unit Labor Cost Trends, 2000.* Washington, DC: Bureau of Labor Statistics.

US Department of Labor. 2002. *Comparative Civilian Labor Force Statistics, Ten Countries, 1959–2001.* Washington, DC: Bureau of Labor Statistics.

USDA. 2010. *Food Availability per Capita Data System*; available at http://www.ers.usda.gov/data-products/food-availability-%28per-capita%29-data-system.aspx#.Ubdgo5ziLKE

Vance, J. Jr. 1970. *The Merchant's World: The geography of wholesaling.* Englewood Cliffs, NJ: Prentice Hall.

Venables, A.J. 2006. Shifts in economic geography and their causes, *Federal Reserve Bank of Kansas City Economic Review* 31, 61–85.

Verhulst, A. 1999. *The Rise of Cities in North-West Europe.* Cambridge: Cambridge University Press.

Vernon, R. 1966. International investment and international trade in the product cycle, *Quarterly Journal of Economics* 80, 190–207.

Vogel, D. 2001. *The Regulation of GMOs in Europe and the United States: A case-study of contemporary European regulatory politics*, paper prepared for a workshop on trans-Atlantic differences in GMO regulation sponsored by the Council on Foreign Relations.

Vogel, E. 1980. *Japan as Number 1: Lessons for America.* New York: Harper & Row.

Wachtel, H.M. 1987. Currency without a country: the global funny money game, *The Nation* 245, December 26, 784–790.

Walker, R. 1988. The geographical organization of production-systems, *Society and Space* 6, 377–408.

Wallerstein, I. 1974. *The Modern World-System: Capitalist agriculture and the origins of the European world-economy in the sixteenth century.* New York: Academic Press.

Wallerstein, I. 1979a. *The Capitalist World-Economy.* Cambridge: Cambridge University Press.

Wallerstein, I. 1979b. Underdevelopment and phase-B, in W. Woldfrank (ed.) *The World-System of Capitalism: Past and present.* Beverly Hills, CA: Sage.

Wallerstein, I. 1980. *The Modern World-System II: Mercantilism and the consolidation of the world-economy 1600–1750.* London: Academic Press.

Wallerstein, I. 1984. *The Politics of the World-Economy.* Cambridge: Cambridge University Press.

Wallerstein, I. 1991. *Geopolitics and Geoculture. Essays on the changing world-system*. Cambridge: Cambridge University Press.

Warf, B. 1990. US employment in foreign-owned high-technology firms, *Professional Geographer* 42, 421–432.

Warf, B. 1995. Telecommunications and the changing geographies of knowledge transmission in the late 20th century, *Urban Studies* 32, 361–378.

Warf, B. 2001. Segueways into cyberspace: multiple geographies of the digital divide, *Environment and Planning B: Planning and Design* 28, 3–19.

Warf, B. 2007. Embodied information, actor-networks, and global value-added services, in J. Bryson and P. Daniels (eds.) *The Handbook of Service Industries*. London: Edward Elgar.

Warman, A. 1980. *"We Come to Object": The peasants of Morelos and the national state*. Baltimore, MD: Johns Hopkins University Press.

Warren, B. 1980. *Imperialism, Pioneer of Capitalism*. London: Verso.

Wasylenko, M. 1991. Empirical evidence on interregional business location decisions and the role of fiscal incentives in economic development, in H.W. Herzog and A.W. Schlottmann (eds.) *Industry Location and Public Policy*. Knoxville, TN: University of Tennessee Press.

Wasylenko, M. 1997. Taxation and economic development: the state of the economic literature, *New England Economic Review*, March/April, 37–52.

Watts, M. 1994. Life under contract: contract farming, agrarian restructuring and flexible accumulation, in P.D. Little and M.J. Watts (eds.) *Living Under Contract: Contract farming and agrarian transformation in sub-Saharan Africa*. Madison, WI: University of Wisconsin Press.

Webb, M.C. 1991. International economic structures, government interests, and international coordination of macroeconomic adjustment policies, *International Organization* 45, 309–342.

Webber, M.J. and Rigby, D.L. 1996. *The Golden Age Illusion: Rethinking postwar capitalism*. New York: Guilford Press.

Wei, Y.D. 2000. *Regional Development in China: States, globalization, and inequality*. London and New York: Routledge.

Wells, L.T. 1972. International trade: the product lifecycle approach, in L.T. Wells (ed.) *The Product Life-Cycle and International Trade*. Boston, MA: Harvard University Press.

Wells, P. and Rawlinson, M. 1992. New procurement regimes and the spatial distribution of suppliers: the case of Ford in Europe, *Area* 24, 380–390.

Wheatley, A. 2009. World Bank cuts China 2009 GDP forecast to 6.5 percent, *Reuters*, March 18; available at http://www.reuters.com/article/2009/03/18/us-china-economy-worldbank-idUSTRE52H0PU20 090318

White, G. 1987. Cuban planning in the mid-1980s: centralization, decentralization and participation, *World Development* 15, 153–161.

WHO (World Health Organization). 2012. *Children: Reducing mortality*, Fact Sheet 178. September; available at http://www.who.int/mediacentre/factsheets/fs178/en/

Wigen, K. 1992. The geographic imagination in early modern Japanese history, *Journal of Asian Studies* 51, 3–29.

Williams, A.M. 1994. *The European Community*, 2nd edn. Oxford: Blackwell.

Williams, C.H. (ed.) 1982. *National Separatism*. Cardiff: University of Wales Press.

Williams, D.E. 1992. Motives for retailer internationalisation: their impact, structure and implications, *Journal of Marketing Management* 8, 269–285.

Williams, G. 1981. *Third World Political Organizations*. Montclair, NJ: Allenheld, Osmun.

Williams, K., Cutler, T., Williams, J., and Haslam, C. 1987. The end of mass production?, *Economy and Society* 16, 405–439.

Williams, R. 1983. *The Year 2000: A radical look at the future and what we can do to change it*. New York: Pantheon.

Williams, R.G. 1986. *Export Agriculture and the Crisis in Central America*. Chapel Hill, NC: University of North Carolina Press.

Williams, W.A. 1981. Radicals and regionalism, *Democracy* 1, 87–98.

Williamson, J.G. 1965. Regional inequality and the process of national development, *Economic Development & Cultural Change* 13, 3–45.

Williamson, O.E. 1985. *The Economic Institutions of Capitalism*. New York: Free Press.

Wilson, P. 2011. *Maintaining US Leadership in Semiconductors*, presentation at AAAS Annual Meeting, Washington, DC, February 18; available at http://www.aaas.org/spp/rd/presentations/20110218 PatrickWilson.pdf

Wise, M. and Gibb, R. 1993. *Single Market to Social Europe*. London: Longman.

Wolf, E.R. 1968. *Peasant Wars of the Twentieth Century*. New York: Harper & Row.

Wolf, E.R. 1982. *Europe and the People without History*. Berkeley, CA: University of California Press.

Wong, K.Y. 1987. China's special economic zone experiment: an appraisal, *Geografiska Annaler*, B 69, 27–40.

Woolley, L. 1963. The urbanisation of society, in J. Hawkes and L. Woolley (eds.) *History of Mankind. Vol. 1: Part 2: The beginnings of civilisation*. Paris: UNESCO.

World Bank. 1979. *World Trade and Output of Manufactures*. Staff Working Paper, January.

World Bank. 1982. *World Development Report 1982*. New York: Oxford University Press.

World Bank. 1983. *World Development Report 1983*. New York: Oxford University Press.

World Bank. 1991. *World Development Report 1991*. New York: Oxford University Press.

World Bank. 1992. *World Development Report 1992*. New York: Oxford University Press.

World Bank. 1995. *World Development Report 1995: Workers in an integrating world*. Oxford: Oxford University Press.

World Bank. 1996. *World Development Report 1996: From plan to market*. New York: Oxford University Press.

World Bank. 1998. *MIGA: The first ten years*. Washington, DC: Multilateral Investment Guarantee Agency.

World Bank. 2000. *World Development Report 2000/2001: Attacking poverty*. Oxford: Oxford University Press.

World Bank. 2001a. *A Study of Alternative Special and Differential Arrangements for Small Economies*. Washington, DC: World Bank.

World Bank. 2001b. *Global Development Finance 2001*. Washington, DC: World Bank.

World Bank. 2001c. *World Development Indicators 2001*. Washington, DC: World Bank.

World Bank. 2001d. *World Development Report 2001*. Washington, DC: World Bank.

World Bank. 2002. *Globalization, Growth, and Poverty: Building an inclusive world economy*. New York: Oxford University Press.

World Bank. 2006. *World Development Indicators 2006*. Washington, DC: World Bank.

World Bank. 2007. *World Development Report 2006: Development and the next generation*. Oxford: Oxford University Press.

World Bank. 2008. *Special Economic Zones: Performance, lessons learned, and implications for zone development*. Washington, DC: World Bank Group.

World Bank. 2011. *Migration and Remittances Factbook, 2011*, 2nd edn. Washington, DC: World Bank.

World Bank. 2012. *Migration and Remittance Brief*, April 23. Washington, DC: World Bank.

World Commission on Environment and Development. 1987. *Our Common Future*. New York: Oxford University Press.

World Resources Institute. 1996. *World Resources 1996–97*. New York: Oxford University Press.

World Resources Institute. 2000. *World Resources 2000–2001: People and ecosystems: The fraying web of life*. Oxford: Elsevier Science and World Resources Institute; available at http://www.wri.org/wr2000/

Wortman, S. and Cummings, R.W. 1978. *To Feed This World*. Baltimore, MD: Johns Hopkins University Press.

WTO (World Trade Organization). 2001. *International Trade Statistics 2001*. Geneva: WTO.

WTO. 2004. *Developing Countries in the WTO Services Negotiations*. Staff working paper ERSD-2004–06. Geneva: Economic Research and Statistics Division.

WTO. 2006. *World Trade Report 2006: Exploring the links between subsidies, trade and the WTO.* Geneva: WTO; available at http://www.wto.org/english/res_e/booksp_e/anrep_e/world_trade_report 06_e.pdf

WTO. 2012a. *World Trade 2011, Prospects for 2012: Trade growth to slow in 2012 after strong deceleration in 2011.* Press release, April 12. Geneva: WTO; available at http://www.wto.org/english/ news_e/pres12_e/pr658_e.htm

WTO. 2012b. *International Trade Statistics 2012.* Geneva: WTO.

WWF (World Wide Fund for Nature). 2000. *Living Planet Report 2000.* Gland: WWF.

WWF. 2006. *Living Planet Report 2006.* Gland: WWF.

WWF. 2010. *Living Planet Report 2010: Biodiversity, biocapacity and development.* Gland: WWF.

Yang, D. 1994. Reform and the restructuring of central-local relations, in D.S.G. Goodman and G. Segal (eds.) *China Deconstructs: Politics, trade and regionalism.* London: Routledge.

Yang, F. 2004. Services and metropolitan development in China: the case of Guangzhou, *Progress in Planning* 61, 181–209.

Yapa, L. 1979. Ecopolitical economy of the Green Revolution, *Professional Geographer* 31, 371–376.

Yapa, L. 1980. The concept of the basic goods theory, in R.L. Singh and R.P.B. Singh (eds.) *Rural Habitat Transformation in World Frontiers.* Varanasi: National Geographical Society of India.

Yates, R.L. 1959. *Forty Years of Foreign Trade.* London: Allen & Unwin.

Yusuf, F. 2001. The East Asian miracle at the millennium, in J.E. Stiglitz and S. Yusuf (eds.) *Rethinking the East Asian Miracle.* Washington, DC and New York: World Bank and Oxford University Press.

Ziegler, C. 2007. *Favored Flowers: Culture and economy in a global system.* Durham, NC: Duke University Press.

Zimbalist, A. 1987. Cuban industrial growth, 1965–84, *World Development* 15, 83–93.

Zimbalist, A. and Eckstein, S. 1987. Patterns of Cuban development, *World Development* 15, 5–22.

Zinnov. 2005. *Outsourcing: Product innovation from India*; available at RealInnovation.com

Zysman, J. 1983. *Government, Markets and Growth: Financial systems and the politics of industrial change.* Ithaca, NY: Cornell University Press.

Index

Note: Page numbers in **bold** type refer to **figures**. Page numbers in *italic* type refer to *tables*. Page numbers in blue type refer to definitions in the Glossary

Abu Dhabi 338
AC (Andean Community) 361, 365
accumulation, primitive 97
Achebe, C. 226
actor-network theory 192
Adelman, I. 276
ADLI (agriculture-demand-led-industrialization) 276
advanced capitalism 141, 145–59, 179–81, 187–90, 194–7, 211, 403
Advanced Research Projects Agency (ARPA) 150
advertising 9, 316, 319, 336
affluence 146, 152–3
Africa 5, *160*, *199*, *224*, 254, *255*, *335*
  colonization 219, **220**
  Sub-Saharan 254–8, 274, 308, *326*, *335*
  telegraph system 221, **222**
African Union (AU) 357, 403
AFTA (Arab Free Trade Area) 361, 364
agency, human 94
agglomeration 136, 166, 187–94, 201, 205, 295–9, 403
  economies 161, 172, 204, 207, 386
  urban 90
Agnew, J.A. 225, 385
Agreement on Textiles and Clothing (ATC) 293
agribusiness 263–7, 308, 371, 376, 403
Agricultural Bank of China (ABC) 339
Agricultural Development Bank (ADB) 253–4
agricultural regions, world 38–9
Agricultural Revolution 96–7
agriculture 126–7, 233–5, 263, 276–7, 313–15, 346–8, 370–3
  arable land, world's 36
Asia 248, 270–2
  Bangladesh 246, 258–9, 271, 272
  beef boom, Central America 268–70, 375
  capitalization of 245, 262–73
  cash crops 113, 254, *255*, 404
  China 256, 259–62
  coffee 248, 250, 253, 267
  collectivization of 262, 300, 301, 405
commercial 228, 247, 251, 254, 257–9, 269, 273
  commercialization of 270–3
  commodities 17
  concerns 245–73
  cultivable land **37**
  development 95–6, 260–2, 276
  drug crops 268
  employment 38–9, 247–8, 272–3, 325
  export-oriented **40**, 111, 235, 253, 256, 266
  fallow 96, 267, 409
  food 256, 257
  food production 96–7, 245–9
  genetically modified organisms (GMOs) 5, 375–6
  Green Revolution 245–7, 253, 257–60, 270–2, 375, 411
  hearth areas 96–7, 100, 411
  imports 250
  improvements 98, 131
  intensive 36
  investment 264
  labor 254, 270
  latifundia 246, 413
  Latin America 36, 246, 247, 250, 251–4, 258, 273
  Malaysia 249, 258, 270
  monoculture 113, 214, 267, 414
  Neolithic period 96, 416
  organization 246–8, 273
  patterns 37–40
  periphery, in 246–51
  production 5, 30, **40**, 101, 104, 119–20, 255, 272
  profitability 8
  rice 101, 271
  scientific 270–3
  seed 96
  settled 97
  share cropping 246, 258–9, 267, 272, 418
  shifting cultivation 96, 409, 418–19
  specialization 12

subsistence economics 6, 18, 267, 419
Taiwan 259, 260–2, 271, 283, 304
technology in 270–2
unsustainable 30
women and 246–7, 256, 273
world 37–8
see also land
agriculture-demand-led-industrialization (ADLI)
276
aid 29, 237, 283
bilateral 59, 279
food 256
foreign 236
international 58–9
multilateral 279
organizations 250, 333
regional 385
state 386
AIDS *see* HIV/AIDS
al Qaeda 389
alliances, strategic (among firms) 86, 91, 164–6,
169, 280, 419
Amin, A. 396
and Robins, K. 205
Amsden, A. 74
Andean Common Market (ANCOM) 361
Andean Community (AC) 361, 365
Andean Pact (1997) 365
antitrust legislation/laws 68, 82, 156
apartheid 66
APEC (Asia Pacific Economic Cooperation) 294,
362
Arab Common Market 361
Arab Free Trade Area (AFTA) 361, 364
Arab Spring 5
Argentina 221, **223**
aristocracy, European 226
Aronson, J.D., and Cowhey, P.F. 84
ARPA (Advanced Research Projects Agency)
150
ASEAN (Association of South East Asian
Nations) 49, 362–4
Asia 102, 109, 147, 165, 168, *199*, *224*, 232,
258–60
agriculture 248, 270–2
telegraph system 221, **222**
Asia Pacific Economic Cooperation (APEC) 294,
362
Asian financial crisis (1997) 239–41, 280, 288,
365
Asian Tigers 67, 202, 403, 404
assets 53
corporate 53
industrial, European 125

Association of South East Asian Nations
(ASEAN) 49, 362–4
ATC (Agreement on Textiles and Clothing)
293
Atlantic system of trade (1650–1850) 218,
219
AU (African Union) 357, 403
austerity measures, national 133
Australia *342*, *359*
autarky 50, 301, 304, 403
communist 310
economic 106
authoritarianism 364
automobile industry
foreign in US 184, **184**
labor process 290, **290**
autonomy
Asia 109
farms, of 263
financial institutions, of 71
Japanese economy 130
lack of, Asian towns 102
markets, of 71
states, of 138

baby-boomers 153
back-office functions 90, 173, 186, 403
backwash effects 136, 210, 363, 377, 403
Bairoch, P. 215
Baldwin, R.E., and Meier, G.M. 20–1
Bangladesh
agriculture 246, 258–9, 271, 272
clothing industry 154, 167, 292
footwear industry 234
banking 120, 137
commercial 316
crisis 369
investment 316
offshore *331*, 339–40, 343–4, **343**, 349
transnational 339, 349
bankruptcy 120, 395
Barbie 16
Barnet, R., and Cavanagh, J. 54, 153
barriers 135
investment 361–2
non-tariff 359–60, 416
removal of 370
tariff 363
trade 159, 170, 241, 293, 353, 360–1, 364
barter 97
terms of trade 230–1, *232*, 235, 240
Basel Ban Amendment (e-waste) 33
Basques 124–5, 380, 387, 388, 389, 395
batch production 169, 403

Bayoumi, T., and Lipworth, G. 294
beef boom 268–80, 375
Better Life Index (OECD) 26
biodiversity 377
biofuels 264
biotechnology 196, 199, 375, 386
Black Death (bubonic plague) 102
Blinder, A.S. 14
Bluestone, B., and Harrison, B. 146, 158
Boeing 787 Dreamliner **165**
Bogor Declaration (APEC, 1994) 362
boosterism 126, 188, 211
Borchert, J. 126, 190
borderless world 83
Boserup, E. 96, 97
Bowler, I. 370
BPO *see* business process outsourcing
brain drain 42–5, 107, 277
branch-plant industrialization 183, 201, 391,
    403–4
Brazil 45, *266*, *284*, *286*, *335*, *359*
Bretton Woods Agreement (1944) 54, 57, 70, 80,
    130, 158, 354, 404
BRICs 108, 308, *30*, *311*, 404 *see also* Brazil;
    Russia; India; China
Britain *see* United Kingdom
brokerage system, colonial 226
Brown, M.B. 250
Brundtland Report (*Our Common Future*, 1987)
    32
Bryson, J.R., *et al* 318
budgets 302, 372
    deficits 70, 180, 283
Bundesbank, Germany (central bank) 78
bureaucracy 77, 136, 299–302, 306, 363, 379,
    388
    imperial 101
    national 81, 139–40
    state 101
Bush, G.W. 384, 389, 397
business
    climate 78, 382
    organization 154
business process outsourcing (BPO) 173, 186–7,
    187, 320–2, 344–9, 403–4
business services 90, 171, 172, 173, 186, 190–1,
    192–3, 197, 316, 319, 355, 404
    international finance and 53–5
    internationalization 344–7, 349

CACM (Central American Common Market) 361
call centers 173, 183, 192, 324, 325, 346
Canada *278*, *335*, *359*
CAP *see* Common Agricultural Policy (EU)

capital 57–8, 119–21, 136–8, 248, 251–60, 308,
    322
    accumulation of 108–9, 117, 124, 367
    circulation of 225
    commercial 137
    costs 384
    crisis 169, 278
    development 101
    drain 363
    exports 214, 285
    financial 80
    fixed 201, 384
    flight 57, 238, 240, 404
    flows 57, 65, 161, 210, 242, 339
    foreign 232, 254, 264, 279, 303–4, 320
    global 190, 242, 340, 384, 394
    globalization of 237–9
    goods 146, 233, 293, 367, 404
    growth 169, 216–18, 278
    human 7, 100
    imperial 131
    imports 122, 131
    investment 106, 134, 159, 182, 218, 264, 285
    local 232
    movement 20, 70, 357, 361
    natural, Earth's 37
    physical 7
    pool of 171
    productive 279
    redeployment of 159
    risk 188
    "rooting capital" 382–3
    shortage of 364
    start-up 56
    state 126
    venture 86, 193
capitalism 65–6, 130, 138, 246, 254, 300–3,
    308–10, 353–4
    advanced 141, 145–59, 179–81, 187–90,
    194–7, 211, 403
    agricultural 252, 260–2
    British 138
    competitive 7–9, 138, 242, 399–400, 406
    contemporary 89
    corporate 397
    democratic 394
    development of 101, 262
    disorganized 9–10, 181, 408
    European 214
    evolution of 8–12
    expansion of 214–16
    exports within 267
    foreign 242
    free market 9, 353

French 138
global 67, 130, 145, 177, 180, 216, 250–1
industrial 93, 109, 119–21, 126–7, 130–2,
    143–5, 212
liberal 213, 242–4
local 242
market 229, 242
merchant 7, 93, 99–113, 116–17, 143, 414
monopolistic 131
organized 7–9, 76, 137–44, 279, 399–400,
    417
pre-capitalist institutions 229
rural 264
transformation of 138, 399
world-systems 159, 300
capitalization 253, 270–3, 404
    agriculture, of 245, 262–73
captive outsourcing 321, 404, 417
carbon fuels 5
CARICOM (Caribbean Community and
    Common Market) 361
carrying capacity 37, 404
cartels 138, 181, 234, 243,
cash crops 113, 254, 255, 404
casino economy 54
Cassa del Mezzogiorno (Southern Development
    Agency) 381
Castells, M. 191–3, 199, 303
    and Hall, P. 179
Castro, A.N. de 111
Cavanagh, J., and Barnet, R. 54, 153
Celtic Tiger (Ireland) 202, 404–5
CEMAC (Economic and Monetary Community
    of Central Africa) 361
Central American beef boom 268–80
Central American Common Market (CACM)
    361
Central Europe see Europe, east-central
central places 190, 302, 405
central place systems 405
centralization
    agriculture, of 263
    corporate 155, 379–80, 405
    EU 400
centres of innovation 193–4
Cereal Partners Worldwide (CPW) 164
change 113
    climate 35, 234, 265, 400
    economic 3–19, 8, 96, 101–2, 113, 126–30,
        209, 351–5
    geopolitical 115, 143
    market 234–5
    political 6, 308
    regional 180–2, 370

spatial 1, 6–12, 6, 18, 116–17, 142
technological 6, 29, 118, 143, 147
urban 180–2
Chaunu, P. 102
Chavez, H. 243
child labor 167
China, People's Republic of 45, 243, 249, 262,
    335, 339, 359
    agriculture 256, 259–62
    Barbie 16
    business process outsourcing (BPO) 320, 322
    clothing industry 167, 292, 293
    communism 308, 310
    early innovatory economic change failure
        101–2
    economic and demographic growth 304, 305
    GDP (industry and services) 306, 307
    hearth area 96
    land reform 262
    poverty 260, 307
    rice 101, 271
    special economic zones (SEZs) 4, 46, 304–6
    Tiananmen Square (1989) 308
    women, and work 16
China Construction Bank (CCB) 339
Chinese Central Bank 310
Chinese Communist Party (CCP) 308
Chisholm, M. 142
Cipolla, C. 102, 112
class (social) 33
    middle 33, 137, 300
    struggle 12, 304
    working 253
classical world, urbanization 98–9, 99
client states 141, 405
climate 97, 142
    change 35, 234, 265, 400
Cline, W.R. 281
clothing industry 154, 167–8, 168, 290, 290,
    292, 293
clustering (spatial)
    corporate 89–90, 403
    exports, of (Italy) 392
clusters (competitiveness, national), France
    199–200
CMEA see Council for Mutual Economic
    Assistance
"codetermination" (in economic democracy)
    395
coffee 248, 250, 253, 267
Cohen, A. 387
Cohen, S.B. 355
cohesion 63, 372–3
    economic (of the state) 384

Cold War (1947–91) 58–9, 74–5, 133, 243, 310, 388–90, 394
  effects 237
collaborative ventures *see* strategic alliances
collectivization of agriculture 262, 300, 301, 405
colonialism 126, 141, 177, 226–30, 243, 254, 259
  competitive 218–19
  end 230–3, 237
  Japanese 74, 283
  market 225
  new 214
  post- 228, 241
  power 227
  pre-colonial institutions 229
  pre-capitalist 214
  and urban settlement patterns **114**, 115
colonization 63, 100–1, 109, 141, 215–16, **216**, 220
  Africa 219, **220**
  overseas 108
  Western 246
COMECON (Council for Mutual Economic Assistance) 301, 405
COMESA (Common Market for Eastern and Southern Africa) 361
Comisso, E. 395
command economy 242, 300, 405
commercial agriculture 228, 247, 251, 254, 257–9, 269, 270–3
commodities 62, 68, 80, 217–18, 231, 246, 267
  agricultural 17
  diversification 235
  economic 216
  exports 221–2, 252
  flows 129
  prices 225, 234, 250, 265
  primary 233–7, 240, 243–5, 252–3, 336, 366, 417
  trade in 79, 104
commodity chains 15–17, **15**, 167–8, 248, 333–4, 348, 405
commodity concentration of exports, index of 52, **53**
Common Agricultural Policy (CAP) (EU) 367, 372
  effects 370–2
common market 361, 364, 365, 372, 405
Common Market for Eastern and Southern Africa (COMESA) 361
communication 50, 227, 243, 303, 316, 347
  complex (by workers) 193
  costs **88**
  improvements 126

innovation 52
internal (national) 134
international 88
national 134
networks 218
poor 119
services 346
technology 83, 117, 149, 315, 319, 330–2, 339
communism 125, 299–301, 304–6, 310, 354, 388
  anti- 354
  spread 133
Communist Party 300, 394
  Chinese 308
comparative advantage 81, 112, 115, 119, 129, 141, 143, 194, 214–16, 281, 362–3, 405
competition 170
  capitalist 7–9, 138, 242, 399–400, 406
  colonial 218–19
  economic 382
  international 170
  market 10
  competition 382–6
  structural 181
  zero-sum 78
competitive advantage 12, 62, 79, 82, 90, 133, 154, 192, 204, 216, 294, 302, 308, 323, 346, 361, 363, 400, 405
competitive disadvantage 120, 405, 406
competitiveness clusters, France 199, **200**
concentration, corporate 155–6, 406
conflict 277
  armed 120
  economic 218
  ethnic 218, 386–7, 391
  global 388
  political 277, 310
  regional 400
conservatism 180, 276, 302
consolidation
  merchant capitalism, of 106–7
  spatial/regional 179, 183, 187–94
consumer demand 153–5, 164, 201, 202, 229, 251
consumer durables 406
consumer goods 233, 367, 406
consumer non-durables 406
consumer services 13, 406
consumption 3, 64
  conspicuous 122
  demand 153–5, 201, 229, 251
  domestic 252
  energy 35–6, **35**
  integration of 244
  market 81, 89, 145, 158, 354

mass 152, 162, 354
national 304
patterns 152–5
per capita 301
spread 241
containerization 406
Cook, P., and Kirkpatrick, C. 238–9
Cooper, F. 229
cordon sanitaire 301, 406
core *see* core countries
core countries 116–44, 159–61, 179–212, 240–1, 353–5, 363–6, 375–7
"fixed industrial" 213
spatial reorganization of 182–7
core-periphery patterns 20–1, 26, 140, 205, 206, 210, 211, 215, 237, 35, 398
core-periphery polarization/contrasts 20, 35, 4, 139, 143, 205, 210, 240–2, 320–3
corporations see transnational corporations (TNCs)
corporate centralization 155, 379–80, 405
corporate concentration 155–6, 406
corporate control centres 190–2
corporate restructuring 147, 174, 188, 201
corruption perceptions index 63, **63**
Council for Mutual Economic Assistance (COMECON) 301, 405
Cowhey, P.F., and Aronson, J.D. 84
CPW (Cereal Partners Worldwide) 164
creative destruction 10, 208
credit policy 273
crisis 139, 145, 355
Asian financial (1997) 239–41, 280, 288, 365
banking 369
capital 169, 278
debt 54, 58, 78–80, 139, 239, 244, 397
feudal system 102–4
fiscal 78, 383
Fordism, of 146–7, 174, 211
global financial (2007–8) 56–7, 68–70, 239, 288, 330, 337–8, 369
political (Japan) 136
pre-2008 global financial 68
profit, of 281
crops 247, 254–7, 266, 271–2, 370–2, 375–6
drug 268
exports of 225, 246, 259, 264, 270
imports of 270
industrial 217
market 267
production of 254–5, 273
rotation of 102, 257, 267
water-intensive 265
yields 260

crossover system of trade 224, 225, 230, 243, 406
crude oil 77, 171
cultivable land **37**
cultivation
slash-and-burn 96
shifting 96, 409, 418–19
cultural economy 204–5, 400, 406
cultural integration 228–30
cumulative causation model (Myrdal) 196, 207–9, 210, **210**, 406
currency 68, 159, 367–9
floating 70
global 410
manipulation of 123
customs unions 360, 361, 406

DCs *see* developed countries
dead zones, marine 30–1
debt 4, 29, 43, 70, 142, 239–40, 311
crisis 54, 58, 78–80, 139, 239, 244, 397
cycle 270
foreign 58
global 244
government 54, **55**
growth of 282
international 55, 239
national 202
per capita 2
reduction in 202
repayment 58, 238, 269
short-term 56
trap 55–8, 241, 406–7
decentralization 179, 182–3, 201, 211–12, 379–80, 385, 394–7
economic 184, 394
governmental 389
inter-metropolitan 186–8
metropolitan 185–8
regional 183–5, 188
Russian government 388, **389**
spatial 183–7
decolonization 215, 243, 278
decommodification 250
defetishization 250–1
deforestation 30–6, 269
deindustrialization 180–1, 195–7, 215, 298, 304, 311, 407
demand 153
consumer 153–5, 164, 201, 202, 229, 251
for education 319
elasticity of 55, 142, 231, 245, *252*, 318, 336, 408
patterns 152–5

democracy *73*, 126, 226, 229, 299
  associative 395, 396
  capitalist 394
  economic 18, 351, 394–7
  electoral 64
  participatory 396
  political 308, 380
  social 242
Demographic Transition (model) 40–1, **41, 42**, 45, 407
demographics 6, 23, 40–5, 113, 143, 305
dependence 12, 20–1, 36, 51–2, 59–60, 63, 90, 133, 142, 146, 235, 241, 310, 337, 356–7, 407
dependency *see* dependence
deprivation index, UK 206, **207**
deregulation 56, 77, 171–2, 181, 339
design-intensive industry *203*
deskilling 149, 407
destabilization, economic 158, 355
Deutsche Bank 54, 172, 266
developed countries (DCs) 39–46, 51–9, 231–40, 274–82, 308–17, 320–30, 335–40
  economic change 8, **8**
  exports with less-developed countries (LDCs) **236**, 237
development 25, *48*, 365, 373, 386
  agricultural 95–6, 260–2, 276
  alternate models of 242–3
  capital 101
  commercial 227
  economic 113–15, 124–6, 208–11, 239–40, 302–4, 380–6, 398–400
  endogenous 385
  global 241
  impasse 242
  industrial 122, 130–1, 208, 258, 276, 299–300, 365
  international 365
  local 381–2
  metropolitan 136
  national 230–3, 237, 242, 386
  patterns of 373
  political-economic 388
  regional 194, 363, 381–2
  rural-agricultural 276
  strategies 230–3
  sustainable 24, 32–4, 237, 244, 420
  theory 356
  uneven 64, 79, 180, 209–10, 355, 357, 369, 399, 421
  urban 106
developmental plasticity 271

devolution 380, 390
  regional 139, 380, 390, 418
diagonal integration 155–6, 407
Diamandi, Z., *et al* 152
differential of contemporaneousness 152, 407
diffuse industrialization 183, 407
digital divide, the 33, 198, 407–8
diminishing returns 100, 107, 113, 117, 143, 408
disasters, natural 5, 35, 272, 337
disease 32, 217
  absence 24
  outbreak 337
disintegration
  vertical 86, *163*, 164, 201, 421
disintermediation 339
dislocation 122–3
disorganized capitalism 9–10, 181, 408
division of labor
  international 56, 412
  sexual 229, 418
Dobb, M. 102
Doha Development Round of the GATT (WTO) 360
Donahue, J.D. 396
droughts 98, 272
drug crops 268
Duignan, P., and Gann, L. 227–8

e-finance 340, *342*
e-waste 33
Earth Summit (UN, 1992) 32–4
earthquakes 272
EC *see* European Union (EU)
ECB (European Central Bank, EU) 76, 202, 369
ecological footprint 26, 29, 37–9, **39**, 408
economic
  change 3–19, **8**, 96, 101–2, 113, 126–30, 209, 351–5
  cohesion (of the state) 384
  conflict 218
  competition 382
  decline 196, 285, 303, 380
  destabilization 158, 355
  growth 74–6, 129–33, 137–9, 208–10, 235–7, 240–2, 394–5
  organization 6–12, **6**, 100, 118–20, 129–30, 139–43, 169, 299
  policy 225, 361, 380
  specialization 137
Economic Community of West African States (ECOWAS) 361
Economic and Monetary Community of Central Africa (CEMAC) 361

economies
  external 86, 89, 197, 209
  market 78, 91, 207, 303
  national 65, 91–3, 202, 215, 322, 358, 386
  scale, of 117, 143, 203, 211, 262–3, 357, 408
  scope, of 85, 147–52, 169, 203, 400, 408
Economist 36, 173, 317
economy 1
  aestheticization of the 204
  commodities, and 216
  continental 127
  cultural 204–5, 400, 406
  decentralized 184, 394
  democratic 18, 351, 394–7
  development, and 113–15, 124–6, 208–11,
    239–40, 302–4, 380–6, 398–400
  difference, global 399
  European 125
  evolution (of industrial) 138
  global 17, 145–75, 218–25
  increasing pace of the **69**
  industrial 47, 116, 138–9
  inequality within 207–11
  informal and development level of 326, **327**
  information 171–2
  integration of the 50, 90, 351–3, 360–6, 373,
    379
  interdependence of the 91, 355
  international 157
  liberal 9
  management of the 308
  marginal 110
  political 131, 206
  power of 91, 110–12, 124, 355, 375
  prosperity, and 23
  public 118, 139–41
  reforms, and 304, 308, 320
  regional 93, 196, 294, 364, 383
  reorganization of the 299–300
  resource-oriented 243
  rights and the 386
  shifting fortunes of the 61, **62**
  stable 202
  transformation in the 306
  urban 222
  well-being and the 129, 179–80, 187, 381
ecotourism 338
ECOWAS (Economic Community of West
    African States) 361
ECSC (European Coal and Steel Community)
    366, 372
education 23–6, 44, 198, 202, 228–9, 321–4, 348
  colonial 229
  demand for 319

  free 63
  higher 193
  low achievement and 206
  reforms of 132
  technical 170
EEA (European Economic Area) 362
EEC (European Economic Community) (now EU)
    366–7
EFTA (European Free Trade Association) 360–2
egalitarian liberalism 9, 76–7
egalitarian political-economic structure 259–60,
    394
EIA (Energy Information Association) 29–30,
    34–5
EIB (European Investment Bank, of the EU) 372
elasticity of demand 55, 142, 231, 245, 318, 336,
    408
electronic finance (e-finance) 340, *342*
electronic waste (e-waste) 33
electronics industry 201, 291, **291**, 292, 297,
    297, *298*
Emilia-Romagna *see* Third Italy
empires 95, 97
  trading 141, 353
employee stock option plans (ESOPs) 395
employment 206, 286, 297, 313–14, **314**, 340,
    348, 384
  agriculture 38–9, 247–8, 272–3, 325
  creation 285
  expansion 393
  formal-sector 326
  industrial 295
  insourcing and outsourcing of (US) 14, **14**
  manufacturing *279*
  UK 194, **195**
  USA 194–5, **196**
encapsulation (of goods and materials) 317, 348,
    408–9
End Poverty campaign (UN, by 2015) 24
energy 35–7
  consumption 35–6, **35**
  nonrenewable 35
Energy Information Administration (US, EIA)
    29–30, 34–5
England *see* United Kingdom (UK)
enterprise zones (EZs) 381, 409
entrepôts 109, 112, 190, 228, 409
environment 29–32
  activism and 276
  degradation of 5, 32, 269, 303, 338
  pollution 30–2, 260, 307, 338, 400
  protection of 147, 338
environmentalism 338
EPZs *see* export processing zones

equity markets 68
equity, social 367
ERDF (European Regional Development Fund, of the EU) 373
ESF (European Social Fund, of the EU) 373
ESOPs (employee stock option plans) 395
ethnic conflict 218, 386–7, 391
ethnic nationalism 387, 391
EU *see* European Union
euro-zone countries, government debt, and 54, 55
eurodollars 17, 54, 171, 238, 239, 409
Europe
    in 1875 **121**
    core and periphery **124**
    east-central 99, 113, 121, 122
    growth rates **123**
    political control (1500–1950) 213–14, **214**
    second-wave industrialization 119–20
    third-wave industrialization 120–1
European Central Bank (ECB, of the EU) 76, 202, 369
European Coal and Steel Community (ECSC) 366, 372
European Economic Area (EEA) 362
European Economic Community (EEC, now EU) 366–7
European Free Trade Association (EFTA) 360–2
European Investment Bank (EIB, of the EU) 372
European Regional Development Fund (ERDF, of the EU) 373
European Social Fund (ESF, of the EU) 373
European Union (EU) 78–9, 160–1, 232–3, 306, 353, 360–74
    economy 125
    enlargement 367, **368**
    euro (single European currency) 54, **55**
    external effects 373–7
    industrialization 62
    members 42, 363–4, 367, 368–70
    polarization 370
    regional inequality 205–6, **206**
    regional policy 373, **374**
    Single European Act (SEA) 372
    Single European Market (1992) 367–9
    Structural Adjustment Program 70, 419
    trade 254
    US foreign direct investment (FDI) 53
exchange 398
    economic and technological, circuits of 398
    foreign 49, 52, 253, 283, 337–40, 349
    market 65, 106
    rates 54, 238–9
    stock 54
    unequal 142, 250, 421
expansion 106–7, 130, 141, 225
    dynamic 110
    global 214
    industrial 120
    overseas 109
exploitation
    economic 66
export processing zones (EPZs) *48*, 49, 234, 288–91, 294, 295, 311, 331, 332, 348, 399, 409
exporters, intra- vs inter-regional connectedness *52*
exports *48*, **51**, 52–3, *160–1*, 243–5, 253, 272
    agricultural **40**, 111, 235, 253, 256, 266
    capital 214, 285
    capitalist 267
    commodities 221–2, 252
    commodity concentration index 52, **53**
    crops 225, 246, 259, 264, 270
    developed and less-developed countries **236**, 237
    earnings 248
    expansion 74, 247
    foreign 391
    growth 282
    manufactures 241, 284–5, 306
    market 235, 254, 281, 310–11, 391
    services 322–5, 329–30, **330**
    shares 392
    world *224*, 329
external control 183–7, 211, 363, 409
external economies 86, 89, 197, 209
externalization 164, 181, 188, 201
EZs *see* enterprise zones

factors of production 12, 91, 149, 155, 216, 405, 408, 409, 414, 415, 420
fair trade 248–51
Fair Trade (Brown) 250
Fair Trade Labeling Organizations International (FLO) 250
fallow agriculture 96, 267, 409
famines 5, 271, 371
FAO (UN Food and Agriculture Organization) 36–7, 246, 256
farming *see* agriculture
fascism 386
favouritism, political 278, 387
FDA (US Food and Drug Administration) 376
FDI *see* foreign direct investment
federalism 390, 395–6
Fertile Crescent 97

feudal system crisis 102–4
feudalism 6, 18, 99–104, 130–2, 246, 260, 409
   spread 95
fiber-optic cable network, submarine 323, **323**
filesharing, peer-to-peer (P2P) 157
finance 129
   electronic (e-finance) 340, *342*
   infrastructure 340
   international 53–5
   patterns of 45–59
   system of 68–70
finance, insurance, and real estate (FIRE) 316,
      410
financial centers 71, 90, 172, 340, 349, 367
   offshore 17, 416
financial crisis, global 56–7, **57**, 68–70, 239, 288,
      330, 337–8, 369
First World 230
fiscal crisis 78, 383
fiscal policy 135, 355–7, 380
fiscal reform 382
Fisher-Clark thesis 313–17, 320
Five-Year Plans (Soviet) 300–2
Flanagan, W., and Gugler, J. 222
flexibility 147
   corporate 166–70
   flexible accumulation 410, 415
   flexible production 79, 162–6, 181, *182*,
      410
   market 77, 173
FLO (Fair Trade Labeling Organizations
      International) 250
floods 98, 217, 270–2
food 257
   aid 256
   deficit 257
   production 96–7, 245–9
Food and Agriculture Organization (UN FAO)
      36–7, 246, 256
Food and Drug Administration (US FDA) 376
footwear industry 234
Ford, H. 5, 9, 126
Fordism 9–12, 80, 125–30, 143–5, 169–70,
      201–3, 354–6
   crisis 146–7, 174, 211
   disintegration 399
foreign direct investment (FDI) 53, 67–9, 225,
      243, 264–5, 279–82, 292–4, 339–40, 348
   British (1860–1959) 219, **220**
   inflows **69**
   foreign policy 75, 361, 367
forestry 30, 36
fossil fuels 34
fracking 5, 29, 410

fragmentation 182
   of forests 30
   political 379
France *220*, **220**, 286, *341–2*
   capitalism 138
   competitiveness clusters 199, **200**
franchises 248, 268, 410
Franco, F. 380
free markets 9, 77, 298, 353, 367, 397
free trade 52, 63, 70, 91, 139, 250, 256
   associations 361, 364, 410
   internal 362
   zones 48–9, 77
freshwater supplies 36, **38**
Fridell, G. 250
Friedmann, J. 190–1
   and Weaver, C. 129
fuels 5, 34, 36

G7 countries 308, *309*
G8 countries 58, 70
G77 countries 365–6, 373
Galbraith, J.K. 146
Gann, L., and Duignan, P. 227–8
Gardiner, K. 202
gateway ports 110, 126
GATS (General Agreement on Trade in Services)
      324
GATT *see* General Agreement on Tariffs and
      Trade
GCC (Gulf Cooperation Council) 361
GDP *see* gross domestic product
Geertz, C. 387
General Agreement on Tariffs and Trade (GATT)
      71, 233, 268, 282, 358–62, 368, 372
General Agreement on Trade in Services (GATS)
      324
genetically modified organisms (GMOs) 5,
      375–6
Genuine Progress Indicator (GPI) 26
geoengineering 34, 410
geographical path dependence 90, 398–9, 410
geographical tiers 63–4, 91
geopolitics 61, 74, 283, 351–5, 377
   change 115, 143
   rivalry 75
George, S. 366–7
Gereffi, G. 333
Germany 78, *220*, 286, *335*, *341–2*
Ghana 256
Glasmeier, A. 172
global assembly systems 15, 162, 164, 172, 420
global cities, world cities 90, 136, 154, 190–2,
      191, 349, 420, 421

global commodity chains *see* commodity chains
global currencies 239, 410
global financial crisis (2007–8) 56–7, 68–70, 239, 288, 330, 337–8, 369
global foreign exchange market 17, *17*
global highway, transportation and communications costs **88**
global interdependence 50, 141–2
global office 171–2
"global" reach of transnational corporations (TNCs) 49
global shift 67, 73–4
global shopping mall 153
global sourcing 233–4, 263, 275, 281, 410
global warming 32, 34, 410
globalization 13–18, 53–4, 77–81, 90–3, 192–4, 396–400
  capitalist 190, 237–9, 242, 340, 384, 394
  economic 17, 145–75, 218–25
  export-oriented 275
  food production and 263
  growth 388, 391
  patterns 20–60, 159–74
  process 61
  of supply 354
  violent response to 389
GMOs (genetically modified organisms) 5, 375–6
GNI *see* gross national income
GNP (gross national product) 411
Golden Age, 16th century 108
goods 11, 62, *161*
  capital 146, 233, 293, 367, 404
  consumer 233, 367, 406
  demand for 127–8
  exports 275
  flow 160, 242, 379
  intermediate 233, 412, 417
  manufactures 233, 317
  movement 357, 361
  positional 152, 417
  producer 233, 417
government
  debt 54, **55**
  intervention 381–4
  Russian decentralization 388, **389**
Graham, S., and Martin, S. 398
grassroots movement 351, 391–6
Great Britain *see* United Kingdom
Great Depression (1929–34) 9, 123, 129, 225, 230, 253
Great Pacific Garbage Patch 32
Greater East Asian Co-Prosperity Sphere 133

Greek Empire 98
Green Revolution 245–7, 253, 257–60, 270–2, 375, 411
Grindle, M.S. 267
gross domestic product (GDP) 21–3, 50–2, **50**, 134–5, 157–9, 272–4, 284–8, 303–10
  per capita **22**, **23**, **118**, 202, 205, 305, 313, **314**
gross national income (GNI) 21, 25, 49, 59, 130, 326, 411
gross national product (GNP) 411
gross value added (GVA) 411
growth 12, 64, *123*, 189
  economic 74–6, 129–33, 137–9, 208–10, 235–7, 240–2, 394–5
  Europe, rates of **123**
  globalization and 388, 391
  income 241, 318
  industrial 122, 129, 298, 301
  manufacturing 304
  market 67, 125, 332–3
  metropolitan 172
  national 76
  output, of 118
  outsourcing, of 348
  poles 210–11, 307, 411
  population, of 102, 132, 159, 259, 265, 272, 283–5
  postwar 133–7
  regional 129, 298
  services 318–19
  trade 50–1, **50**, 79, 233
  urban 106
Guatemala 267–8
Gugler, J., and Flanagan, W. 222
Gulf Cooperation Council (GCC) 361
Gutenberg, J. 107
GVA (gross value added) 411

Hall, P. 199
  and Castells, M. 179
Hall, T.D. 64
Hamilton, F.E.I. 127
Hanseatic League 104–5, **105**
Hansen, G., and Prescott, E. 116
Happy Planet Index (HPI) 26–7, **27**
Harrison, B. 86
  and Bluestone, B. 146, 158
Harvey, D.W. 354
HDI (UN Human Development Index) 25–8, **27**, **28**, *284*
health 23–6, 139, 229, 241, 313, 323, 348
  economic 140

poor 206
public 140
hearth areas, agricultural 96–7, 100, 411
hegemony 64–5, 91, 411
　American 354
　British 214
　political 235
Henry, N., and Pinch, S. 189
Henshel, H.B. 158
Hicks, J.R. 210
high-tech manufacturing 80, 82, 86, 89, 136,
　　152, 169, 172, 197, 203, 211, 380
high technology complexes 5, 89, 169, 201,
　　297
hinterlands
　gateway ports, of 110, 126
　metropolitan 90, 222
　regional 119, 161, 306
Hirschman, A. 210
Hirst, P. 395
HIV/AIDS 5, 24, 41, 160
Hodge, J.S., and Martin, R.L. 385
Holland, S. 370
"hollowing-out" of national government 77,
　　181
Hollywood 204–5
homeland, European 110, 115
homesourcing 183
homogenization 379, 399
　spatial 174
Honduras 247, 269, 290
Hong Kong 74, 202, 240, 283, 291, 306
　Barbie 16
　clothing industry 292–3
　East Asian G4 281
　electronics industry 292
　foreign direct investment (FDI) 292–4
　industrial complexes 297
　manufacturing 304
　newly industrializing economy (NIE) 288,
　　　232, 236, 279–80, 288
　services 326, 348
　tourism 337
　world city 90, 191
Hopkins, A.G. 226
horizontal integration 155–6, *163*, 411
Howells, J. 317
HPI (Happy Planet Index) 26–7
human agency 94
human capital 7, 100
Human Development Index (UN HDI) 25–8,
　　27, 28, *284*
Hundred Years War (1337–1453) 102
hunger, food deficit 257

hunter-gatherers 95–7
hydraulic fracturing *see* fracking
hydroelectricity 34
Hymer, S. 86–7

Iceland, and global financial crisis 56, **57**
ILO (International Labour Organization) 41–2,
　　326
IMF *see* International Monetary Fund
immigration 46
　brain drain (flow of highly-skilled workers)
　　　42–5, 107, 277
　to Central American plantations 217
　within China 46
　illegal, US 364
　increase to EU 5
　remittances to families 43
　top destinations 42
　to US 46, 125
　see also migration
imperialism 8, 63–4, 112, 141, 365, 411
　anti-imperialist rhetoric 235
　Chinese 101
　European 226
　Japanese 131
　local "imperialisms" 219
　neo- 375
　power 12, 130
　Spanish 112
imports **51**, 160, *161*, 237, 258, 293, *331*
　agricultural 250
　capital 122, 131
　crops 270
　foreign 74, 299
　manufactured 311
　market 235
　penetration and 147, 148, 412
　quotas 233, 293, 306, 359
　services 322–5, 329–30, **330**
　substitution and 115, 275–6, 279, 295, 412
income 25, 55, 70, 307
　distribution 137, 283, 316
　growth 241, 318
　inequality 5, 242
　interregional 208
　low 206
　national 240
　per capita 205, 209, 241, 308, 318–20, 325–6,
　　　348
　polarization 18
　security 139
　tax 78
increasing returns to scale 12, 204, 211, 399,
　　408, 412

incrementalism (in resource allocation in the
    USSR) 302
independence 70, 259, 380
    national 386
    political 388
index of commodity concentration of exports
    52, **53**
India 45, **187**, 286, *342*, *359*
industrial capitalism 93, 109, 119–21, 126–7,
    130–2, 143–5
industrial complexes, new 295–9
industrial districts, new 166, 170
industrial evolution 294–5
industrial growth 122, 129, 298, 301
industrial policy 70, *73*
Industrial Revolution 41, 101, 112–13, 116–18,
    127, 138–41, 144, 214–15, 301
industrial spaces 201
    new 179, 196–7, *203*
    old 194–6, 201
industrialization 17–18, 116–26, 129–32, 141–5,
    224–5, 241–3, 276, 298–300
    American 125–30, 144
    branch-plant 183, 201, 391, 403–4
    diffuse 183, 407
    EU 67
    export-oriented 282
    first-wave (Britain) 119–21, 126
    Japanese 130–7, 144
    limits to 280–5
    national 276, 277
    in periphery 280, 294–5, 311, 379
    progress 274–312
    proto- 100, 119, 130
    rationale for 277–80
    second-wave (Europe) 119–20
    Soviet model 245, 299–304
    spread of 119, 120, 144
    stimuli 275–80
    third-wave (Europe) 120–1
    three tests of 286–8, **287**
    trade-led 333
industry 194
    decline of 194
    design-intensive *203*
    development of 122, 130–1, 208, 258, 276,
        299–300, 365
    economy and 47, 116, 138–9
    evolution of 294–5
    infant 232, 295, *296*, 412
    investment and *73*, 262, 276
    labor-intensive 212, 218, 234, 273, 290, 295,
        333
    mass-markets and 113

patterns of 45–59
production 7, 8, 45, 54, 77, 161, 194, 218,
    263, 275, 300, 303, 412, 414, 416
promotion of 232
relocation of 141
technology-intensive *296*
inequality 207
    economic 207–11
    inter-regional **208**
    regional 205–11, 307
    spatial 383–4
infant industries 232, 295, *296*, 412
inflation *56–7*, 146, 171, 208, 235, 300, 322,
    412
informal economy 326–7
information 161
    economy 150, 171–2, 179, 205, 211, 412
information technology (IT) 319, 330, 339,
    344
Information and Technology Agreement (WTO,
    1996) 362
Information Technology Association of America
    (ITAA) 322
infrastructure 141, 292, 323–4, 383
    investment in 139, 250, 322
    well-developed *203*
initial advantage 11, 127, 139, 209, 210, 216,
    281, 302, 346, 405, 412
innovation 50, 85, 100, 106, 119–21, 197, 386
    centers of 193–4
    communications and 52
    introduction of 147
    Japanese 133
    policy 170
    political crisis 136
    production and 128
    technological 29, 96, 107, 125–6, 138, 185,
        253
    waves 295
instability 239–40, 283
    political 292
institutional integration 369
institutionalization of savings 339
integration 363
    business and 126
    consumption of 244
    cultural 228–30
    diagonal 155, *156*, 407
    economic 50, 90, 351–3, 360–6, 373, 379
    horizontal 155–6, *163*, 411
    institutional 369
    international 353–78
    logic and 356–7
    political 95, 126, 351

production, of 244
regional 360, 364–5
spatial 221
supranational 353–78, **358**
territorial 129
vertical 80, 154–6, *163*, 195, 393, 421
intellectual property (IP) 320
protection 324
interdependence 70, 125, 230, 351, 356
dynamics of 213–44
economic 91, 355
global 50, 141–2
government and 137
international 47
political 355
systematic 136
intermediate goods 233, 412, 417
intermediation 71
inter-metropolitan decentralization 182–3
intermodal transportation 150, 412
internal economies of scale 117, 187, 412
Internal Revenue Service (US IRS) 55
international and supranational organizations 81,
    353–78, 396
international corporate alliances (ICAs) 84
international division of labor 56, 412
International Labour Organization (ILO) 41–2,
    326
International Monetary Fund (IMF) 57–8, 68–70,
    73, 202, 258, 282, 357
structural adjustment programs 70, 419
International Telecommunications Union (ITU)
    198
international terrorism *see* terrorism
internationalization 50–4, 74, 145, 172, 237,
    243–4, 330
business services, of 344–7, 249
finance, of 339–44
retailing, of 332–4
services, of 332–47
Internet users **149**, 150
intervention, government, by 381–4
investment 80–2, 158, 202, 218–19, 306–8, 348,
    360
agriculture and 264
banking and 316
barriers to 361–2
British foreign 219, **220**
capital 106, 134, 159, 182, 218, 264, 285
direct 225, 262, 294–7, 375
flows 294, 379–80, 383
foreign direct (FDI) 67–9, 218–20, **220**, 243,
    252, 264–5, 275, 279–82, 292–4, 298, 324,
    331–3, 339–40, 348

indirect 262
industrial 73, 262, 276
infrastructure 139, 250, 322
international 54, 202, 340, 354
inward 382, 383
national 76
overseas 108, 330
portfolio 225, 264
private 239, 384
productive 277, 373
professional 171, 339
regulation of 71
restricted 136
spatial distribution of 143
strategic 115
technological 146
transnational 292–4, 330–1
irrigation 98, 259, 265
IRS (US Internal Revenue Service) 55
Islam 242, 244, 338, 389
isolationism (US) 50
isotropic plain 401, 413
Italy 111, *341, 342*
export shares 391, **392**
ITU (International Telecommunications Union)
    198

Jacobs, J. 97, 106, 115
Japan 63, 132–3, 286, *335, 342, 359*
colonialism 74, 283
electronics industry 297, **297**
industrialization and 130–7, 144
innovation and 133
METI (Ministry of Economy, Trade and
    Industry) 75, 84, 135
MITI (Ministry of International Trade and
    Industry) 75, 84, 135
modernization and 131
Pacific Corridor, megapolitan Japan 135–6
see also Tokyo
Japan Development Bank (DBJ) 135
Jessop, B. 181
JIT production (just-in-time production) 164,
    281, 333, 413
Joekes, S.P. 285
Johnston, R. 20, 94, 98, 104, 113, 187, 384
Journal of Economic Geography 189
just-in-time (JIT) production 164, 281, 333, 413
justice (during colonial period) 229

Karmarker, U. 322
Katouzian, H. 242
Kedourie, E. 387
keiretsu 84, 135, 413

Kenya 256
Keynes, J.M. 9, 130
Keynesianism 9, 77, 130, 181, 413
Khrushchev, N. 301
kin-ordered system 97, 413
kiosk economy (in Russia) 303
Kirkpatrick, C., and Cook, P. 238–9
knowledge economy 10, 413
Kondratiev cycles 413
Korea, Republic of *see* South Korea
Korean War (1950–3) 133
Kornhauser, D. 133
Kraft Foods 68, 156, 376
Krebs, G. 208
Krugman, P. 211
Kuznets cycles 413

labor 10–12, 62, 180, 217, 245, 251–60, 300
    agricultural 254, 270
    automobile industry 290, **290**
    cheap 158, 167, 186, 210, 281
    child 167
    costs 146–7, 159, 183, 188, 201, 280, 292
    cultural division of 387
    demands 109, 270
    division of 12–18, 55–6, 135, 180–2, 201–5,
        229, 348, 362
    exploitation of 21
    field 109
    flexible 164
    flows 210
    force 85, 125, 170, 245–6, 263, 267, 308
    free (during colonial period) 228
    global 216
    intensity (in mass production) 83
    international 55–6
    involuntary 33, 413
    local 241
    manual 109
    markets 143, 161, 181, 203, 229–30, 298,
        319
    migrant 167–8, 259
    militant 149
    movements 357
    organized 388
    pools 86, 166, 209
    power 99, 142, 216
    productivity 245
    relations 85
    rigidity 203
    skilled 8–10, 111, 135, 159, 174, 179, 183
    slave 65, 109
    surplus 183
    unions 9, 77, 117, 144–6, 170
    unskilled 13, 159, 183, 281, 285, 293
    wage 117, 228–9, 243, 246–8, 256, 267, 421
labor-intensive industry 212, 218, 234, 273, 290,
    295, 333
Labour Party (UK) 394
LAIA (Latin American Integration Association)
    362
laissez-faire policy 8, 139, 400
land 7, 245, 246, 247, 251–60, 267
    arable 35–7
    cheap 210
    clearing 30
    costs 188, 306
    cultivable **37**
    grabs 264–5
    marginal 248
    privatization of 262
    redistribution of 260, 262
    reforms 74, 245, 259–62, 273, 282, 283
    tenure 38, 413
landholding 251–3, 258–60
Langley, L.D. 230
Lash, S., and Urry, J. 138, 181
latifundia 246, 413
Latin America 214, 215, 227–8, 230, 283
    agribusiness 264–6
    agriculture 36, 246, 247, 250, 251–4, 258,
        273
    arable land 36
    banks 339
    capital outflow 238–9
    drug crops 268
    export processing zones (EPZs) 291
    food production 272
    foreign direct investment (FDI) 225
    Green Revolution 411
    growth 41, 241
    Happy Planet Index 26
    Human Development Index (HDI) 25
    industrialization 276
    instability 29
    Internet users *199*
    land reform 260, 262–3, 273
    latifundia 413
    manufacturing 279, *279*, 295
    maquiladoras 4, 290, 414
    markets 158, 375
    migration in 246, 253, 267–8
    new agricultural country 262–3
    newly industrializing economy (NIE) 411,
        288
    retailers 333
    services 326, 348
    trade 49, 142, *224*

Latin American Integration Association (LAIA) 362
law 308, 367
  public 100
  taxation 395
law of diminishing returns 100, 107, 113, 117, 143
LDCs *see* less developed countries
less developed countries (LDCs) 21–5, 31–6, 39–45, 50–9, 224–5, 231–51, 268–93, 320–49
  exports with developed countries (DCs) **236**, 237
Levy, F., and Murnane, R. 192
liberalism 77
  classical 77
  economic 9
  egalitarian 9, 76–7
liberalization 256
  service and 324
  of trade 233, 275, 281, 333, 361–2, 369
Light, I. 131
Lipworth, G., and Bayoumi, T. 294
Little, I. 245
livestock 246, 255, 266–7, 370
Living Planet Report 2012 (WWF) 37
loans 57, 80, 240–2
  cheap 295, 382
  repayments 239
localism 377
localization 91, 391, 400
localization economies 16, 207, 209, 386, 406, 414
locational hierarchies 161–2
Lomé Conventions (EU) 373–4
London 18, 124, 206, 217, 227, 349, 354
  capital of core country 168
  financial center 71, 90, 340, 367
  terrorism 390
  world city 191, **191**, 349, 420, 421
Long Depression (1883–96) 215–16
Lonsdale, J. 229
Los Angeles 90, 168, 186, 196, 421
Los Angeles School 170
Los Angeles Times 16
Lugard, F. (Lord) 226

M&As (mergers and acquisitions) 339
Maastricht Treaty (1992) 367–8
MacFarquhar, R. 310
machinofacture 8, 117–25, 143–4, 414
macroeconomics 70, 130, 235, 240–2, 357–8, 385, 388
  policymaking 75–6

Malassis, L. 246
Malaysia 41, 64, 232, 364
  electronics industry 201, 291, **291**, 292, 297
  agriculture 249, 258, 270
  food production 249
  foreign direct investment (FDI) 333
  manufacturing 274
  newly industrializing economy (NIE) 21, 281
  transnational corporations (TNCs) 291
  women's employment 291
Malecki, E. 193–4
Mankiw, N.G. 14
manor, medieval 102, **103**
manufacturing 17–18, 183–4, 277–9, 286–8, **286**, 295, 298–9, 306–10, 315–21,
  clothing **168**
  economy and 308
  employment *279*
  exports 241, 284–5, 306
  global 47, 278, 299, 311
  goods 233, 317
  growth 304
  high-tech 80, 82, 86, 89, 136, 152, 169, 172, 197, 203, 211, 380
  imports 311
  labor-intensive 46
  output 288, **289**
  outsourcing 173
  production index **134**
  spread (Europe) 91
  traditional 313
  value added (MVA) 45–6, 274, 298
  world 46, 54
Manufacturing Belt (US) 78, 127–9, 196, 212
maquiladoras 4, 290, 299, 414
marine dead zones 30, **31**
"market access" 79–91
market power 82, 333
marketing 282, 316, 317, 319
markets *48*, 122, 180, *252*, 375
  autonomous 71
  capitalist 229, 242
  changes and 234–5
  colonial 225
  common 361, 364–5, 372
  competition and 10
  consumer 81, 89, 145, 158, 354
  deregulation 172
  discreet 344
  domestic 48, 145, 258, 311, 391
  dominance and 216
  economic 78, 91, 207, 303
  equity 68
  evolving 21

exchange and 65, 106
expanded (in Asia) 132
export 235, 254, 281, 310–11, 391
external 89
failure and 77
financial 70
flexibility and 77, 173
foreign 84–5, 266, 280, 304, 310, 330–3, 346–8
free 9, 77, 298, 353, 367, 397
global 3, 79, 89–90
growth of 67, 125, 332–3
homogenization of 153
import 235
internalized 82
international 3, 130, 299
knowledge of 323
labor 143, 161, 181, 203, 229–30, 298, 319
local 203, 256, 363
mass 113, 128, 152, 167, 281, 354
national 17, 117, 129, 172
niche 153
oligopolistic structures 324
overseas 117
protection and 247
regional 13, 117
regulation and 63, 137, 181, 370
saturation of 152, 155
stagnation of 241
stock 129, 133, 239, 340, **343**
well-developed 119
Markusen, A. 197, 201
Marshall, A. 166
Marshall Plan (1948) 58, 125, 130, 158, 354, 414
Martin, R.L. 189
and Hodge, J.S. 385
Martin, S., and Graham, S. 398
Marxism 250–1
mass consumption *see* Fordism
mass production *see* Fordism
mass markets 113, 128, 152, 167, 281, 354
Massey, D. 12, 384
materialism 153
Mauritius 288
Meat Import Act (US, 1979) 269
mechanization 8, 246–7, 253, 263
mediation (of world economy by national governments) 71–4
medical tourism 336
medieval manor 102, **103**
megacities 192
Meier, G.M., and Baldwin, R.E. 20–1
mercantilism *see* merchant capitalism

merchant capitalism 7, 93, 99–113, **105**, 116–17, 143, 414
growth 109–11
rise 105
transoceanic rim settlements 109–10, **110**
mergers 156–7, 197
and acquisitions (M&As) 339
Metcalf, D. 263
METI (Ministry of Economy, Trade and Industry, Japan) 75, 84, 135
metropolitan decentralization 182–3
metropolitan growth 172
Mexico 4, 13, 42, 58, 159, 236, 239, 253–4, 264, 267, 335
export processing zone (EPZ) 291
less developed country (LDC) 34
maquiladoras 299, 414
and NAFTA 361–2, 364
Meyer, D.R. 128
MFA (Multifiber Agreement) 233, 293
MFN (most favored nation) 360–2, 415
middle class 33, 137, 300
migrant labor 167–8, 259
migration 5, 14, 40–5, 126
external flows within Sub-Saharan Africa (for labor) 254, 256, **257**
in Guatemala 267–8
international 277, 390
in Latin America 246, 253, 267–8
long-distance 256, 267
rural-to-urban 45, 246, 253, 277
seasonal 268
from Yemen 298
see also immigration
militarism 100, 115, 143
militarization 237
military 120, 225, 253
growth 277–8
power 365
security and 301
Millennium Development Goals (UN) 24
minifundia 246
minimum wage 291
mining 315
Ministry of Economy, Trade and Industry (METI, Japan) 75, 84, 135
Ministry of International Trade and Industry (MITI, Japan) 75, 84, 135
minisystems 95, 414
MITI (Ministry of International Trade and Industry, Japan) 75, 84, 135
MNCs (multinational corporations) 68, 179
mobilization, political 387
mode of production 6, 409, 414

mode of regulation 413, 414
modernism 137
modernity 112
modernization 132, 275–7, 299, 311, 371, 387
    Japanese 131
    pace of 120
Mondragon 395
monetarism 58, 414
monetary policy 75, 353–61, 369
monoculture 113, 214, 267, 414
monopoly 78, 182, 267, 324, 400
    capitalist 131
    state power and 308–9
most favored nation (MFN) 360–2, 415
Motherland (colonial) 219, 228
Multifiber Agreement (MFA) 233, 293
multinational corporations (MNCs) 68, 179
multinationality 82
multiplier effects 276, 283, 291, 377, 415
Murnane, R., and Levy, F. 192
Murphy, J.M. 375
MVA (manufacturing value added) 45–6, 274, 298
Myrdal, G. 207, 210
    cumulative causation model 210, **210**

NAFTA see North American Free Trade Agreement
Nairn, T. 377
nanotechnology 196
Napoleonic Wars (1803–15) 120
nation-states 353, 356–7, 386, 415
    sovereign 367
nationalism 79, 136–7, 229–30, 253, 306, 377, 387–9
    active 253
    economic 356
    ethnic 387, 391
    political 390
    resurgence of 368
nationalist separatism 386–91, 396, 415
nationalization (of the means of production) 301
NATO (North Atlantic Treaty Organization) 64, 158
natural disasters 5, 35, 272, 337
natural resources 26, 35, 125, 302
    distribution of 29, 143
    extraction of 34
neo-Fordism 410, 415
neoclassicism 14, 415
neocolonialism 131
neoliberalism 76–7, 181, 415
Neolithic period 96, 416

Nestlé 49, 85, 164, 248, 263
Netherlands 112, *342*
networking, social 322–3, 401
networks, production 85–6, 87
New Deal (US) 9, 129, 416
New Economic Policy (Russia, 1921) 300
new industrial complexes 295–9
new industrial districts 166, 170
New International Division of Labor (NIDL) 86–7, 162, 278, 311, 322, 384–5, 416
New International Economic Order (NIEO) (Group of 77) 365–6, 373
New York 78, 90, 194, 203, 354
    clothing industry 154, 167–8
    control center 190–1
    financial centre 340, 349, 367
    international terrorism 389–90
    Three Mile Island 30
    world city 90, 154, 420, 421
New Zealand 76, 110, 142, 164, 268, 375, 416
Newby, H. 263
newly industrializing economies (NIEs) 3, 21, 41, 54, 74, 157–8, 213–15, 232, 235–7, 240–3, 279–89, 293–5, 310–11, 411, 416
    share in LDC exports 288, **289**
NGOs (nongovernmental organizations) 23, 33, 251, 265
NIDL (New International Division of Labor) 86–7, 162, 278, 311, 322, 384–5, 416
NIEO (New International Economic Order) (Group of 77) 365–6, 373
NIEs see newly industrializing economies
Nigerian Civil War (1967–70) 235
Nike 154, 333
non-governmental organizations (NGOs) 23, 33, 251, 265
non-profit organizations 247
non-tariff barriers 359–60, 416
non-tradable services 316, 330
North America see United States of America
North American Free Trade Agreement (NAFTA) 49, 79, 290–1, 361–2, 364
North Atlantic Treaty Organization (NATO) 64, 158

OAU (formerly African Union) 357, 403
Öberg, S., and Sachar, A. 152
O'Brien, R. 70
ODA (official development assistance) 59
OECD see Organization for Economic Cooperation and Development
OEEC (Organization for European Economic Cooperation) 130
official development assistance (ODA) 59

offshore banking *331*, 339–40, 343–4, **343**, 349
offshore financial centers 17, 416
offshore outsourcing **187**, 292, **293**, 416
offshoring 320, *321*, 416
oil 16, 34, 35, 188, 213, 238, 241, 338, 398
  crude 77, 171
  embargo (1973) 355
OPEC 57–9, 153, 238, 243, 282, 416–17
  prices 80, 145, 146, 153, 171, 235, 239, 241, 277, 303
oligarchy 396
oligopolistic market structures 324
On Hollywood (Scott) 204
OPEC *see* Organization of the Petroleum Exporting Countries
Opium Wars (1839–42 & 1856–60) 268
organization 144
  agricultural 246–8, 273
  business 154
  economic 6–12, 100, 118–20, 129–30, 139–43, 169, 299
  spatial 127, 131
Organization of African Unity (OAU), now African Union (AU) 357, 403
Organization for Economic Cooperation and Development (OECD) 25–6, **45**, 146, 202, 310–11, *321*, 416
Organization for European Economic Cooperation (OEEC) 130
Organization of the Petroleum Exporting Countries (OPEC) 57–9, 153, 238, 243, 282, 416–17
  oil embargo (1973) 355
organizations
  aid 250, 333
  international and supranational 81, 353–78, 396
organized capitalism 7–9, 76, 137–44, 279, 399–400, 417
Our Common Future, see Brundtland Report
outsourcing 173, 193, 197, 313–15, 319–25, 344–6, 417
  anti- 173
  captive 321, 404, 417
  external 321
  growth in 348
  internal 321
  manufactures and 173
  offshore **187**, 292, **293**, 416
  services and 319–24, 329, 348
overaccumulation 417
Owen, N. 369
Oxfam 250, 333

Pacific Corridor, megapolitan Japan 135–6
Paris Club 58
path dependence *see* geographical path dependence
patriarchy 131
patronage 278
Peace of Westphalia (1648) 66
peasantry 104, 131–2, 246–8, 252–5, 260, 270, 299–300
Peck, J., and Tickell, A. 77
peer-to-peer (P2P) filesharing 157
peripheral countries 124–5, 140–3, 157–9, 210–12, 231, 244–8, 322–5, 357, 363–6, 375–7
  and agriculture 246–51
  contemporary 251
  disparity and 235–7
  industrialization and 280, 294–5, 311, 379
  transformation of 213–44
periphery *see* peripheral countries
Perroux, F. 210
Perry, M.C. (Admiral) 131
Petras, J. 299
Philippines 236, *249*, *359*
phonesourcing 183
Pinch, S., and Henry, N. 189
plantations 65, 246–55, 259, 264, 268, 273
pluralism 182
polarization 210, 240–1, 283
  core-periphery 20, 35, 4, 139, 143, 205, 210, 240–2, 320–3
  European Union 370
  income 18
  regional 377
  social 398
  spatial 363, 366, 370–3
policy 273, 365
  competition 382–6
  credit 273
  economic 225, 361, 380
  fiscal 135, 355–7, 380
  foreign 75, 361, 367
  industrial 70, *73*
  market-access 244
  monetary 75, 353–61, 369
  regional 140, 180, 351, 363, 366, 372–3, **374**, 380–1, 384–6, 396N
  social (Nordic) 63
  tariffs 71
  taxation 256, 355
political climate 102, 140
political economy 131, 206
political integration *see* integration
political mobilization 387

political stability 234, 400
political crisis (Japan) 136
politics 50
    change and 6, 308
    conflict and 277, 310
    democratic 308, 380
    division and 388
    favoritism 278, 387
    fragmentation and 379
    hegemony and 235
    independent 388
    instability and 237
    interdependent 355
    lobbying and 282
    nationalist 390
    power and 38, 129, 216, 226, 230, 254, 355
    reforms and 133
    stability and 234, 400
Pollard, S. 118–22
pollution 30–2, 260, 307, 338, 400
    environmental 30–2, 260, 307, 338, 400
    water 260, 400
Popke, J. 250
population 29, 140, 274
    density, international 40
    expansion, international 41
    growth 102, 132, 159, 259, 265, 272, 283–5
    increase 32, 241, 255–7
    movements 125
    patterns 29–45
positional goods 152, 417
post-Fordism (neo-Fordism) 410, 415
poverty 32, 36, 107, 125, 251, 305–7, 329
    alleviation of 329
    decline in 241, 260
    mass 395
power 77, 198, 228–9, 271, 390
    balance of 205, 311
    colonial 227
    decentralization of 242
    devolution of 139
    economic 91, 110–12, 124, 355, 375
    hegemonic 64
    imperial 12, 130
    labor 99, 142, 216
    market 82, 333
    military 365
    national 181
    political 38, 129, 216, 226, 230, 254, 355
    purchasing 58, 111, 142, 146, 240–1
    state and the 67–8
PPP (purchasing power parity) 205, 418
Prebisch, R. 356

preindustrial foundations (of the world economy) 95–115
Prescott, E., and Hansen, G. 116
price elasticity of demand 55
primacy/primate city 222, 295, 417
primary commodities 233–7, 240, 243–5, 252–3, 336, 366, 417
primary production 13, 83, 122, 231, 277, 417
primitive accumulation 97
private sector 316
privatization 68, 256, 319, 417
    land 262
    large-scale 320
procurement (corporate) 335, *345*
producer cartels 234
producer goods 233, 417
producer services 82, 171–2, 203, 205, 206, 211, 294, 308, 316, 319, 332, 367, 404, 410, 417
product development 166
product life cycle **83**
production 3, 11, 282
    agricultural 5, 30, **40**, 101, 104, 119–20, 255, 272
    batch 169, *403*
    capital-intensive 16, 21
    chains 195
    collapse 112
    costs 281
    crops and 254–5, 273
    decentralization of 89
    development of 85
    factors of 12, 91, 149, 155, 216, 405, 408, 409, 414, 415, 420
    flexible 79, 162–6, 181, *182*, 410
    flows 65
    food and 96–7, 245–9
    forces 7
    global 162
    hierarchy 174
    industrial 7, 8, 45, 54, 77, 104, 161, 194, 218, 263, 275, 300, 303, 412, 414, 416
    innovation and 128
    integration of 244
    just-in-time (JIT) 164, 281, 333, 413
    lifecycle and 82–3, 417
    mass 82–3, 152, 162, 170, 354
    mode of 6, 409, 414
    networks 85–6, 87
    per capita *249*, 255
    primary 13, 83, 122, 231, 277, 417
    scale and 393
    services and 171, 329, 417
    specialization of 231
    technology and 151, 162

profit 8, 21, 53, 61, *252*, 311, 322, *331*
    agricultural 8
    crisis and 281
    cycles 280–1
    decline 146, 278
    transfer 43
Progressive Era (US) 144, 417
property rights 77
propulsive industries *203*, 210–11,
prosperity 1, 21, 110–12, 125, 212, 317, 355
    economic 23
    national 386
    pre-war (WWII, Europe) 125
    relative 381
protectionism 236, 256, 281, 322, 360, 373
proto-industrialization 119
public economy 118, 139, 140–1
public law 100
public relations 319
public sector 316, 319, 348
public-private partnerships 77, 181, 382, 418
purchasing power 58, 111, 142, 146, 240–1
purchasing power parity (PPP) 205, 418
Putin, V. 243, 388

quality of life 25, 26, 34, 199, 241

R&D *see* research and development
railways 121–2, 127, 132, 150, 221, 227
Ranger, T. 226
rank redistribution 7, 418
raw materials 12–15, 158, 183, 214–15, *232*,
    234, 237
Reagan, R. 76, 180, 384, 397
Reaganomics 58, 418
Reardon, T., *et al* 335
reconstruction
    postwar (WWII) 133–7
redeployment, international, corporate 161–2
reforms 299, 304, 372
    economic 304, 308, 320
    educational 132
    fiscal 382
    land 74, 245, 259–62, 273, 283
    liberal 242
    political 133
    social 133
    trade 279
regime of accumulation 410, 414, 415, 418
regional change 180–2, 370
regional cumulative causation *see* cumulative
    causation
regional decentralization 183–5, 188
regional devolution 139, 380, 390, 418

regional dispersal 13
regional economic motors *see* regional motors
regional inequality 205–11
    China 307
    core economies 205–6
    European Union 205–6, **206**
national economic development and 208–9
regional linkages 294–5
regional motors 79–81, 85–6, 89, 161, 295, 383,
    399, 400
regional policy 140, 180, 351, 363, 366, 372–3,
    374, 380–1, 384–6, 396N
regional specialization 13, 215–18, 243, 247,
    302
regionalism 18, 380–6, 396, 418
    economic 93, 196, 294, 364, 383
    formal 381
regulation 67, 71
    government 308, 324
    market 63, 137, 181, 370
    mode of 413, 414
    monetary system and 118
    state 242–3
    wages and 118
regulationist theories 414
Reich, R. 3
religion 112, 242, 368, 388
relocation, industrial 141
remittance flows 43, **43**, **44**
reorganization 179
    corporate 179, 186
    economic 299–300
    spatial 179–212
research and development (R&D) 82–4, 169–70,
    184–5, 188–90, 193–4, 199–201, 347–9
    global service providers 346, **347**
    United States 184–5, **185**
resource curse 235, 418
resource grabbing 5, 418
resource-based industry 298
resources *163*
    flows 307
    natural 26, 29, 34, 35, 125, 143, 302
    patterns 29–45
    shale gas 29, **30**, **31**
    undiscovered 29, 421
restructuring, corporate 147, 174, 188, 201
retailing 88, 332–4
returns to scale, increasing 12, 204, 211, 399,
    408, 412
revenue 235, 383
    federal 140
Reynolds, R. 95, 109, 268–9
Ricardo, D. 216

rice 101, 271
Rigby, D.L., and Webber, M.J. 279
rights 386
  economic 386
  political 386
  property 77
  protection of 23–4
Robins, K., and Amin, A. 205
Rokkan, S., and Urwin, D. 387
Roman Empire 98–9
Rondinelli, D.A. 222
Ruggiero, R. 360
rural consolidation 99–100
rural-to-urban migration 45, 246, 253, 277
Russia 7, 29, 45, 58, 70, 93, 159, 164, *243*, 275, 303–4, 339, 379
  BRICs 308, *309*, 404,
  government decentralization 388, **389**
  kiosk economy 303
  New Economic Policy (1921) 300
  Tsarist 299
Russian Federation 36, 42, 240
Russian Revolution (1917) 66

Sachar, A., and Öberg, S. 152
SACU (South African Customs Union) 361
SADC (South African Development Community) 361
San Francisco
  high-technology industries 197
  services 186
  transnational corporations (TNCs) 264
  world city 191
Sassen, S. 191
savings, institutionalization of 339
scale economies 117, 143, 203, 211, 262–3, 357, 408
schismogenesis 387
Schumpeter, J.A. 10
Scott, A. 89, 161, 166, 169, 201–4, 291, 297
SEA (Single European Act) of the EU 372
Second World 230
security 23–4, 319, 324, 340, 390
  income 139
  international 357
  military 301
  national 301, 357, 400
  social 380
Segal, G. 310
self-sufficiency 275, 304
semi-peripheral countries 21–2, 141–5, 152–3, 322–5, 339, 353, 356–7
semi-periphery *see* semi-peripheral countries
semiconductor industry, labor process 290, **290**

Sen, A. 257
Sender, J., and Smith, S. 229, 235
separatism 79, 351, 380
  nationalist 386–91, 396, 415
serfdom 299
service providers **315**
  global R & D 346, **347**
services 11, *161*, 186
  annual growth **328**, 329
  consumer 13, 406
  employment and GDP 313, **314**
  encapsulation 317, **318**
  exports 322–5, 329–30, **330**
  financial 315
  global 313–49
  growth 318–19
  imports 322–5, 329–30, **330**
  internationalization 332–47
  outsourcing 319–24, 329, 348
  producer 82, 171–2, 203, 205, 206, 211, 294, 308, 316, 319, 332, 367, 404, 410, 417
  production 171, **329**, 417
  trade performance 329
  workers in 326, **327**
  see also business services
settlement, urban patterns and colonialism **114**, 115
sexual divisions of labor 229, 418
SEZs (Special Economic Zones) in China 4, 46, 304–6
shale gas resources 29, **30**, **31**
share cropping 246, 258–9, 267, 272, 418
shifting cultivation shifting 96, 409, 418–19
shopping mall, global 153
Silicon Valley (California) 90, 172, 189, 197–201, 212, 297, 386
Singapore 159, 240, 271, 294, 409
  Asian Tiger 67, 202, 403
  banking 340
  East Asian G4 281
  electronics industry 292, 297, *298*
  export processing zone (EPZ) 288, 291
  foreign direct investment (FDI) 53
  growth rate 47
  intra-ASEAN trade 365
  medical tourism 336
  newly industrializing economy (NIE) 3, 21, 41, 54, 74, 232, 279–80, 283–4, 288
  services 326, 329, 348
  technopole 199
  transnational corporations (TNCs) 49, 292
  world city 90, 191
Single European Act (SEA) of the EU 372
Single European Market (1992) of the EU 367–9

slash-and-burn cultivation 96
slave labor 65, 109
slavery 6–7, 18, 214
SMBs (small and medium businesses) 152
Smith, A. 14, 216, 276
Smith, D. 302
Smith, S., and Sender, J. 229, 235
Smith, T. 226
social class *see* class (social)
social equity 367
social networking 322–3, 401
social policy, Nordic 63
socialism 6–7, 18, 21, *261*, 299–303, 308, 419
    industrial 302
    Soviet-style 397
    state 301–4, 388, 394
soil erosion 32–4, 267, 270
sourcing, global 233–4, 263, 275, 281, 410
South Africa 66, *335*
South African Customs Union (SACU) 361
South America
    River Plate region road and railways 221, **223**
    see also Latin America
South Korea 159, 239, 265, *335*, *359*
    export processing zones (EPZs) 288
    export-orientated industrial development
        296
    industries 295
    land reform 259, 260, 262, 304
    newly industrializing country (NIE) 74, 283
    Samsung Group 157
Southern African Development Community
    (SADC) 361
sovereignty 367, 388, 419
    national/state 66, 357
Soviet model of industrialization 245, 299–304
Soviet Union 299–304
    Five-Year Plans 300–2
    socialism 397
    spheres of influence 237, **238**
    see also Russia; Russian Federation
space-process relationship within firms (Hymer)
    419
Spain 286
    economic power and 111
    imperialism and 112
spatial change 1, 6–12, **6**, 18, 116–17, 142
spatial decentralization *see* decentralization
spatial division of labor 95, 113, 142–3, 419
spatial homogenization 174, 379, 399
spatial inequality 383–4
spatial polarization 363, 366, 370–3
spatial reorganization 179–212
spatial transaction costs *88*, *89*, *91*, 399

Special Economic Zones (SEZs), China 4, 46,
    304–6
specialization 166, 170, 221
    agricultural 12
    economic 137
    flexible 166, 263
    international 23
    local 174
    production 231
    regional 13, 215–18, 243, 247, 302
splintering urbanism 398
spread effects (Myrdal) 362, 419
stability, political 234, 400
stagflation 123, 146, 153, 158, 355, 419
stagnation 64, 110, 133, 269, 280, 311
    markets and 241
    prices, of 231
Stalin, J. 300, 301, 303
states 66–79, 419
    aid and 386
    autonomous 138
    bureaucracy and 101
    capital and 126
    client 141, 405
    hierarchy of 67
    nation 353, 356–7, 367, 386, 415
    new 230, **231**
    policy and 63
    power and 67–8
    quasi- 74
    regulation and 242–3
    socialist 301–4, 388, 394
    sovereignty and 66
    system of 62–3, 91
    territorial 78
    welfare 64, 79, 139, 180–1
statism 301–3
steamships 218
    routes 221, **221**
Stein, P. 225
stock markets 129, 133, 239, 340, **343**
Stokes, E. 226
strategic alliances (among firms) 86, 91, 164- 6,
    169, 280, 419
structural adjustment programs (IMF and World
    Bank) 70, 419
Studwell, J. 240, 259
Sub-Saharan Africa 254–8, 274, 308, *326*, *335*
    external flows within (for labor) 254, 256,
        257
subcontracting 61,140, 167, 172, 173, 275, 277,
    280, 410
submarine fiber-optic cable network 323, **323**
subsistence economics 6, 18, 267, 419

suburbanization 186
Sunbelt, the 141,186, 188,196, 379
supply side 130, 181, 332, 418, 419
supranational integration 357–78, **358**, 379
supranational organizations 81, 353–78, 396
supranational political union 31, 365, 370, 420
sustainable development 24, 32–4, 237, 344, 420
Sutcliffe, R.B. 286
Swindell, K. 256

Taiwan 237, 240, 274
    agriculture 259, 260–2, 271, 283, 304
    APEC, and 362
    Asian Tiger 67, 202, 403
    Barbie 16
    export-oriented industrial development 276,
        279, 283–4
    export processing zones (EPZs) 288, 291
    foreign direct investment (FDI) 292
    newly industrializing economy (NIE) 3, 74,
        213, 232, 236, 283–4
    protectionism 281
    special economic zones (SEZs) 306
    transnational corporations (TNCs) 159, 280,
        292
tariffs 62, 73, 133–5, 139, 360–2, 420
    average 358, 359, *359*
    barriers and 363
    policy and 71
taxation 131–2, 183, 202, 276, 298–300, *331*,
        371, 380–3
    codes 4
    concessions 295
    discriminatory 324
    incentives 170
    income of 77–8
    laws 395
    policy 256, 355
    sales and 332
Taylor, F.W. 9
Taylor, P.J. 191–2
Taylorism 9, 126–30, 420
tea 108, 215, 216, 217–18
technology 10–12, 18, 50, 136, 180, 323–4, 354,
        365
    access to 38, 281–2
    advances in 5
    agriculture 270–2
    breakthroughs 150–1, 313
    changes 6, 29, 118, 143, 147
    communications and 83, 117, 149, 315, 319,
        330–2, 339
    exchange 398
    flows 383

high- 203
    improvements in 98, 131–2
    industrial 121, 127–8, 132
    information and 319, 330, 339, 344
    innovation and 29, 96, 107, 125–6, 138, 185,
        253
    investment 146
    production 151, 162
    promise of 375
    systems 399, 420
    transport and 80, 83, 149, 180, 391
technology-intensive industry 296
technopoles 197–200, 420
telecommunications 73, 147, *298*, 323–4, 357
    technology 180
telegraph system, Asia and Africa 221, **222**
telemarketing 322
Tempest, R. 16
Terman, F. 197
terms of trade 230–1, *232*, 235, 240
territory 79, 81, 143, 399
    colonial 226
    conquest and 214
    division of 62
    expansion and 98–101, 107–9, 113, 126,
        301
    integration of 129
    policy and 133
    states and 78
terrorism 75, 337, 389–90
textiles, labor processes and 290, **290**
Thatcher, M. 76, 180, 384–5, **385**, 397
Third Italy 86, 90, 166, 169–70, 203, 391
Third Reich 356
Third World 230, 237–8
Three Mile Island (New York) 30
Thrift, N.J. 145
Tiananmen Square (1989) 308
Tickell, A., and Peck, J. 77
time-space compression 180, 351, 420
TNCs *see* transnational corporations
Tokyo 5, 90, 131, 154, 157–8, 161
    financial centre 172, 340, 349, 367
    high-technology industries 169
    technopole 199
    terrorism 389
    Tokaido Megalopolis 135–6
    world city 90, 136,190–1, 420
    topography 142, 198
tourism 315, 316–18, 331
    international 334–8, 349
    medical 336
tourists, international 337, **337**
Toxic Threads report (Greenpeace 2012) 155

trade **51**, 59, 63, 73, 75, 80–2, 85, 97, 108, 112, 134, 142, 218–19, 231, 232, 240, 277, 359, 370
  agreements 123
  Andean Pact 365
  anti- 235
  Atlantic system 218, **219**
  barriers 159, 170, 241, 293, 353, 360–1, 364
  bilateral 360, 373
  blocs 49–50, 420
  commercial 228
  commodities and 79, 104
  cooperation and 362
  creation 362, 366, 369, 377
  crossover system of 224, 225, 230, 243, 406
  decline 112, 225
  deficits 16, 161, 171, 224, 239, 310, 355
  diversion 294, 366, 373–7, 420
  ethical 250
  European 254
  fair 248–51
  flows 50, 159, 233, 306
  foreign 20, 51, 111, 306, 365
  free 48–9, 52, 63, 70, 77, 91, 139, 250, 256, 361, 362, 364, 41
  global *see* international
  growth 50–1, **50**, 79, 233
  increase 106, 237, 319
  intensification 224
  international 9, 29, 49–58, 104, 133, 141, 202, 216, 224, 233, 319–21, 329–30, 348, 365
  interregional 172–3
  intra-ASEAN 365
  intra-regional 52, 127
  intra-EU 363, 369
  liberalization and 233, 275, 281, 333, 361–2, 369
  merchandise growth and GDP 50–1, **50**
  multilateral 225
  negotiations 82
  patterns 49–53, 104, 109, 363–4, 373
  policy 68, 71, 242, 256
  preference associations 362, 420
  realignment 224–5
  reform 279
  regional 50, 104, 360
  regulation 71, 353
  relations 160, 375
  restraints 227, 360
  resurgence 100, 104–6
  retail 316, 347
  routes 106
  Soviet 301
  specialization 233
  support 48
  surplus 16, 160–1
  terms of 230–1, *232*, 235, 240
  transoceanic 112
  triangular system of 218–19
  world 50–1, 62, 79–81, 232, 238, 268–70, 363
trade unions 137, 229
trading blocs 49–50, 68, 79, 91, 360, 400, 420
trading empires 141, 353
trading stations 109
traditional manufacturing industry 313
traditionalism 215, 387
transactions 169
  costs and 88–9
  volume of 171
transnational banking 339, 349
transnational corporations (TNCs) 155–9, 242–5, 260–4, 267–9, 272–5, 278–82, 291–2, 394–6
  "global" reach 49
  roles 233–4
  state-owned 73
transnational investment patterns 292–4, 330–1
transoceanic rim settlements, mercantile era 109–10, **110**
transport 11–13, 129, 139–41, 218, 221, 303, 316–18
  costs **88**, 211, 233, 400
  global *see* international
  improvements 186
  infrastructure 283, 307
  intermodal 150, 412
  international 88, 150
  modern 228
  routes 219–22
  steamship 218
  technology 80, 83, 149, 180, 391
Treaty of European Union (Maastricht Treaty 1992) 367–8
Turkey 111, *342*
Turnock, D. 302

UK *see* United Kingdom
UNCED (United Nations Conference on Environment and Development) 32–4
UNCTAD *see* United Nations (UN), Conference on Trade and Development
underdevelopment 21, 23, 65, 192, 215, 357
undiscovered resources 29, 421
UNDP (United Nations Development Programme) 25–8, 36, 43
unemployment 77, 146, 202, 206, 281, 385, 395

unequal exchange 142, 250, 421
uneven development 64, 79, 180, 209–10, 355,
    357, 369, 399, 421
UNFPA (United Nations Population Fund)
    247
unionization 65, 85, 167–9, 183, 336
unions
    labor 9, 77, 117, 144–6, 170
    organization and 285
    trade 137, 229
United Kingdom (UK) 148, 195, *220*, *278*, 286,
    335, *341–2*
    British foreign investment 219, **220**
    capitalism and 138
    deindustrialization and **148**
    deprivation index 206, **207**
    economic power 112
    employment 194, **195**
    first-wave industrialization 119–21,
    hegemony 214
    Labour Party 394
    regional aid 384, **385**
    see also London
United Nations (UN) 40–1, 45, 64, 198, 246,
    256, 260
    Conference on Environment and Development
        (UNCED) 32–4
    Conference on Trade and Development
        (UNCTAD) 47, 51–3, 159, 321, 346, 365–6,
        373
    Development Programme (UNDP) 25–8, 36,
        43
    Food and Agriculture Organization (UN FAO)
        36–7, 246, 256
    General Assembly 358, 366
    Human Development Index (HDI) 25–8, **27**,
        28, *284*
    Millennium Development Goals (MDGs) 24
    Population Fund (UNFPA) 247
    Security Council 358
United States of America (USA) 50, 160, 196,
    248, *278–9*, 286, *341–2*, 359
    automobile industry, foreign 184, **184**
    capitalism and 138, 218
    Census Bureau *156*
    Defense Department 150
    employment 194–5, **196**
    Food and Drug Administration (FDA) 376
    hegemony and 354
    income, per capita, regional 208, **209**
    industrialization 125–30, 144
    insourcing and outsourcing 14, **14**
    Internal Revenue Service (IRS) 55
    Islam 242, 244, 338, 389

Manufacturing Belt (1880s -1960s) 78, **128**,
    129
military **76**, 156
New Deal 9, 129, 416
plant size 393, **393**
product life cycles **83**
regional per capita incomes 208, **209**
Research & Development (R&D) 184–5, **185**
shale gas resources 29, **30**
Silicon Valley (California) 90, 172, 189,
    197–201, 212, 297, 386
spheres of influence 237, **238**
see also Los Angeles; New York; San Francisco
unproven reserves 29, 421
urban settlement patterns, and colonialism **114**,
    115
urbanism, splintering 398
urbanization 97–9, 126, 135, 190, 222, 229
    classical world 98–9, **99**
Urry, J., and Lash, S. 138, 181
Uruguay Round (GATT) 359–60
Urwin, D., and Rokkan, S. 387
USSR *see* Soviet Union

venture capital 86, 193
Vernon, R. 82
vertical disintegration 86, *163*, 164, 201, 421
vertical integration 80, 154–6, *163*, 195, 393,
    421
violence 228, 269
    rural 253

Wachtel, H.M. 238
WAEMU (West African Economic and Monetary
    Union) 361
wages 140, 167, 291
    labor and 117, 228–9, 243, 246–8, 256, 267,
        421
    regulation and 118
Wal-Mart 332–6, 348
Wallerstein, I. 21, 95, 192
war 111, 133, 228, 268, 272, 300, 306
    resources 307
    see also World War I; World War II
warehousing 129
Warf, B. 192, 198
Warman, A. 267
Warren, B. 311
waste 33, 37
water *11*, 29, 32, 35–7, 56, 122, 270, 411
    fresh 29, 36, **38**, 265
    pollution 260, 400
water power 118, 119
WDI (World Development Indicators) 23

wealth 18, 38, 104, 198, 271, 398
  redistribution 367
Weaver, C., and Friedmann, J. 129
Webber, M.J., and Rigby, D.L. 279
welfare 78, *182*, 189, 395–6
  corporate 76
  government (state) 64, 79, 139, 180–1
  human 32, 400
  policy 137
  services 355
  social 34, 147, 383
welfare state 64, 79, 139, 145, 180–1
well-being 26, 40, 139
  economic 129, 179–80, 187, 381
West African Economic and Monetary Union
    (WAEMU) 361
Western colonization 246
Western Europe, output growth (500–1990)
    118
Westphalian system 66–7
wholesaling 88, 129, 172, 316, 347
  merchant 173
Williams, C.H. 391
Williams, D.E. 332
Williams, G. 366
Williams, R.G. 269
Williams, W.A. 396
Williamson, J.G. 208
Wolf, E.R. 217–18
women
  and agriculture 246–7, 256, 273
  and work 16, 291
workfare state 181
working class 253
World Bank 39–40, 57–8, 68–70, 236, 283, 326,
    357–8
  structural adjustment programs 70, 419
  World Development Indicators (WDI) 23

world cities 90, 136, 154, 190–2, 349, 420, 420,
    421
  system **191**
World Commission on Environment and
    Development (Brundtland Report) 32
World Development Indicators (WDI) (World
    Bank) 23
world empires 95, 97
world products 153
World Research Institute (WRI) 149
world system *see* world-system
World Trade Organization (WTO) 64, 71, 251,
    292–4, 358–60, 372, 376
World War I (1914–18) 12, 132, 299, 301, 356
  disruption 122
  post- 125
World War II (1939–45) 63–4, 145, 204, 230–1,
    266–7, 353, 358
  post- 71, 81, 225
  recovery 123–5
World Wide Fund for Nature (WWF) 37
world-system **22**, 95, 108, 422
  capitalist 159, 300
  European-based 100–12
world-systems theory 21–2, 192
WRI (World Research Institute) 149
WTO *see* World Trade Organization
WWF (World Wide Fund for Nature) 39

xenophobia 131, 356, 422

Yavorosky, B. 26, 33, 56, 265
Yemen, migration from 298
Youndé Convention (EU, 1963) 373
  see also Lomé Conventions
zaibatsu 133–5, 144
Zambia 256
zero-sum competition 78